HYDRODYNAMICS AND SOUND

This book is designed for a first graduate course in fluid dynamics. It focuses on knowledge and methods that find application in most branches of fluid mechanics and aims to supply a theoretical understanding that will permit sensible simplifications to be made in the formulation of problems and enable the reader to develop analytical models of practical significance. The study of simplified model problems can be used to guide experimental and numerical investigations. The first part (Chapters 1–4) is concerned entirely with the incompressible flow of a homogeneous fluid. Chapters 5 and 6 deal with dispersive waves and acoustics.

Professor Howe is in the Department of Aerospace and Mechanical Engineering at Boston University. He received his PhD in Continuum Mechanics from Imperial College, London. He has published more than 200 technical papers on fluid mechanics and acoustics and is the author of four books, including *Acoustics of Fluid-Structure Interactions* and *Theory of Vortex Sound*, both published by Cambridge University Press.

Hydrodynamics and Sound

M. S. HOWE

Boston University

CAMBRIDGE UNIVERSITY PRESS
Cambridge, New York, Melbourne, Madrid, Cape Town,
Singapore, São Paulo, Delhi, Mexico City

Cambridge University Press
The Edinburgh Building, Cambridge CB2 8RU, UK

Published in the United States of America by Cambridge University Press, New York

www.cambridge.org
Information on this title: www.cambridge.org/9781107410671

First published 2007
First paperback edition 2012

A catalogue record for this publication is available from the British Library

Library of Congress Cataloguing in Publication Data
Howe, M. S.
Hydrodynamics and sound / M. S. Howe
p.   cm.
Includes bibliographical references and index.
ISBN-13: 978-0-521-86862-4 (hardback)
ISBN-10: 0-521-86862-9 (hardback)
1. Hydrodynamics – Mathematical models.   2. Fluid dynamics – Mathematical models.
3. Waves.   4. Sound.   I. Title.
QC151.H69   2006
532′. 5015118 – dc22   2006023800

ISBN 978-0-521-86862-4 Hardback
ISBN 978-1-107-41067-1 Paperback

In memoriam James Lighthill

Contents

CONTENTS

Preface

Fluid mechanics impinges on practically all areas of human endeavour. But it is not easy to grasp its principles and ramifications in all of its diverse manifestations. Industrial applications usually require the numerical solution of the equations of motion of a fluid on a very large scale, perhaps coupled in a complicated manner to equations describing the response of solid structures in contact with the fluid. There has developed a tendency to regard the subject as defined solely by its governing equations whose treatment by numerical methods can furnish the solution of any problem.

There are actually many practical problems that are not yet amenable to full numerical evaluation in a reasonable time, even on the fastest of present-day computers. It is therefore important to have a proper theoretical understanding that will permit sensible simplifications to be made when formulating a problem. As in most technical subjects such understanding is acquired by detailed study of highly simplified 'model problems'. Many of these problems fall within the realm of classical fluid mechanics, which is often criticised for its emphasis on ideal fluids and potential flow theory. The criticism is misplaced, however: For example, potential flow methods provide a good first approximation to airfoil theory, and 'free-streamline' theory (pioneered in its modern form by Chaplygin) permits the two-dimensional modelling of complex flows involving separation and jet formation.

There is a certain body of knowledge and methods that finds application in most branches of fluid mechanics. This book aims to supply this basic material and to present the most important theoretical methods that will enable the reader to develop analytical models of practical significance. Such analyses can be used to guide more detailed experimental and numerical investigations. The first part (Chapters 1–4) is concerned entirely with the incompressible flow of a homogeneous fluid. It was written for the Boston University introductory graduate-level course 'Advanced Fluid Mechanics'. The remaining chapters, 5 and 6, deal with dispersive waves and acoustics and are unashamedly inspired by James Lighthill's masterpiece *Waves in Fluids*.

M. S. Howe

1

Equations of Motion

1.1 The fluid state

Consider a fluid that can be regarded as continuous and locally homogeneous at all levels of subdivision. At any time t and position $\mathbf{x} = (x_1, x_2, x_3)$ the state of the fluid is defined when the velocity $\mathbf{v}$ and any two thermodynamic variables are specified. A fluid in unsteady motion, in which temperature and pressure vary with position and time, cannot strictly be in thermodynamic equilibrium, and it will be necessary to discuss how to define the thermodynamic properties of the small individual *fluid particles* of which the fluid may be supposed to consist.

The distinctive fluid property possessed by both liquids and gases is that these fluid particles can move freely relative to one another under the influence of applied forces or other externally imposed changes at the boundaries of the fluid. Five scalar partial differential equations are required for determining these motions. They are statements of conservation of mass, momentum, and energy, and they are to be solved subject to appropriate *boundary* and *initial* conditions, dependent on the problem at hand. This book is concerned with the use of these equations to formulate and analyse a wide range of model problems whose solutions will help the reader to understand the intricacies of fluid motion.

1.2 The material derivative

Let v_i denote the component of the fluid velocity $\mathbf{v}$ in the x_i direction of the fixed rectangular coordinate system (x_1, x_2, x_3) and consider the rate at which any function $F(\mathbf{x}, t)$ varies *following the motion* of a fluid particle. Suppose the particle is at $\mathbf{x}$ at time t, and at $\mathbf{x} + \delta\mathbf{x}$ a short time later at time $t + \delta t$, where $\delta\mathbf{x} = \mathbf{v}(\mathbf{x}, t)\delta t +, \ldots,$ where the terms omitted vanish more rapidly than δt as $\delta t \to 0$. Then the value of F at the new position of the fluid particle is

$$F(\mathbf{x} + \delta\mathbf{x}, t + \delta t) = F(\mathbf{x}, t) + v_j \delta t \frac{\partial F}{\partial x_j}(\mathbf{x}, t) + \delta t \frac{\partial F}{\partial t}(\mathbf{x}, t) + \cdots +,$$

where the repeated suffix j implies summation over $j = 1, 2, 3$. The limiting value of the ratio

$$\frac{F(\mathbf{x} + \delta\mathbf{x}, t + \delta t) - F(\mathbf{x}, t)}{\delta t} \quad \text{as } \delta t \to 0$$

is called the material (or 'Lagrangian') derivative of F. It is denoted by DF/Dt, and

$$\frac{DF}{Dt} = \frac{\partial F}{\partial t} + v_j \frac{\partial F}{\partial x_j} \equiv \frac{\partial F}{\partial t} + \mathbf{v} \cdot \nabla F. \tag{1.2.1}$$

DF/Dt measures the time rate of change of F as seen by an observer moving with the fluid particle that occupies position $\mathbf{x}$ at the current time t.

1.3 Conservation of mass: Equation of continuity

A fluid particle of volume V and mass density ρ has a total mass of ρV. This cannot change as the particle moves around in the fluid, and therefore satisfies

$$\frac{D(\rho V)}{Dt} = 0,$$

so that

$$\frac{1}{\rho} \frac{D\rho}{Dt} + \frac{1}{V} \frac{DV}{Dt} = 0. \tag{1.3.1}$$

Now DV/Dt is the rate at which the volume of the fluid particle increases and is ultimately equal to $V\mathrm{div}\,\mathbf{v}$ when $V \to 0$. This is a consequence of the following integral definition of the divergence:

$$\mathrm{div}\,\mathbf{v} = \lim_{V \to 0} \frac{1}{V} \oint_S \mathbf{v} \cdot d\mathbf{S} \tag{1.3.2}$$

where the integration is over the closed material surface S forming the boundary of V, on which the vector surface element $d\mathbf{S}$ is directed *out* of V. Hence, using definition (1.2.1) of D/Dt, mass conservation equation (1.3.1) can be transformed into any of the following equivalent forms of the *equation of continuity*

$$\left. \begin{array}{l} \frac{1}{\rho} \frac{D\rho}{Dt} + \mathrm{div}\,\mathbf{v} = 0, \\[2mm] \frac{\partial \rho}{\partial t} + \mathrm{div}(\rho \mathbf{v}) = 0, \\[2mm] \frac{\partial \rho}{\partial t} + \frac{\partial}{\partial x_j}(\rho v_j) = 0. \end{array} \right\} \tag{1.3.3}$$

In the special case of an *incompressible* fluid, the density ρ of a fluid particle cannot change, although it may be different for different fluid particles. Therefore both $D\rho/Dt = 0$ and $DV/Dt = 0$, and the continuity equation reduces to

$$\mathrm{div}\,\mathbf{v} = 0. \tag{1.3.4}$$

This represents a kinematic (or *geometric*) constraint on possible motions of the fluid.

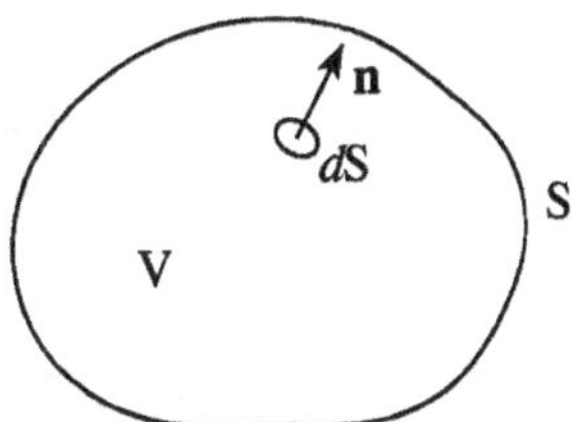

Figure 1.4.1

1.4 Momentum equation

The momentum equation is derived by consideration of the rate of change of momentum of a fluid particle (Figure 1.4.1) subject to the effects of the normally applied pressure p on its bounding surface S, the normal and tangential viscous stresses on S, and any body force (such as gravity) $\mathbf{F}$ per unit volume. Let σ_{ij} denote the *viscous stress tensor*, defined such that the *traction* force per unit area on a surface element of S with unit normal n_i is $\sigma_{ij}n_j$. Then the i component of the momentum equation for a small fluid element of volume V becomes

$$\rho V \frac{Dv_i}{Dt} = \oint_S \left(-p\delta_{ij} + \sigma_{ij} \right) n_j dS + VF_i,$$

where the unit normal $\mathbf{n}$ is directed out of V. The surface integral can be transformed into an integral over the interior volume of the fluid element by application of the divergence theorem,

$$\oint_S \mathcal{F}(\mathbf{x}) n_j dS = \int_V \frac{\partial \mathcal{F}}{\partial x_j}(\mathbf{x}) d^3\mathbf{x}, \tag{1.4.1}$$

where $\mathcal{F}(\mathbf{x})$ is any scalar or vector field. Thus, as $V \to 0$,

$$\oint_S \left(-p\delta_{ij} + \sigma_{ij} \right) n_j dS \to V \frac{\partial}{\partial x_j} \left(-p\delta_{ij} + \sigma_{ij} \right),$$

and the momentum equation becomes

$$\rho \frac{Dv_i}{Dt} = -\frac{\partial p}{\partial x_i} + \frac{\partial \sigma_{ij}}{\partial x_j} + F_i. \tag{1.4.2}$$

The viscous stress is caused by the molecular diffusion of momentum between neighbouring fluid particles and is non-zero only when neighbouring particles are in relative motion; σ_{ij} must therefore depend on the velocity gradient.

1.4.1 Relative motion of neighbouring fluid elements

Let v_i denote the velocity at $\mathbf{x}$. The velocity $v_i + \delta v_i$ at a neighbouring point $\mathbf{x} + \delta\mathbf{x}$ at the same time is given to first order in $\delta\mathbf{x}$ by

$$\delta v_i = \delta x_j \frac{\partial v_i}{\partial x_j} \equiv \delta x_j e_{ij} + \delta x_j \xi_{ij}, \tag{1.4.3}$$

where

$$e_{ij} = \frac{1}{2}\left(\frac{\partial v_i}{\partial x_j} + \frac{\partial v_j}{\partial x_i}\right), \quad \xi_{ij} = \frac{1}{2}\left(\frac{\partial v_i}{\partial x_j} - \frac{\partial v_j}{\partial x_i}\right) \tag{1.4.4}$$

are, respectively, the symmetric and antisymmetric components of $\partial v_i/\partial x_j$.

The diagonal elements ($i = j$) of the 3×3 antisymmetric tensor ξ_{ij} are zero; the remaining six elements satisfy $\xi_{ij} = -\xi_{ji}$, and are therefore determined by three independent quantities ω_1, ω_2, and ω_3, say. We can then write

$$\xi_{ij} = -\frac{1}{2}\epsilon_{ijk}\omega_k, \tag{1.4.5}$$

where ϵ_{ijk} is the *alternating tensor* whose components are zero unless i, j, and k are all different, and then $\epsilon_{ijk} = \pm 1$ according to whether i, j, and k are or are not in cyclic order. We obtain an explicit representation of ξ_{ij} by identifying the i, j component of ξ_{ij} with the element in the ith row and jth column of a 3×3 matrix, i.e.,

$$\xi_{ij} = -\frac{1}{2}\begin{bmatrix} 0 & \omega_3 & -\omega_2 \\ -\omega_3 & 0 & \omega_1 \\ \omega_2 & -\omega_1 & 0 \end{bmatrix}.$$

By equating corresponding terms on the two sides of Eq. (1.4.5), we see that ω_i is just the ith component of the *vorticity* vector $\boldsymbol{\omega} = \text{curl } \mathbf{v}$,

$$\omega_1 = \frac{\partial v_3}{\partial x_2} - \frac{\partial v_2}{\partial x_3}, \quad \omega_2 = \frac{\partial v_1}{\partial x_3} - \frac{\partial v_3}{\partial x_1}, \quad \omega_3 = \frac{\partial v_2}{\partial x_1} - \frac{\partial v_1}{\partial x_2},$$

and that

$$\delta x_j \xi_{ij} = -\frac{1}{2}\epsilon_{ijk}\delta x_j \omega_k = \frac{1}{2}(\boldsymbol{\omega} \wedge \delta\mathbf{x})_i.$$

The symmetric array e_{ij} is called the *rate of strain* tensor. The sum of the diagonal elements,

$$e_{11} + e_{22} + e_{33} \equiv e_{kk} = \text{div } \mathbf{v},$$

is independent of the orientation of the coordinate axes. The contribution of e_{ij} to the relative velocity δv_i of Eq. (1.4.3) can be written as

$$\delta x_j e_{ij} = \frac{1}{2}\frac{\partial}{\partial \delta x_i}\left(e_{jk}\delta x_j \delta x_k\right),$$

where the differentiation is with respect to the displacement δx_i (in terms of which e_{jk} is constant).

Hence the velocity of the fluid at $\mathbf{x} + \delta\mathbf{x}$ relative to that at $\mathbf{x}$ can be written as

$$\delta\mathbf{v} = \frac{1}{2}\nabla\left(e_{ij}\delta x_i \delta x_j\right) + \frac{1}{2}\boldsymbol{\omega} \wedge \delta\mathbf{x}, \tag{1.4.6}$$

where the gradient is taken with respect to $\delta\mathbf{x}$. The term in $\boldsymbol{\omega}$ represents relative motion as a *rigid-body rotation*, at angular velocity $\frac{1}{2}\boldsymbol{\omega}$, with no distortion of the fluid particle.

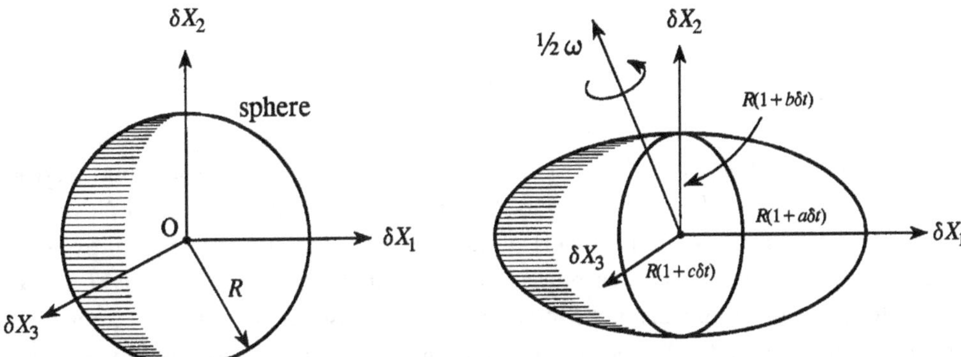

Figure 1.4.2

The gradient term, however, represents an *irrotational* distortion of the fluid element, in the direction of the normal at $\delta\mathbf{x}$ to the quadric surface:

$$\frac{1}{2}e_{ij}\delta x_i \delta x_j = C,$$

where C is a constant whose value is chosen to make the surface pass through the point $\delta\mathbf{x}$. By means of a suitable rotation of the local coordinate axes at $\mathbf{x}$, from δx_i to δX_i, say, we can transform the quadric to the normal form:

$$\frac{1}{2}\left[a(\delta X_1)^2 + b(\delta X_2)^2 + c(\delta X_3)^2\right] = C, \tag{1.4.7}$$

where a, b, and c are called the principal rates of strain; their sum is an invariant of the coordinate transformation that satisfies

$$a + b + c = e_{kk} \equiv \operatorname{div}\mathbf{v}.$$

The distortion produced by e_{ij} can now be seen to be one of *pure strain*. For, if the fluid element at $\mathbf{x}$ was initially a sphere of radius R (Figure 1.4.2), after time δt it is deformed without rotation into an ellipsoid with semi-axes $R(1 + a\delta t)$, $R(1 + b\delta t)$, and $R(1 + c\delta t)$, respectively, along the directions of the principal axes δX_1, δX_2, δX_3, with change in volume equal to

$$\frac{4}{3}\pi R^3(a + b + c)\delta t \equiv \frac{4}{3}\pi R^3 \operatorname{div}\mathbf{v}\delta t,$$

which vanishes when the fluid is incompressible.

1.4.2 Viscous stress tensor

There is no relative motion between neighbouring points of a fluid particle in solid-body rotation. Therefore there can be no viscous force $\sigma_{ij}n_j$ on a surface element with normal $\mathbf{n}$ that separates these points. Thus in a first approximation we assume that σ_{ij} is a *linear* combination of those gradients $\partial v_i/\partial x_j$ of the velocity that represent a purely straining motion of the fluid, that is, of relative motion defined by the strain tensor e_{ij}.

Put

$$e_{ij} = \left(e_{ij} - \frac{1}{3}e_{kk}\delta_{ij}\right) + \frac{1}{3}e_{kk}\delta_{ij}.$$

The first term on the right-hand side represents a straining motion involving no net change in volume [with principal axes of strain a, b, c of the corresponding quadric (1.4.7) satisfying $a + b + c = 0$], whereas the second term describes an isotropic dilatation of a fluid element (so that a spherical fluid particle remains spherical but expands or contracts to a new size). Such physically different straining motions might be expected to make essentially distinct contributions to the viscous stress tensor. If the fluid properties are assumed to be locally isotropic (independent of the orientation of local coordinate axes at any point in the fluid), we can therefore set

$$\sigma_{ij} = 2\eta\left(e_{ij} - \frac{1}{3}e_{kk}\delta_{ij}\right) + \eta'e_{kk}\delta_{ij}, \tag{1.4.8}$$

where η and η' are called, respectively, the **shear** and **bulk** coefficients of viscosity. They generally vary with both the pressure and temperature and with position in the fluid. The bulk coefficient of viscosity η' vanishes for monatomic gases, and in this case (and for most liquids, such as water) the fluid is said to be 'Stokesian', with

$$\sigma_{ij} = 2\eta\left(e_{ij} - \frac{1}{3}e_{kk}\delta_{ij}\right). \tag{1.4.9}$$

When velocity gradients are present the fluid cannot be in strict thermodynamic equilibrium, and thermodynamic variables, such as the pressure and density, require special interpretation. For a fluid in non-uniform motion, it is usual to define the density ρ and internal energy e per unit mass in the usual way, such that ρ and ρe are the mass and internal energy per unit volume, respectively. The pressure and all other thermodynamic quantities are then defined by means of the same functions of ρ and e that would be used for a system in thermal equilibrium. However, the *thermodynamic pressure* $p = p(\rho, e)$ so defined is then no longer the sole source of normal stress in the fluid. We obtain the mean normal stress at $\mathbf{x}$ by averaging $(p\delta ij - \sigma_{ij})n_i n_j$ over all possible orientations of a unit vector $\mathbf{n}$ at $\mathbf{x}$. We do this by evaluating the following integral over the surface of the unit sphere,

$$\frac{1}{4\pi}\oint n_i n_j dS = \frac{1}{3}\delta_{ij},$$

and multiplying by $(p\delta ij - \sigma_{ij})$, to find

$$\text{mean normal stress} = p - \eta'e_{kk} \equiv p - \eta'\operatorname{div}\mathbf{v}.$$

The mean normal stress therefore differs from the thermodynamic pressure p if the bulk coefficient of viscosity η' is non-zero. This happens in a fluid whose molecules possess rotational (or other internal) degrees of freedom whose relaxation time (during which thermal equilibrium is re-established after, say, a compression of the fluid) is large relative to the equilibration time of the translational degrees of freedom.

For example, when a diatomic gas is compressed ($\mathrm{div}\,\mathbf{v} < 0$) the temperature must rise, but the corresponding increase in the rotational energy lags slightly behind that of the translational energy; the thermodynamic pressure $p = (\gamma - 1)\rho e$ (γ = specific-heat ratio) accordingly is smaller than the actual pressure $p - \eta'\mathrm{div}\,\mathbf{v}$ by an amount equal to $-\eta'\mathrm{div}\,\mathbf{v}$.

It may be shown (Landau & Lifshitz 1987) that, whereas the thermodynamic pressure differs from the mean normal stress by a term linear in $\mathrm{div}\,\mathbf{v}$, the corresponding departure of the thermodynamic entropy s (per unit mass) from the true entropy is proportional at least to the square of such gradients, and the difference is usually small in practice. This can be deduced from a consideration of thermodynamic relation (1.5.4) given in the next section.

1.4.3 Navier–Stokes equation

When the variations of η and η' can be neglected, the substitution of Eq. (1.4.8) into momentum equation (1.4.2) yields the *Navier–Stokes* equation:

$$\rho\frac{D\mathbf{v}}{Dt} = -\nabla p + \eta\nabla^2\mathbf{v} + \left(\eta' + \frac{1}{3}\eta\right)\nabla\,\mathrm{div}\,\mathbf{v} + \mathbf{F}. \tag{1.4.10}$$

By means of the vector identity $\mathrm{curl}\,\mathrm{curl} = \nabla\,\mathrm{div} - \nabla^2$, we can also write

$$\rho\frac{D\mathbf{v}}{Dt} = -\nabla p - \eta\,\mathrm{curl}\,\omega + \left(\eta' + \frac{4}{3}\eta\right)\nabla\,\mathrm{div}\,\mathbf{v} + \mathbf{F}. \tag{1.4.11}$$

1.4.4 The Reynolds equation and Reynolds stress

We obtain an equation for the rate of change of momentum density ρv_i by adding continuity equation (1.3.3) multiplied by v_i to momentum equation (1.4.2), and writing the result in the form

$$\frac{\partial(\rho v_i)}{\partial t} = -\frac{\partial \pi_{ij}}{\partial x_j} + F_i, \tag{1.4.12}$$

where

$$\pi_{ij} = p\delta_{ij} + \rho v_i v_j - \sigma_{ij} \tag{1.4.13}$$

is the called the *momentum flux tensor*.

Equation (1.4.12) is the *Reynolds equation*. By integrating it over the interior volume V of a *fixed* control surface S and applying the divergence theorem, we can write

$$\frac{\partial}{\partial t}\int_V \rho v_i\,d^3\mathbf{x} = -\oint_S \pi_{ij}\,dS_j + \int_V F_i\,d^3\mathbf{x}$$

$$\equiv \oint_S \left(-p\,dS_i - \rho v_i v_j\,dS_j + \sigma_{ij}\,dS_j\right) + \int_V F_i\,d^3\mathbf{x}, \tag{1.4.14}$$

where the surface element dS_i is directed *out* of V. The terms in the surface integral on the second line respectively represent the flux of i momentum through S into V produced by the surface pressure p, by the *Reynolds stress* $-\rho v_i v_j$ (by the convection of momentum ρv_i per unit volume by the normal component of the velocity v_j), and by the action of frictional forces on S.

1.5 The energy equation

The energy equation is derived from a consideration of the total energy of the fluid: the kinetic energy of the gross fluid motions and the thermodynamic 'internal' energy. The equation governs the dissipation of mechanical energy and its transformation into heat.

Consider a small fluid element of volume V bounded by a surface S with unit *outward* normal $\mathbf{n}$ (Figure 1.4.1). The kinetic and internal energies per unit volume are equal respectively to $\frac{1}{2}\rho v^2$ and ρe, and the total energy of the fluid in V is $E = \rho V(\frac{1}{2}v^2 + e)$. Changes in E are produced by the work done by the pressure and viscous frictional forces on the boundary S, by the flux of heat energy through S by molecular diffusion, and by the work performed by the body force $\mathbf{F}$ within V. Because the mass in V is conserved $[D(\rho V)/Dt = 0]$, we can write

$$V\rho \frac{D}{Dt}\left(\frac{1}{2}v^2 + e\right) = \oint_S \left(-pn_i + \sigma_{ij}n_j\right) v_i\,dS + \oint_S \kappa \frac{\partial T}{\partial x_j} n_j\,dS + VF_i v_i,$$

where T is the temperature and κ is the thermal conductivity of the fluid. Using the divergence theorem (for small V), dividing through by V, and expanding divergence derivatives on the right-hand side by the product rule, we find

$$\rho \frac{D}{Dt}\left(\frac{1}{2}v^2\right) + \rho \frac{De}{Dt} = -v_i \frac{\partial p}{\partial x_i} - p\,\text{div}\,\mathbf{v} + v_i \frac{\partial \sigma_{ij}}{\partial x_j} + \sigma_{ij}\frac{\partial v_i}{\partial x_j} + \frac{\partial}{\partial x_j}\left(\kappa \frac{\partial T}{\partial x_j}\right) + F_i v_i.$$

This is greatly simplified by subtraction of the product of v_i and momentum equation (1.4.2) to obtain

$$\rho \frac{De}{Dt} = -p\,\text{div}\,\mathbf{v} + \sigma_{ij}\frac{\partial v_i}{\partial x_j} + \frac{\partial}{\partial x_j}\left(\kappa \frac{\partial T}{\partial x_j}\right). \tag{1.5.1}$$

We obtain a more useful form of this equation by first noting, from continuity equation (1.3.3) and from definitions (1.4.4) of e_{ij} and (1.4.8) of σ_{ij}, that

$$\text{div}\,\mathbf{v} = -\frac{1}{\rho}\frac{D\rho}{Dt}, \quad \sigma_{ij}\frac{\partial v_i}{\partial x_j} = \sigma_{ij}e_{ij} \equiv 2\eta\left(e_{ij} - \frac{1}{3}e_{kk}\delta_{ij}\right)^2 + \eta'(\text{div}\,\mathbf{v})^2,$$

so that Eq. (1.5.1) becomes

$$\rho \frac{De}{Dt} - \frac{p}{\rho}\frac{D\rho}{Dt} = 2\eta\left(e_{ij} - \frac{1}{3}e_{kk}\delta_{ij}\right)^2 + \eta'(\text{div}\,\mathbf{v})^2 + \text{div}\left(\kappa\nabla T\right). \tag{1.5.2}$$

The left-hand side can be expressed in terms of the specific entropy s of the fluid by application of the first law of thermodynamics to unit mass of fluid:

$$de = Tds - pdV.$$

If V is the volume occupied by unit mass, then $\rho V = 1$,

$$dV = d\left(\frac{1}{\rho}\right) = -\frac{1}{\rho^2}d\rho,$$

and therefore

$$Tds = de - \frac{p}{\rho^2}d\rho. \tag{1.5.3}$$

Hence energy equation (1.5.2) becomes

$$\rho T\frac{Ds}{Dt} = 2\eta\left(e_{ij} - \frac{1}{3}e_{kk}\delta_{ij}\right)^2 + \eta'(\operatorname{div}\mathbf{v})^2 + \operatorname{div}(\kappa\nabla T). \tag{1.5.4}$$

The quantity $\rho TDs/Dt$ is the time rate of change following the fluid particles of the heat gained per unit volume of fluid. The term $2\eta(e_{ij} - \frac{1}{3}e_{kk}\delta_{ij})^2 + \eta'(\operatorname{div}\mathbf{v})^2 > 0$ is the rate of production of heat by frictional dissipation of macroscopic motions, i.e., the rate at which mechanical energy is dissipated per unit volume of the fluid; $\operatorname{div}(\kappa\nabla T)$ is the rate at which heat energy is gained per unit volume by molecular diffusion.

1.5.1 Alternative treatment of the energy equation

Let us use the identity

$$(\mathbf{v}\cdot\nabla)\mathbf{v} = \boldsymbol{\omega}\wedge\mathbf{v} + \nabla\left(\frac{1}{2}v^2\right) \tag{1.5.5}$$

to write momentum equation (1.4.2) in the form

$$\rho\frac{\partial v_i}{\partial t} + \rho\frac{\partial}{\partial x_i}\left(\frac{1}{2}v^2\right) + \frac{\partial p}{\partial x_i} = -\rho(\boldsymbol{\omega}\wedge\mathbf{v})_i + \frac{\partial\sigma_{ij}}{\partial x_j} + F_i.$$

Take the scalar product with v_i and use continuity equation (1.3.3) to obtain

$$\frac{\partial}{\partial t}\left(\frac{1}{2}\rho v^2\right) + \operatorname{div}\left(\rho\mathbf{v}\frac{1}{2}v^2\right) + \mathbf{v}\cdot\nabla p = v_i\frac{\partial\sigma_{ij}}{\partial x_j} + F_i v_i. \tag{1.5.6}$$

This result is further transformed by introduction of the *enthalpy* w, defined by

$$w = e + \frac{p}{\rho}, \tag{1.5.7}$$

in terms of which the first law (1.5.3) becomes

$$dw = Tds + \frac{dp}{\rho}. \tag{1.5.8}$$

Then a simple calculation shows that

$$\mathbf{v}\cdot\nabla p = \frac{\partial}{\partial t}(\rho e) + \operatorname{div}(\rho\mathbf{v}w) - \rho T\frac{Ds}{Dt},$$

and therefore that Eq. (1.5.6) becomes

$$\frac{\partial}{\partial t}\left(\frac{1}{2}\rho v^2 + \rho e\right) + \frac{\partial}{\partial x_j}\left[\rho v_j\left(w + \frac{1}{2}v^2\right) - v_i\sigma_{ij}\right] = \rho T\frac{Ds}{Dt} - \sigma_{ij}\frac{\partial v_i}{\partial x_j} + F_i v_i. \quad (1.5.9)$$

This equation shows how the overall energy of the fluid is coupled to the production of heat within the fluid and the work done by viscous stresses and the body force $\mathbf{F}$. For an isentropic, inviscid fluid,

$$\frac{\partial}{\partial t}\left(\frac{1}{2}\rho v^2 + \rho e\right) + \operatorname{div}\left[\rho\mathbf{v}\left(w + \frac{1}{2}v^2\right)\right] = \mathbf{F}\cdot\mathbf{v}. \quad (1.5.10)$$

Let this equation be integrated over the interior V of a *fixed* control surface S:

$$\frac{\partial}{\partial t}\int_V\left(\frac{1}{2}\rho v^2 + \rho e\right)d^3\mathbf{x} = -\oint_S\left[\mathbf{v}\left(\frac{1}{2}\rho v^2 + \rho e\right) + p\mathbf{v}\right]\cdot d\mathbf{S} + \int_V \mathbf{F}\cdot\mathbf{v}\,d^3\mathbf{x},$$

where the surface element $d\mathbf{S}$ is directed out of V. This equates the rate of increase of energy inside S to the sum of its rate of convection across S by the flow velocity $\mathbf{v}$, and to the rates of working of the ambient pressure on S and the body force in V. In a viscous fluid the surface integral is augmented by the contribution

$$\oint_S v_i\sigma_{ij}n_j\,dS,$$

which represents the rate of working by frictional forces on the boundary S. In addition, the remaining terms on the right-hand side of Eq. (1.5.9) (other than the body force $\mathbf{F}$) represent the net energy gain within S by heat addition. Indeed,

$$\sigma_{ij}\frac{\partial v_i}{\partial x_j} = 2\eta\left(e_{ij} - \frac{1}{3}e_{kk}\delta_{ij}\right)^2 + \eta'(\operatorname{div}\mathbf{v})^2$$

$$= \text{rate of frictional heating per unit volume.}$$

Therefore, if $\mathbf{Q} = -\kappa\nabla T$ is the *heat flux* vector, so that the rate at which heat flows into S is just $-\oint_S\mathbf{Q}\cdot d\mathbf{S}$, then

$$\rho T\frac{Ds}{Dt} = \sigma_{ij}\frac{\partial v_i}{\partial x_j} - \operatorname{div}\mathbf{Q}. \quad (1.5.11)$$

This is just Equation (1.5.4).

1.5.2 Energy equation for incompressible flow

When the flow is incompressible, the energy equation is merely a linear combination of the continuity and momentum equations. However, the special case of flow subject to a conservative body force $\mathbf{F} = \rho\nabla\Phi(\mathbf{x})$ is of particular interest. By using the third form of continuity equation (1.3.3), we can write Eq. (1.5.6) as

$$\frac{\partial}{\partial t}\left(\frac{1}{2}\rho v^2 + \rho\Phi\right) + \frac{\partial}{\partial x_j}\left[v_j\left(p - \rho\Phi + \frac{1}{2}\rho v^2\right) - v_i\sigma_{ij}\right] = -2\eta e_{ij}^2. \quad (1.5.12)$$

For an incompressible fluid of uniform density $\rho = \rho_o = $ constant, the term $\rho\Phi$ in the time derivative can be omitted.

The term $2\eta e_{ij}^2$ on the right-hand side of Eq. (1.5.12) is the rate at which mechanical energy is dissipated per unit volume by the action of viscosity. It is non-zero in any region where the fluid particles are being *strained*, i.e., where material line elements suffer extension or contraction. Thus the only case in which motion can occur without dissipation is that in which the fluid as a whole translates and rotates as a rigid body.

1.6 Summary of governing equations

The coefficients η, η', k are functions of the temperature and pressure and generally vary with position and time in the flow. In many cases these variations are small enough to be neglected, however, and this will be assumed to be the case in all of the applications to be discussed below.

INCOMPRESSIBLE FLOW For an incompressible fluid (such as water, or air at low *Mach* numbers) energy equation (1.5.4) is not required for determining the motion. The governing equations are

$$\text{continuity,} \qquad \operatorname{div}\mathbf{v} = 0; \tag{1.6.1}$$

$$\text{momentum,} \qquad \rho\frac{D\mathbf{v}}{Dt} = -\nabla p + \eta\nabla^2\mathbf{v} + \mathbf{F}; \tag{1.6.2}$$

$$\text{state,} \qquad \frac{D\rho}{Dt} = 0. \tag{1.6.3}$$

The third equation can be dispensed with in a fluid of uniform density, $\rho = \rho_0$, say.

COMPRESSIBLE FLOW For compressible flow all of the equations must be retained:

$$\text{continuity,} \qquad \operatorname{div}\mathbf{v} + \frac{1}{\rho}\frac{D\rho}{Dt} = 0; \tag{1.6.4}$$

$$\text{momentum,} \qquad \rho\frac{D\mathbf{v}}{Dt} = -\nabla p - \eta\operatorname{curl}\boldsymbol{\omega} + \left(\eta' + \frac{4}{3}\eta\right)\nabla\operatorname{div}\mathbf{v} + \mathbf{F}; \tag{1.6.5}$$

$$\text{energy,} \qquad \rho T\frac{Ds}{Dt} = 2\eta\left(e_{ij} - \frac{1}{3}e_{kk}\delta_{ij}\right)^2 + \eta'(\operatorname{div}\mathbf{v})^2 + \operatorname{div}\left(\kappa\nabla T\right); \tag{1.6.6}$$

$$\text{state,} \qquad p = p(\rho, T), \quad s = s(\rho, T), \ \text{etc.} \tag{1.6.7}$$

Equations of state (1.6.7) permit any thermodynamic variable to be expressed in terms of any two variables, such as the density and temperature, although in applications it may be more convenient to use other such equations.

For example, if it is permissible to assume *homentropic* flow, the equation $s = $ constant provides a relation between the thermodynamic variables that enables the motion to be determined from the equations of continuity and momentum together with the equation of state $p = p(\rho, s)$, and the energy equation is then ignored. In more general situations, in which s is variable, it is necessary to retain the energy equation to account for coupling between macroscopic motions and the internal energy of the fluid.

1.7 Boundary conditions

The physical continuity of the fluid requires that the normal component of velocity at an impermeable boundary should be equal to that of the boundary.

Let a moving impermeable boundary S be specified by the equation $f(\mathbf{x}, t) = 0$, where $f > 0$ in the fluid, and let $\mathbf{u}$ denote the velocity of point $\mathbf{x}$ on the surface at time t. After a short interval δt we must have

$$f(\mathbf{x} + \mathbf{u}\delta t, t + \delta t) = 0, \quad \text{and therefore} \quad \frac{\partial f}{\partial t} + \mathbf{u} \cdot \nabla f = 0.$$

Because ∇f is parallel to the surface normal at $\mathbf{x}$ *directed into the fluid*, it follows that the normal component of velocity at $\mathbf{x}$, which must equal that of the fluid, is given by

$$v_n = \frac{\mathbf{u} \cdot \nabla f}{|\nabla f|} = -\frac{\partial f/\partial t}{|\nabla f|}.$$

This is equivalent to $\mathbf{u} \cdot \nabla f = \mathbf{v} \cdot \nabla f$, and the condition to be satisfied on the moving surface can also be expressed in the form

$$\frac{Df}{Dt} \equiv \frac{\partial f}{\partial t} + \mathbf{v} \cdot \nabla f = 0. \tag{1.7.1}$$

This expresses the *kinematic* condition that continuity of the fluid prevents the formation of 'holes' between the fluid and the boundary, so that the relative velocity of a fluid particle on the boundary must be in the tangential direction or must vanish.

All real fluids possess viscosity, which imposes an additional condition on the velocity. In most cases, experiment confirms that the fluid must satisfy the 'no-slip' condition, that the relative velocities of the surface and fluid must vanish on S. Exceptions occur, however, for example, for a rarefied gas for which a finite amount of tangential slip may be possible, and also when a real fluid is approximated by an *ideal* fluid, whose viscosity is assumed to vanish. In the latter case no condition can be imposed on the tangential component of velocity, and the fluid must be allowed to 'slide' over the surface without exerting any tangential stress.

PROBLEMS 1

1. The total entropy in an infinite fluid medium is $\int \rho s\, d^3\mathbf{x}$, where the integration is over the whole of the fluid. Use energy equation (1.5.4) and the divergence theorem to show that

$$\frac{\partial}{\partial t} \int \rho s\, d^3\mathbf{x} = \int \frac{2\eta}{T} \left(e_{ij} - \frac{1}{3} e_{kk}\delta_{ij} \right)^2 d^3\mathbf{x} + \int \frac{\eta'}{T} (\operatorname{div} \mathbf{v})^2 d^3\mathbf{x} + \int \frac{k(\nabla T)^2}{T^2} d^3\mathbf{x}.$$

Hence deduce that, because the total entropy can only increase, the coefficients η, η', and k must each be positive.

2. Consider the equilibrium of a small rectangular parallelopiped of homogeneous fluid. Show by taking moments about an edge that the viscous stress tensor σ_{ij} must be symmetric. Hence deduce relation (1.4.8) of σ_{ij} in terms of e_{ij}.

3. Show that in an unbounded irrotational flow

$$\int \left(e_{ij} - \frac{1}{3} e_{kk} \delta_{ij} \right)^2 d^3\mathbf{x} = \frac{2}{3} \int (\operatorname{div} \mathbf{v})^2 d^3\mathbf{x}.$$

4. Show that in an unbounded incompressible fluid the net rate of dissipation of mechanical energy can be expressed in the form

$$2\eta \int e_{ij}^2 d^3\mathbf{x} = \eta \int \boldsymbol{\omega} \cdot \boldsymbol{\omega}\, d^3\mathbf{x}.$$

5. Show that the material derivative $D\mathcal{A}/Dt$ of an arbitrary vector field $\mathcal{A}$ can be expressed in the invariant form

$$\frac{D\mathcal{A}}{Dt} = \frac{\partial \mathcal{A}}{\partial t} + \frac{1}{2}\Big[\nabla(\mathbf{v} \cdot \mathcal{A}) + \boldsymbol{\omega} \wedge \mathcal{A} - \mathbf{v} \wedge \operatorname{curl} \mathcal{A} - \operatorname{curl}(\mathbf{v} \wedge \mathcal{A}) + \mathbf{v} \operatorname{div} \mathcal{A} - \mathcal{A} \operatorname{div} \mathbf{v} \Big].$$

Deduce that

$$\frac{D\mathbf{v}}{Dt} = \frac{\partial \mathbf{v}}{\partial t} + \nabla \left(\frac{1}{2} v^2 \right) + \boldsymbol{\omega} \wedge \mathbf{v}.$$

2

Potential Flow of an Incompressible Fluid

2.1 Ideal fluid

An ideal fluid is inviscid and does not conduct heat. If the fluid is incompressible and of uniform density ρ_o the motion is governed by the continuity and momentum equations:

$$\operatorname{div} \mathbf{v} = 0; \quad \frac{D\mathbf{v}}{Dt} = -\frac{1}{\rho_o} \nabla p + \frac{\mathbf{F}}{\rho_o}. \tag{2.1.1}$$

If the fluid is compressible we shall generally assume also that it is *homentropic*, with $s = s_o = \text{constant}$. Then $\rho = \rho(p, s_o)$ may be regarded as a function of the pressure alone. When the explicit dependence on s_o is suppressed, the equations of motion are

$$\operatorname{div} \mathbf{v} + \frac{1}{\rho}\frac{D\rho}{Dt} = 0; \quad \frac{D\mathbf{v}}{Dt} = -\nabla \int \frac{dp}{\rho(p)} + \frac{\mathbf{F}}{\rho}. \tag{2.1.2}$$

The order of magnitude of the ratio of the pressure and Reynolds stress terms to the viscous stresses in the Navier–Stokes equation $\sim \mathrm{Re} = v\ell/\nu$, where v and ℓ are characteristic velocities and length scales, respectively, of the problem at hand, $\nu = \eta/\rho$ is the *kinematic* viscosity, and Re is called the Reynolds number. The ideal fluid approximation requires at the very least that $\mathrm{Re} \gg 1$, although it may not be possible to satisfy this condition at all points of the flow.

2.2 Kelvin's circulation theorem

Let C be a closed material contour drawn in the fluid (Fig. 2.2.1); this moves with the fluid and therefore always passes through the same fluid particles. The **circulation** Γ around C is defined by

$$\Gamma = \oint_C \mathbf{v} \cdot d\mathbf{x}.$$

14

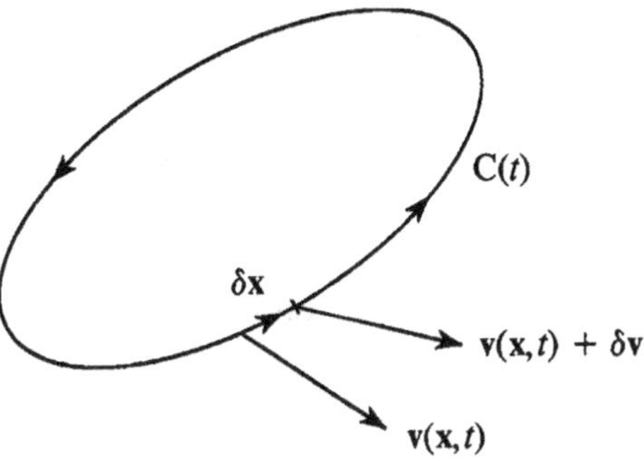

Figure 2.2.1

Stokes' theorem permits this to be expressed in terms of the vorticity:

$$\Gamma = \oint_C \mathbf{v} \cdot d\mathbf{x} = \int_S \operatorname{curl} \mathbf{v} \cdot d\mathbf{S} \equiv \int_S \omega \cdot d\mathbf{S}, \qquad (2.2.1)$$

where S is any two-sided surface bounded by C, and the sense of integration along C is positive with respect to the direction of the normal associated with the surface element $d\mathbf{S}$. When the fluid is ideal (compressible or incompressible) and the body force *per unit mass* $\mathbf{F}/\rho$ is *conservative*, so that there exists a potential function Φ such that

$$\mathbf{F} = \rho \nabla \Phi, \qquad (2.2.2)$$

the flow will evolve in such a way that the circulation around the moving contour remains constant:

$$\frac{D\Gamma}{Dt} = \frac{D}{Dt} \oint_C \mathbf{v} \cdot d\mathbf{x} = 0. \qquad (2.2.3)$$

This is **Kelvin's circulation theorem**. We prove it by taking the scalar product of the vector length element $d\mathbf{x}$ with the momentum equation of Eqs. (2.1.1) or (2.1.2) and integrating around C. To do this we use the definition of an integral as the limit of a sum $\sum_C \mathbf{v} \cdot \delta\mathbf{x} \to \oint_C \mathbf{v} \cdot d\mathbf{x}$ of discrete elements of arc on C of vector length $\delta\mathbf{x}$. For such an element, $D(\delta\mathbf{x})/Dt$ is just the difference in the velocities of the end points of $\delta\mathbf{x}$:

$$\frac{D}{Dt}\delta\mathbf{x} = \delta\mathbf{v}.$$

For a compressible fluid the momentum equation of Eqs. (2.1.2) and Eq. (2.2.2) yield

$$\sum_C \frac{D\mathbf{v}}{Dt} \cdot \delta\mathbf{x} \equiv \frac{D}{Dt} \sum_C \mathbf{v} \cdot \delta\mathbf{x} - \sum_C \mathbf{v} \cdot \delta\mathbf{v} = \sum_C \delta\mathbf{x} \cdot \nabla \left(-\int \frac{dp}{\rho} + \Phi \right),$$

and therefore, as $\delta\mathbf{x} \to 0$,

$$\frac{D}{Dt} \int_C \mathbf{v} \cdot d\mathbf{x} = \int_C d\left(-\int \frac{dp}{\rho} + \Phi + \frac{1}{2}v^2 \right) \equiv 0,$$

because the net change in the value of $\left(-\int \frac{dp}{\rho} + \Phi + \frac{1}{2}v^2\right)$ around C must vanish. The proof for incompressible flow is the same except that $\int \frac{dp}{\rho}$ is replaced with $\frac{p}{\rho_o}$.

The validity of Kelvin's theorem depends on continuity of pressure. On the other hand, it is easily seen that conservation of circulation is a necessary and sufficient condition for the pressure to be continuous. Indeed, if a velocity field $\mathbf{v}(\mathbf{x}, t)$ has been found by some means, then integration of the momentum equation (as in the proof of Kelvin's theorem) shows that $\mathbf{v}$ will represent a possible motion of the fluid only if

$$\frac{D}{Dt} \int_{\mathbf{a}(t)}^{\mathbf{b}(t)} \mathbf{v} \cdot d\mathbf{x} = \left(-\int \frac{dp}{\rho} + \Phi + \frac{1}{2}v^2\right)_{\mathbf{a}(t)}^{\mathbf{b}(t)}, \qquad (2.2.4)$$

where the integration is along a material path between any two points $\mathbf{a}(t)$ and $\mathbf{b}(t)$ (which move with the fluid). This in turn implies that the pressure is continuous only if the value of the left-hand side does not depend on the path between $\mathbf{a}$ and $\mathbf{b}$. Two such arbitrary paths form a closed material contour around which the circulation must be invariant.

It may be concluded that if any body of ideal fluid is initially moving 'irrotationally' ($\boldsymbol{\omega} = 0$) it will remain in that state for all time. Otherwise, according to Eq. (2.2.1), at some later time it would be possible to find a material circuit within the fluid body around which the circulation would cease to vanish. This means that motion produced from rest in an ideal fluid, say, by a moving boundary, is irrotational and remains irrotational. The implications of this conclusion will be considered in the remainder of this chapter for *incompressible* flow governed by Equations (2.1.1).

The proof of Kelvin's theorem requires the fluid to be inviscid. In practice this means that the theorem will be valid to a good approximation in a real fluid in those regions where viscous stresses are not important, for example, away from boundary layers and regions of high shear.

2.3 The velocity potential

If the vorticity is initially zero in an ideal fluid it must remain zero. The vanishing of curl $\mathbf{v}$ implies the existence of a velocity potential φ such that

$$\mathbf{v} = \nabla\varphi.$$

2.3.1 Bernoulli's equation

A first integral of the momentum equation [the second of Eqs. (2.1.2)] is obtained in terms of φ provided the body force $\mathbf{F} = \rho\nabla\Phi$. The vector identity

$$(\mathbf{v} \cdot \nabla)\mathbf{v} = \nabla\left(\frac{1}{2}v^2\right) + \boldsymbol{\omega} \wedge \mathbf{v}, \qquad (2.3.1)$$

with $\omega = 0$, is first used to write the momentum equation in the form

$$\nabla\left(\frac{\partial\varphi}{\partial t} + \int\frac{dp}{\rho} + \frac{1}{2}v^2 - \Phi\right) = 0.$$

The integral of this is called *Bernoulli's* equation:

$$\frac{\partial\varphi}{\partial t} + \int\frac{dp}{\rho} + \frac{1}{2}v^2 - \Phi = f(t), \tag{2.3.2}$$

where $f(t)$ is an arbitrary function of the time. This is the irrotational form of Eq. (2.2.4). The velocity potential φ is defined only up to an arbitrary function of time, so that $f(t)$ can usually be set equal to a constant or zero.

INCOMPRESSIBLE FLOW When the irrotational motion is incompressible the continuity equation div $\mathbf{v} = 0$ is equivalent to *Laplace's* equation,

$$\nabla^2\varphi = 0,$$

and Bernoulli's equation becomes

$$\frac{\partial\varphi}{\partial t} + \frac{p}{\rho_o} + \frac{1}{2}v^2 - \Phi = f(t). \tag{2.3.3}$$

In many applications the solution of Laplace's equation is governed entirely by the kinematic, or geometric, properties of the flow (as opposed to the *dynamics*), for example, by prescribed values of the normal component of velocity on a surface bounding the flow. Bernoulli's equation is then used to evaluate the pressure when the velocity potential has been so determined. However, in problems involving 'free surfaces', the pressure and velocity potential are coupled on the boundaries and must be found by the simultaneous solution of the equations.

A time-independent body force potential can often be discarded by balancing it against a mean pressure gradient. Thus, if the x_2 axis is vertically upwards, the gravitational potential can usually be approximated by $\Phi = -gx_2$ (where g is the acceleration due to gravity), and we can put $p = p' + p_o(x_2)$, where

$$\nabla\left(\frac{p_o}{\rho_o} + gx_2\right) = 0, \quad \text{so that} \quad p_o = \text{constant} - \rho_o gx_2. \tag{2.3.4}$$

Bernoulli's equation then determines the variations of the pressure relative to the local mean value p_o and may be written in the reduced form

$$\frac{p}{\rho_o} = -\frac{\partial\varphi}{\partial t} - \frac{1}{2}v^2 + f(t), \tag{2.3.5}$$

where the prime on p' has been dropped.

Equation (2.3.5) is used to calculate the pressure distribution in a general irrotational flow. For *steady* flow the pressure changes only if the velocity $\mathbf{v}$ changes with position and $p = \text{constant} - \frac{1}{2}\rho_o v^2$. This is called the *dynamic pressure*. For rapidly varying, time-dependent flows the pressure fluctuations are usually dominated by $-\rho_o\partial\varphi/\partial t$, and this is frequently referred to as the *transient pressure*.

2.3.2 Impulsive pressure

When a solid body immersed in a stationary, incompressible ideal fluid is suddenly set into motion, irrotational flow starts *everywhere* instantaneously. A fluid particle located at a distant point $\mathbf{x}$ is accelerated from rest by an infinitely large pressure pulse of infinitesimal duration, which is transmitted throughout the whole of the fluid from the surface of the solid. If the solid begins to move at time $t = 0$, the impulsive pressure $\hat{\varpi}$, say, is defined by

$$\hat{\varpi} \equiv \lim_{\delta t \to 0} \int_0^{\delta t} p(\mathbf{x}, t)dt = -\rho_o \lim_{\delta t \to 0} \left[\int_0^{\delta t} \frac{\partial \varphi}{\partial t} dt + \frac{1}{2} \int_0^{\delta t} v^2 dt - \int_0^{\delta t} f(t)dt \right].$$

The velocity potential vanishes at $t = 0$ but increases to the finite value characteristic of the final motion in the fluid produced by the impulsive start. The velocity $\mathbf{v}$, however, is finite during the starting motion, and the second term in the brackets therefore vanishes as $\delta t \to 0$. A finite contribution from the function $f(t)$ is of no practical significance, because it represents a quantity whose gradient is zero and cannot contribute to the impulsive start of the motion.

It follows from this that the actual fluid motion at any instant can be regarded as generated instantaneously *from a state of rest* by the application throughout the fluid of an

$$\text{impulsive pressure} \equiv \hat{\varpi} = -\rho_o \varphi(\mathbf{x}, t). \tag{2.3.6}$$

In the same way, a distributed impulsive pressure equal to $+\rho_o \varphi(\mathbf{x}, t)$ applied throughout the fluid at time t will *completely stop the fluid motion*.

The relationship between impulsive pressure and changes in the velocity potential can also be expressed in the following manner. Let $\mathbf{v}$, $\mathbf{v}'$ and φ, φ' denote the velocities and velocity potentials, respectively, just before and just after the application of an impulse $\hat{\varpi}$. Because impulse measures the change of momentum it follows (in the absence of extraneous impulses) that

$$\rho_o(\mathbf{v}' - \mathbf{v}) = -\nabla \left[\lim_{\delta t \to 0} \int_0^{\delta t} p(\mathbf{x}, t)dt \right] = -\nabla \hat{\varpi}.$$

However, $\mathbf{v}' - \mathbf{v} = \nabla \varphi' - \nabla \varphi \equiv \nabla[\varphi]$ (where $[\ ]$ denotes the jump in value),

$$\therefore \quad \hat{\varpi} = -\rho_o[\varphi].$$

The actual motion of the fluid (with a single-valued velocity potential) could therefore be produced from rest instantaneously by impulses properly applied throughout the fluid. Similarly, such a motion could be brought to rest by application of an impulsive pressure distribution.

2.3.3 Streamlines and intrinsic equations of motion

A curve drawn in the fluid at a given instant of time that is tangential to the velocity vector at each of its points is called a *streamline*. The vector line element $d\mathbf{x}$ on such a

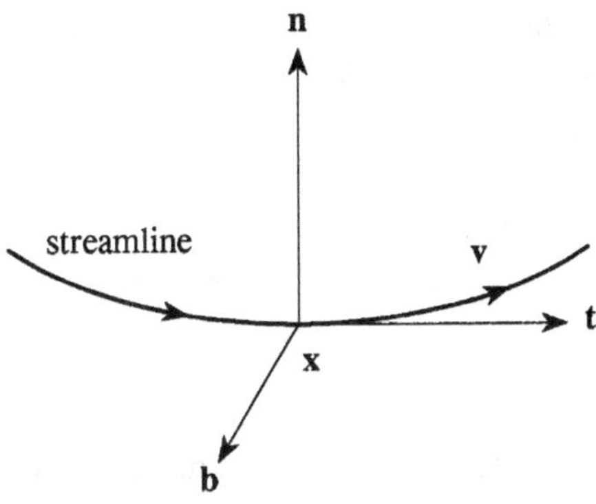

Figure 2.3.1

curve is parallel to $\mathbf{v}(\mathbf{x}, t)$, so that the family of streamlines is determined at any instant by the solutions of the equations

$$\frac{dx_1}{v_1(\mathbf{x}, t)} = \frac{dx_2}{v_2(\mathbf{x}, t)} = \frac{dx_3}{v_3(\mathbf{x}, t)}, \quad t = \text{constant.} \tag{2.3.7}$$

When $\mathbf{v}(\mathbf{x}, t)$ is known as a function of $\mathbf{x}$ for a given value of t, the integration of these equations from any given point $\mathbf{x} = \mathbf{x}_o$ will yield the equations for the streamline through $\mathbf{x}_o$ at time t.

A streamline coincides with the path traced out by the motion of a fluid particle only in the special case of *steady* flow. In general, the streamlines change continuously with time and they do not coincide with the particle paths.

To analyse the motion of a fluid particle along a streamline of a steady flow, it is convenient to introduce at the current point $\mathbf{x}$ on the streamline a local right-handed triad of mutually perpendicular unit vectors $(\mathbf{t}, \mathbf{n}, \mathbf{b})$ with $\mathbf{t}$ parallel to the streamline at $\mathbf{x}$, $\mathbf{n}$ in the direction of the *principal normal* to the streamline, and $\mathbf{b} = \mathbf{t} \wedge \mathbf{n}$ in the direction of the *binormal* (Figure 2.3.1). The triad rotates as a solid body as the particle moves along the streamline, such that if s denotes distance measured along the streamline, then

$$\frac{d\mathbf{t}}{ds} = \mathcal{D} \wedge \mathbf{t}, \quad \frac{d\mathbf{n}}{ds} = \mathcal{D} \wedge \mathbf{n}, \quad \frac{d\mathbf{b}}{ds} = \mathcal{D} \wedge \mathbf{b},$$

$$\text{where} \quad \mathcal{D} = \frac{\mathbf{b}}{\mathcal{R}} + \tau\mathbf{t}. \tag{2.3.8}$$

$\mathcal{D}$ is called the *Darboux vector*, $\mathcal{R}$ is the radius of curvature of the streamline at $\mathbf{x}$, and τ is the *torsion*, which measures the rate of 'twist' of the streamline.

Let $v(t)\mathbf{t}$ denote the velocity of a 'marked' fluid particle on the streamline at $s = s(t)$ at time t. The momentum equation (with the body force absorbed into the pressure) is then

$$\rho_o \frac{D\mathbf{v}}{Dt} \equiv \rho_o \left(\frac{dv}{dt}\mathbf{t} + v\frac{D\mathbf{t}}{Dt} \right) = -\nabla p,$$

$$\text{i.e.} \quad \rho_o \frac{dv}{dt} = -\frac{\partial p}{\partial s}, \quad \frac{\rho_o v^2}{\mathcal{R}} = -\frac{\partial p}{\partial n}, \tag{2.3.9}$$

where n denotes distance measured along the principal normal $\mathbf{n}$ to the streamline. These are the *intrinsic* equations of motion. The second of (2.3.9) applied to flow over a body whose local radius of curvature $\sim\mathcal{R}$ shows that the pressure gradient normal to the surface $\sim\rho_o v^2/\mathcal{R}$, which implies, for example, that the pressure change across a thin *viscous boundary layer* (thin compared with $\mathcal{R}$) formed on the surface will be small compared with the value of $\rho_o v^2$ in the irrotational region just outside the boundary layer.

2.3.4 Bernoulli's equation in steady flow

In steady flow the family of streamlines that pass through a fixed closed contour forms the boundary of a 'stream tube', on the surface of which the normal component of velocity is zero. Integration of the time-independent form of the continuity equation $\mathrm{div}(\rho\mathbf{v}) = 0$ over any section of the tube of finite length shows that the *mass flux* $\rho v\mathcal{A}$ is constant, where $\mathcal{A} = \mathcal{A}(\mathbf{x})$ is the local value of the cross-sectional area of the tube and v is the mean flow velocity. For incompressible motion the *volume flux* $= v\mathcal{A}$ is constant along the tube.

Similarly, integration of the steady form of inviscid energy equation (1.5.10) over an interior section of a stream tube, with $w = \int dp/\rho$ and $\mathbf{F} = \rho\nabla\Phi$, reveals that

$$\rho v\mathcal{A}\left(\int \frac{dp}{\rho} + \frac{1}{2}v^2 - \Phi\right) = \text{constant}.$$

Because the mass flux $\rho v\mathcal{A}$ is constant, we see that the energy equation reduces to the steady form of Bernoulli's equation.

In incompressible flow, for example, the relation

$$p + \frac{1}{2}\rho_o v^2 - \rho_o\Phi = \text{constant}$$

represents the balance of flow energy between the kinetic energy $\frac{1}{2}\rho_o v^2$ per unit volume and the potential energy $p - \rho_o\Phi$.

EXAMPLE 1. TRANSIENT RESPONSE TO IMPULSIVE ACTION The initial perturbation velocity produced by an impulse is always irrotational, whatever be the initial state of the fluid motion.

The velocity and its spatial gradients are finite during the impulse, and

$$\frac{D}{Dt} \sim \frac{\partial}{\partial t}.$$

Therefore integration of the momentum equation across the impulse yields

$$[\mathbf{v}] = -\nabla\left(\frac{1}{\rho_o}\int p\,dt\right),$$

showing that the change in velocity is described by a velocity potential.

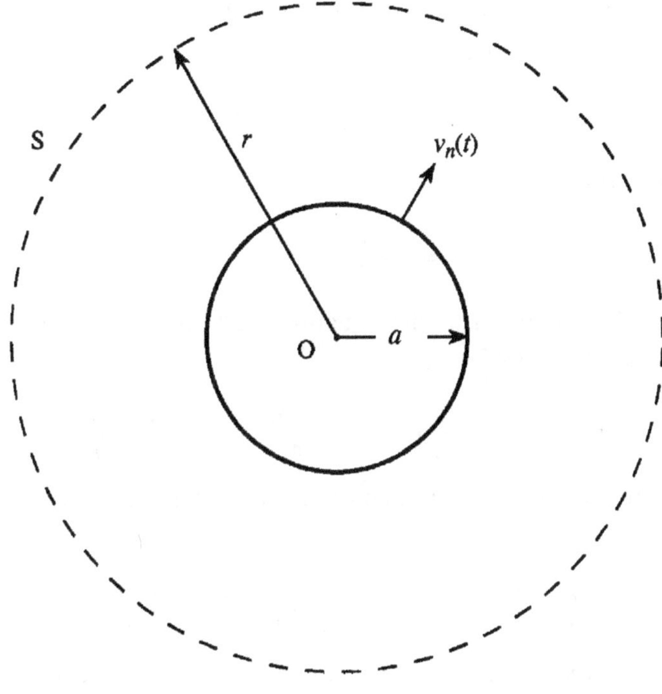

Figure 2.4.1

2.4 Motion produced by a pulsating sphere

In an infinite incompressible fluid of uniform density, irrotational motion is produced by moving boundaries. The simplest example is the radially symmetric flow generated by pulsations in the volume of a solid sphere (Figure 2.4.1). Let the centre of the sphere be at the origin, and let its radius at time t be denoted by $a(t)$, so that its normal velocity $v_n(t) = da(t)/dt$. The velocity potential therefore satisfies

$$\left.\begin{array}{ll} \nabla^2\varphi = 0, & r > a(t) \\[2mm] \partial\varphi/\partial r = v_n(t), & r = a(t) \end{array}\right\} \quad \text{where} \quad r = |\mathbf{x}|.$$

Because the motion is obviously radially symmetric,

$$\nabla^2\varphi \equiv \frac{1}{r^2}\frac{\partial}{\partial r}\left(r^2\frac{\partial}{\partial r}\right)\varphi = 0, \quad r > a.$$

Hence

$$\varphi = \frac{A}{r} + B,$$

where $A \equiv A(t)$, $B \equiv B(t)$ are functions of t. $B(t)$ can be discarded because the pressure fluctuations must vanish as $r \to \infty$. The condition that $\partial\varphi/\partial r = v_n$ when $r = a$ then yields

$$\varphi = -\frac{a^2 v_n(t)}{r}, \quad r > a(t). \tag{2.4.1}$$

At any time t the volume flux $q(t)$ of fluid is the same across any closed surface enclosing the sphere. Evaluating it for any sphere S of radius $r > a$ we find

$$q(t) = \oint_S \nabla\varphi \cdot d\mathbf{S} = 4\pi a^2 v_n(t),$$

and we may therefore write

$$\varphi = \frac{-q(t)}{4\pi r}, \quad r > a(t). \tag{2.4.2}$$

The function $f(t) \equiv 0$ in Bernoulli's equation (2.3.5) because φ, $p \to 0$ as $r \to \infty$. Then the pressure is given by

$$p = -\rho_0 \frac{\partial\varphi}{\partial t} - \frac{1}{2}\rho_0(\nabla\varphi)^2 = \frac{\rho_0}{4\pi r}\frac{dq}{dt}(t) - \frac{\rho_0 q^2(t)}{32\pi^2 r^4}.$$

In this formula the orders of magnitude of the two components of the pressure are, respectively,

$$\frac{\rho_0}{4\pi r}\frac{dq}{dt}(t) \sim \frac{3\rho_0\omega^2 a^3}{r}, \quad \frac{\rho_0 q^2(t)}{32\pi^2 r^4} \sim \frac{\rho_0\omega^2 a^6}{2r^4},$$

where ω is a characteristic frequency of the volume pulsations of the sphere. The second term is important only in the **near field**; it decreases rapidly with increasing distance r and can be neglected in the **far field** where $r \gg a$.

2.5 The point source

It is often useful to introduce an artificial generalization of the continuity equation by inserting a **volume source** distribution $q(\mathbf{x}, t)$ on the right-hand side. In the general case of a compressible fluid we would then write

$$\frac{1}{\rho}\frac{D\rho}{Dt} + \operatorname{div}\mathbf{v} = q(\mathbf{x}, t); \tag{2.5.1}$$

for an incompressible, irrotational flow this becomes

$$\nabla^2\varphi = q(\mathbf{x}, t). \tag{2.5.2}$$

The distribution q is the rate of increase of fluid volume per unit volume of the fluid and might represent, for example, the effect of volume pulsations of a small body in the fluid, as we shall now see.

The incompressible motion generated by a volume point source of strength $q(t)$ at the origin corresponds to the case $q(\mathbf{x}, t) \equiv q(t)\delta(\mathbf{x})$, where

$$\delta(\mathbf{x}) = \delta(x_1)\delta(x_2)\delta(x_3)$$

is the three-dimensional δ function. The velocity potential satisfies

$$\nabla^2\varphi = q(t)\delta(\mathbf{x}). \tag{2.5.3}$$

The solution must be radially symmetric and decay as $r = |\mathbf{x}| \to \infty$. Therefore

$$\varphi = \frac{A}{r} \quad \text{for } r > 0. \tag{2.5.4}$$

To find A we integrate Eq. (2.5.3) over the interior of a sphere of radius $r = R > 0$ and use the divergence theorem $\int_{r<R} \nabla^2 \varphi \, d^3\mathbf{x} = \oint_S \nabla\varphi \cdot d\mathbf{S}$, where S is the surface of the sphere. Then

$$\oint_S \nabla\varphi \cdot d\mathbf{S} \equiv \left(\frac{-A}{R^2}\right)(4\pi R^2) = q(t).$$

Hence $A = -q(t)/4\pi$ and $\varphi = -q(t)/4\pi r$, which agrees with solution (2.4.2) for the sphere with the same volume outflow in the region $r > a = $ radius of the sphere. This indicates that when we are interested in modelling the effect of a pulsating sphere at large distances $r \gg a$, it is permissible to replace the sphere with a point source (a 'monopole') of the same strength $q(t) = $ rate of change of the volume of the sphere. This conclusion is valid for any pulsating body, not just a sphere.

The point-source solution $\varphi = -q(t)/4\pi r$ is strictly valid only for $r > 0$, where it satisfies $\nabla^2\varphi = 0$. What happens as $r \to 0$, where its value is actually undefined? To answer this question we write the solution in the form

$$\varphi = \lim_{\epsilon \to 0} \frac{-q(t)}{4\pi (r^2 + \epsilon^2)^{\frac{1}{2}}}, \quad \epsilon > 0, \quad \text{in which case} \quad \nabla^2\varphi = \lim_{\epsilon \to 0} \frac{3\epsilon^2 q(t)}{4\pi (r^2 + \epsilon^2)^{\frac{5}{2}}}.$$

The last limit is just equal to $q(t)\delta(\mathbf{x})$. Indeed, when ϵ is small, $3\epsilon^2/4\pi (r^2 + \epsilon^2)^{\frac{5}{2}}$ is also small except close to $r = 0$, where it attains a large maximum $\sim 3/4\pi\epsilon^3$. Therefore, for any smoothly varying 'test' function $f(\mathbf{x})$ and any volume V enclosing the origin,

$$\lim_{\epsilon \to 0} \int_V \frac{3\epsilon^2 f(\mathbf{x}) d^3\mathbf{x}}{4\pi (r^2 + \epsilon^2)^{\frac{5}{2}}} = f(0) \lim_{\epsilon \to 0} \int_{-\infty}^{\infty} \frac{3\epsilon^2 d^3\mathbf{x}}{4\pi (r^2 + \epsilon^2)^{\frac{5}{2}}} = f(0) \int_0^{\infty} \frac{3\epsilon^2 r^2 dr}{(r^2 + \epsilon^2)^{\frac{5}{2}}} = f(0),$$

where the value of the last integral is independent of ϵ. This is the defining property of the three-dimensional δ function.

Thus the correct interpretation of the solution

$$\varphi = \frac{-1}{4\pi r} \quad \text{of } \nabla^2\varphi = \delta(\mathbf{x}) \tag{2.5.5}$$

for a *unit* point source ($q = 1$) is

$$\frac{-1}{4\pi r} = \lim_{\epsilon \to 0} \frac{-1}{4\pi (r^2 + \epsilon^2)^{\frac{1}{2}}}, \quad r \geq 0, \tag{2.5.6}$$

where

$$\nabla^2\left(\frac{-1}{4\pi r}\right) = \lim_{\epsilon \to 0} \nabla^2\left[\frac{-1}{4\pi (r^2 + \epsilon^2)^{\frac{1}{2}}}\right] = \lim_{\epsilon \to 0} \frac{3\epsilon^2}{4\pi (r^2 + \epsilon^2)^{\frac{5}{2}}} = \delta(\mathbf{x}). \tag{2.5.7}$$

2.6 Free-space Green's function

The free-space Green's function $G(\mathbf{x}, \mathbf{y})$ is the solution of Laplace's equation generated by a unit point source located at the point with position vector $\mathbf{y}$. We obtain the formula for G from solution (2.5.5) [with interpretation (2.5.6)] for a source at $\mathbf{x} = 0$ simply by replacing r with $|\mathbf{x} - \mathbf{y}|$. In other words, if

$$\nabla^2 G = \delta(\mathbf{x} - \mathbf{y}) \tag{2.6.1}$$

where $G \to 0$ as $|\mathbf{x}| \to \infty$, then

$$G(\mathbf{x}, \mathbf{y}) = \frac{-1}{4\pi |\mathbf{x} - \mathbf{y}|}. \tag{2.6.2}$$

This is the velocity potential of a radially symmetric flow from the point $\mathbf{y}$.

Green's function is the fundamental building block for constructing solutions of inhomogeneous Laplace equation (2.5.2),

$$\nabla^2 \varphi = q(\mathbf{x}, t),$$

where the source field $q(\mathbf{x}, t)$ is assumed to generate a disturbance that decays at large distances from the source region.

The continuous distribution $q(\mathbf{x}, t)$ can be regarded as an array of point sources of the type on the right-hand side of Eq. (2.6.1), because

$$q(\mathbf{x}, t) = \int_{-\infty}^{\infty} q(\mathbf{y}, t)\delta(\mathbf{x} - \mathbf{y})d^3\mathbf{y}.$$

The solution for each constituent source of strength $q(\mathbf{y}, t)\delta(\mathbf{x} - \mathbf{y})d^3\mathbf{y}$ is $q(\mathbf{y}, t)G(\mathbf{x}, \mathbf{y})d^3\mathbf{y}$, so that, by combining these individual contributions, we obtain

$$\varphi(\mathbf{x}, t) = \int_{-\infty}^{\infty} q(\mathbf{y}, t)G(\mathbf{x}, \mathbf{y})d^3\mathbf{y} \tag{2.6.3}$$

$$= \frac{-1}{4\pi} \int_{-\infty}^{\infty} \frac{q(\mathbf{y}, t)}{|\mathbf{x} - \mathbf{y}|}d^3\mathbf{y}. \tag{2.6.4}$$

This integral represents the potential at $\mathbf{x}$ as the *linear superposition* of contributions from sources at positions $\mathbf{y}$.

Observe that changes in the velocity potential $\varphi(\mathbf{x}, t)$ at an arbitrary point $\mathbf{x}$ in the fluid produced by changes with time t of the source strength $q(\mathbf{y}, t)$ are felt *instantaneously*. This is a property of an incompressible fluid; in reality such changes at $\mathbf{x}$ would occur after a time delay $\sim |\mathbf{x} - \mathbf{y}|/c_o$ equal to the time required for a sound wave to propagate from $\mathbf{y}$ to $\mathbf{x}$ at the *speed of sound* c_o.

2.7 Monopoles, dipoles, and quadrupoles

A volume source $q(t)\delta(\mathbf{x})$ of the type considered in equation §2.5 as a model for a pulsating sphere is also called a point *monopole*. We have seen that the velocity potential produced by a monopole of strength $q(t)$ concentrated at the origin is given by

$$\varphi(\mathbf{x}, t) = \frac{-q(t)}{4\pi |\mathbf{x}|}. \tag{2.7.1}$$

THE POINT DIPOLE Let $\mathbf{f} = \mathbf{f}(t)$ be a time-dependent vector; then a 'source' in Laplace equation (2.5.2) of the form

$$q(\mathbf{x}, t) = \mathrm{div}\Big[\mathbf{f}(t)\delta(\mathbf{x})\Big] \equiv \frac{\partial}{\partial x_j}\Big[f_j(t)\delta(\mathbf{x})\Big] \tag{2.7.2}$$

is called a *point dipole* (located at the origin).

The velocity potential that is due to the dipole can be calculated from (2.6.4), as follows:

$$\varphi(\mathbf{x}, t) = \frac{-1}{4\pi} \int_{-\infty}^{\infty} \frac{\partial}{\partial y_j}\Big[f_j(t)\delta(\mathbf{y})\Big]\frac{d^3\mathbf{y}}{|\mathbf{x} - \mathbf{y}|}.$$

Integrate by parts with respect to each y_j [recalling that $\delta(\mathbf{y}) = 0$ at $y_j = \pm\infty$], and note that

$$\frac{\partial}{\partial y_j}\frac{1}{|\mathbf{x} - \mathbf{y}|} = -\frac{\partial}{\partial x_j}\frac{1}{|\mathbf{x} - \mathbf{y}|}.$$

Then

$$\varphi(\mathbf{x}, t) = \frac{-1}{4\pi} \int_{-\infty}^{\infty} f_j(t)\delta(\mathbf{y})\frac{\partial}{\partial x_j}\left(\frac{1}{|\mathbf{x} - \mathbf{y}|}\right) d^3\mathbf{y}$$

$$= \frac{-1}{4\pi}\frac{\partial}{\partial x_j} \int_{-\infty}^{\infty} f_j(t)\delta(\mathbf{y})\frac{d^3\mathbf{y}}{|\mathbf{x} - \mathbf{y}|},$$

i.e.

$$\varphi(\mathbf{x}, t) = -\frac{\partial}{\partial x_j}\left[\frac{f_j(t)}{4\pi |\mathbf{x}|}\right] = \frac{x_j f_j(t)}{4\pi |\mathbf{x}|^3} \equiv \frac{f(t)\cos\theta}{4\pi r^2}, \tag{2.7.3}$$

where θ is the angle between $\mathbf{x}$ and $\mathbf{f}(t)$ and $r = |\mathbf{x}|$.

The same procedure shows that for a *distributed* dipole source of the type $q(\mathbf{x}, t) = \mathrm{div}\,\mathbf{f}(\mathbf{x}, t)$ on the right-hand side of Equation (2.5.2), the velocity potential becomes

$$\varphi(\mathbf{x}, t) = \frac{-1}{4\pi}\frac{\partial}{\partial x_j} \int_{-\infty}^{\infty} \frac{f_j(\mathbf{y}, t)}{|\mathbf{x} - \mathbf{y}|}d^3\mathbf{y}. \tag{2.7.4}$$

A point dipole at the origin orientated in the direction of a unit vector $\mathbf{n}$ is entirely equivalent to two point monopoles of equal but opposite strengths placed a short distance apart on opposite sides of the origin on a line through the origin parallel to $\mathbf{n}$. For example, if $\mathbf{n}$ is parallel to the positive x_1 direction and the sources are distance ϵ apart, the two monopoles would be

$$q(t)\delta\left(x_1 - \frac{\epsilon}{2}\right)\delta(x_2)\delta(x_3) - q(t)\delta\left(x_1 + \frac{\epsilon}{2}\right)\delta(x_2)\delta(x_3)$$

$$\approx -\epsilon q(t)\delta'(x_1)\delta(x_2)\delta(x_3) \equiv -\frac{\partial}{\partial x_1}\left[\epsilon q(t)\delta(\mathbf{x})\right] = -\mathrm{div}\left[\epsilon q(t)\mathbf{n}\delta(\mathbf{x})\right]. \qquad (2.7.5)$$

The vector direction of the dipole in this definition is always reckoned to be that of the vector $\mathbf{n}$ from the negative monopole $-q$ (the 'sink') to the $+q$ monopole.

QUADRUPOLES A source distribution involving two space derivatives is equivalent to a combination of four monopole sources (whose net volume source strength is zero) and is called a quadrupole. A general quadrupole is a source of the form

$$q(\mathbf{x}, t) = \frac{\partial^2 T_{ij}}{\partial x_i \partial x_j}(\mathbf{x}, t) \qquad (2.7.6)$$

in Eq. (2.5.2). The argument above leading to expression (2.7.4) can be applied twice to show that the corresponding velocity potential is given by

$$\varphi(\mathbf{x}, t) = \frac{-1}{4\pi}\frac{\partial^2}{\partial x_i \partial x_j}\int_{-\infty}^{\infty}\frac{T_{ij}(\mathbf{y}, t)}{|\mathbf{x} - \mathbf{y}|}\,d^3\mathbf{y}. \qquad (2.7.7)$$

At large distances from the source region it is clear from (2.7.1), (2.7.3), and (2.7.7) that in the *far field*, where $|\mathbf{x}| \to \infty$, the magnitudes of the respective velocity potentials decay as

$$\mathrm{monopole} \sim \frac{1}{|\mathbf{x}|}, \quad \mathrm{dipole} \sim \frac{1}{|\mathbf{x}|^2}, \quad \mathrm{quadrupole} \sim \frac{1}{|\mathbf{x}|^3}.$$

Thus the far field of a combination of different source types will be dominated by the potential of the *lowest-order multipole*.

2.7.1 The vibrating sphere

Let a rigid sphere of radius a execute small-amplitude oscillations at speed $U(t)$ in the x_1 direction [Figure 2.7.1(a)]. Take the coordinate origin at the mean position of the centre. In §2.11 we shall derive the velocity potential for the motion induced in an incompressible ideal fluid and show that it is equivalent to that produced by a point-volume dipole of strength $2\pi a^3 U(t)$ at its centre directed along the x_1 axis, determined by the solution of

$$\nabla^2\varphi = -\frac{\partial}{\partial x_1}\left[2\pi a^3 U(t)\delta(\mathbf{x})\right]. \qquad (2.7.8)$$

By analogy with (2.7.2) and (2.7.3), we have

$$\varphi(\mathbf{x}, t) = \frac{\partial}{\partial x_1}\left[\frac{2\pi a^3 U(t)}{4\pi |\mathbf{x}|}\right] = \frac{-a^3 x_1 U(t)}{2|\mathbf{x}|^3}. \qquad (2.7.9)$$

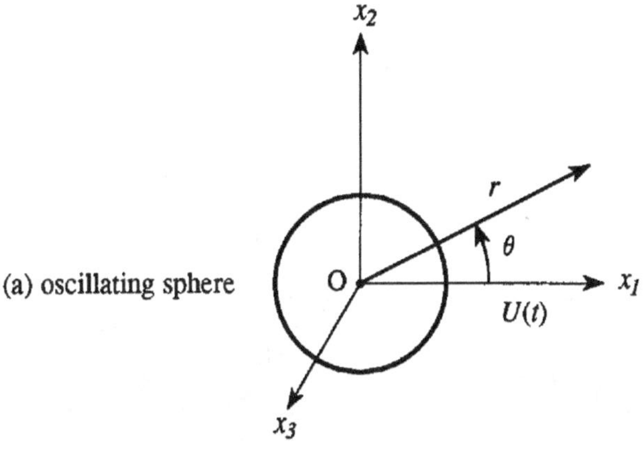

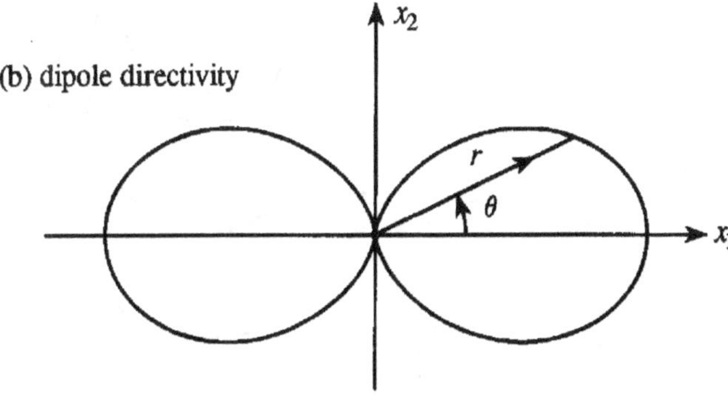

Figure 2.7.1

By setting $r = |\mathbf{x}|$ and $x_1 = r\cos\theta$, we can also write this as

$$\varphi(\mathbf{x}, t) = \frac{-a^3 U(t)\cos\theta}{2r^2}. \tag{2.7.10}$$

On the surface $r = a$ of the moving sphere, the sphere and fluid should have the same velocity $U(t)\cos\theta$ in the radial (i.e. *normal*) direction. This is readily seen to be satisfied by our solution, because

$$\frac{\partial\varphi}{\partial r} = \frac{a^3 U(t)\cos\theta}{r^3} = U(t)\cos\theta \quad \text{on } r = a.$$

For small-amplitude oscillations of the sphere the fluid particles move back and forth along path lines that closely approximate the dipole pattern shown in Figure 2.7.2. At each instant the effective dipole vector $\mathbf{n}$ is in the direction of motion of the sphere, the advancing 'front' of which behaves as a source and the 'rear' as a sink.

In the far field of the sphere the pressure

$$p \approx -\rho_0\frac{\partial\varphi}{\partial t} = \frac{\rho_0 a^3\cos\theta}{2r^2}\frac{dU}{dt}, \quad r \gg a.$$

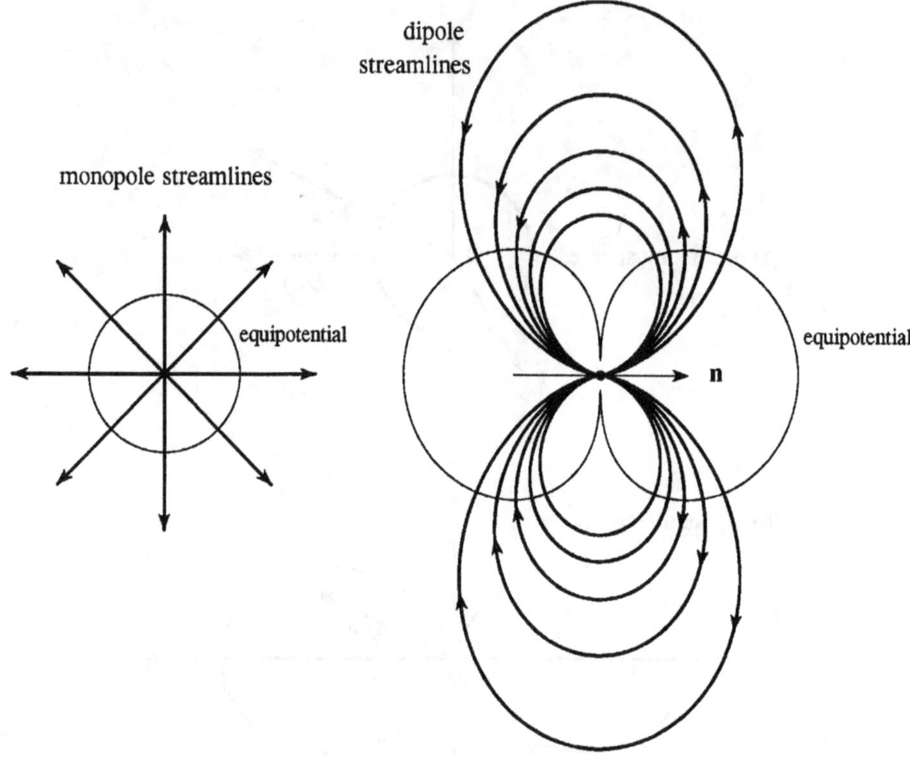

Figure 2.7.2

The mean-square pressure in the far field is therefore

$$< p^2 > \approx \frac{\rho_o^2 a^6}{4r^4} \left\langle \left(\frac{dU}{dt} \right)^2 \right\rangle \cos^2 \theta,$$

where the angle brackets $< >$ denote a time average. The dependence on θ determines the *directivity* of the pressure fluctuations. For the dipole it has the 'figure-of-eight' pattern illustrated in Figure 2.7.1(b), with peaks in directions parallel to the dipole axis ($\theta = 0, \pi$) (the curve should be imagined to be rotated about the x_1 axis).

2.7.2 Streamlines

The streamlines of the outflow from a point monopole are evidently parallel to the radius vector from the source, as illustrated in Figure 2.7.2. Because the streamline element $d\mathbf{x}$ is parallel to $\mathbf{v} = \nabla\varphi$, the streamlines are locally orthogonal to the *equipotential* surfaces $\varphi(\mathbf{x}, t) = $ constant, which for a monopole are spheres centred on the source point. The streamlines for a dipole are also illustrated in the figure. Suppose the dipole is formed [as in (2.7.5) with velocity potential (2.7.3)] by neighbouring equal and opposite monopoles orientated in the direction $\mathbf{n}$, which we temporarily take to coincide with the x_1 direction, or by the oscillating sphere of Figure 2.7.1(a) [with φ given

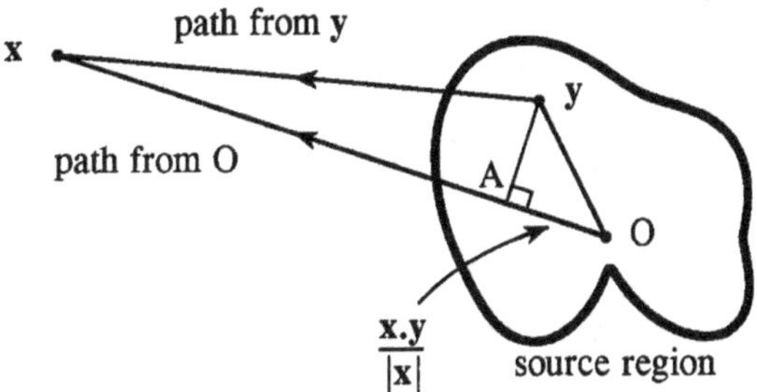

Figure 2.7.3

by (2.7.10)]; then the streamlines are determined by the axisymmetric polar form of Eqs. (2.3.7):

$$\frac{dr}{\partial\varphi/\partial r} = \frac{r\,d\theta}{\partial\varphi/r\partial\theta}, \quad \text{i.e.,} \quad \frac{dr}{r} = \frac{2\cos\theta\,d\theta}{\sin\theta}. \tag{2.7.11}$$

Therefore the streamlines consist of a family of re-entrant space curves, each of which has the polar representation

$$r = \text{constant} \times (\sin^2\theta), \tag{2.7.12}$$

where r is the radial distance from the dipole and θ is the angle between this ray and the $\mathbf{n}$ direction. The flow is outwards from the positive monopole and inwards towards the sink. In this case the equipotentials are the surfaces $r = \text{constant} \times |\cos\theta|^{\frac{1}{2}}$.

2.7.3 Far field of a monopole distribution of zero strength

It is frequently difficult to deduce the multipole nature of a source distribution by inspection. The far-field approximation,

$$\varphi(\mathbf{x},t) = \frac{-1}{4\pi}\int_{-\infty}^{\infty}\frac{q(\mathbf{y},t)}{|\mathbf{x}-\mathbf{y}|}d^3\mathbf{y} \approx \frac{-1}{4\pi|\mathbf{x}|}\int_{-\infty}^{\infty}q(\mathbf{y},t)d^3\mathbf{y}, \quad |\mathbf{x}|\to\infty, \tag{2.7.13}$$

fails if the net volume source strength $\int_{-\infty}^{\infty}q(\mathbf{y},t)d^3\mathbf{y}\equiv 0$, as is often the case in applications.

Suppose that $q(\mathbf{x},t)\neq 0$ only within a finite source region (Figure 2.7.3) and take the coordinate origin O within the region. When $|\mathbf{x}|\to\infty$ and $\mathbf{y}$ lies within the source region (so that $|\mathbf{x}|\gg|\mathbf{y}|$),

$$|\mathbf{x}-\mathbf{y}| \equiv \left(|\mathbf{x}|^2 - 2\mathbf{x}\cdot\mathbf{y} + |\mathbf{y}|^2\right)^{\frac{1}{2}} = |\mathbf{x}|\left(1 - \frac{2\mathbf{x}\cdot\mathbf{y}}{|\mathbf{x}|^2} + \frac{|\mathbf{y}|^2}{|\mathbf{x}|^2}\right)^{\frac{1}{2}}$$

$$\approx |\mathbf{x}|\left[1 - \frac{\mathbf{x}\cdot\mathbf{y}}{|\mathbf{x}|^2} + O\left(\frac{|\mathbf{y}|^2}{|\mathbf{x}|^2}\right)\right],$$

i.e., $\quad |\mathbf{x}-\mathbf{y}| \approx |\mathbf{x}| - \dfrac{\mathbf{x}\cdot\mathbf{y}}{|\mathbf{x}|} \quad$ when $\dfrac{|\mathbf{y}|}{|\mathbf{x}|}\ll 1. \tag{2.7.14}$

Also,

$$\frac{1}{|\mathbf{x} - \mathbf{y}|} \approx \frac{1}{\left(|\mathbf{x}| - \frac{\mathbf{x} \cdot \mathbf{y}}{|\mathbf{x}|}\right)} \approx \frac{1}{|\mathbf{x}|}\left(1 + \frac{\mathbf{x} \cdot \mathbf{y}}{|\mathbf{x}|^2}\right),$$

$$\therefore \quad \frac{1}{|\mathbf{x} - \mathbf{y}|} \approx \frac{1}{|\mathbf{x}|} + \frac{\mathbf{x} \cdot \mathbf{y}}{|\mathbf{x}|^3} \quad \text{when} \quad \frac{|\mathbf{y}|}{|\mathbf{x}|} \ll 1. \tag{2.7.15}$$

Thus, when $\int_{-\infty}^{\infty} q(\mathbf{y}, t)d^3\mathbf{y} \equiv 0$, we can derive a non-trivial estimate of the far-field value of the first integral in (2.7.13) by replacing $1/|\mathbf{x} - \mathbf{y}|$ with the right-hand side of (2.7.15):

$$\varphi(\mathbf{x}, t) = \frac{-1}{4\pi} \int_{-\infty}^{\infty} \frac{q(\mathbf{y}, t)}{|\mathbf{x} - \mathbf{y}|} d^3\mathbf{y}$$

$$\approx \frac{-1}{4\pi |\mathbf{x}|} \int_{-\infty}^{\infty} q(\mathbf{y}, t)\left(1 + \frac{\mathbf{x} \cdot \mathbf{y}}{|\mathbf{x}|^2}\right) d^3\mathbf{y}$$

$$= \frac{-x_j}{4\pi |\mathbf{x}|^3} \int_{-\infty}^{\infty} y_j q(\mathbf{y}, t)d^3\mathbf{y}, \quad |\mathbf{x}| \to \infty.$$

This shows that the far field is actually equivalent to that produced by a dipole source whose strength $\int_{-\infty}^{\infty} y_j q(\mathbf{y}, t)d^3\mathbf{y}$ is the first moment of the original source distribution.

If the dipole moment $\int_{-\infty}^{\infty} y_j q(\mathbf{y}, t)d^3\mathbf{y}$ should also vanish, the expansion of the integrand in powers of $\mathbf{y}/|\mathbf{x}|$ must be carried to higher order; the next term yields a *quadrupole* far field whose strength involves the second-moment integrals $\int_{-\infty}^{\infty} y_i y_j q(\mathbf{y}, t)d^3\mathbf{y}$ and decays as $1/|\mathbf{x}|^3$ as $|\mathbf{x}| \to \infty$.

2.8 Green's formula

Solution (2.6.4) of inhomogeneous Laplace equation (2.5.2) in an infinite fluid will now be generalized to include the influence of moving or stationary solid bodies immersed in the flow. The general procedure is also applicable to more general problems involving vorticity and is also the basis for the numerical computation of the flow. A system of mathematical **control surfaces** is introduced; the surfaces can be deformed to coincide with the surfaces of the moving or stationary bodies. We start by establishing a transformation formula for surface and volume integrals that is used repeatedly in problems of this kind.

2.8.1 Volume and surface integrals

Let V be the fluid *outside* a closed control surface S (Figure 2.8.1) defined by the equation

$$f(\mathbf{x}) = 0, \quad \text{where} \quad \begin{cases} f(\mathbf{x}) > 0 \text{ for } \mathbf{x} \text{ in } V \\ f(\mathbf{x}) < 0 \text{ for } \mathbf{x} \text{ inside } S \end{cases}, \tag{2.8.1}$$

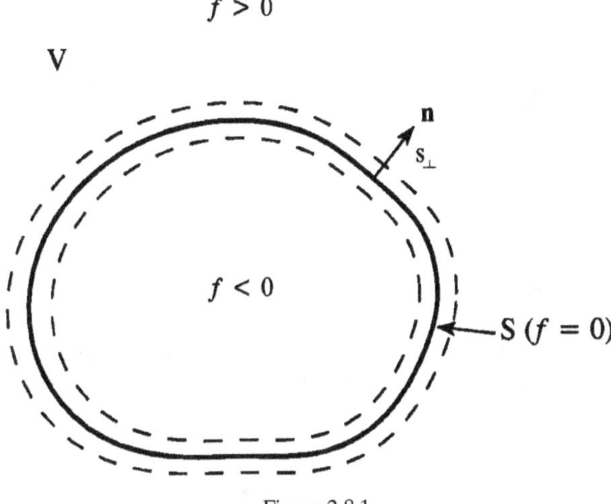

Figure 2.8.1

and consider the *Heaviside unit function*

$$H(f) = \begin{cases} 1 \text{ for } \mathbf{x} \text{ in V} \\ 0 \text{ for } \mathbf{x} \text{ inside S} \end{cases}.$$

Then, for an arbitrary function $\Phi(\mathbf{x})$ defined in V and on S,

$$\int_{-\infty}^{\infty} \Phi(\mathbf{x}) \nabla H d^3 \mathbf{x} = \oint_S \Phi(\mathbf{x}) \mathbf{n} dS \equiv \oint_S \Phi(\mathbf{x}) \, d\mathbf{S} \tag{2.8.2}$$

$$\text{or} \quad \int_{-\infty}^{\infty} \Phi(\mathbf{x}) \frac{\partial H}{\partial x_j} d^3 \mathbf{x} = \oint_S \Phi(\mathbf{x}) n_j dS \equiv \oint_S \Phi(\mathbf{x}) \, dS_j, \tag{2.8.3}$$

where $H \equiv H(f)$ and $\mathbf{n}$ is the unit normal on S directed into V.

PROOF

$$\nabla H(f) \equiv \delta(f) \nabla f \tag{2.8.4}$$

is non-zero only on S, where ∇f is in the direction of $\mathbf{n}$. The volume integral is therefore confined to the region between the inner and outer faces of a shell of infinitesimal thickness (between the broken line surfaces in Figure 2.8.1) that just encloses S, and in which the volume element is

$$d^3 \mathbf{x} = ds_{\perp} dS,$$

where $s_{\perp} = 0$ on S and $s_{\perp}$ is measured parallel to $\mathbf{n}$. Because $f = 0$ on S, we can write, for small values of $s_{\perp}$,

$$f = \left(\frac{\partial f}{\partial s_{\perp}} \right)_S s_{\perp},$$

where

$$\left(\frac{\partial f}{\partial s_{\perp}} \right)_S \equiv |\nabla f| > 0$$

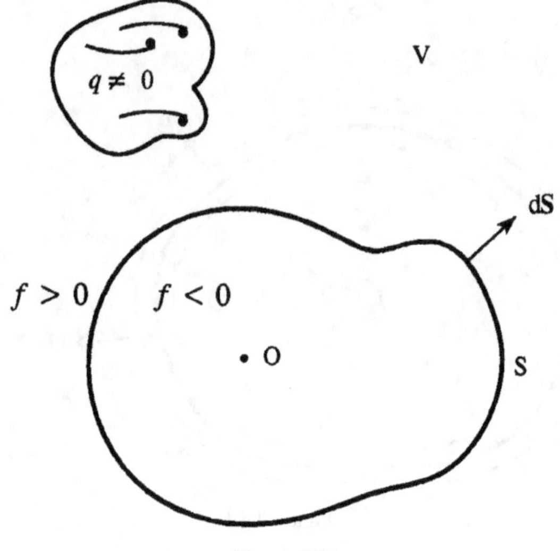

Figure 2.8.2

is evaluated on S,

$$\therefore \quad \delta(f) = \delta\left(|\nabla f|s_\perp\right) \equiv \frac{\delta(s_\perp)}{|\nabla f|}.$$

Hence

$$\int_{-\infty}^{\infty} \Phi(\mathbf{x})\nabla H d^3\mathbf{x} \equiv \int_{-\infty}^{\infty} \Phi(\mathbf{x})\nabla f\delta(f)d^3\mathbf{x} = \int_{-\infty}^{\infty} \Phi(\mathbf{x})\frac{\nabla f}{|\nabla f|}\delta(s_\perp)ds_\perp dS$$

$$= \oint_S \Phi(\mathbf{x})\mathbf{n}dS,$$

because

$$\mathbf{n} = \frac{\nabla f}{|\nabla f|}. \qquad\qquad\qquad \text{Q.E.D.}$$

2.8.2 Green's formula

Green's formula is a formal representation of the solution of inhomogeneous Laplace equation (2.5.2) in a region bounded by a system of arbitrary fixed or moving surfaces. To fix ideas, we shall consider a single closed surface S defined as previously by an equation $f(\mathbf{x}) = 0$, such that $f(\mathbf{x}) > 0$ in the fluid region V outside S (Figure 2.8.2). The surface may enclose a solid body or merely constitute a control surface used to isolate a fixed region of space containing both solid bodies and fluid or just fluid (perhaps a fluid region undergoing, say, a chemical reaction within which the flow cannot be regarded as irrotational).

To derive Green's formula Eq. (2.5.2) is multiplied by $H \equiv H(f)$. The identity

$$H\nabla^2\varphi \equiv \nabla \cdot (H\nabla\varphi) - \nabla H \cdot \nabla\varphi \equiv \nabla^2(H\varphi) - \nabla \cdot (\varphi\nabla H) - \nabla H \cdot \nabla\varphi \qquad (2.8.5)$$

then permits Eq. (2.5.2) to be written as

$$\nabla^2(H\varphi) = Hq + \mathrm{div}\,(\varphi\nabla H) + \nabla H \cdot \nabla\varphi. \qquad (2.8.6)$$

This is the differential form of Green's formula. The relation $\nabla H = \nabla f\delta(f)$ implies that the second and third terms on the right-hand side may respectively be regarded as *dipole* and *monopole* 'sources' distributed over the control surface S. These sources, together with the prescribed source distribution q in V, formally determine the velocity potential φ *in the exterior region* V [where $H(f) \equiv 1$]. However, Eq. (2.8.6) is valid throughout all space, including the region enclosed by S where $H(f)$ vanishes. The surface distribution of monopoles and dipoles have the following interpretation:

The control surface S will in general enclose fluid, possibly also solid bodies, and may or may not contain vortical or other non-irrotational disturbances; the surface dipole and monopole sources represent the influence of this region on the motion in V; in other words, the aggregate effect of the dipole and monopole sources accounts for the presence of solid bodies and other flow inhomogeneities within S and also for the interaction of the motion generated by the source q outside S with the fluid and solid bodies in S.

Because Eq. (2.8.6) is valid throughout all space, including the region within S, the solution that decays at large distances from S is found from general solution (2.6.4) of the Laplace equation in the absence of boundaries, by use of special form (2.7.4) for dipole sources. When account is taken of the transformation formulae (2.8.2) and (2.8.3), this yields **Green's formula:**

$$H\varphi(\mathbf{x}, t) = -\int_V \frac{q(\mathbf{y}, t)}{4\pi|\mathbf{x} - \mathbf{y}|}d^3\mathbf{y} - \frac{\partial}{\partial x_j}\oint_S \frac{\varphi(\mathbf{y}, t)n_j dS}{4\pi|\mathbf{x} - \mathbf{y}|} - \oint_S n_j \frac{\partial\varphi}{\partial y_j}(\mathbf{y}, t)\frac{dS}{4\pi|\mathbf{x} - \mathbf{y}|}.$$
$$(2.8.7)$$

Note that, because $H(f) \equiv 0$ inside S, the sum of the three integrals on the right-hand side must also vanish when field point $\mathbf{x}$ is within S.

Green's formula implies that $\varphi \sim O(1/|\mathbf{x}|)$ as $|\mathbf{x}| \to \infty$ provided that $\int_V q(\mathbf{y}, t)d^3\mathbf{y} \neq 0$. Otherwise φ decays typically as $1/|\mathbf{x}|^2$ at large distances from S, because

$$\oint_S n_j \frac{\partial\varphi}{\partial y_j}(\mathbf{y}, t)dS \equiv 0$$

in an incompressible flow, except, for example, when S coincides with the surface of a body whose volume varies as a function of time. However, naive estimates of the orders of magnitude of the surface integral terms in (2.8.7) as $|\mathbf{x}| \to \infty$ can sometimes be misleading when S is so large that the surface integrals extend into the far field.

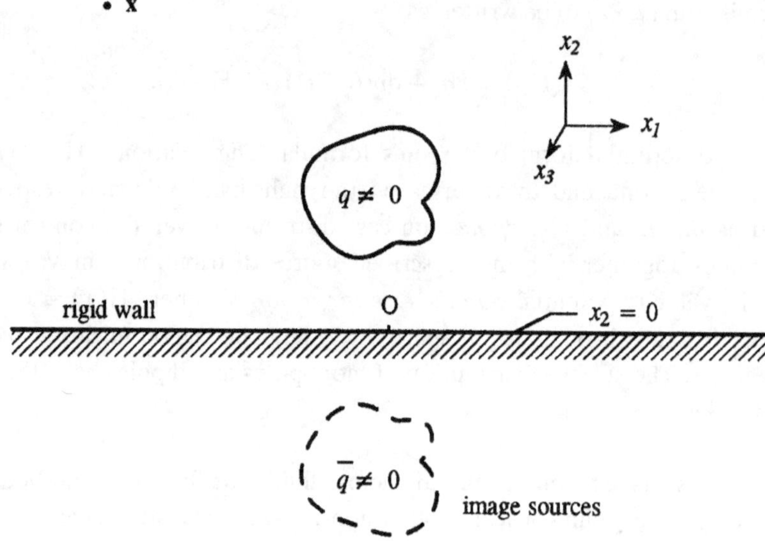

Figure 2.8.3

2.8.3 Sources adjacent to a plane wall

Consider a source distribution $q(\mathbf{x}, t)$ in $x_2 > 0$ adjacent to the infinite rigid wall $x_2 = 0$ in Figure 2.8.3. Let us apply Green's formula (2.8.7) to calculate the velocity potential at the point $\mathbf{x}$ in the fluid, where $H(f) \equiv H(x_2) = 1$:

$$\varphi(\mathbf{x}, t) = -\int \frac{q(\mathbf{y}, t)\, d^3\mathbf{y}}{4\pi |\mathbf{x} - \mathbf{y}|} + \oint_{y_2=0} \frac{x_2 \varphi(y_1, 0, y_3, t)\,dy_1 dy_3}{4\pi |\mathbf{x} - \mathbf{y}|^3}, \tag{2.8.8}$$

there being no contribution from the final monopole integral of (2.8.7) because $\partial\varphi/\partial y_2 = 0$ on the wall. The surface integral represents a dipole velocity potential that would normally be expected to decay as $1/|\mathbf{x}|^2$ as $|\mathbf{x}| \to \infty$.

However, by applying Green's formula (2.8.7) at the *image* $\bar{\mathbf{x}} = (x_1, -x_2, x_3)$ in the wall of the field point $\mathbf{x}$, where $H(f) = 0$, we find

$$0 = -\int \frac{q(\mathbf{y}, t)\, d^3\mathbf{y}}{4\pi |\bar{\mathbf{x}} - \mathbf{y}|} - \oint_{y_2=0} \frac{x_2 \varphi(y_1, 0, y_3, t)\,dy_1 dy_3}{4\pi |\bar{\mathbf{x}} - \mathbf{y}|^3}. \tag{2.8.9}$$

The surface integral is equal but opposite in sign to that in solution (2.8.8), because $|\mathbf{x} - \mathbf{y}| \equiv |\bar{\mathbf{x}} - \mathbf{y}|$ when $y_2 = 0$. Therefore, when (2.8.8) and (2.8.9) are added, the solution at $\mathbf{x}$ becomes

$$\varphi(\mathbf{x}, t) = -\int \frac{q(\mathbf{y}, t)\, d^3\mathbf{y}}{4\pi |\mathbf{x} - \mathbf{y}|} - \int \frac{q(\mathbf{y}, t)\, d^3\mathbf{y}}{4\pi |\bar{\mathbf{x}} - \mathbf{y}|}. \tag{2.8.10}$$

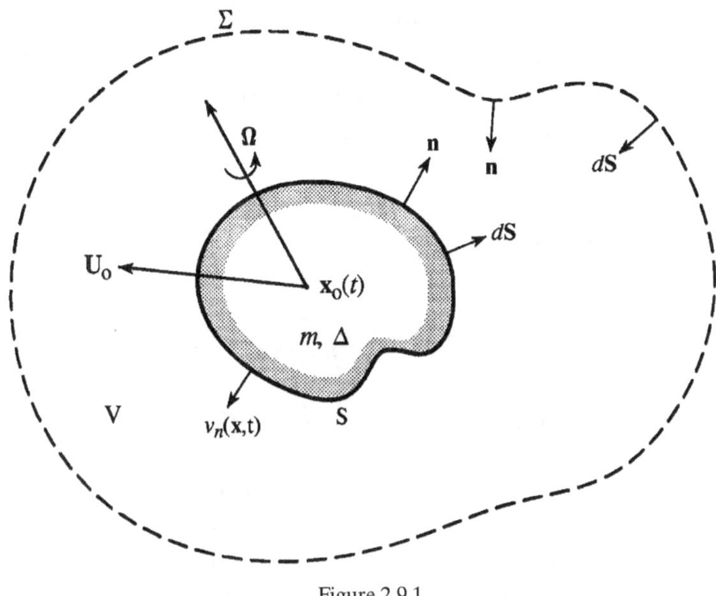

Figure 2.9.1

However, $|\bar{\mathbf{x}} - \mathbf{y}| \equiv |\mathbf{x} - \bar{\mathbf{y}}|$, so that the surface integral in solution (2.8.8) is now seen to represent exactly the potential produced by a system of images in the wall of the original distribution $q(\mathbf{x}, t)$.

Thus the aggregate contribution from the surface dipoles in (2.8.8) is actually equal to a monopole field whose strength is identical to that of the original source distribution $q(\mathbf{x}, t)$. What we have shown, in fact, is that the field produced by a distribution of sources next to the wall can be calculated by the well-known 'method of images'. This result is a warning, however, that care should be exercised when Green's formula is applied to surfaces that extend into the far field. The naive interpretation of the surface integral as a dipole fails because the surface integral *diverges* when $|\mathbf{x} - \mathbf{y}|^3$ is approximated by $|\mathbf{x}|^3$ and $\int q(\mathbf{y}, t)\, d^3\mathbf{y} \neq 0$.

2.9 Determinancy of the motion

Consider the flow produced in an infinite, ideal fluid by arbitrary motion from rest of a solid body with surface S when there are no sources in the fluid ($q \equiv 0$). In Figure 2.9.1 the body has volume Δ and moves with velocity

$$\mathbf{U} = \mathbf{U}_o + \mathbf{\Omega} \wedge [\mathbf{x} - \mathbf{x}_o(t)], \tag{2.9.1}$$

where $\mathbf{U}_o = d\mathbf{x}_o/dt$ is the velocity of its centre of volume $\mathbf{x}_o(t)$ and $\Omega(t)$ is its angular velocity.

Irrotational motion with velocity $\mathbf{v} = \nabla\varphi$ satisfies the kinematic boundary condition

$$\frac{\partial \varphi}{\partial x_n} = U_n \text{ on S},$$

where x_n is a local coordinate in the normal direction on S and U_n is the normal component of velocity on S, but fails to satisfy the physical no-slip condition on the tangential component of velocity. The latter is satisfied implicitly by imagining the presence of a tangential 'vortex sheet' on S through which the tangential fluid velocity adjusts to that of S (an approximation that is *exact*, however, for the initial motion started impulsively from rest).

There are no sources within the instantaneous region V occupied by the fluid, where $\nabla^2 \varphi = 0$. Therefore the kinetic energy T_o of the fluid is

$$T_o = \frac{1}{2}\rho_o \int_V (\nabla\varphi)^2 d^3\mathbf{x} = \frac{1}{2}\rho_o \int_V \left[\text{div}(\varphi\nabla\varphi) - \varphi\nabla^2\varphi \right] d^3\mathbf{x}$$

$$= -\frac{1}{2}\rho_o \oint_S \varphi \frac{\partial\varphi}{\partial x_n} dS \equiv -\frac{1}{2}\rho_o \oint_S \varphi U_n dS, \qquad (2.9.2)$$

where the divergence theorem has been used to obtain the second line (there is no contribution from the surface Σ at 'infinity' in Figure 2.9.1, where, according to the note following Green's formula (2.8.7), $\varphi \sim O[1/|\mathbf{x}|^2)]$. Formula (2.9.2) therefore implies that if S is suddenly brought to rest ($U_n \to 0$) the motion everywhere in the fluid ceases instantaneously, because $\int_V (\nabla\varphi)^2 d^3\mathbf{x}$ can vanish only if $\nabla\varphi \equiv 0$. This unphysical behavior is never observed in a real fluid because (i) no fluid is perfectly incompressible, and 'signals' generated by changes in the boundary conditions propagate at the finite speed of sound, and (ii) diffusion of *vorticity* from the boundary supplies irrecoverable kinetic energy to the fluid.

It is evident that the formula

$$T_o = \frac{1}{2}\rho_o \int_V (\nabla\varphi)^2 d^3\mathbf{x} = -\frac{1}{2}\rho_o \oint_S \varphi \frac{\partial\varphi}{\partial x_n} dS$$

for the irrotational kinetic energy in the region V bounded by S remains valid when the surface S is an arbitrary control surface instead of a rigid boundary, provided that, when V extends to infinity, $\varphi \to 0$ faster than $1/|\mathbf{x}|$.

Green's formula (2.8.7) determines the velocity potential for the general problem involving a source distribution $q(\mathbf{x}, t)$ in terms of q and the values of φ and $\partial\varphi/\partial x_n$ on S. The surface data are redundant, however, because a knowledge of either φ *or* $\partial\varphi/\partial x_n$ on S is actually sufficient to completely determine φ in V. We can prove this and also establish the uniqueness of the solution of the potential flow problem by a simple extension of the procedure used previously to evaluate the kinetic energy. Suppose $\varphi_1(\mathbf{x}, t)$ and $\varphi_2(\mathbf{x}, t)$ are two velocity potentials satisfying $\nabla^2\varphi = q$ and the same boundary conditions on S. If S is the inner boundary of an infinite fluid, as in Figure 2.9.1, we may also assume that $\varphi \to 0$ as $|\mathbf{x}| \to \infty$. Then (2.9.2) implies for $\Phi = \varphi_1(\mathbf{x}, t) - \varphi_2(\mathbf{x}, t)$ that

$$\int_V (\nabla\Phi)^2 d^3\mathbf{x} = -\oint_S \Phi \frac{\partial\Phi}{\partial x_n} dS \equiv -\oint_S (\varphi_1 - \varphi_2)\left(\frac{\partial\varphi_1}{\partial x_n} - \frac{\partial\varphi_2}{\partial x_n}\right) dS. \qquad (2.9.3)$$

However, the right-hand side vanishes identically because $\partial\varphi_1/\partial x_n = \partial\varphi_2/\partial x_n = U_n$ on S; this is possible only if $\nabla\Phi \equiv \nabla\varphi_1 - \nabla\varphi_2 = 0$ everywhere in V, that is, only if the velocity fields predicted by the two solutions are equal. Hence

$$\varphi_1 = \varphi_2 + f_o(t),$$

where $f_o(t)$ is an arbitrary function of the time that vanishes if the fluid extends to infinity, but otherwise has no physical significance.

It follows that, if $\partial\varphi/\partial x_n$ is prescribed on S, then φ is determined everywhere to within an arbitrary function of the time, including at points on S, where the functional form of φ in the dipole surface integral of Green's formula (2.8.7) cannot be specified independently. Indeed, given any solution with φ and $\partial\varphi/\partial x_n$ known on S, then any other solution with the same value of $\partial\varphi/\partial x_n$ on S must also have the same value of φ on S, whose variation on S cannot therefore be prescribed. Hence, Green's formula (2.8.7) is strictly an alternative, *integral equation* representation of the partial differential equation $\nabla^2\varphi = q$; when the motion of S is known in terms of the normal velocity, the equation provides a useful prediction of the velocity potential only when φ is also known on S or can be estimated by some other means. The principal application of Green's formula is to the numerical solution of potential flows. The first step in such calculations consists of solving a system of linear algebraic equations for the values of φ at a discrete set of points on S; these equations are obtained by application of a discretized version of the integral formula at these points. The results of this calculation of φ on S can then be used to determine φ within the fluid.

EXAMPLE 1. D'ALEMBERT'S PARADOX The uniqueness theorem implies that the kinetic energy of an ideal incompressible fluid (at rest at infinity) is constant when the flow is produced by *steady* translational motion of a rigid body at constant velocity $\mathbf{U}_o$. Thus the moving body does no work on the fluid and therefore experiences *no drag*; this is D'Alembert's paradox. The conclusion is obviously unchanged when the body is placed at rest in a uniform flow at velocity $\mathbf{U}_o$.

2.9.1 Fluid motion expressed in terms of monopole or dipole distributions

Green also showed that his formula (2.8.7), which expresses the solution in terms of monopole and dipole distributions on S, can generally be replaced with one involving either monopoles or dipoles alone. To do this we imagine the region $\bar{V}$ within the surface S of Figure 2.8.2 to be filled with fluid. Let the motion of this fluid be determined by the velocity potential $\bar{\varphi}$ (when S is a 'closed surface', as in Figure 2.8.2, the existence of $\bar{\varphi}$ requires that $\oint_S \partial\bar{\varphi}/\partial x_n \, dS = 0$). Green's formula (2.8.7) for this problem becomes

$$H(-f)\bar{\varphi}(\mathbf{x}, t) = -\frac{\partial}{\partial x_j} \oint_S \frac{\bar{\varphi}(\mathbf{y}, t)\bar{n}_j \, dS}{4\pi|\mathbf{x} - \mathbf{y}|} - \oint_S \bar{n}_j \frac{\partial\bar{\varphi}}{\partial y_j}(\mathbf{y}, t)\frac{dS}{4\pi|\mathbf{x} - \mathbf{y}|}, \qquad (2.9.4)$$

where $H(-f) = 1$ in $\bar{V}$ and $H(-f) = 0$ in the exterior region V, and where $\bar{n}_j = -n_j$ is the unit surface normal directed into $\bar{V}$. By applying this formula at the point $\mathbf{x}$ in V we find

$$0 = \frac{\partial}{\partial x_j} \oint_S \frac{\bar{\varphi}(\mathbf{y}, t)n_j \, dS}{4\pi|\mathbf{x} - \mathbf{y}|} + \oint_S n_j \frac{\partial\bar{\varphi}}{\partial y_j}(\mathbf{y}, t)\frac{dS}{4\pi|\mathbf{x} - \mathbf{y}|}.$$

When this is added to (2.8.7) for $\mathbf{x}$ in V [where $H = H(f) = 1$] it follows that $\varphi(\mathbf{x}, t)$ can be written as

$$\varphi(\mathbf{x}, t) = -\int_V \frac{q(\mathbf{y}, t)}{4\pi |\mathbf{x} - \mathbf{y}|} d^3\mathbf{y} - \frac{\partial}{\partial x_j} \oint_S \frac{n_j (\varphi - \bar{\varphi})(\mathbf{y}, t) dS}{4\pi |\mathbf{x} - \mathbf{y}|}$$

$$- \oint_S \frac{\mathbf{n} \cdot (\nabla\varphi - \nabla\bar{\varphi})(\mathbf{y}, t) dS}{4\pi |\mathbf{x} - \mathbf{y}|}. \tag{2.9.5}$$

We are free to define the value of $\bar{\varphi}$ or of $\mathbf{n} \cdot \nabla\bar{\varphi}$ on S in any convenient manner. If we take $\bar{\varphi} = \varphi$ on S the solution in V becomes

$$\varphi(\mathbf{x}, t) = -\int_V \frac{q(\mathbf{y}, t)}{4\pi |\mathbf{x} - \mathbf{y}|} d^3\mathbf{y} - \oint_S \frac{\mathbf{n} \cdot (\nabla\varphi - \nabla\bar{\varphi})(\mathbf{y}, t) dS}{4\pi |\mathbf{x} - \mathbf{y}|}. \tag{2.9.6}$$

More particularly, for $\mathbf{x}$ in either region we can write

$$H(f)\varphi(\mathbf{x}, t) + H(-f)\bar{\varphi}(\mathbf{x}, t) = -\int_V \frac{q(\mathbf{y}, t)}{4\pi |\mathbf{x} - \mathbf{y}|} d^3\mathbf{y} - \oint_S \frac{\mathbf{n} \cdot (\nabla\varphi - \nabla\bar{\varphi})(\mathbf{y}, t) dS}{4\pi |\mathbf{x} - \mathbf{y}|}$$

for the *overall* velocity potential. This represents a flow whose tangential component of velocity is continuous across S, but the normal velocity is discontinuous. At any instant the motion could be produced from rest if the body were removed and the surface S replaced with a layer of infinitesimal thickness within which the impulsive pressure $\hat{\varpi} = -\rho_o\varphi = -\rho_o\bar{\varphi}$ is applied. The remaining integral in (2.9.6) is the field produced by a distribution of monopole sources whose strength per unit surface area of S is equal to the jump $\mathbf{n} \cdot (\nabla\varphi - \nabla\bar{\varphi})$ in the normal velocity across S.

Similarly, if we take $\mathbf{n} \cdot \nabla\bar{\varphi} = \mathbf{n} \cdot \nabla\varphi$ on S (which is possible for a closed surface only provided that $\oint_S \mathbf{n} \cdot \nabla\varphi\, dS = 0$) the normal component of velocity is continuous but the tangential component is discontinuous (so that S is equivalent to a vortex sheet; see §4.1.6). The overall motion is then determined by a prescribed normal velocity distribution on S. In V,

$$\varphi(\mathbf{x}, t) = -\int_V \frac{q(\mathbf{y}, t)}{4\pi |\mathbf{x} - \mathbf{y}|} d^3\mathbf{y} - \frac{\partial}{\partial x_j} \oint_S \frac{n_j (\varphi - \bar{\varphi})(\mathbf{y}, t) dS}{4\pi |\mathbf{x} - \mathbf{y}|}, \tag{2.9.7}$$

which expresses the surface effect in terms of a distribution of dipoles whose strength per unit surface area of S is just $\varphi - \bar{\varphi}$, which is proportional to the difference in the surface values of the impulsive pressures on the exterior and interior faces of S. Note that (2.9.7) cannot be applied when S is the surface of a pulsating body for which $\oint_S \mathbf{n} \cdot \nabla\varphi\, dS \neq 0$.

It is clear from relation (2.9.5) that the representation of the influence of S in terms of a combination of monopole and dipole distributions on S is not unique. However, it is easy to show that the separate monopole and dipole formulae (2.9.6) and (2.9.7) are unique.

EXAMPLE 2. The velocity potential of motion produced by a rigid sphere of radius a moving at velocity $\mathbf{U}$ along the x axis is $\varphi = -Ua^3 \cos\theta / r^2$, $r = |\mathbf{x}|$, where the origin is taken at the centre of the sphere (see Figure 2.7.1). We find surface monopole and dipole representations of φ [corresponding respectively to (2.9.6) and (2.9.7)] by noting that the bounded solution of Laplace's equation proportional to $\cos\theta$ in the interior $\bar{V}$

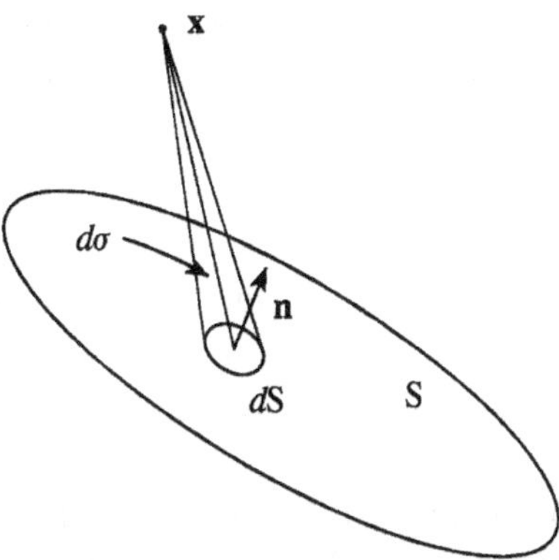

Figure 2.9.2

of the sphere is $\bar{\varphi} = $ constant $(r\cos\theta)$. We choose the constant by taking $\partial\varphi/\partial r = \partial\bar{\varphi}/\partial r$ at $r = a$ for the monopole and $\varphi = \bar{\varphi}$ at $r = a$ for the dipole distribution, which yields the representations

$$\text{monopole,} \quad \varphi(\mathbf{x}) = -\frac{3}{8\pi}\oint_S \frac{\mathbf{n}\cdot\mathbf{U}dS}{|\mathbf{x}-\mathbf{y}|}; \qquad \text{source density} = \frac{3}{2}\mathbf{n}\cdot\mathbf{U};$$

$$\text{dipole,} \quad \varphi(\mathbf{x}) = \frac{3}{8\pi}\frac{\partial}{\partial x_j}\oint_S \frac{n_j\mathbf{y}\cdot\mathbf{U}dS}{|\mathbf{x}-\mathbf{y}|}; \quad \text{dipole density} = -\frac{3}{2}\mathbf{y}\cdot\mathbf{U}.$$

EXAMPLE 3. The potential $\varphi(\mathbf{x})$ produced by a distribution of dipole sources of strength f_n per unit area of a surface S can be represented in the alternative forms

$$\varphi(\mathbf{x}) = -\frac{\partial}{\partial x_j}\oint_S \frac{f_n(\mathbf{y})n_j dS(\mathbf{y})}{4\pi|\mathbf{x}-\mathbf{y}|} = \oint_S \frac{f_n(\mathbf{y})\mathbf{n}\cdot(\mathbf{x}-\mathbf{y})dS(\mathbf{y})}{4\pi|\mathbf{x}-\mathbf{y}|^3} \equiv \oint_S f_n(\mathbf{y})\frac{d\sigma(\mathbf{y})}{4\pi},$$

where $d\sigma$ is the element of solid angle subtended at $\mathbf{x}$ by the surface element dS (for an open surface, with two sides wetted by the fluid, the integration is taken over only the 'upper' side in Figure 2.9.2).

As the field point $\mathbf{x}$ approaches dS from 'above' the solid angle $d\sigma \to 2\pi$. It follows that the jump $[\varphi]$ in the value of φ in passing through S from 'below' is given by

$$[\varphi(\mathbf{x})] = f_n(\mathbf{x})\left[\frac{2\pi}{4\pi} - \frac{(-2\pi)}{4\pi}\right] = f_n(\mathbf{x}).$$

EXAMPLE 4. The velocity potential φ cannot attain a maximum or minimum within the flow because the value of φ at a point is equal to the mean of its values on a small sphere centred on that point. Similarly, $(\nabla\varphi)^2$ cannot attain a maximum *at a point*, although it can vanish.

2.9.2 Determinancy of cyclic irrotational flow

When a flow is irrotational $(\mathrm{curl}\,\mathbf{v} = 0)$ the circulation around any closed contour can be asserted to vanish only if the contour is the boundary of a surface lying entirely within the fluid. Such a contour is said to be *reducible* and can be contracted to a point within the fluid. It cannot be concluded, however, that the circulation is zero around an irreducible contour. Thus it is not permissible to assume that, in irrotational flow, the circulation around a contour enclosing an infinite cylinder is zero, although the circulation around all curves encircling the cylinder just once must be the same. This is an example of flow in a 'doubly connected' domain: The velocity $\nabla\varphi$ is a single-valued function of position, but $\varphi(\mathbf{x})$ is undefined to within an arbitrary constant and increases or decreases by a multiple of the circulation when $\mathbf{x}$ loops once around an irreducible contour.

We have hitherto implicitly assumed in our discussion of potential theory that the fluid occupies a *simply connected* region, i.e. that any closed contour is reducible. In *multiply connected* fluid regions, however, it is always possible to insert at least one 'barrier' in the form of a surface having a closed curve for its (solid) boundary without breaking the domain into a set of disconnected regions. In a doubly connected domain one barrier can be drawn; for example, for fluid bounded internally by an infinite cylinder, the barrier is a surface (such as a half-plane) extending to infinity from, say, a generator of the cylinder. When it is possible to insert $n - 1$ such barriers the fluid region is said to be n-ply connected.

The circulation around a contour that cuts through only one barrier is the same for all such contours, equal to κ, say. This is called the *cyclic constant* for the circuit. In an n-ply connected domain there are $n - 1$ distinct paths that cut a barrier once, with respective cyclic constants $\kappa_1, \kappa_2, \ldots, \kappa_{n-1}$. Of course, in irrotational motion starting from rest there can never be a non-zero circulation around *any* contour. Thus the existence of non-vanishing cyclic constants implies that at some stage during the motion vorticity has been released from the boundary and subsequently swept away, into the far field of the flow, a process that in practice can take place only by means of the intervention of viscous stresses at the boundary (§4.1).

2.9.3 Kinetic energy of cyclic irrotational flow

In a multiply connected domain the velocity potential φ *may* be undefined to within an arbitrary constant, but $\nabla\varphi$ is a single-valued function. In that case the kinetic energy $T = \frac{1}{2}\rho_o \int (\nabla\varphi)^2 d^3\mathbf{x}$ is still well defined. However, to transform this expression into a surface integral of the type in (2.9.2), it is necessary to form an artificial, simply connected region, within which φ can be regarded as single valued, by introducing $n - 1$ barriers. Then

$$T = \frac{1}{2}\rho_o \int (\nabla\varphi)^2 d^3\mathbf{x}$$

$$= \frac{1}{2}\rho_o \int \left[\operatorname{div}(\varphi\nabla\varphi) - \varphi\nabla^2\varphi\right] d^3\mathbf{x}$$

$$= -\frac{1}{2}\rho_o \oint_S \varphi\frac{\partial\varphi}{\partial x_n} dS - \frac{1}{2}\rho_o \sum_j \oint_{S_j} [\varphi]\frac{\partial\varphi}{\partial x_n} dS, \qquad (2.9.8)$$

where S is the solid boundary and the summation is over the $n - 1$ barrier surfaces S_j; for each j the integral is taken over one side of S_j, where $[\varphi] = \kappa_j$ is the jump in the value of φ in crossing S_j in the direction of x_n. Hence the required modification of (2.9.2) is

$$T = -\frac{1}{2}\rho_o \oint_S \varphi \frac{\partial \varphi}{\partial x_n} dS - \frac{1}{2}\rho_o \sum_j \kappa_j \oint_{S_j} \frac{\partial \varphi}{\partial x_n} dS. \tag{2.9.9}$$

DETERMINANCY OF THE MOTION When the cyclic constants κ_1, κ_2, ..., are given, let φ_1, φ_2 be two possible solutions of Laplace's equation having the same cyclic constants. Then $\Phi = \varphi_1 - \varphi_2$ is a velocity potential of an *acyclic* motion for which $\partial \Phi / \partial x_n = 0$ on S and is therefore unique and equal to a constant. Hence the two flows described by φ_1 and φ_2 are the same.

2.10 The kinetic energy

Consider a solid body initially at rest and immersed in a stationary, ideal incompressible fluid. When the body is set into motion it generates irrotational flow throughout the fluid by the transmission of impulsive pressures from its surface. If the body subsequently stops moving, the motion everywhere ceases instantaneously. Kelvin showed that the kinetic energy of this irrotational flow is smaller than that of any other possible motion of the fluid that is consistent with the boundary conditions. Any other possible flow must have the same normal velocity on the surface of the body, and it must also contain regions where the vorticity $\omega = \text{curl } \mathbf{v} \neq 0$, because the irrotational motion is unique.

Let the body have surface S and be in an infinite expanse of fluid (Figure 2.10.1), and let the irrotational motion be described by the velocity potential φ, so that

$$\frac{\partial \varphi}{\partial x_n} = U_n \text{ on S,}$$

where U_n is the normal component of velocity. The kinetic energy T_o of the fluid is calculated as in §2.9:

$$T_o = -\frac{1}{2}\rho_o \oint_S \varphi \frac{\partial \varphi}{\partial x_n} dS \equiv -\frac{1}{2}\rho_o \oint_S \varphi U_n dS. \tag{2.10.1}$$

If S is suddenly brought to rest ($U_n \to 0$) the motion everywhere also stops immediately, because $\int_V (\nabla\varphi)^2 d^3\mathbf{x}$ can vanish only if $\nabla\varphi \equiv 0$. We have already noted (§2.9) that this never happens in a real fluid, first because an instantaneous response must actually be delayed by the time required for the pressure pulse to travel from S at the speed of sound, and second because the release of vorticity into the fluid from S supplies the fluid with irrecoverable kinetic energy.

For a real incompressible fluid, the velocity can be written as

$$\mathbf{v} = \nabla\varphi + \mathbf{u},$$

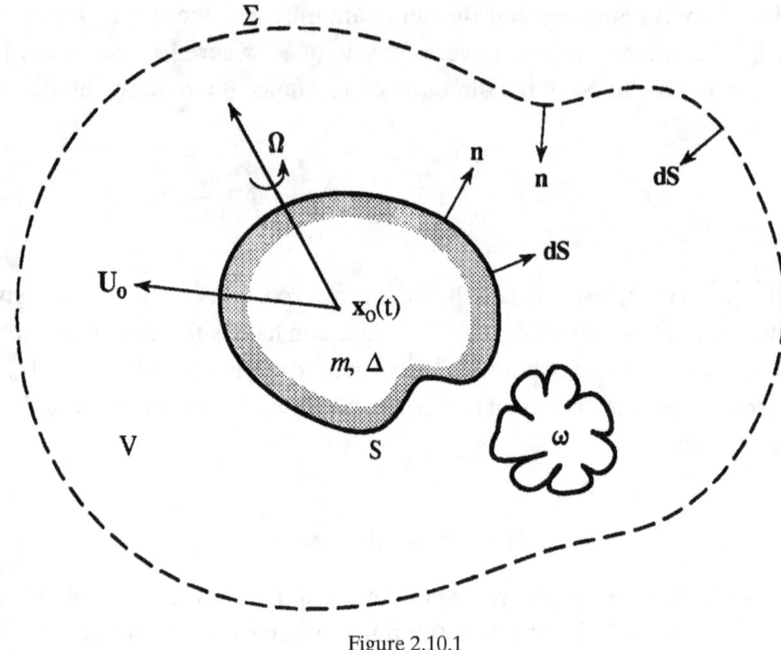

Figure 2.10.1

where $\omega = \text{curl}\,\mathbf{u} \equiv \text{curl}\,\mathbf{v}$, and where the normal velocity component $u_n = \mathbf{n} \cdot \mathbf{u} = 0$ on S.

The kinetic energy is now given by

$$
\begin{aligned}
T &= \frac{1}{2}\rho_o \int_V (\nabla\varphi + \mathbf{u})^2 d^3\mathbf{x} \;=\; \frac{1}{2}\rho_o \int_V \Big[(\nabla\varphi)^2 + 2\nabla\varphi \cdot \mathbf{u} + u^2 \Big] d^3\mathbf{x} \\
&= \frac{1}{2}\rho_o \int_V \Big[(\nabla\varphi)^2 + u^2 \Big] d^3\mathbf{x} + \rho_o \int_V \text{div}(\varphi\mathbf{u}) d^3\mathbf{x} \\
&= \frac{1}{2}\rho_o \int_V \Big[(\nabla\varphi)^2 + u^2 \Big] d^3\mathbf{x} - \rho_o \oint_S \varphi u_n dS \\
&= -\frac{1}{2}\rho_o \oint_S \varphi U_n dS + \frac{1}{2}\rho_o \int_V u^2 d^3\mathbf{x} \;\equiv\; T_o + \frac{1}{2}\rho_o \int_V u^2 d^3\mathbf{x}.
\end{aligned}
\tag{2.10.2}
$$

If the surface stops moving the flow described by the velocity potential φ stops instantaneously, but that associated with the rotational velocity $\mathbf{u}$ continues. The crucial difference between rotational and irrotational flows is that, once established, vortical motions proceed irrespective of whether or not the fluid continues to be driven by impulsive pressures transmitted from the boundary or other external agencies.

Equation (2.10.2) yields **Kelvin's theorem** that $T \geq T_o$: The kinetic energy of the real flow (for which $\mathbf{u} \neq 0$) always exceeds that of the irrotational flow that would be produced by the same surface motion. In other words, the irrotational motion represents the *least* possible disturbance that can be produced in the fluid by the moving body.

2.10.1 Converse of Kelvin's minimum-energy theorem

The fact that a prescribed boundary motion of a solid body produces a unique irrotational flow in an ideal, incompressible fluid is equivalent to the statement that the finite number of generalized coordinates required for specifying the motion of S (six, three of translation and three of rotation) is also sufficient for defining the motion of the fluid. Any additional degrees of freedom in the fluid correspond to the presence of vorticity, which indicates the existence of fluid kinetic energy independent of forcing of the motion by the boundary. It is obvious therefore that the requirement that the kinetic energy of a homogeneous, ideal fluid should be a minimum must necessarily imply that the motion is irrotational. The same conclusion must hold in a real viscous fluid for the initial motion from rest caused by a sudden movement at the boundary. Kelvin gave a direct proof of this proposition by minimizing the kinetic energy $T = \int_V \frac{1}{2}\rho_o v^2 d^3\mathbf{x}$ when $\mathrm{div}\,\mathbf{v} = 0$.

By introducing a Lagrange multiplier λ, we are required to minimize

$$\mathcal{I} = \int_V \left(\frac{1}{2}\rho_o v^2 + \lambda\,\mathrm{div}\,\mathbf{v}\right) d^3\mathbf{x} \quad \text{subject to} \quad \mathrm{div}\,\mathbf{v} = 0. \tag{2.10.3}$$

Then, taking the first variation and integrating by parts, we obtain

$$\delta\mathcal{I} = \int_V (\rho_o\mathbf{v}\cdot\delta\mathbf{v} + \lambda\,\mathrm{div}\,\delta\mathbf{v})\,d^3\mathbf{x} = \int_V (\rho_o\mathbf{v} - \nabla\lambda)\cdot\delta\mathbf{v}\,d^3\mathbf{x} - \oint_S \lambda\,\delta v_n dS,$$

provided that λ is finite in fluid unbounded at infinity, where $\delta\mathbf{v} < O(|\mathbf{x}|^{-2})$ as $|\mathbf{x}| \to \infty$. The latter condition is satisfied when the normal velocity on S is prescribed, so that $\delta v_n \equiv 0$. Hence, equating to zero the coefficient of $\delta\mathbf{v}$ in the remaining volume integral, we arrive at the required result:

$$\rho_o\mathbf{v} = \nabla\lambda,$$

where the second of Equations (2.10.3) shows that $\nabla^2\lambda = 0$. Hence the initial motion is described by the velocity potential $\lambda(\mathbf{x})/\rho_o$, and the Lagrange multiplier is equal but opposite in sign to the impulsive pressure $\varpi = -\lambda$ required for establishing the motion from rest. The possibly unsatisfactory condition that λ should be finite at infinity can be replaced with the requirement that the motion at infinity is prescribed, so that $\delta v_n = 0$ on a surface at infinity.

2.10.2 Energy of motion produced by a translating sphere

The velocity potential of flow produced by a rigid sphere of radius a moving at speed U in the x_1 direction is $\varphi = -a^3 U \cos\theta/2r^3$, when the coordinate system is taken as in Figure 2.10.2, with the origin at the centre of the sphere. The kinetic energy of the flow is therefore

$$T_o = -\frac{1}{2}\rho_o \oint_S \varphi \frac{\partial\varphi}{\partial r}\,dS = \frac{\pi\rho_o U^2 a^3}{2} \int_0^\pi \cos^2\theta \sin\theta\,d\theta = \frac{1}{2}M'U^2,$$

$$\text{where} \quad M' = \frac{2\pi a^3\rho_o}{3} = \frac{1}{2}\,\text{mass of the displaced fluid.} \tag{2.10.4}$$

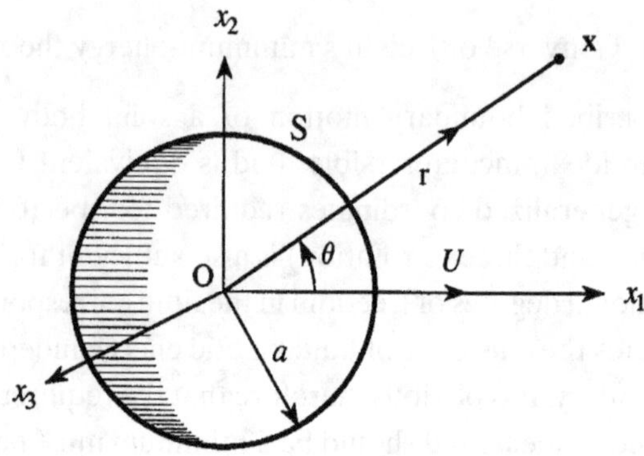

Figure 2.10.2

M' is called the 'added mass' of the sphere and represents the effective mass of the fluid that is dynamically linked to translational motion of the sphere.

Suppose the sphere is accelerated in the x_1 direction by the application of a force $F_1(t)$. The rate of working of the force $F_1 U$ must equal the rate of increase of the kinetic energy of the sphere and the surrounding fluid, so that

$$\frac{d}{dt}\left[\frac{1}{2}(M + M')U^2\right] = F_1 U,$$

where M is the mass of the sphere. Hence the equation of motion of the sphere can be written as

$$M\frac{dU}{dt} = -M'\frac{dU}{dt} + F_1,$$

which shows that the motion of the sphere is opposed by the surface pressures on S whose net effect is equivalent to reaction force $-M' dU/dt$.

EXAMPLE 1. The velocity potential of the motion with respect to fixed coordinate axes $\mathbf{x}$ can be expressed in the form

$$\varphi = -\frac{a^3[x_1 - x_{o1}(t)]U(t)}{2|\mathbf{x} - \mathbf{x}_o(t)|^3}, \quad \text{where} \quad U = \frac{dx_{o1}}{dt}, \quad \mathbf{x}_o = (x_{o1}, 0, 0).$$

Bernoulli's equation $p/\rho_o = -\partial\varphi/\partial t - \frac{1}{2}(\nabla\varphi)^2$ then shows that the pressure on S is given by

$$\frac{p}{\rho_o} = \frac{a}{2}\frac{dU}{dt}\cos\theta + \frac{9}{16}U^2\cos 2\theta - \frac{1}{16}U^2,$$

and therefore that the net force exerted on the sphere by the fluid is

$$-\int_S p\cos\theta\, dS = -M'\frac{dU}{dt} \quad \text{in the } x_1 \text{ direction.}$$

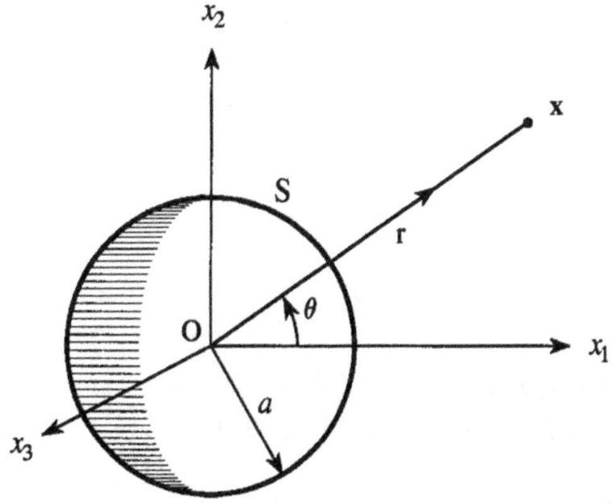

Figure 2.11.1

2.11 Problems with spherical boundaries

Consider a sphere of radius a (Figure 2.11.1) whose centre O is taken as the coordinate origin. Introduce spherical polar coordinates (r, θ, ϕ), where θ is the angle between the vector $\mathbf{x}$ from O and the positive x_1 axis, such that

$$x_1 = r \cos\theta, \quad x_2 = r \sin\theta \cos\phi, \quad x_3 = r \sin\theta \sin\phi.$$

The corresponding vector line element $d\mathbf{x}$ and gradient operator with components respectively parallel to unit vectors in the directions of increasing r, θ, ϕ are

$$d\mathbf{x} = (dr, r\,d\theta, r\sin\theta\,d\phi),$$

$$\nabla = \left(\frac{\partial}{\partial r}, \frac{1}{r}\frac{\partial}{\partial\theta}, \frac{1}{r\sin\theta}\frac{\partial}{\partial\phi} \right),$$

(2.11.1)

and the Laplacian operator is

$$\nabla^2 = \frac{1}{r^2}\frac{\partial}{\partial r}\left(r^2 \frac{\partial}{\partial r} \right) + \frac{1}{r^2 \sin\theta}\frac{\partial}{\partial\theta}\left(\sin\theta \frac{\partial}{\partial\theta} \right) + \frac{1}{r^2 \sin^2\theta}\frac{\partial^2}{\partial\phi^2}.$$

(2.11.2)

2.11.1 Legendre polynomials

We shall examine first *axisymmetric* potential flows that do not depend on the azimuthal angle ϕ, governed by the axisymmetric Laplace equation

$$\frac{\partial}{\partial r}\left(r^2 \frac{\partial\varphi}{\partial r} \right) + \frac{1}{\sin\theta}\frac{\partial}{\partial\theta}\left(\sin\theta \frac{\partial\varphi}{\partial\theta} \right) = 0, \quad 0 \le \theta \le \pi.$$

(2.11.3)

A separable solution,

$$\varphi = R(r)\Theta(\theta),$$

in which $R(r)$ and $\Theta(\theta)$ are, respectively, functions of r and θ alone, satisfies

$$\frac{1}{R}\frac{d}{dr}\left(r^2 \frac{dR}{dr} \right) = -\frac{1}{\sin\theta\,\Theta}\frac{d}{d\theta}\left(\sin\theta \frac{d\Theta}{d\theta} \right) = \lambda,$$

where λ is constant. R and Θ are therefore determined by

$$r^2 \frac{d^2 R}{dr^2} + 2r \frac{dR}{dr} - \lambda R = 0,$$

$$\frac{1}{\sin\theta} \frac{d}{d\theta} \left(\sin\theta \frac{d\Theta}{d\theta} \right) + \lambda\Theta = 0, \ 0 \le \theta \le \pi.$$

$$(2.11.4)$$

The *Legendre polynomials* $\Theta = P_n(\cos\theta)$ are solutions of the second of these equations that are bounded in $0 \le \theta \le \pi$ provided $\lambda = n(n+1)$ for positive integral values of n. $P_n(\cos\theta)$ is a polynomial of the order of n in $\cos\theta$, normalized such that $P_n(1) = 1$; the following special cases are noted for future reference:

$$P_0(\cos\theta) = 1, \quad P_1(\cos\theta) = \cos\theta, \quad P_2(\cos\theta) = \frac{3}{2}\cos^2\theta - \frac{1}{2}. \qquad (2.11.5)$$

The Legendre polynomials satisfy the orthogonality relations

$$\int_0^\pi P_n(\cos\theta) P_m(\cos\theta) \sin\theta \, d\theta = \begin{cases} \frac{2}{2n+1} & (n = m) \\ 0 & (n \ne m) \end{cases}, \qquad (2.11.6)$$

which permit an arbitrary function $f(\theta)$ defined in $0 \le \theta \le \pi$ to be expanded in the form

$$f(\theta) = \sum_{n=0}^\infty \alpha_n P_n(\cos\theta), \quad \text{where } \alpha_n = \left(n + \frac{1}{2}\right) \int_0^\pi f(\theta) P_n(\cos\theta) \sin\theta \, d\theta. \qquad (2.11.7)$$

When $\lambda = n(n+1)$ the first of Equations (2.11.4) has the solution $R = A_n r^n + B_n/r^{n+1}$, where A_n and B_n are arbitrary constants. The general, bounded solution of axisymmetric Laplace equation (2.11.3) is therefore

$$\varphi(r, \theta) = \sum_{n=0}^\infty \left(A_n r^n + \frac{B_n}{r^{n+1}} \right) P_n(\cos\theta), \quad 0 \le \theta \le \pi. \qquad (2.11.8)$$

We must take $A_n = 0$, $n \ge 0$ if the solution is required to vanish as $r \to \infty$; for a bounded solution in a region including $r = 0$ we must take $B_n = 0$, $n \ge 0$.

EXAMPLE 1. Find the velocity potential of the motion produced when the sphere of Figure 2.11.1 is rigid and translates in the x_1 direction at speed U.

The potential is axisymmetric and of the general form of (2.11.8) with $A_n = 0$. The normal component of velocity on S is equal to $\partial\varphi/\partial r$ evaluated at $r = a$, where it must be equal to $U_n = U\cos\theta$. This condition can be satisfied when only the term $n = 1$ is retained in the expansion, so that

$$\varphi = \frac{B_1}{r^2} P_1(\cos\theta) \equiv \frac{B_1 \cos\theta}{r^2}, \quad r \ge a,$$

and

$$\frac{\partial\varphi}{\partial r} = \frac{-2B_1 \cos\theta}{r^3} = U\cos\theta \ \text{ when } r = a;$$

$$\therefore \ \varphi = -\frac{Ua^3 \cos\theta}{2r^2}, \quad r > a. \qquad (2.11.9)$$

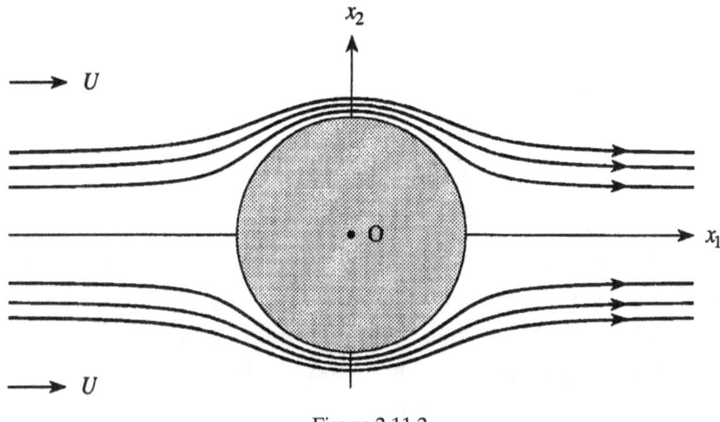

Figure 2.11.2

EXAMPLE 2. Find the velocity potential of uniform flow at speed U in the x_1 direction past a stationary, rigid sphere of radius a whose center is at the origin.

The potential is axisymmetric and must satisfy $\varphi \to Ux_1$ as $r \to \infty$. Because $x_1 = r\cos\theta \equiv r P_1(\cos\theta)$, we put

$$\varphi = Ur P_1(\cos\theta) + \sum_{n=0}^{\infty} \frac{B_n}{r^{n+1}} P_n(\cos\theta).$$

The normal component of velocity $\partial\varphi/\partial r$ vanishes on the sphere at $r = a$:

$$\therefore \quad \sum_{n=0}^{\infty} \frac{(n+1)B_n}{a^{n+2}} P_n(\cos\theta) = U P_1(\cos\theta), \quad 0 < \theta < \pi.$$

Hence, all of the B_n vanish except for $B_1 = Ua^3/2$, and

$$\varphi = Ur\cos\theta \left(1 + \frac{a^3}{2r^3}\right). \tag{2.11.10}$$

This can also be derived from solution (2.11.9) for a sphere translating to the right at speed U. We first reverse the direction of motion of the sphere by replacing U with $-U$ (so that φ becomes $+Ua^3\cos\theta/2r^2$) and then superpose a velocity U on the fluid and sphere, which adds the potential $Ux_1 = Ur\cos\theta$.

The streamlines are determined [from (2.3.7)] by the equations

$$\frac{dr}{\left(1 - \frac{a^3}{r^3}\right)\cos\theta} = \frac{r\,d\theta}{-\left(1 + \frac{a^3}{2r^3}\right)\sin\theta} = \frac{d\phi}{0}.$$

Each streamline lies in a plane of constant ϕ and satisfies

$$\frac{2\cos\theta\,d\theta}{\sin\theta} = \left(\frac{1}{r} - \frac{3r^2}{r^3 - a^3}\right)dr, \quad \text{i.e.,} \quad \frac{r}{r^3 - a^3} = \text{constant}\,(\sin^2\theta).$$

The streamline pattern is shown in Figure 2.11.2.

EXAMPLE 3. A rigid sphere of radius a and mass M is placed at the centre of a hollow spherical shell of radius $b > a$, and the space between them is filled with an ideal, incompressible fluid of density ρ_o. The shell executes translational oscillations at velocity $u(t)$ parallel to the x_1 axis. Show that if gravity is neglected and the amplitude of the motion is *small*, the inner sphere moves in the x_1 direction at velocity

$$v(t) = \frac{3m_o b^3\, u(t)}{2M(b^3 - a^3) + m_o(2a^3 + b^3)},\tag{2.11.11}$$

where m_o is the mass of fluid displaced by the inner sphere.

Take the coordinate origin at the centre of the shell and assume that the equations of motion can be *linearised*. Then boundary conditions on the shell and on the surface of the sphere can be applied at their respective undisturbed positions $r = b,\ a$. The velocity potential of the fluid motion can be taken in the form

$$\varphi = \left(Ar + \frac{B}{r^2} \right) \cos\theta, \quad a < r < b,$$

subject to the conditions

$$\left(\frac{\partial\varphi}{\partial r} \right)_{r=a} = v\cos\theta, \quad \left(\frac{\partial\varphi}{\partial r} \right)_{r=b} = u\cos\theta.$$

These supply

$$A = \frac{ub^3 - va^3}{b^3 - a^3}, \quad B = \frac{(u-v)a^3 b^3}{2(b^3 - a^3)}.$$

Using the linearised Bernoulli equation, we find that the net force on the sphere (in the x_1 direction) is

$$F_1 = 2\pi a^2 \rho_o \left(a\frac{dA}{dt} + \frac{1}{a^2}\frac{dB}{dt} \right) \int_0^\pi \cos^2\theta \sin\theta\, d\theta = m_o \left(\frac{dA}{dt} + \frac{1}{a^3}\frac{dB}{dt} \right).$$

Substituting into the equation of motion $Mdv/dt = F_1$ and integrating, we find

$$Mv = m_o \left(A + \frac{1}{a^3}B \right),$$

which leads directly to (2.11.11).

EXAMPLE 4. Find the velocity potential of the flow produced by rotation of the rigid prolate spheroid

$$\frac{x^2}{a^2(1+\epsilon)} + \frac{y^2}{a^2} + \frac{z^2}{a^2} = 1, \quad \epsilon \ll 1,$$

at constant angular velocity $\boldsymbol{\Omega} = (0, 0, \Omega)$ about the z axis (Figure 2.11.3).

When $\epsilon \ll 1$ the major axis has length $2a\sqrt{1+\epsilon} \approx 2a(1 + \tfrac{1}{2}\epsilon)$, and the surface S of the spheroid differs in shape from a sphere of radius a by $O(\epsilon)$. We therefore expand all quantities to first order in ϵ.

Suppose first that the coordinates are fixed relative to the rotating body. In this frame the velocity potential φ does not depend on time. Because the velocity at $\mathbf{x}$ *on* the spheroid is $\boldsymbol{\Omega} \wedge \mathbf{x}$, the condition to be satisfied by φ on S is

$$\mathbf{n} \cdot \nabla\varphi = \mathbf{n} \cdot \boldsymbol{\Omega} \wedge \mathbf{x}.\tag{2.11.12}$$

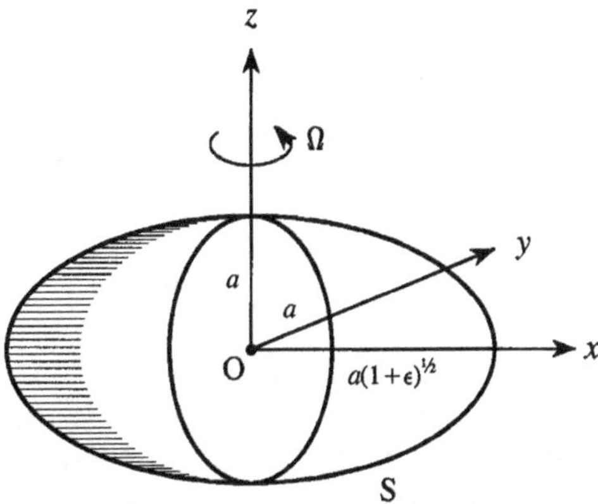

Figure 2.11.3

When $\epsilon \ll 1$,
$$\mathbf{n} \approx \frac{\left[x(1-\epsilon), y, z\right]}{\sqrt{|\mathbf{x}|^2 - 2\epsilon x^2}} = \frac{\mathbf{x} - \epsilon(x, 0, 0,)}{\sqrt{|\mathbf{x}|^2 - 2\epsilon x^2}},$$

$$\therefore \quad \mathbf{n} \cdot \mathbf{\Omega} \wedge \mathbf{x} \approx \frac{-\epsilon(x, 0, 0) \cdot \mathbf{\Omega} \wedge \mathbf{x}}{\sqrt{|\mathbf{x}|^2 - 2\epsilon x^2}} \approx \frac{-\epsilon(x, 0, 0) \cdot \mathbf{\Omega} \wedge \mathbf{x}}{|\mathbf{x}|} = \frac{\epsilon \Omega x y}{|\mathbf{x}|}.$$

The normal velocity on S is a small quantity of order ϵ; it vanishes when $\epsilon = 0$, when the spheroid reduces to a sphere. Thus *correct to $O(\epsilon)$* normal velocity condition (2.11.12) can actually be applied on the surface $r = a$ of the sphere. Then, if the polar coordinates (r, θ, ϕ) are defined by

$$x = r \sin\theta \cos\phi, \quad y = r \sin\theta \sin\phi, \quad z = r \cos\theta,$$

condition (2.11.12) becomes

$$\frac{\partial \varphi}{\partial r} \approx \frac{\epsilon \Omega x y}{|\mathbf{x}|} \approx \frac{\epsilon \Omega a \sin^2\theta \sin 2\phi}{2} \quad \text{on} \quad r = a. \tag{2.11.13}$$

We must now find the solution of Laplace's equation

$$\frac{1}{r^2} \frac{\partial}{\partial r}\left(r^2 \frac{\partial \varphi}{\partial r}\right) + \frac{1}{r^2 \sin\theta} \frac{\partial}{\partial \theta}\left(\sin\theta \frac{\partial \varphi}{\partial \theta}\right) + \frac{1}{r^2 \sin^2\theta} \frac{\partial^2 \varphi}{\partial \phi^2} = 0 \tag{2.11.14}$$

that decays as $r \to \infty$ and satisfies (2.11.13). By substitution into (2.11.13) we find that a trial solution proportional to $r^n \sin^2\theta \sin 2\phi$ will satisfy the equation provided that

$$n^2 + n - 6 = 0, \quad \text{i.e., provided that} \quad n = -3, \, 2,$$

$$\therefore \quad \varphi = \left(Ar^2 + \frac{B}{r^3}\right) \sin^2\theta \sin 2\phi.$$

We take $A = 0$ to ensure the solution is bounded as $r \to \infty$; then condition (2.11.13) yields

$$\varphi = \frac{-\epsilon \Omega a^5 \sin^2\theta \sin 2\phi}{6r^3} = \frac{-\epsilon \Omega a^5 x y}{3|\mathbf{x}|^5} \equiv \frac{-\epsilon \Omega a^5}{9} \frac{\partial^2}{\partial x \partial y}\left(\frac{1}{|\mathbf{x}|}\right). \tag{2.11.15}$$

This is identical with the velocity potential generated by the point quadrupole

$$\frac{4\pi\epsilon\Omega a^5}{9}\frac{\partial^2}{\partial x\partial y}\delta(\mathbf{x})$$

at the centre of the spheroid, i.e., with the solution of

$$\nabla^2\varphi = \frac{4\pi\epsilon\Omega a^5}{9}\frac{\partial^2}{\partial x\partial y}\delta(\mathbf{x}).$$

To complete the solution we must be express (2.11.15) in terms of a fixed, non-rotating coordinate system $\mathbf{x} = (x_1, x_2, x_3)$, say. Let the x and y axes respectively coincide with the fixed x_1 and x_2 axes at time $t = 0$. Then, provided the spherical polar angles θ, ϕ are now considered to be defined with respect to the fixed coordinate system, the polar form of solution (2.11.15) with respect to fixed axes becomes

$$\varphi = \frac{-\epsilon\Omega a^5 \sin^2\theta \sin 2(\phi - \Omega t)}{6r^3}, \tag{2.11.16}$$

which is in a form suitable for the evaluation of the pressure from Bernoulli's equation. Alternatively, because

$$x_1 = x\cos\Omega t - y\sin\Omega t,$$
$$x_2 = x\sin\Omega t + y\cos\Omega t,$$

we can also write

$$\varphi = \frac{-\epsilon\Omega a^5}{18}\frac{\partial^2}{\partial x_i\partial x_j}\left[\frac{T_{ij}(t)}{|\mathbf{x}|}\right], \quad \text{where } T_{ij} = \begin{bmatrix} -\sin 2\Omega t & \cos 2\Omega t & 0 \\ \cos 2\Omega t & \sin 2\Omega t & 0 \\ 0 & 0 & 0 \end{bmatrix}, \tag{2.11.17}$$

which is the solution of

$$\nabla^2\varphi = \frac{2\pi\epsilon\Omega a^5}{9}\frac{\partial^2}{\partial x_i\partial x_j}\left[T_{ij}(t)\delta(\mathbf{x})\right].$$

The relatively weak quadrupole behaviour is unusual for the far field of a rotating body, which normally produces a dipole disturbance. It is a consequence of the special symmetry properties of the prolate spheroid, *not* of the assumption that $\epsilon \ll 1$. The full solution for a rotating spheroid with ϵ specified arbitrarily is known (Lamb 1932) and exhibits the same quadrupole behaviour as $r \to \infty$.

2.11.2 Velocity potential of a point source in terms of Legendre polynomials

A unit point source is located on the z axis at $(0, 0, h)$. Introduce spherical polar coordinates as indicated in Figure 2.11.4, where $z = r\cos\theta$. Then the velocity potential generated by the source at distance r from the origin is

$$\varphi = \frac{-1}{4\pi\sqrt{x^2 + y^2 + (z-h)^2}} \equiv \frac{-1}{4\pi\sqrt{r^2 - 2rh\cos\theta + h^2}},$$

which satisfies the axisymmetric form of Laplace's equation (2.11.3).

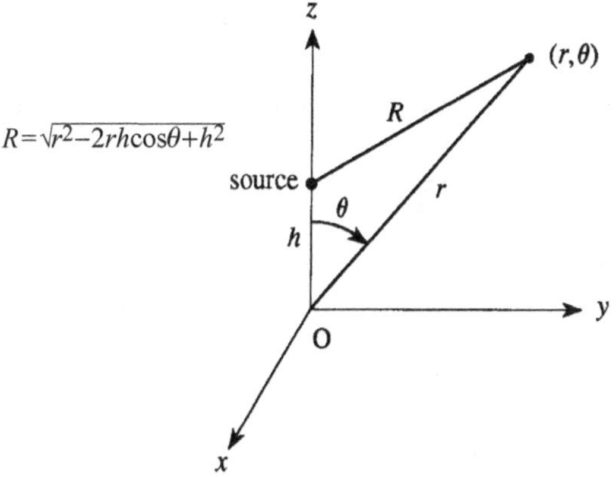

Figure 2.11.4

We can expand φ in ascending or descending powers of r/h depending on whether $r < h$ or $r > h$. To do this we first consider

$$\frac{1}{\sqrt{r^2 - 2rh\cos\theta + h^2}} = \begin{cases} \dfrac{1}{h}\displaystyle\sum_{n=0}^{\infty}\left(\dfrac{r}{h}\right)^n \mathcal{P}_n(\cos\theta), & r < h \\[2ex] \dfrac{1}{r}\displaystyle\sum_{n=0}^{\infty}\left(\dfrac{h}{r}\right)^n \mathcal{P}_n(\cos\theta), & h < r \end{cases} \qquad (2.11.18)$$

where symmetry requires the expansion coefficients $\mathcal{P}_n(\cos\theta)$ to be the same in both series. It is clear by inspection that $\mathcal{P}_n(\cos\theta)$ is a polynomial in $\cos\theta$ of the order of n. Substitution into Equation (2.11.3) shows that each term in these series must separately satisfy Laplace's equation. Therefore $\mathcal{P}_n(\cos\theta) = \alpha_n P_n(\cos\theta)$, where α_n is a constant.

Now when $\theta = 0$ the expansions become

$$\frac{1}{|r - h|} = \begin{cases} \dfrac{1}{h}\displaystyle\sum_{n=0}^{\infty}\left(\dfrac{r}{h}\right)^n \mathcal{P}_n(1), & r < h \\[2ex] \dfrac{1}{r}\displaystyle\sum_{n=0}^{\infty}\left(\dfrac{h}{r}\right)^n \mathcal{P}_n(1), & h < r \end{cases},$$

which can be true only if $\mathcal{P}_n(1) = 1$. Thus because $P_n(1) = 1$ we must have $\alpha_n = 1$, i.e., $\mathcal{P}_n(x) \equiv P_n(x)$. Hence we arrive at the following important representation of the potential of a unit point source:

$$\frac{-1}{4\pi\sqrt{r^2 - 2rh\cos\theta + h^2}} = \begin{cases} \dfrac{-1}{4\pi h}\displaystyle\sum_{n=0}^{\infty}\left(\dfrac{r}{h}\right)^n P_n(\cos\theta), & r < h \\[2ex] \dfrac{-1}{4\pi r}\displaystyle\sum_{n=0}^{\infty}\left(\dfrac{h}{r}\right)^n P_n(\cos\theta), & h < r \end{cases}. \qquad (2.11.19)$$

EXAMPLE 5. Find the potential of the flow produced by a point source of strength q situated at a distance h from the centre of a rigid sphere of radius $a < h$.

Let the source lie on the x_1 axis, and let θ denote the angle between the radius vector $\mathbf{x}$ from the centre of the sphere and the positive x_1 axis. Using (2.11.8) with $A_n = 0$, we can write the potential of the source in the presence of the sphere as

$$\varphi = \frac{-q}{4\pi\sqrt{r^2 - 2rh\cos\theta + h^2}} + \sum_{n=0}^{\infty} \frac{B_n}{r^{n+1}} P_n(\cos\theta), \quad r > a.$$

The coefficients B_n are found by use of the first expansion of (2.11.19) for $r < h$ and application of the normal velocity condition $\partial\varphi/\partial r = 0$ on $r = a$, which gives

$$B_0 = 0, \quad B_n = \frac{-qna^{2n+1}}{4\pi(n+1)h^{n+1}r^{n+1}}, \quad n \ge 1.$$

Therefore the required potential is

$$\varphi = \frac{-q}{4\pi\sqrt{r^2 - 2rh\cos\theta + h^2}} - \sum_{n=1}^{\infty} \frac{qna^{2n+1}}{4\pi(n+1)h^{n+1}r^{n+1}} P_n(\cos\theta), \quad r > a. \quad (2.11.20)$$

2.11.3 Interpretation in terms of images

The influence of the sphere on the velocity potential is represented by the infinite series in (2.11.20). The series can be recast in terms of the 'inverse point' at $r = \bar{h} \equiv a^2/h$ within the sphere on the ray between the centre of the sphere and the source, as follows:

$$-\sum_{n=1}^{\infty} \frac{qna^{2n+1}}{4\pi(n+1)h^{n+1}r^{n+1}} P_n(\cos\theta)$$

$$= -\frac{(qa/h)}{4\pi r} \sum_{n=0}^{\infty} \left(\frac{\bar{h}}{r}\right)^n \frac{nP_n(\cos\theta)}{(n+1)}$$

$$= -\frac{(qa/h)}{4\pi r} \sum_{n=0}^{\infty} \left(\frac{\bar{h}}{r}\right)^n P_n(\cos\theta) + \frac{(qa/h)}{4\pi r} \sum_{n=0}^{\infty} \left(\frac{\bar{h}}{r}\right)^n \frac{P_n(\cos\theta)}{n+1}$$

$$= \frac{-(qa/h)}{4\pi(r^2 - 2r\bar{h}\cos\theta + \bar{h}^2)^{\frac{1}{2}}} + \frac{(qa/h)}{4\pi r} \sum_{n=0}^{\infty} \left(\frac{\bar{h}}{r}\right)^n \frac{P_n(\cos\theta)}{n+1}. \quad (2.11.21)$$

Now, for $\lambda < r$,

$$\frac{1}{\sqrt{r^2 - 2r\lambda\cos\theta + \lambda^2}} = \sum_{0}^{\infty} \frac{\lambda^n}{r^{n+1}} P_n(\cos\theta),$$

$$\therefore \int_0^{\bar{h}} \frac{d\lambda}{\sqrt{r^2 - 2r\lambda\cos\theta + \lambda^2}} = \sum_{0}^{\infty} \left(\frac{\bar{h}}{r}\right)^{n+1} \frac{P_n(\cos\theta)}{n+1}.$$

Hence, when this is used in (2.11.21), overall potential (2.11.20) becomes

$$\varphi = \frac{-q}{4\pi(r^2 - 2rh\cos\theta + h^2)^{\frac{1}{2}}} - \frac{(qa/h)}{4\pi(r^2 - 2r\bar{h}\cos\theta + \bar{h}^2)^{\frac{1}{2}}}$$

$$+ \frac{(q/a)}{4\pi} \int_0^{\bar{h}} \frac{d\lambda}{\sqrt{r^2 - 2r\lambda\cos\theta + \lambda^2}}. \quad (2.11.22)$$

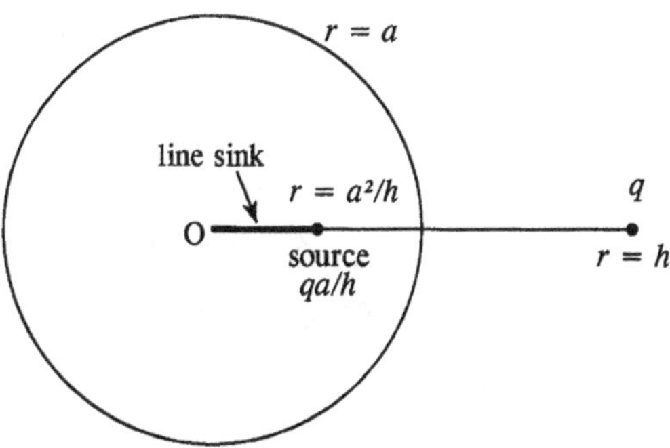

Figure 2.11.5

This shows that the effect of the sphere is equivalent to the potential produced by an image point source of strength qa/h at the inverse point $r = \bar{h}$ together with a 'line sink' stretching along the radius from the centre of the sphere to the inverse point of strength q/a per unit length (Figure 2.11.5).

2.12 The Stokes stream function

Flows that are axisymmetric with respect to some preferred direction are conveniently treated by use of cylindrical polar coordinates. If the x axis is taken as the axis of symmetry, we write

$$\mathbf{x} = (x, \varpi \cos\phi, \varpi \sin\phi), \tag{2.12.1}$$

where ϖ is the perpendicular distance from this axis and ϕ is the polar angle. For axisymmetric flow the velocity has components v_x and v_ϖ in the x and ϖ directions, respectively, which depend on only x, ϖ, and possibly also on time t.

The equation of continuity for an incompressible fluid then becomes

$$\frac{\partial}{\partial x}(\varpi v_x) + \frac{\partial}{\partial \varpi}(\varpi v_\varpi) = 0. \tag{2.12.2}$$

This is satisfied identically by

$$v_x = \frac{1}{\varpi}\frac{\partial \psi}{\partial \varpi}, \quad v_\varpi = -\frac{1}{\varpi}\frac{\partial \psi}{\partial x}, \tag{2.12.3}$$

where $\psi(x, \varpi)$ is called the *Stokes stream function*.

The physical significance of ψ can be understood by reference to Figure 2.12.1. Consider two points, 1 at (x_1, ϖ_1, ϕ) and 2 at (x_2, ϖ_2, ϕ), in the same meridian plane $\phi =$ constant, and let C be any plane curve joining these points in the plane. A surface of revolution between 1 and 2 is generated by the rotation of C about the x axis. The net volume flux of fluid through this surface, from left to right, is $2\pi \int_C \varpi \mathbf{v} \cdot \mathbf{n}\,ds$, where s

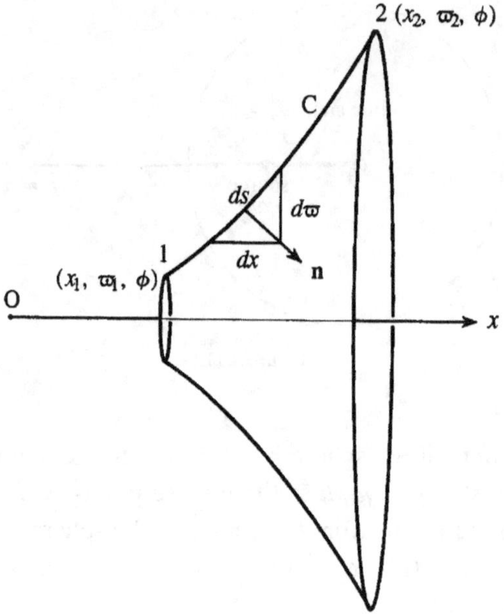

Figure 2.12.1

is the arc length measured along C from 1 to 2 and $\mathbf{n} = (d\varpi/ds, -dx/ds, 0)$ is the unit normal on C. Hence, using (2.12.3), we obtain

$$\text{volume flux} = 2\pi \int_C \varpi \mathbf{v} \cdot \mathbf{n} \, ds = 2\pi \int_C \left(\frac{\partial \psi}{\partial \varpi} d\varpi + \frac{\partial \psi}{\partial x} dx \right)$$

$$= 2\pi \int_C d\psi = 2\pi (\psi_2 - \psi_1).$$

Thus 2π times the change in the value of ψ between any two points is equal to the volume flux across the annular surface that we obtain by rotating around the axis of symmetry *any* curve joining those points. In particular, if we take point 1 on the axis of symmetry and put $\psi = 0$ there, then

$$\psi(x, \varpi) = \text{volume flux per radian}.$$

Also, on C,

$$\mathbf{v} \cdot \mathbf{n} = \frac{1}{\varpi} \left(\frac{\partial \psi}{\partial \varpi} \frac{d\varpi}{ds} + \frac{\partial \psi}{\partial x} \frac{dx}{ds} \right) = \frac{1}{\varpi} \frac{\partial \psi}{\partial s},$$

i.e. we obtain the velocity in any given direction by differentiating at right angles *to the left*.

The streamlines in axisymmetric flow lie in planes of constant ϕ and satisfy

$$\frac{dx}{v_x} = \frac{d\varpi}{v_\varpi}, \quad \text{i.e.,} \quad \frac{dx}{\partial \psi/\partial \varpi} = \frac{-d\varpi}{\partial \psi/\partial x},$$

$$\therefore \quad \frac{\partial \psi}{\partial x} dx + \frac{\partial \psi}{\partial \varpi} d\varpi \equiv d\psi = 0,$$

$$\text{i.e.} \quad \psi = \text{constant on a streamline}.$$

The rotation of a streamline about the x axis forms an axisymmetric stream surface, on which $\psi = $ constant.

The vorticity ω in axisymmetric flow can have only one component $\omega = \omega_\phi \mathbf{i}_\phi$, where $\mathbf{i}_\phi$ is a unit vector in the azimuthal direction, and

$$\omega_\phi = \frac{\partial v_\varpi}{\partial x} - \frac{\partial v_x}{\partial \varpi},$$

i.e.

$$\frac{1}{\varpi}\left(\frac{\partial^2 \psi}{\partial x^2} + \frac{\partial^2 \psi}{\partial \varpi^2} - \frac{1}{\varpi}\frac{\partial \psi}{\partial \varpi}\right) = -\omega_\phi. \tag{2.12.4}$$

Therefore, when the flow is irrotational, ψ must satisfy

$$\frac{\partial^2 \psi}{\partial x^2} + \frac{\partial^2 \psi}{\partial \varpi^2} - \frac{1}{\varpi}\frac{\partial \psi}{\partial \varpi} = 0. \tag{2.12.5}$$

Note that this is *not* the same as the Laplace equation satisfied by the velocity potential φ, which for axisymmetric flow becomes

$$\frac{\partial^2 \varphi}{\partial x^2} + \frac{\partial^2 \varphi}{\partial \varpi^2} + \frac{1}{\varpi}\frac{\partial \varphi}{\partial \varpi} = 0. \tag{2.12.6}$$

The Stokes stream function has dimensions of velocity $\times$ length^2, whereas the dimensions of φ are velocity $\times$ length.

2.12.1 Stream function examples

Figure 2.12.2 illustrates the three simple cases of uniform flow in the x direction [Figure 2.12.2(a)], a point source of strength q at the origin [Figure 2.12.2(b)], and a uniform line source of strength q per unit length on the x axis between points A and B [Figure 2.12.2(c)].

For the point source, the velocity in the spherically radial direction r is $q/4\pi r^2$, so that when we differentiate to the *left* of the radial direction,

$$\frac{1}{\varpi}\frac{1}{r}\frac{\partial \psi}{\partial \theta} \equiv \frac{1}{r^2 \sin\theta}\frac{\partial \psi}{\partial \theta} = \frac{q}{4\pi r^2},$$

$$\therefore \quad \psi = \text{constant} - \frac{q\cos\theta}{4\pi}.$$

In Figure 2.12.2(b) the constant has been chosen to make $\psi = 0$ on the streamline along the positive x axis. We can also derive the result by noting that a spherical cap of angle θ subtends a spherical angle $\sigma = 2\pi r^2(1 - \cos\theta)$ at the source, and the flux $2\pi\psi$ through the cap is therefore equal to $q\sigma/4\pi r^2$.

DIPOLE SOURCE The stream function for a point source can also be written as

$$\psi = -\frac{\partial}{\partial x}\left(\frac{q}{4\pi}r\right) + \text{constant}.$$

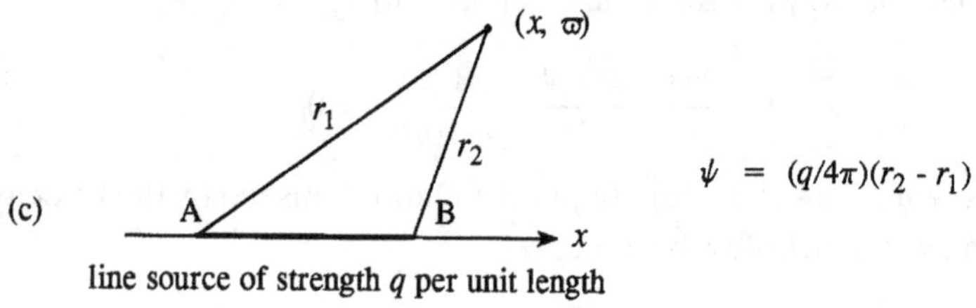

Figure 2.12.2

For a dipole at the origin orientated in the x direction, it follows by linear superposition that

$$\psi = -\frac{\partial^2}{\partial x^2}\left(\frac{q}{4\pi}r\right) = \frac{-q\,\sin^2\theta}{4\pi r}.$$

For a **sphere** moving at speed U along the x axis, with the origin at its centre,

$$\varphi = -\frac{Ua^3\cos\theta}{2r^2} = \frac{\partial}{\partial x}\left(\frac{Ua^3}{2r}\right),$$

$$\therefore\ \psi = \frac{\partial^2}{\partial x^2}\left(\frac{Ua^3}{2}r\right) = \frac{Ua^2\sin^2\theta}{2r}. \tag{2.12.7}$$

This is a particular instance of the general relation that, for constant A,

$$\psi = A\frac{\partial^{n+1}r}{\partial x^{n+1}} \quad \text{corresponds to} \quad \varphi = A\frac{\partial^n}{\partial x^n}\left(\frac{1}{r}\right). \tag{2.12.8}$$

The polar spherical equation of the streamlines for the sphere is $r = Ua^3\sin^2\theta/2\psi$. They are plotted in Figure 2.12.3: $\psi = 0$ corresponds to the x axis of symmetry and the surface of the sphere.

2.12.2 Rankine solids

The ideal flow produced by a point source q at the origin in the presence of a uniform flow at speed U in the x direction is given by

$$\psi = \frac{1}{2}U\varpi^2 - \frac{q}{4\pi}(1+\cos\theta) \equiv \frac{1}{2}U\varpi^2 - \frac{q}{4\pi}\left[1 + \frac{x}{(\varpi^2+x^2)^{\frac{1}{2}}}\right],$$

$$\varphi = Ux - \frac{q}{4\pi(\varpi^2+x^2)^{\frac{1}{2}}}.$$

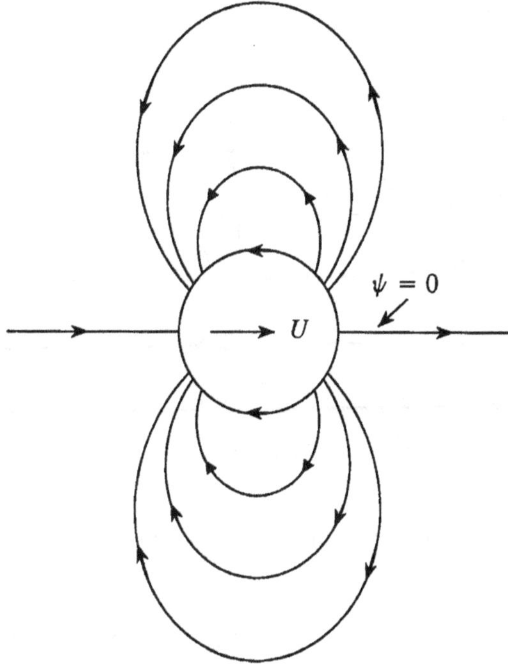

Figure 2.12.3

The outflow from the source on the negative x axis just cancels the mean flow velocity U at the point x where $\partial\varphi/\partial x = U - q/4\pi x^2 = 0$. Let this stagnation point be at $x = -a/2$; then $q = \pi a^2 U$, and the stream function becomes

$$\psi = \frac{1}{2}U\left\{\varpi^2 - \frac{a^2}{2}\left[1 + \frac{x}{(\varpi^2 + x^2)^{\frac{1}{2}}}\right]\right\}. \tag{2.12.9}$$

The streamlines are illustrated in Figure 2.12.4. The stream function $\psi = 0$ on the negative x axis and on the semi-infinite, cigar-shaped dividing surface that starts at the stagnation point A where the opposing flow velocities of the mean stream and the source flow are equal. The flows within and outside this surface are essentially distinct, and the surface may be replaced with a solid boundary without affecting the flow. The

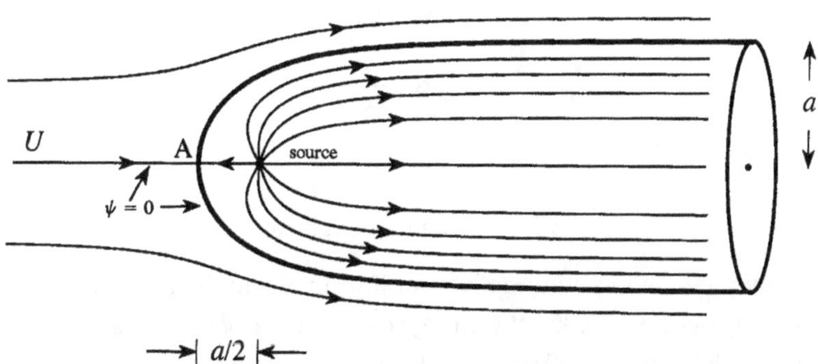

Figure 2.12.4

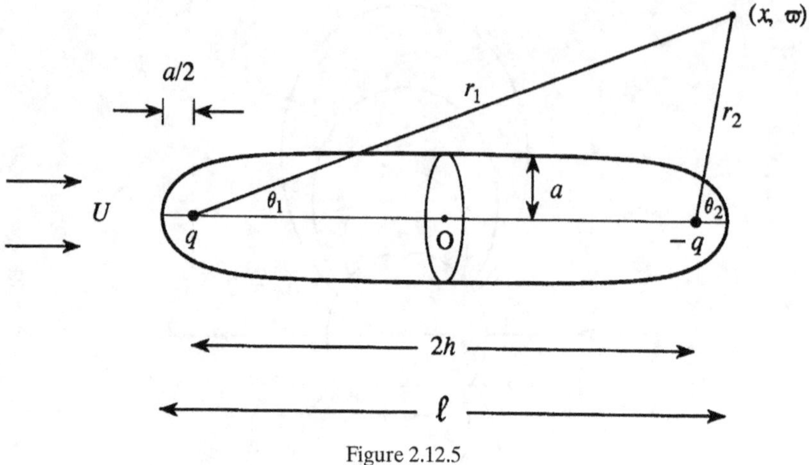

Figure 2.12.5

boundary defines the surface of a Rankine solid, whose shape is given in this case by the formula

$$\varpi^2 = \left(a^2 - x^2 + x\sqrt{2a^2 + x^2}\right)/2, \quad x > -a/2. \tag{2.12.10}$$

Far downstream of the nose the solid becomes circular cylindrical of radius a. At these points the flow velocity both inside and outside the dividing surface is asymptotically equal to U. The volume flux within the solid is therefore $\pi a^2 U$, which is, of course, exactly the rate at which fluid volume is being generated by the source.

2.12.3 Rankine ovoid

We obtain uniform flow past a finite, axisymmetric solid surface by placing a source q at $x = -h$ and a sink of equal strength sink at $x = +h$ on the x axis (Figure 2.12.5). Then

$$\psi = \frac{1}{2}U\varpi^2 + \frac{q}{4\pi}(\cos\theta_2 - \cos\theta_1), \quad \varphi = Ux + \frac{q}{4\pi}\left(\frac{1}{r_2} - \frac{1}{r_1}\right). \tag{2.12.11}$$

If we take $q = \pi a^2 U$ (as for the semi-infinite Rankine solid) then the overall length $\ell \approx 2h + a$, provided that $a < h/3$, and the central region is again cylindrical of radius a. The equation of the surface is easily shown to be approximately

$$\varpi^2 \approx \frac{a^2 - (h - |x|)^2}{2} + \frac{(h - |x|)}{2}\sqrt{(h - |x|)^2 + 2a^2}, \quad \text{for } \frac{a}{h} < \frac{1}{3}.$$

2.12.4 Drag in ideal flow

An isolated body moving irrotationally at uniform velocity in an unbounded fluid experiences no drag because the kinetic energy of the fluid does not change, and the body does no work on the fluid (see §2.9). For a Rankine ovoid it is evident from the fore–aft symmetry [and from expressions (2.12.11)] that the net pressure force on the body must vanish. However, the same conclusion can be deduced also for semi-infinite bodies, as we shall now verify for the Rankine solid of Figure 2.12.4.

The net excess pressure on the dividing surface (2.12.10) is evaluated from Bernoulli's equation

$$p = \frac{1}{2}\rho_0 U^2 - \frac{1}{2}\rho_0 v^2,$$

in terms of which the net drag (in the positive x direction) is

$$F_D = 2\pi \int_0^a p(\varpi)\,\varpi\,d\varpi. \tag{2.12.12}$$

From stream function (2.12.9) the velocity $\mathbf{v}$ is first found in component form:

$$v_x = U\left[1 + \frac{a^2 x}{4(\varpi^2 + x^2)^{\frac{3}{2}}}\right], \quad v_\varpi = \frac{Ua^2\varpi}{4(\varpi^2 + x^2)^{\frac{3}{2}}}.$$

We express these in terms of ϖ by making use of the surface formula $\psi = 0$, which supplies the alternative equivalent relations,

$$\frac{x}{(\varpi^2 + x^2)^{\frac{1}{2}}} = \frac{2\varpi^2}{a^2} - 1, \quad \varpi^2 + x^2 = \frac{a^2}{4(1 - \varpi^2/a^2)},$$

from which it follows that

$$v^2 = U^2\left(\frac{4\varpi^2}{a^2} - \frac{3\varpi^4}{a^4}\right).$$

Hence the surface pressure becomes

$$p = \frac{\rho_0 U^2}{2}\left(1 + \frac{3\varpi^4}{a^4} - \frac{4\varpi^2}{a^2}\right).$$

This is plotted in Figure 2.12.6 as a function of the distance s from the nose, given by

$$s = \frac{a}{2} + x \equiv \frac{a}{2}\left[1 + \frac{2\varpi^2/a^2 - 1}{(1 - \varpi^2/a^2)^{\frac{1}{2}}}\right].$$

The excess pressure is positive for $\varpi/a < \sqrt{3}$, i.e., in the forward, shaded region near the nose where $s < (a/2)(1 - 1/\sqrt{6}) \approx 0.3a$. To the rear of this point $v^2 > U^2$ and the excess pressure is negative, causing this region of the surface to experience a net *suction* force in the negative x direction. This suction force is just sufficient to overcome the drag in the nose region, because

$$F_D = 2\pi \int_0^a p(\varpi)\,\varpi\,d\varpi = \pi\rho_0 U^2 \int_0^a \left(1 + \frac{3\varpi^4}{u^4} - \frac{4\varpi^2}{a^2}\right)\varpi\,d\varpi \equiv 0.$$

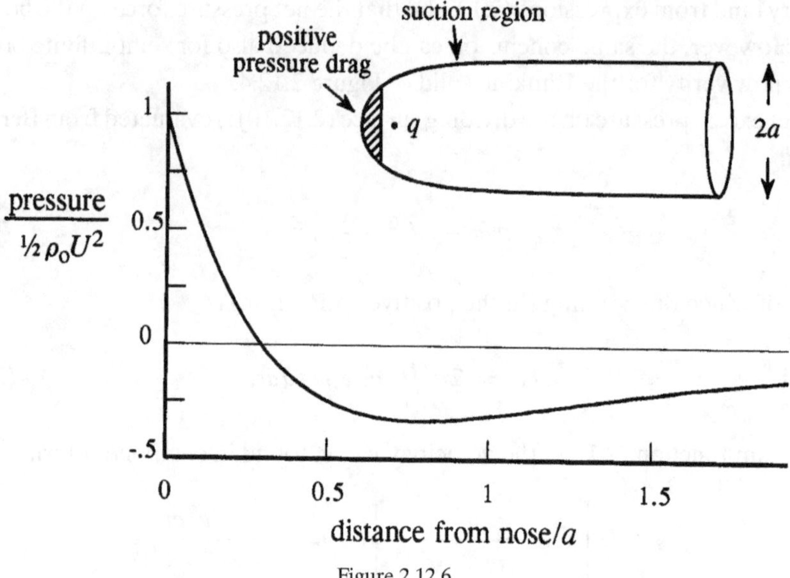

Figure 2.12.6

The velocity potential of any semi-infinite body that is ultimately cylindrical far down-stream of the nose can always be represented by a suitable distribution of monopole sources in the nose region. A more general proof of the vanishing of the net drag can be made in this case by use of the time-independent form of Reynolds equation (1.4.12). With body forces and viscous stresses ignored, this equation is integrated over the fluid region V bounded externally by a large surface Σ (which is pierced far downstream by the cylindrical tail of the body) and internally by the portion of the surface S contained within Σ. The net force F_i on S is then determined by application of the divergence theorem

$$F_i \equiv -\oint_S p\, n_i\, dS = \oint_\Sigma (p\delta_{ij} + \rho_0 v_i v_j)\, n_j\, dS, \qquad (2.12.13)$$

where the unit normal **n** is directed *into* the region V. The integral over Σ is evaluated, and shown to vanish as Σ recedes to infinity, by expression of the excess pressure and velocity in terms of the asymptotic velocity potential of the monopole source distribution that represents the nose profile of the body.

2.12.5 Axisymmetric flow from a nozzle

An ideal fluid travelling at constant velocity in a duct of uniform cross section is ob-viously moving irrotationally. When the duct terminates it is natural to assume that the continuation of the irrotational flow outside the duct is similar to that shown in Figure 2.12.7(a). Thus, *ideal* flow from a circular cylindrical 'nozzle' (into a nominally stationary ambient fluid) might be expected to be in the form of a circular cylindrical

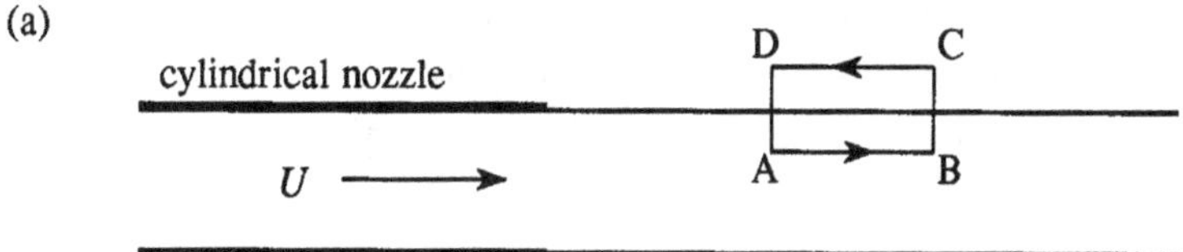

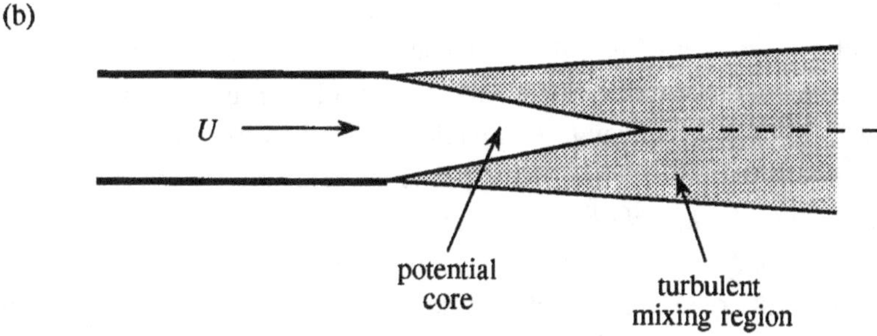

Figure 2.12.7

jet whose cross section is just the continuation of the nozzle. However, this flow is *not* irrotational outside the nozzle. To see this we calculate the circulation $\Gamma = \oint_{\mathrm{ABCD}} \mathbf{v} \cdot d\mathbf{x}$ around the infinitesimal rectangular contour ABCD (see figure) in a plane parallel to the jet, with sides $\mathrm{AB} = \delta x$ and $\mathrm{BC} = \delta \varpi$ and lying partly within and partly outside the jet. Then, $\mathbf{v} \cdot d\mathbf{x} = U dx$ on the arm AB of the contour but is zero elsewhere, so that $\Gamma = U\delta x$. Hence, because $\Gamma = \omega_\phi \delta x \delta \varpi$, the vorticity within the contour $\omega_\phi = U/\delta \varpi$, provided the points B and C lie on opposite sides of the jet boundary.

In these circumstances the vorticity is confined to a 'shear layer' of infinitesimal thickness at the edge of the jet, across which the velocity changes discontinuously. This is a *vortex sheet* with the vorticity distribution

$$\omega = \omega_\phi \mathbf{i}_\phi \equiv U\delta(\varpi - R)\mathbf{i}_\phi, \tag{2.12.14}$$

where R is the jet and nozzle radius. We can otherwise derive this result trivially by taking the curl of the velocity distribution:

$$\mathbf{v} = U\mathbf{i}_x \mathrm{H}(R - \varpi),$$

where $\mathbf{i}_x$ is a unit vector in the x direction. The quantity $\Gamma/\delta x = U$ is the circulation per unit length of the jet and is called the strength of the vortex sheet.

The ideal jet flow of Figure 2.12.7(a) is often a useful model at high Reynolds number, at which it is permissible to regard the fluid as 'inviscid'. However, we must then postulate a source of the vorticity. Evidently vorticity appears at the duct exit, and the jet can actually be considered to consist of the flow associated with a succession of vortex rings 'shed' continuously from the circular lip of the nozzle. It will be seen in Chapter 4 that the release of vorticity into a flow from a surface can occur only if the viscosity is non-zero. In practice viscous effects are usually important only in regions of high shear, especially near solid boundaries where the fluid velocity must reduce to that of the boundary, usually through a very thin layer of fluid (a 'boundary layer') adjacent to the surface. Therefore it is possible that, whereas it is quite permissible to neglect viscous stresses within the body of the flow, they become very important at surfaces where, in particular, they act to release vorticity into the main flow.

We shall see in the next subsection (Figure 2.12.8) that the velocity turns out to be infinite at a nozzle lip of infinitesimal thickness when the exit flow is strictly irrotational. In a real fluid, viscous stresses at the boundary must act to keep the velocity finite. However, for a wide range of problems it is possible to obtain approximate, inviscid solutions that incorporate the principal effects of viscous wall stresses by invoking the *Kutta–Joukowski hypothesis*, that vorticity is shed into the flow from the edge (in the form of discrete vortices or continuous vortex sheets) in a sufficient amount to remove the singularities that are otherwise predicted by ideal (potential) flow theory. In the case of flow from a cylindrical nozzle the required vortex sheet strength is simply the jet velocity U.

Considerations discussed in Chapter 4 reveal further that the ideal picture of nozzle flow in Figure 2.12.7(a) can never occur in practice. First, viscosity causes vorticity to be released from the lip by a process of molecular diffusion, and this mechanism continues within the body of the fluid. Thus a vortex sheet cannot persist within a viscous flow because diffusion immediately spreads vorticity into the formerly irrotational regions on either side of the ideal interface in such a way that at *short* distances x downstream of the nozzle the width of the shear layer $\sim \sqrt{vx/U}$. Second, and very much more important, a vortex sheet is unstable to small disturbances; this causes the sheet to break up into a random distribution of vorticity and the flow in the shear layer to become 'turbulent'. The convection (or 'diffusion') of vorticity by the turbulent velocity field (which is superimposed on the local mean jet flow) greatly increases the rate of growth in the width of the shear layer, so that the exit flow is more typically like that pictured in Figure 2.12.7(b). The shear layer turbulence promotes 'mixing' of the jet with ambient fluid, which is drawn into the flow and results in an overall increase in mass flux. The inner boundary of the 'mixing region' is roughly conical in shape and (for incompressible flow) extends four or five nozzle diameters downstream; it is called the *potential core* of the jet. At large downstream distances the lateral growth of the jet becomes independent of the shape and size of the nozzle. Dimensional analysis then implies that the mean radius R_ϖ of the turbulent jet is given by

$$R_\varpi \approx \alpha x, \qquad \alpha \sim 0.22, \qquad (2.12.15)$$

where α is a universal constant (determined by experiment) and x is measured from the nozzle exit plane, so that the angle of spread of the jet is ultimately about $25°$.

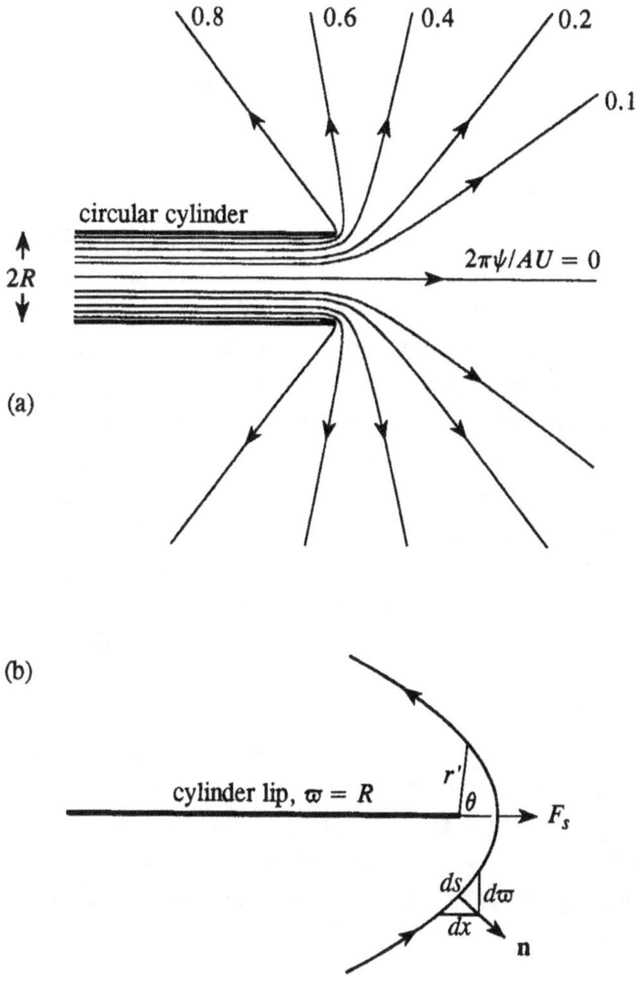

Figure 2.12.8

2.12.6 Irrotational flow from a circular cylinder

If the exit flow is entirely irrotational, fluid flowing out of the nozzle that was initially in contact with the inner wall of the cylinder remains 'attached' to the cylinder outside. The streamlines then have the axisymmetric pattern shown in Figure 2.12.8(a), which have been plotted by use of the formula (Howe 1998)

$$\frac{2\pi\psi(x,\varpi)}{\mathcal{A}U} = \frac{\varpi^2}{2R^2} - \frac{\varpi}{\pi R}\int_0^\infty I_1\left(\lambda\frac{\varpi}{R}\right)\sqrt{\frac{2K_1(\lambda)}{I_1(\lambda)}}\sin\left\{\lambda\left[\frac{x}{R} + \mathcal{F}(\lambda)\right]\right\}\frac{d\lambda}{\lambda}, \quad \varpi < R,$$

$$= \frac{1}{2} - \frac{\varpi}{\pi R}\int_0^\infty K_1\left(\lambda\frac{\varpi}{R}\right)\sqrt{\frac{2I_1(\lambda)}{K_1(\lambda)}}\sin\left\{\lambda\left[\frac{x}{R} + \mathcal{F}(\lambda)\right]\right\}\frac{d\lambda}{\lambda}, \quad \varpi > R,$$

$$(2.12.16)$$

where

$$\mathcal{F}(\lambda) = \frac{1}{\pi} \int_0^\infty \frac{\ln[K_1(\mu)I_1(\mu)/K_1(\lambda)I_1(\lambda)]}{\mu^2 - \lambda^2} \, d\mu,$$

and I_1 and K_1 are Bessel functions.

The flow at large distances from the exit resembles that from a point source of strength $q = \mathcal{A}U$ at the opening, where $\mathcal{A} = \pi R^2$ is the cross-sectional area of the cylinder. This source flow has no overall momentum, yet the uniform flow within the cylinder has x momentum equal to $\rho_o \mathcal{A}U$ per unit length. This momentum disappears at a rate $\rho_o \mathcal{A}U^2$ at the nozzle exit, which requires the application of a retarding force equal to $-\rho_o \mathcal{A}U^2$ in the x direction. Part of this force is supplied by the *negative* excess pressure (or *suction*) far upstream within the cylinder, which, according to Bernoulli's equation, is smaller by $\frac{1}{2}\rho_o U^2$ than the ambient mean pressure. Therefore the pressure force (in the x direction) applied to the rear end of a long slug of fluid flowing out of the nozzle is $-\frac{1}{2}\rho_o U^2 \mathcal{A}$. The remaining contribution to the retarding force, $F_s = -\frac{1}{2}\rho_o U^2 \mathcal{A}$, is supplied by suction at the circular lip of the nozzle, where the flow velocity becomes very large (infinite when the thickness of the cylinder wall is infinitesimal), and Bernoulli's equation predicts a large, negative pressure. To verify this, it is necessary to calculate the flow near the lip, where $x \sim 0$, $\varpi \sim R$.

The integral representations (2.12.16) of the flow are dominated by contributions from the high 'wavenumber' regions of the integrands ($\lambda \gg 1$) when $x \sim 0$, $\varpi \sim R$. When λ is large the main contribution to the integral defining $\mathcal{F}(\lambda)$ is from the neighbourhood of $\mu = \lambda$, which yields $\mathcal{F}(\lambda) \sim -\pi/4\lambda$. Using this and the large argument approximations for the Bessel functions, we can then easily show that, very close to the lip (and ignoring additive constants),

$$\varphi \approx U\sqrt{\frac{Rr'}{\pi}} \sin\left(\frac{\theta}{2}\right), \quad \psi \approx -UR^{\frac{3}{2}}\sqrt{\frac{r'}{\pi}} \cos\left(\frac{\theta}{2}\right), \tag{2.12.17}$$

where $r' = \sqrt{x^2 + (\varpi - R)^2}$ and (r', θ) are local polar coordinates in a meridian plane of constant ϕ with origin at the lip, as in Figure 2.12.8(b). These approximations are valid only very close to the lip, where the stream surface $\psi = \text{constant} \sim +0$ has the parabolic section indicated in the figure, which collapses down onto the interior and exterior walls of the cylinder as $\psi \to 0$.

As $\psi \to 0$, the net force F_s applied to the fluid at the lip is evidently in the x direction and equal to that on the stream surface. If s denotes distance measured along a streamline [as in Figure 2.12.8(b)] then

$$F_s = 2\pi R \int_{\psi \to 0} -\frac{1}{2}\rho_o(\nabla\varphi)^2 n_x ds, \tag{2.12.18}$$

where $n_x = d\varpi/ds$ is the x component of the unit normal $\mathbf{n}$. We readily calculate from (2.12.17) that, on $\psi = \text{constant}$,

$$(\nabla\varphi)^2 = \frac{U^2 R}{4\pi r'}, \quad n_x ds \equiv d\varpi = r'd\theta, \quad \text{where} \quad -\pi < \theta < \pi,$$

$$\therefore \quad F_s = 2\pi R \int_{-\pi}^{\pi} -\frac{\rho_o U^2 R}{8\pi} \, d\theta = -\frac{1}{2}\rho_o U^2 \mathcal{A}.$$

The suction force at the rim of the cylinder is finite because the integrand in Eq. (2.12.18) is unbounded in the neighbourhood of the lip as $\psi \to 0$. However, this is not necessary for the practical realisation of this force. Any flow that remains 'attached' to a rounded edge (such as the leading edge of an airfoil; see §3.3.4) experiences suction; and, for example, leading-edge suction makes an important contribution to maintaining flapping winged flight and fish locomotion. It is also unnecessary to assume in Eq. (2.12.18) that the flow is steady, as the contribution to F_s from the transient pressure $-\rho_o \partial\varphi/\partial t$ vanishes identically.

2.12.7 Borda's mouthpiece

Irrotational flow from the cylinder is reversible, and the inward and outward flows are subject to the same suction forces both within the duct and at the cylinder lip. An irrotational flow of this kind that oscillates back and forth as a function of time represents a good approximation to the reciprocating flow through the duct exit produced by a (long-wavelength, plane) sound wave incident from within the duct. In that case the opening behaves as an oscillating point source, alternately producing outward and inward radial flows in the ambient fluid. On the other hand, the corresponding 'Kutta condition' jet flow of Figure 2.12.7(a) cannot be reversed. The ambient streamlines for *steady* flow *into* the cylinder typically exhibit the radially symmetric pattern expected for flow into a sink [Figure 2.12.9(a)]. However, 'separation' usually occurs at the inlet; vorticity is shed from the lip into a continuous, axisymmetric vortex sheet in an amount sufficient to remove the pressure and velocity singularities at the lip. The motion is unstable and usually becomes turbulent with random vorticity rapidly filling the whole cross section of the cylinder in the downstream region. The vortex sheet model is often useful near the entrance, however, especially in cases involving the flow of liquid in which the vortex sheet separates the liquid from vapour or a gas (such as air) of relatively negligible density.

The entrance flow arrangement of this type illustrated in Figure 2.12.9(a) is called 'Borda's mouthpiece'. The stream surface in contact with the outer cylindrical wall of the duct separates tangentially at the lip forming an axisymmetric vortex sheet that becomes the edge of an internal jet within the cylinder. The jet ultimately becomes circular cylindrical of uniform cross-sectional area $\sigma\mathcal{A}$ and flow speed U, say, where $\mathcal{A}$ is the area of the duct and σ is the 'contraction ratio'. If the excess pressure far downstream within the duct is assumed to vanish, Bernoulli's equation implies that the

(a)
Σ
p_0
circular cylinder
F_s
σA
$p = 0$
U
F_s
(b)
Σ
p_0
circular aperture
F_s
σA
$p = 0$
U
F_s

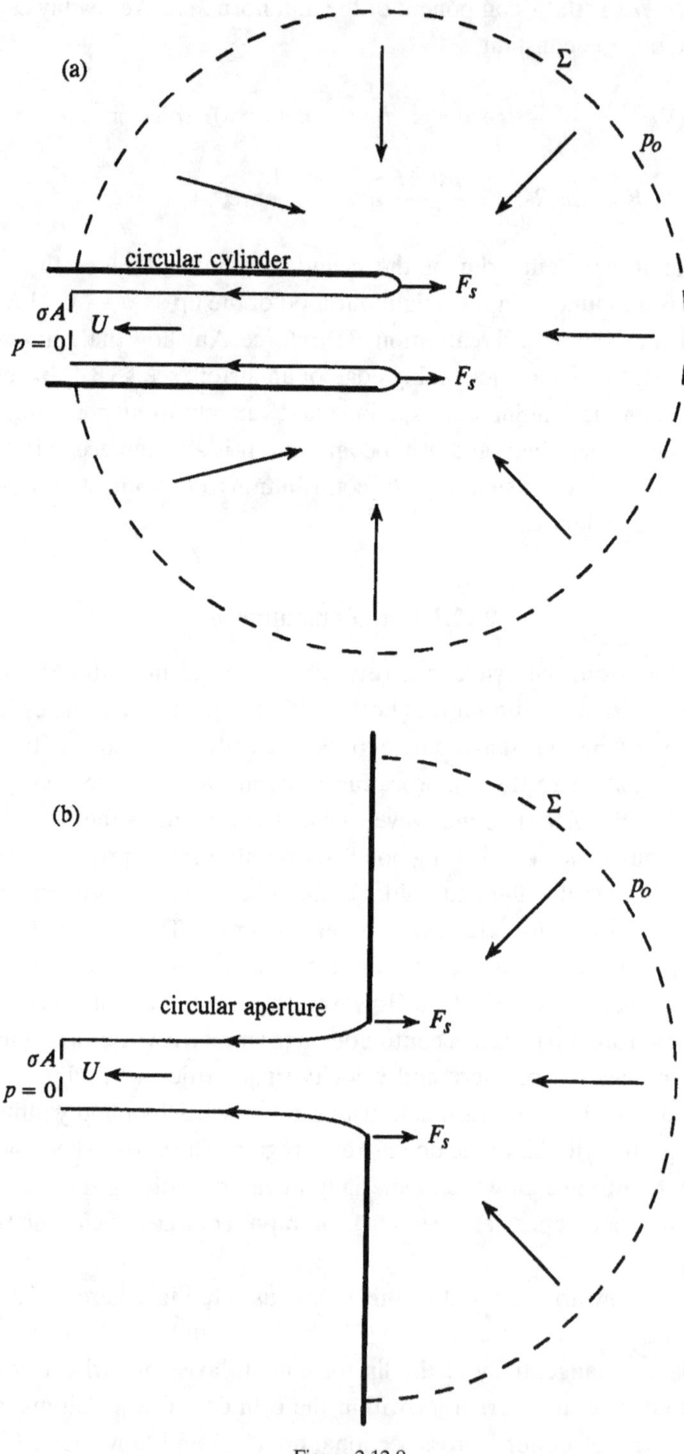

Figure 2.12.9

pressure in the ambient fluid at large distances from the mouth is $p_o = \frac{1}{2}\rho_o U^2$. Because the excess pressure vanishes at the vortex sheet, Bernoulli's equation also shows that the flow speed at the jet boundary is everywhere equal to U.

Conservation of momentum in steady flow can be used to relate the x component of the momentum flux of the jet, $-\rho_o U^2 \sigma \mathcal{A}$, to the pressure forces acting on the fluid. This is done by integration of the steady-state form of Reynolds equation (1.4.12) over the fluid region V in Figure 2.12.9(a) bounded by a closed surface Σ consisting of a sphere at infinity, the outer surface of the cylinder, the axisymmetric vortex sheet at the edge of the jet, and a cross section of the jet where the flow is uniform. When body forces are ignored, application of the divergence theorem yields

$$\oint_{\Sigma} (p\delta_{ij} + \rho_o v_i v_j) n_j \, dS = 0, \tag{2.12.19}$$

where $\mathbf{n}$ is the unit normal on Σ. Hence, because the velocity vanishes on the sphere at infinity and the normal component of velocity vanishes at the edge of the jet,

$$p_o - \rho_o U^2 \sigma = \frac{F_s}{\mathcal{A}}, \tag{2.12.20}$$

where the force F_s represents the effect of suction on the flow around the lip. However, the suction force vanishes identically in Borda's mouthpiece, so that Eq. (2.12.20) can be combined with Bernoulli's equation $p_o = \frac{1}{2}\rho_o U^2$ to deduce that the contraction ratio $\sigma = \frac{1}{2}$.

A similar argument using momentum integral formula (2.12.19) can be applied to the steady flow shown in Figure 2.12.9(b) through a circular aperture in a wall. When the mean excess pressure p_o in the neighbourhood of the aperture is uniform (on a scale that is large compared with the aperture radius) we again obtain relation (2.12.20) involving the suction force F_s. However, F_s cannot now be discarded, because the high-speed exit flow produces a significant pressure decrease on the wall near the aperture. Use of the Bernoulli relation $p_o = \frac{1}{2}\rho_o U^2$ now yields

$$\sigma = \frac{1}{2} - \frac{F_s}{\rho_o U^2 \mathcal{A}}.$$

Thus, because $F_s \leq 0$, this result implies that the contraction ratio $\sigma \geq \frac{1}{2}$ for a wall aperture; experiment indicates that $\sigma \approx 0.6 - 0.64$ (see §3.7).

2.13 The incompressible far field

It is often necessary to evaluate the far-field behaviour of the motion produced by a localised source distribution. The sources may be assumed to be contained within a closed 'control surface' S; inside S solid bodies or other inhomogeneities 'interact' with the fluid. The motion in the 'source-free' region V outside S is irrotational and its amplitude decreases with increasing distance from S. This far-field motion is typically of either monopole or dipole type, irrespective of the details of the interactions occurring within S. In exceptional cases both the monopole and dipole fields may vanish, and the leading-order representation of the far field would then typically be a quadrupole.

2.13.1 Deductions from Green's formula

Let the source region have characteristic dimension ℓ, and let the control surface S be defined by $f(\mathbf{x}, t) = 0$, with $f > 0$ in the source-free region V. Green's formula (2.8.7) determines the velocity potential $\varphi(\mathbf{x}, t)$ at $\mathbf{x}$ in V in terms of conditions on S:

$$\varphi(\mathbf{x}, t) = -\frac{\partial}{\partial x_j} \oint_S \frac{\varphi(\mathbf{y}, t) n_j dS}{4\pi |\mathbf{x} - \mathbf{y}|} - \oint_S n_j \frac{\partial \varphi}{\partial y_j}(\mathbf{y}, t) \frac{dS}{4\pi |\mathbf{x} - \mathbf{y}|}, \qquad (2.13.1)$$

where the unit normal $\mathbf{n}$ on S is directed into V [in which $q(\mathbf{x}, t) \equiv 0$]. Take the coordinate origin O within S, and consider the behaviour of these integrals when $|\mathbf{x}| \to \infty$.

Then $|\mathbf{y}| \sim \ell \ll |\mathbf{x}|$ when $\mathbf{y}$ lies on S and, because

$$\frac{\partial}{\partial y_i} \left(\frac{1}{|\mathbf{x} - \mathbf{y}|} \right) = -\frac{\partial}{\partial x_i} \left(\frac{1}{|\mathbf{x} - \mathbf{y}|} \right),$$

we can write

$$\frac{1}{|\mathbf{x} - \mathbf{y}|} = \frac{1}{|\mathbf{x}|} - y_i \frac{\partial}{\partial x_i} \left(\frac{1}{|\mathbf{x}|} \right) + \frac{y_i y_j}{2} \frac{\partial^2}{\partial x_i \partial x_j} \left(\frac{1}{|\mathbf{x}|} \right) + \cdots +,$$

and thereby develop Eq. (2.13.1) into the 'multipole expansion'

$$\varphi(\mathbf{x}, t) = -\frac{1}{4\pi |\mathbf{x}|} \oint_S \frac{\partial \varphi}{\partial y_n} dS + \frac{1}{4\pi} \frac{\partial}{\partial x_i} \left(\frac{1}{|\mathbf{x}|} \right) \oint_S \left(y_i \frac{\partial \varphi}{\partial y_n} - n_i \varphi \right) dS$$

$$- \frac{1}{4\pi} \frac{\partial^2}{\partial x_i \partial x_j} \left(\frac{1}{|\mathbf{x}|} \right) \oint_S \left(\frac{y_i y_j}{2} \frac{\partial \varphi}{\partial y_n} - y_i n_j \varphi \right) dS + \cdots +, \qquad (2.13.2)$$

$$= \text{monopole} + \text{dipole} + \text{quadrupole} + \cdots +,$$

where

$$\frac{\partial \varphi}{\partial y_n}(\mathbf{y}, t) = n_i \frac{\partial \varphi}{\partial y_i}(\mathbf{y}, t).$$

As $|\mathbf{x}| \to \infty$ only the first non-zero term in expansion (2.13.2) is significant. The monopole will normally vanish unless fluid volume is being created or destroyed within S. We therefore retain only the first two terms and write

$$\varphi(\mathbf{x}, t) \approx -\frac{Q}{4\pi |\mathbf{x}|} + \text{div} \left(\frac{\mathbf{I}}{4\pi |\mathbf{x}|} \right), \qquad |\mathbf{x}| \gg \ell, \qquad (2.13.3)$$

where

$$Q = \oint_S \frac{\partial \varphi}{\partial y_n} dS \; = \; \text{monopole source strength},$$

$$\mathbf{I} = \oint_S \left(\mathbf{y} \frac{\partial \varphi}{\partial y_n} - \mathbf{n}\varphi \right) dS \; = \; \text{dipole source strength.} \qquad (2.13.4)$$

The vector $\mathbf{I}(t)$ is called the specific **total impulse** of the source region. When $Q \equiv 0$ (for example, when S coincides with the surface of a solid of fixed volume) the far field is dominated by the dipole, and

$$\varphi(\mathbf{x}, t) \approx \operatorname{div}\left[\frac{\mathbf{I}(t)}{4\pi |\mathbf{x}|} \right] = -\frac{x_j I_j}{4\pi |\mathbf{x}|^3}, \quad \text{for} \quad |\mathbf{x}| \gg \ell. \qquad (2.13.5)$$

2.13.2 Far field produced by motion of a rigid body

Consider a rigid body of volume Δ and surface S in an unbounded ideal fluid at rest at infinity. Suppose the centre of volume $\mathbf{x}_o$ of S translates at velocity $\mathbf{U}_o(t)$ and that the body rotates with angular velocity $\mathbf{\Omega}(t)$, as in Figure 2.9.1, so that the velocity at $\mathbf{x}$ within the body is

$$\mathbf{U} = \mathbf{U}_o + \mathbf{\Omega} \wedge (\mathbf{x} - \mathbf{x}_o). \qquad (2.13.6)$$

Then in the integral of Eq. (2.13.4) we can put

$$\frac{\partial \varphi}{\partial x_n} = \left[\mathbf{U}_o + \mathbf{\Omega} \wedge (\mathbf{x} - \mathbf{x}_o) \right] \cdot \mathbf{n}. \qquad (2.13.7)$$

Hence, by application of the divergence theorem,

$$\mathbf{I} = \Delta \mathbf{U}_o - \oint_S \mathbf{n}\varphi \, dS. \qquad (2.13.8)$$

The remaining integral can be expressed in a standardised form that is characteristic of the shape and size of the body by the introduction of a set of elementary velocity potentials, each of which defines a simple translational or rotational motion of the body. We define φ_i^* to be the velocity potential of the fluid motion produced when the body translates at unit speed in the i direction without rotation and let χ_i^* be the potential produced when the body rotates (without translation) at unit angular velocity about an axis in the i direction through the centre of volume $\mathbf{x}_o(t)$. Both of these functions decrease at least as fast as $1/|\mathbf{x}|^2$ at large distances from S, and their definitions imply that

$$\frac{\partial \varphi_i^*}{\partial x_n} = n_i, \quad \frac{\partial \chi_i^*}{\partial x_n} = [(\mathbf{x} - \mathbf{x}_o) \wedge \mathbf{n}]_i \quad \text{on S}, \qquad (2.13.9)$$

where $\mathbf{n}$ is the unit normal directed into the fluid. The functional forms of φ_i^* and χ_i^* are determined entirely by the *shape* of S and may therefore be assumed to be known for any given body.

The actual velocity potential φ for arbitrary motion of S can be written as a linear combination of the φ_i^* and χ_i^*. Thus, if the body moves as in Figure 2.9.1, with its centre of volume $\mathbf{x}_o(t)$ translating at velocity $\mathbf{U}_o$ and with a rotation at angular velocity $\boldsymbol{\Omega}$, normal velocity (2.3.7) on S can be written as

$$\frac{\partial\varphi}{\partial x_n} = U_{oj}n_j + \Omega_j[(\mathbf{x} - \mathbf{x}_o) \wedge \mathbf{n}]_j.$$

According to Eq. (2.13.9), the velocity potential satisfying this condition is

$$\varphi = U_{oj}\varphi_j^* + \Omega_j\chi_j^*, \tag{2.13.10}$$

where the right-hand side is summed over the repeated suffix j.

Hence the ith component of specific total impulse (2.13.8) becomes

$$I_i = \Delta U_{oi} - U_{oj}\oint_S n_i\varphi_j^* dS - \Omega_j\oint_S n_i\chi_j^* dS. \tag{2.13.11}$$

2.13.3 Inertia coefficients

The integrals in Eq. (2.13.11) depend on only the shape and size of the body, and they are called inertia coefficients. These coefficients are defined more generally by

$$\hat{M}_{ij} \equiv \hat{M}_{ji} = -\oint_S n_i\varphi_j^* dS \equiv -\oint_S \frac{\partial\varphi_i^*}{\partial x_n}\varphi_j^* dS,$$

$$\hat{B}_{ij} \equiv \hat{B}_{ji} = -\oint_S \chi_i^*\frac{\partial\chi_j^*}{\partial x_n} dS \equiv -\oint_S \chi_i^*[(\mathbf{x} - \mathbf{x}_o) \wedge \mathbf{n}]_j dS, \tag{2.13.12}$$

$$\hat{C}_{ij} = -\oint_S \chi_j^*\frac{\partial\varphi_i^*}{\partial x_n} dS = -\oint_S \varphi_i^*\frac{\partial\chi_j^*}{\partial x_n} dS \equiv -\oint_S \varphi_i^*[(\mathbf{x} - \mathbf{x}_o) \wedge \mathbf{n}]_j dS.$$

Note that $\hat{M}_{ij}$, $\hat{C}_{ij}$, and $\hat{B}_{ij}$ have the dimensions of volume, volume $\times$ length, and volume $\times$ length^2, respectively. We prove the symmetry of $\hat{M}_{ij}$ and $\hat{B}_{ij}$ by recalling that φ_i^* and χ_i^* are solutions of Laplace's equation that decrease at least as fast as $1/|\mathbf{x}|^2$ as $|\mathbf{x}| \to \infty$ and therefore, for example, $\hat{M}_{ij} = \hat{M}_{ji}$ because the divergence theorem implies that

$$\oint_S \left(\frac{\partial\varphi_i^*}{\partial x_n}\varphi_j^* - \varphi_i^*\frac{\partial\varphi_j^*}{\partial x_n}\right) dS = \int_{\text{fluid}} \left(\varphi_i^*\nabla^2\varphi_j^* - \varphi_j^*\nabla^2\varphi_i^*\right) d^3\mathbf{x} \equiv 0.$$

Specific total impulse (2.13.11) may now be cast in the form

$$I_i = \Delta U_{oi} - U_{oj}\oint_S n_i\varphi_j^* dS - \Omega_j\oint_S n_i\chi_j^* dS \equiv (\Delta\delta_{ij} + \hat{M}_{ij})U_{oj} + \hat{C}_{ij}\Omega_j, \tag{2.13.13}$$

and the velocity potential in the far field, approximation (2.13.5), becomes

$$\varphi(\mathbf{x}, t) \approx \frac{\partial}{\partial x_i}\left[\frac{I_i(t)}{4\pi|\mathbf{x}|}\right] = -\frac{x_i}{4\pi|\mathbf{x}|^3}\left[(\Delta\delta_{ij} + \hat{M}_{ij})U_{oj} + \hat{C}_{ij}\Omega_j\right], \quad |\mathbf{x}| \to \infty. \tag{2.13.14}$$

2.13.4 Pressure in the far field

At large distances from the body the unsteady pressure $p \approx -\rho_o\partial\varphi/\partial t$. In general the inertia coefficients depend on the orientation of the body. For a rotating body they should first be calculated in a reference frame fixed relative to the body, with respect to which they are constant. Then

$$p(\mathbf{x}, t) \approx \frac{x_i \rho_o}{4\pi |\mathbf{x}|^3} \left[(\Delta \delta_{ij} + \hat{M}_{ij}) \frac{dU_{oj}}{dt} + \hat{C}_{ij} \frac{d\Omega_j}{dt} \right], \quad |\mathbf{x}| \to \infty, \tag{2.13.15}$$

where the time derivatives are determined by

$$\frac{d\mathbf{U}_o}{dt} = \frac{\partial \mathbf{U}_o}{\partial t} + \mathbf{\Omega} \wedge \mathbf{U}_o, \quad \frac{d\mathbf{\Omega}}{dt} = \frac{\partial \mathbf{\Omega}}{\partial t},$$

where $\partial/\partial t$ represents the time rate of change relative to the rotating axes.

EXAMPLE 1. INERTIA COEFFICIENTS FOR A SPHERE For a sphere of radius a (with centre at the origin)

$$\varphi_i^* = \frac{-a^3 x_i}{2|\mathbf{x}|^3}, \quad \therefore \quad \hat{M}_{ij} = \frac{a^3}{2} \oint_S \frac{n_i n_j dS}{|\mathbf{x}|^2} = \pi a^3 \delta_{ij} \int_0^\pi \cos^2 \theta \sin \theta \, d\theta = \frac{2\pi a^3}{3} \delta_{ij},$$

$$\tag{2.13.16}$$

i.e. $\hat{M}_{ij} = \frac{1}{2} \Delta \delta_{ij}$, and its magnitude is half the volume of the sphere. By symmetry $\hat{B}_{ij} = \hat{C}_{ij} = 0$. Thus the dipole field of a translating sphere is

$$\varphi = -\frac{x_i}{4\pi |\mathbf{x}|^3} \left[(\Delta \delta_{ij} + \frac{1}{2} \Delta \delta_{ij}) U_{oj} \right] \equiv U_{oj} \varphi_j^*(\mathbf{x}).$$

2.14 Force on a rigid body

The distribution of impulsive pressure required to start an irrotational flow instantaneously from a state of rest is $\varpi = -\rho_o \varphi(\mathbf{x}, t)$ (§2.3.2). Changes in the flow produced by unsteady motion of a rigid body are caused by the instantaneous propagation of impulsive pressure throughout the fluid from the body. Because 'force = rate of change of impulse', it follows that the net force $\mathbf{F}(t)$ experienced by the body during this change is given by

$$\mathbf{F} = \rho_o \frac{d}{dt} \oint_S \mathbf{n} \varphi \, dS. \tag{2.14.1}$$

By making use of representation (2.13.10), $\varphi = U_{oj} \varphi_j^* + \Omega_j \chi_j^*$, and the definitions of the inertia coefficients, we can also write

$$F_i = \rho_o \frac{d}{dt} \oint_S \left(U_{oj} n_i \varphi_j^* + \Omega_j n_i \chi_j^* \right) dS = -\frac{d}{dt} \left(U_{oj} M_{ij} + \Omega_j C_{ij} \right), \tag{2.14.2}$$

where

$$M_{ij} = \rho_o \hat{M}_{ij}, \quad C_{ij} = \rho_o \hat{C}_{ij}. \tag{2.14.3}$$

The new coefficient M_{ij} (which has the dimension of mass) is called the added-mass tensor. The quantity $U_{oj} M_{ij}$ evidently represents i momentum supplied to the fluid by surface motion, and the corresponding term in (2.14.2) is the reaction force of the fluid on the body that opposes the motion and is equivalent to an augmentation of its mass by fluid 'dragged along' during unsteady motion. Unless the body is spherical the added-mass coefficient, and therefore the strength of the reaction force, will depend on the direction of motion. The term involving C_{ij} is an analogous augmentation of mass produced by rotation.

From Equation (2.13.8) we can write

$$\rho_o \frac{d\mathbf{I}}{dt} = -\mathbf{F} + m_o \frac{d\mathbf{U}_o}{dt}, \tag{2.14.4}$$

where $m_o = \rho_o \Delta$ is the mass of fluid displaced by the body and $-\mathbf{F}$ is the force exerted on the fluid by the body. Thus $\rho_o d\mathbf{I}/dt$ represents the force required to accelerate the fluid *and* the fluid displaced by the solid body, supposed to move at the velocity $\mathbf{U}_o$ of the centre of volume.

A formal proof of Equation (2.14.1) can be given by use of Bernoulli's equation (2.3.5) and the relation

$$\frac{\partial \mathrm{H}}{\partial t} + \frac{\partial \varphi}{\partial x_j} \frac{\partial \mathrm{H}}{\partial x_j} = 0, \quad \text{where} \quad \mathrm{H} \equiv \mathrm{H}(f),$$

and $f(\mathbf{x}, t) = 0$ is the equation of the surface S of the body, with $f > 0$ in the fluid. Then

$$\frac{d}{dt} \oint_S \mathbf{n}\varphi \, dS = \frac{d}{dt} \int_{-\infty}^{\infty} \varphi \nabla \mathrm{H} \, d^3\mathbf{x} = \int_{-\infty}^{\infty} \left(\frac{\partial \varphi}{\partial t} \nabla \mathrm{H} + \varphi \nabla \frac{\partial \mathrm{H}}{\partial t} \right) d^3\mathbf{x}$$

$$= \oint_S \mathbf{n}\frac{\partial \varphi}{\partial t} \, dS + \int_{-\infty}^{\infty} \left\{ \nabla \left[\varphi \frac{\partial \mathrm{H}}{\partial t} \right] + \frac{\partial}{\partial x_i} \left[\mathrm{H}\nabla\varphi \frac{\partial \varphi}{\partial x_i} \right] \right.$$

$$\left. - \nabla \left[\frac{1}{2}(\nabla\varphi)^2 \mathrm{H} \right] + \frac{1}{2}(\nabla\varphi)^2 \nabla\mathrm{H} \right\} d^3\mathbf{x}$$

$$= \oint_S \mathbf{n}\frac{\partial \varphi}{\partial t} \, dS + 0 + 0 + 0 + \int_{-\infty}^{\infty} \frac{1}{2}(\nabla\varphi)^2 \nabla\mathrm{H} \, d^3\mathbf{x}$$

$$= \oint_S \mathbf{n} \left[\frac{\partial \varphi}{\partial t} + \frac{1}{2}(\nabla\varphi)^2 \right] dS = -\oint_S \frac{p}{\rho_o} \mathbf{n} \, dS \equiv \frac{\mathbf{F}}{\rho_o},$$

where the '0' terms vanish because $\mathbf{v} = \nabla\varphi \sim 1/|\mathbf{x}|^3$ as $|\mathbf{x}| \to \infty$ and because $\partial \mathrm{H}/\partial t$ also vanishes there.

The integral in (2.14.1) is independent of time when the body translates without rotation at constant velocity, and then $\mathbf{F} = 0$. This is a restatement of *d'Alembert's paradox*, that the drag experienced by a body placed in a uniform irrotational stream vanishes identically.

EXAMPLE 1. EQUATION OF MOTION OF A BODY ACCELERATING WITHOUT ROTATION Let a body of mass m with added-mass tensor M_{ij} be subject to an externally applied force $\mathcal{F}$. Then, for motion at velocity $\mathbf{U}(t)$ the equation of motion of the body is

$$m\frac{dU_i}{dt} = -\mathrm{M}_{ij}\frac{dU_j}{dt} + \mathcal{F}_i$$

i.e.,

$$\left(m\delta_{ij} + \mathrm{M}_{ij}\right)\frac{dU_j}{dt} = \mathcal{F}_i.$$

The added-mass tensor determines the effective mass of fluid that accompanies the body in its accelerated motion. The inertia of this fluid, in addition to that of the body, must be overcome by the force $\mathcal{F}$ when the body accelerates. In general, however, a *couple* must also be applied to the translating body to counter a rotational torque simultaneously exerted on the body by the fluid.

EXAMPLE 2. EQUATION OF MOTION OF A RIGID SPHERE From (2.13.16), for a sphere $\mathrm{M}_{ij} = \frac{1}{2}\rho_o\Delta\delta_{if} = \frac{1}{2}m_o\delta_{ij}$, where m_o is the mass of fluid displaced by the sphere. Hence the equation of motion is

$$\left(m + \frac{m_o}{2}\right)\frac{d\mathbf{U}}{dt} = \mathcal{F},$$

where m is the mass of the sphere and $\mathcal{F}$ is the external force applied to the sphere.

EXAMPLE 3. PRESSURE IN THE FAR FIELD At large distances from a moving body the unsteady pressure $p \approx -\rho_o\partial\varphi/\partial t$:

$$p(\mathbf{x},t) \approx -\rho_o\frac{\partial^2}{\partial t\partial x_i}\left[\frac{\mathrm{I}_i(t)}{4\pi|\mathbf{x}|}\right] = \frac{x_i}{4\pi|\mathbf{x}|^3}\frac{d}{dt}\left[(m_o\delta_{ij} + \mathrm{M}_{ij})U_{oj} + \mathrm{C}_{ij}\Omega_j\right], \quad |\mathbf{x}| \to \infty,$$

This may also be set in the form

$$p(\mathbf{x},t) \approx \mathrm{div}\left\{\frac{1}{4\pi|\mathbf{x}|}\left[\mathbf{F}(t) - m_o\frac{d\mathbf{U}_o}{dt}\right)\right], \quad |\mathbf{x}| \to \infty.$$

2.14.1 Moment exerted on a rigid body

The distributed pressure forces on S exert a net moment $\mathbf{M}$ on the body. It is convenient to refer this moment to the centre of volume of the body and to write

$$\mathbf{M} = -\oint_S (\mathbf{x} - \mathbf{x}_o) \wedge \mathbf{n}p\, dS. \tag{2.14.5}$$

It is subsequently shown that

$$\mathbf{M} = \rho_o\frac{d}{dt}\oint_S (\mathbf{x} - \mathbf{x}_o) \wedge \mathbf{n}\varphi\, dS + \rho_o\mathbf{I} \wedge \mathbf{U}_o. \tag{2.14.6}$$

In general $\rho_o\mathbf{I} \wedge \mathbf{U}_o$ is non-zero even when the body moves in *steady translational* motion, when there is no mean drag. The steady pressure distribution on S produces a finite couple that would tend, for example, to cause an elongated body held in a uniform stream to align itself with its long axis perpendicular to the direction of the mean velocity (see Question 24 of Problems 2).

Equation (2.14.5) is established as follows:

$$\frac{d}{dt} \oint_S (\mathbf{x} - \mathbf{x}_o) \wedge \mathbf{n}\varphi \, dS = \frac{d}{dt} \int_{-\infty}^{\infty} (\mathbf{x} - \mathbf{x}_o) \wedge \nabla H\varphi \, d^3\mathbf{x}$$

$$= -\oint_S \mathbf{U}_o \wedge \mathbf{n}\varphi \, dS + \int_{-\infty}^{\infty} (\mathbf{x} - \mathbf{x}_o) \wedge \frac{\partial}{\partial t}(\nabla H\varphi) \, d^3\mathbf{x}$$

$$\equiv -\mathbf{I} \wedge \mathbf{U}_o + \oint_S (\mathbf{x} - \mathbf{x}_o) \wedge \mathbf{n}\left[\frac{\partial \varphi}{\partial t} + \frac{1}{2}(\nabla\varphi)^2\right] dS,$$

$$\therefore \quad \frac{d}{dt} \oint_S (\mathbf{x} - \mathbf{x}_o) \wedge \mathbf{n}\varphi \, dS = -\mathbf{I} \wedge \mathbf{U}_o - \oint_S (\mathbf{x} - \mathbf{x}_o) \wedge \mathbf{n}\frac{p}{\rho_o} \, dS. \qquad \text{Q. E. D.}$$

The third line of this proof is obtained as follows:

$$\int_{-\infty}^{\infty} (\mathbf{x} - \mathbf{x}_o) \wedge \frac{\partial}{\partial t}(\nabla H\varphi) \, d^3\mathbf{x} = \oint_S (\mathbf{x} - \mathbf{x}_o) \wedge \mathbf{n}\frac{\partial \varphi}{\partial t} \, dS + \int_{-\infty}^{\infty} (\mathbf{x} - \mathbf{x}_o) \wedge \varphi\frac{\partial}{\partial t}\nabla H \, d^3\mathbf{x},$$

where, with suffix notation used,

$$\int_{-\infty}^{\infty} \left[(\mathbf{x} - \mathbf{x}_o) \wedge \varphi\frac{\partial}{\partial t}\nabla H\right]_i d^3\mathbf{x}$$

$$\equiv \int_{-\infty}^{\infty} \epsilon_{ijk}(x_j - x_{oj})\varphi\frac{\partial^2 H}{\partial t \partial x_k} d^3\mathbf{x}$$

$$= \int_{-\infty}^{\infty} \epsilon_{ijk}\left\{\frac{\partial}{\partial x_k}\left[(x_j - x_{oj})\varphi\frac{\partial H}{\partial t}\right] + \frac{\partial}{\partial x_l}\left[(x_j - x_{oj})\frac{\partial \varphi}{\partial x_k}\frac{\partial \varphi}{\partial x_l}H\right]\right.$$

$$\left. - \frac{\partial}{\partial x_k}\left[(x_j - x_{oj})\frac{1}{2}(\nabla\varphi)^2 H\right] + (x_j - x_{oj})\frac{1}{2}(\nabla\varphi)^2 \frac{\partial H}{\partial x_k}\right\} d^3\mathbf{x}$$

$$= 0 + 0 + 0 + \oint_S \left[(\mathbf{x} - \mathbf{x}_o) \wedge \mathbf{n}\frac{1}{2}(\nabla\varphi)^2\right]_i dS.$$

EXAMPLE 4. Find the moment exerted on the rigid prolate spheroid,

$$\frac{x^2}{a^2(1+\epsilon)} + \frac{y^2}{a^2} + \frac{z^2}{a^2} = 1, \quad \epsilon \ll 1,$$

rotating at angular velocity $\boldsymbol{\Omega}(t) = (0, 0, \Omega)$ about a fixed axis coinciding with the z axis (see Figure 2.11.3).

The moment M_3 about the z axis is given by (2.14.6) with $\mathbf{U}_o = 0$. The centre of volume coincides with the geometrical centre of the body, so that, in terms of axes rotating with the spheroid (§2.11, Example 4),

$$\varphi = \frac{-\epsilon\Omega a^5 xy}{3|\mathbf{x}|^5}, \quad \mathbf{n} \approx \frac{\mathbf{x} - \epsilon(x, 0, 0)}{\sqrt{|\mathbf{x}|^2 - 2\epsilon x^2}}.$$

Therefore, correct to $O(\epsilon^2)$,

$$\rho_o \oint_S (\mathbf{x} \wedge \mathbf{n})_3\varphi \, dS = \frac{-\epsilon^2 \rho_o\Omega a^5 \mathbf{k}}{3} \oint_{|\mathbf{x}|=a} \frac{x^2 y^2 dS}{|\mathbf{x}|^6} = \frac{-\epsilon^2 m_o\Omega a^2 \mathbf{k}}{15},$$

where m_o is the mass of the displaced fluid and $\mathbf{k}$ is a unit vector in the z direction. Hence

$$M_3 = -\frac{\epsilon^2 m_o a^2}{15}\frac{d\Omega}{dt}.$$

The quantity $\epsilon^2 m_o a^2/15$ is the moment of inertia of the fluid effectively set into motion around the z axis by the rotating body; when the body is subjected to an externally applied moment $\mathcal{M}_3$, say, its angular acceleration is determined by

$$\left(\mathcal{I}_3 + \frac{\epsilon^2 m_o a^2}{15}\right)\frac{d\Omega}{dt} = \mathcal{M}_3,$$

where $\mathcal{I}_3$ is the moment of inertia of the spheroid.

2.15 Sources near solid boundaries

Let us return to the problem of solving

$$\nabla^2 \varphi = q(\mathbf{x}, t) \tag{2.15.1}$$

in an infinite fluid region containing a prescribed source distribution $q(\mathbf{x}, t)$. In the absence of solid boundaries (in *free space*) it was shown in §2.6 that the solution that vanishes in the far field is

$$\varphi(\mathbf{x}, t) = \int_{-\infty}^{\infty} q(\mathbf{y}, \tau) G(\mathbf{x}, \mathbf{y}) d^3\mathbf{y}, \tag{2.15.2}$$

where $G(\mathbf{x}, \mathbf{y})$ is the free-space Green's function

$$G(\mathbf{x}, \mathbf{y}) = \frac{-1}{4\pi |\mathbf{x} - \mathbf{y}|}, \tag{2.15.3}$$

that is, $G(\mathbf{x}, \mathbf{y})$ is the solution of

$$\nabla^2 G = \delta(\mathbf{x} - \mathbf{y}) \tag{2.15.4}$$

that vanishes as $|\mathbf{x}| \to \infty$.

Green's formula (2.8.7) was used in §2.8 to represent the solution of (2.15.1) in the presence of a solid boundary S (Figure 2.15.1) as the sum of the velocity potential produced by the sources in the absence of S and the velocity potentials produced by certain monopole and dipole sources distributed on S, represented by the second and third terms on the right-hand side of Equation (2.8.7). This solution was derived by use of free-space Green's function (2.15.3).

It would be very convenient if the functional form of $G(\mathbf{x}, \mathbf{y})$ could be adjusted so that it *automatically* accounts for the dipole and monopole sources on S, without the need to evaluate surface integrals. To do this we must find a solution of the Green's function equation (2.15.4) that not only decays as $|\mathbf{x}| \to \infty$, but also satisfies the correct boundary conditions on S. The solution φ of (2.15.1) would then be given by formula (2.15.2) with G replaced with the modified Green's function, there being no additional surface integrals to evaluate.

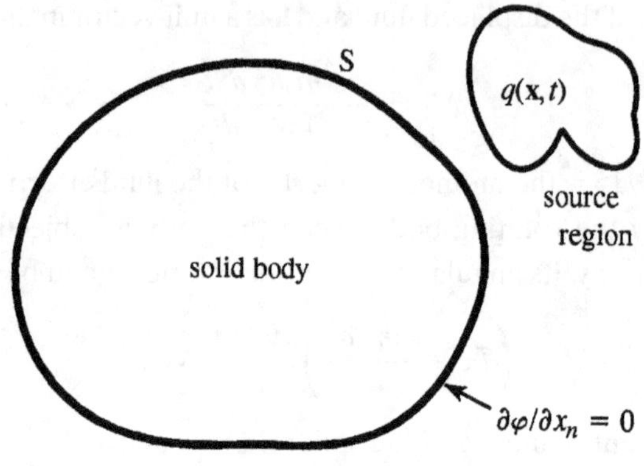

Figure 2.15.1

The main practical difficulty is the calculation of the modified Green's function. Exact analytical representations can be found only for solid bodies of very simple shapes such as spheres and circular cylinders (a series representation of G for a rigid sphere is given by the result of Example 5 of §2.11 by the setting of $q = 1$). However, a relatively simple and general approximate formula can always be found for the modified Green's function when the solution of Equation (2.15.1) is required only *in the far field* where $|\mathbf{x}| \to \infty$. This is called the *far-field Green's function*.

2.15.1 The reciprocal theorem

The calculation of the far-field Green's function is greatly simplified by invoking the *reciprocal theorem*.

Consider the two problems indicated in Figure 2.15.2, of the determination of the velocity potentials of the flows produced by two unit point sources, one at $\mathbf{x} = \mathbf{x}_A$ and the other at $\mathbf{x} = \mathbf{x}_B$ in the presence of a solid body S. Let $G(\mathbf{x}, \mathbf{x}_A)$ and $G(\mathbf{x}, \mathbf{x}_B)$ respectively denote the velocity potentials generated at $\mathbf{x}$, where

$$\nabla^2 G(\mathbf{x}, \mathbf{x}_A) = \delta(\mathbf{x} - \mathbf{x}_A), \tag{2.15.5}$$

$$\nabla^2 G(\mathbf{x}, \mathbf{x}_B) = \delta(\mathbf{x} - \mathbf{x}_B). \tag{2.15.6}$$

In addition we shall permit $G(\mathbf{x}, \mathbf{x}_A)$ and $G(\mathbf{x}, \mathbf{x}_B)$ to satisfy simple mechanical boundary conditions on S, which we take to have the same general *linear* form:

$$\frac{\partial G}{\partial x_n}(\mathbf{x}, \mathbf{x}_A) = \frac{G(\mathbf{x}, \mathbf{x}_A)}{\mathcal{Z}(\mathbf{x})}, \quad \frac{\partial G}{\partial x_n}(\mathbf{x}, \mathbf{x}_B) = \frac{G(\mathbf{x}, \mathbf{x}_B)}{\mathcal{Z}(\mathbf{x})}, \quad \text{for } \mathbf{x} \text{ on S}, \tag{2.15.7}$$

where x_n is measured in the normal direction from S *into the fluid* and $\mathcal{Z}(\mathbf{x})$ is the surface 'impedance'. For a *rigid* surface, $\mathcal{Z}(\mathbf{x}) = \infty$.

In the far field, at large distances from S, both solutions are required to decay in the usual monopole manner,

$$G(\mathbf{x}, \mathbf{x}_A) \sim G(\mathbf{x}, \mathbf{x}_B) \sim -\frac{1}{4\pi |\mathbf{x}|}, \quad |\mathbf{x}| \to \infty, \tag{2.15.8}$$

where it may be supposed that the coordinate origin is in the neighbourhood of S.

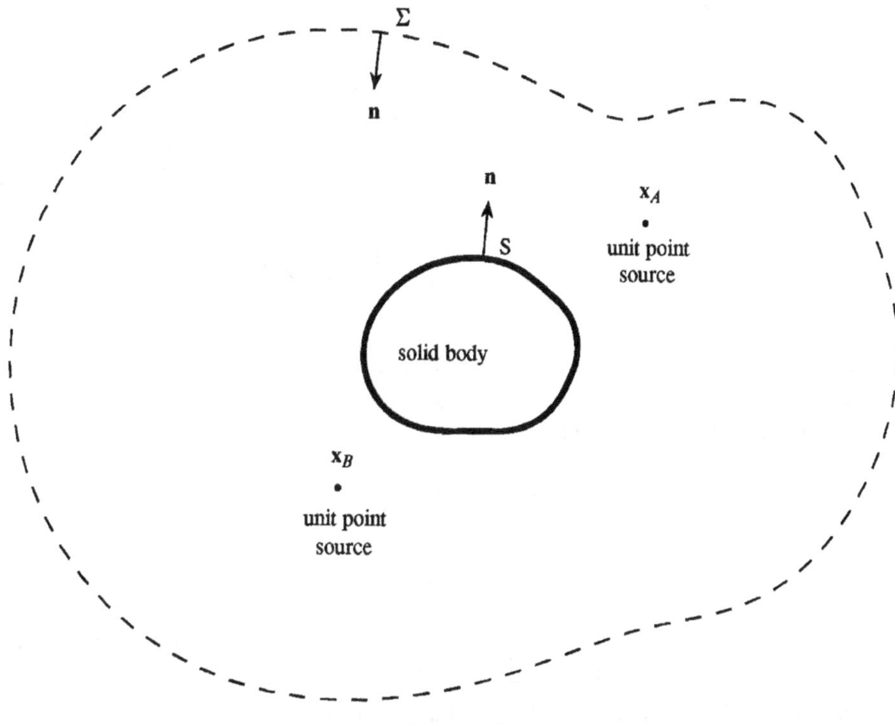

Figure 2.15.2

The **reciprocal theorem** states that

$$G(\mathbf{x}_A, \mathbf{x}_B) = G(\mathbf{x}_B, \mathbf{x}_A), \qquad (2.15.9)$$

that is, the potential at $\mathbf{x}_A$ produced by the point source at $\mathbf{x}_B$ is equal to the potential at $\mathbf{x}_B$ produced by an equal point source at $\mathbf{x}_A$.

PROOF Multiply Equation (2.15.5) by $G(\mathbf{x}, \mathbf{x}_B)$ and Equation (2.15.6) by $G(\mathbf{x}, \mathbf{x}_A)$, subtract the results, and integrate over the region between the surface S and a large surface Σ in the far field. Green's identity

$$G(\mathbf{x}, \mathbf{x}_B)\nabla^2 G(\mathbf{x}, \mathbf{x}_A) - G(\mathbf{x}, \mathbf{x}_A)\nabla^2 G(\mathbf{x}, \mathbf{x}_B)$$
$$= \mathrm{div}\Big[G(\mathbf{x}, \mathbf{x}_B)\nabla G(\mathbf{x}, \mathbf{x}_A) - G(\mathbf{x}, \mathbf{x}_A)\nabla G(\mathbf{x}, \mathbf{x}_B) \Big]$$

and the divergence theorem permit the volume integral of the term obtained from the left-hand sides to be expressed as surface integrals over S and Σ, whereas the integrals involving the δ functions can be evaluated explicitly. This procedure gives

$$\oint_{S+\Sigma} \left[G(\mathbf{x}, \mathbf{x}_A)\frac{\partial G}{\partial x_n}(\mathbf{x}, \mathbf{x}_B) - G(\mathbf{x}, \mathbf{x}_B)\frac{\partial G}{\partial x_n}(\mathbf{x}, \mathbf{x}_A) \right] dS = G(\mathbf{x}_B, \mathbf{x}_A) - G(\mathbf{x}_A, \mathbf{x}_B).$$

The surface integral over S vanishes because of impedance conditions (2.15.7). The surface integral over Σ vanishes because conditions (2.15.8) imply that the integrand is $O(1/|\mathbf{x}|^3)$ on Σ. This proves the theorem.

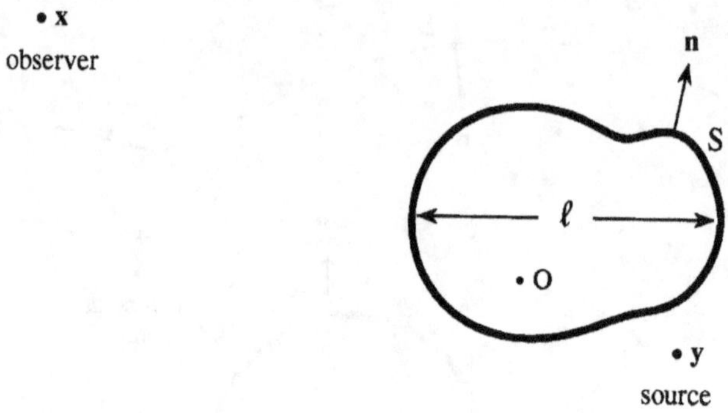

Figure 2.16.1

The result is usually expressed as the simple reciprocal relation

$$G(\mathbf{x}, \mathbf{y}) = G(\mathbf{y}, \mathbf{x}). \qquad (2.15.10)$$

2.16 Far-field Green's function

We are now ready to derive the far-field Green's function $G(\mathbf{x}, \mathbf{y})$ for the problem shown
in Figure 2.16.1, where motion is produced by a point source at $\mathbf{y}$ adjacent to a fixed,
solid body of characteristic diameter ℓ. We have to solve

$$\nabla^2 G(\mathbf{x}, \mathbf{y}) = \delta(\mathbf{x} - \mathbf{y}), \quad \frac{\partial G}{\partial x_n} = 0 \text{ on S}. \qquad (2.16.1)$$

We shall see later that the condition $\partial G/\partial x_n = 0$ on S does not restrict the applicability
of G only to problems involving rigid surfaces. The influence of the body is equivalent to
an additional distribution of monopoles and dipoles on S. The far-field Green's function
determines the net effect of these monopole and dipoles for an observer in the far field,
so obviating the need to evaluate surface integrals.

Take the coordinate origin at O within S. The source point $\mathbf{y}$ is assumed to be close
to S (so that $|\mathbf{y}| \sim \ell$) and the observer at $\mathbf{x}$ is taken to be in the far field, where $|\mathbf{x}| \gg \ell$.

In these circumstances the far-field approximation for $G(\mathbf{x}, \mathbf{y})$ can be found very easily
from the solution of the *reciprocal problem*:

$$\left(\frac{\partial^2}{\partial y_1^2} + \frac{\partial^2}{\partial y_2^2} + \frac{\partial^2}{\partial y_3^2} \right) G(\mathbf{y}, \mathbf{x}) = \delta(\mathbf{y} - \mathbf{x}), \quad \frac{\partial G}{\partial y_n} = 0 \text{ on S}, \qquad (2.16.2)$$

where the source is at the far-field point $\mathbf{x}$ and $G(\mathbf{y}, \mathbf{x})$ is determined as a function of $\mathbf{y}$ close to S. The solution of (2.16.1) is then given by the reciprocal theo-
rem (§2.15) $G(\mathbf{x}, \mathbf{y}) = G(\mathbf{y}, \mathbf{x})$ [the potential $G(\mathbf{x}, \mathbf{y})$ at the far-field point $\mathbf{x}$ produced
by the point source at $\mathbf{y}$ is exactly equal to the potential $G(\mathbf{y}, \mathbf{x})$ produced at the near-
field point $\mathbf{y}$ by an equal point source at the far-field point $\mathbf{x}$].

To solve (2.16.2) put

$$G(\mathbf{y}, \mathbf{x}) = G_o(\mathbf{y}, \mathbf{x}) + G'(\mathbf{y}, \mathbf{x})$$

$$\equiv \frac{-1}{4\pi |\mathbf{x} - \mathbf{y}|} + G'(\mathbf{y}, \mathbf{x}),$$

where $G_o(\mathbf{y}, \mathbf{x})$ is the radially symmetric velocity potential of the flow produced by the point source at $\mathbf{x}$ when the presence of the solid is ignored. The term $G'(\mathbf{y}, \mathbf{x})$ is the additional velocity potential required to account for the modification of this flow in the neighbourhood of S.

When $|\mathbf{x}| \gg \ell$ and $|\mathbf{y}| \sim \ell$, the expansion

$$\frac{1}{|\mathbf{x} - \mathbf{y}|} = \frac{1}{|\mathbf{x}|} \left[1 + \frac{\mathbf{x} \cdot \mathbf{y}}{|\mathbf{x}|^2} + O\left(\frac{\ell^2}{|\mathbf{x}|^2}\right) \right]$$

implies that

$$G_o(\mathbf{y}, \mathbf{x}) \equiv \frac{-1}{4\pi |\mathbf{x} - \mathbf{y}|} \approx \frac{-1}{4\pi |\mathbf{x}|} \left[1 + \frac{x_j y_j}{|\mathbf{x}|^2} + O\left(\frac{\ell^2}{|\mathbf{x}|^2}\right) \right]. \tag{2.16.3}$$

The linear dependence on y_j in this formula represents the first approximation (of the order of $\ell/|\mathbf{x}|$) in a power-series expansion of rapidly decreasing terms that describes the variation of the incident spherically symmetric flow from the source at $\mathbf{x}$ close to the body. Thus, regarded as functions of $\mathbf{y}$, the terms shown explicitly in

$$G_o(\mathbf{y}, \mathbf{x}) = \frac{-1}{4\pi |\mathbf{x}|} - \frac{x_j y_j}{4\pi |\mathbf{x}|^3} + \cdots +$$

$$\equiv \text{constant} + U_j y_j + \cdots +, \quad \text{where} \quad U_j = \frac{-x_j}{4\pi |\mathbf{x}|^3} \tag{2.16.4}$$

can be regarded as the velocity potential of a *uniform flow* at velocity U_j impinging on the solid.

At distances $|\mathbf{y}| \gg \ell$ from S the distortion of this flow produced by the body must be small. Let it be represented by the velocity potential

$$G'(\mathbf{y}, \mathbf{x}) = -U_j \Phi_j(\mathbf{y}), \quad \text{where} \quad \Phi_j(\mathbf{y}) \to 0 \text{ when } |\mathbf{y}| \gg \ell.$$

The function Φ_j satisfies Laplace's equation $\nabla^2 \Phi_j(\mathbf{y}) = 0$ and has the dimensions of length and $\sim \ell$ in order of magnitude. Then

$$G(\mathbf{y}, \mathbf{x}) = G_o(\mathbf{y}, \mathbf{x}) + G'(\mathbf{y}, \mathbf{x}) = \frac{-1}{4\pi |\mathbf{x}|} + U_j \left[y_j - \Phi_j(\mathbf{y}) \right] + \cdots +, \tag{2.16.5}$$

where the terms shown explicitly represent a potential flow past the body.

The rigid surface condition requires that

$$\frac{\partial}{\partial y_n} \left[y_j - \Phi_j(\mathbf{y}) \right] = 0 \text{ on S},$$

that is,

$$\frac{\partial \Phi_j}{\partial y_n} = \frac{\partial y_j}{\partial y_n} \equiv n_i \frac{\partial y_j}{\partial y_i} = n_i \delta_{ij} = n_j \text{ on S}.$$

Hence Φ_j is just the instantaneous velocity potential of the motion that would be produced by translational motion of S as a *rigid body* at unit speed in the j direction, and it therefore coincides with the function $\varphi_j^*(\mathbf{y})$ introduced in §2.13.2 and used in the definition of the inertia coefficients of the body.

SUMMARIZING THESE RESULTS When $\mathbf{x}$ is in the far field and $\mathbf{y}$ is close to the body,

$$G(\mathbf{x}, \mathbf{y}) = \frac{-1}{4\pi |\mathbf{x}|} \left\{ 1 + \frac{x_j}{|\mathbf{x}|^2}\left[y_j - \varphi_j^*(\mathbf{y}) \right] + O\left(\frac{\ell^2}{|\mathbf{x}|^2}\right) \right\}, \quad \mathbf{y} \sim O(\ell), \ |\mathbf{x}| \to \infty. \quad (2.16.6)$$

The first term in the braces represents the contribution from the incident source flow potential $G_o(\mathbf{x}, \mathbf{y})$ evaluated at $\mathbf{y} = 0$. The next term is $O(\ell/|\mathbf{x}|)$ and includes a component $x_j y_j/|\mathbf{x}|^2$ from the incident flow plus a correction $-ix_j\varphi_j^*(\mathbf{y})/|\mathbf{x}|^2$ produced by S.

2.16.1 The Kirchhoff vector

The vector field

$$\mathbf{Y}(\mathbf{y}) \equiv \mathbf{y} - \varphi^*(\mathbf{y}) \qquad (2.16.7)$$

is called the *Kirchhoff vector* for the body; the jth component

$$Y_j(\mathbf{y}) \equiv y_j - \varphi_j^*(\mathbf{y})$$

satisfies Laplace's equation $\nabla^2 Y_j = 0$ with $\partial Y_j/\partial y_n = 0$ on S and can be interpreted as the velocity potential of an incompressible flow past S that has unit speed in the j direction at large distances from S. The function $\varphi_j^*(\mathbf{y})$ decays with distance from S, satisfies

$$\frac{\partial \varphi_j^*}{\partial y_n}(\mathbf{y}) = n_j \ \text{ on S}, \qquad (2.16.8)$$

and is just the instantaneous velocity potential of the motion that would be produced by translational motion of S as a *rigid body* at unit speed in the j direction.

DEFINITION

$$G(\mathbf{x}, \mathbf{y}) = \frac{-1}{4\pi |\mathbf{x}|} \left\{ 1 + \frac{x_j}{|\mathbf{x}|^2}\left[y_j - \varphi_j^*(\mathbf{y}) \right] \right\}, \quad \mathbf{y} \sim O(\ell), \ |\mathbf{x}| \to \infty, \qquad (2.16.9)$$

is called the far-field Green's function for source points $\mathbf{y}$ near the body and observer positions $\mathbf{x}$ in the far field.

In §2.18 a very much more elegant representation of the far-field Green's function is introduced that greatly expands its utility.

2.16.2 Far-field Green's function for a sphere

Let the sphere have radius a and take the coordinate origin O at its center, as illustrated in Figure 2.16.2. We have to determine the Kirchhoff vector whose jth component

$$Y_j(\mathbf{y}) = y_j - \varphi_j^*(\mathbf{y}), \quad \text{for } j = 1, 2, 3,$$

is equal to the velocity potential of incompressible flow past the sphere having unit speed in the j direction at large distances from the sphere. A particular case of this problem

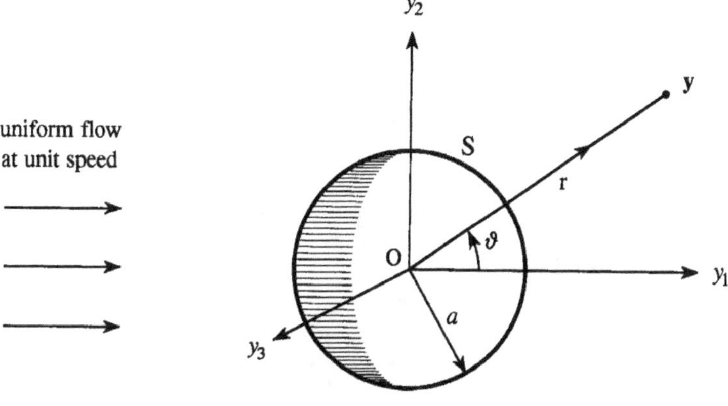

Figure 2.16.2

was solved in §2.11 (Example 2) by use of the expansion of the general solution of the axisymmetric Laplace equation in terms of Legendre polynomials. A direct method of solution is briefly outlined.

Consider the case $j = 1$ shown in the figure. The flow is evidently symmetric about the y_1 axis. Take spherical polar coordinates (r, ϑ, ϕ) with ϑ measured from the positive y_1 axis. Then $y_1 = r \cos \vartheta$, and condition (2.16.8) to be satisfied on the sphere is

$$\frac{\partial \varphi_1^*}{\partial r} = \cos \vartheta \quad \text{at} \quad r = a. \tag{2.16.10}$$

The boundary condition on S suggests that $\varphi_1^*(\mathbf{y})$ has the separable form

$$\varphi_1^* = \Psi(r) \cos \vartheta,$$

which satisfies the axisymmetric Laplace equation

$$\left[\frac{1}{r^2} \frac{\partial}{\partial r} \left(r^2 \frac{\partial}{\partial r} \right) + \frac{1}{r^2 \sin \vartheta} \frac{\partial}{\partial \vartheta} \left(\sin \vartheta \frac{\partial}{\partial \vartheta} \right) \right] \Psi(r) \cos \vartheta = 0,$$

provided that

$$r^2 \frac{d^2 \Psi}{dr^2} + 2r \frac{d\Psi}{dr} - 2\Psi = 0.$$

The solutions of this equation are proportional to r^n, where n is a root of the quadratic equation

$$n^2 + n - 2 = 0, \quad \text{i.e.} \quad n = -2, \ 1.$$

Hence

$$Y_1 \equiv y_1 - \varphi_1^* = r \cos \vartheta - \left(Ar + \frac{B}{r^2} \right) \cos \vartheta, \quad \text{where } A \text{ and } B \text{ are constants.}$$

The condition that $\varphi_1^* \to 0$ as $r \to \infty$ implies that $A = 0$, and condition (2.16.10) supplies $B = -a^3/2$. Therefore,

$$Y_1 = r \cos \vartheta + \frac{a^3}{2r^2} \cos \vartheta \equiv y_1 \left(1 + \frac{a^3}{2r^3} \right).$$

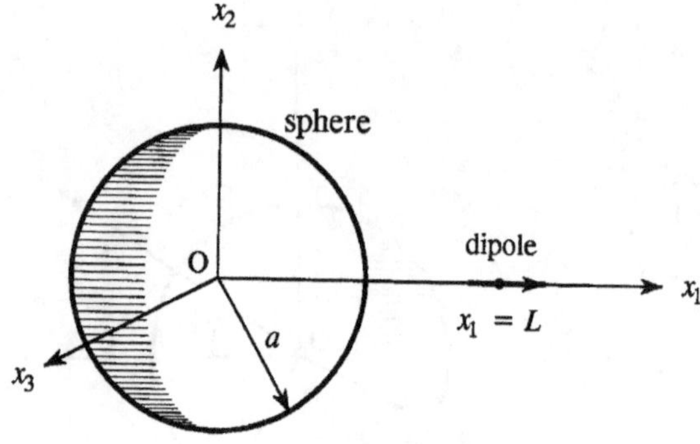

Figure 2.16.3

Because of the symmetry of the sphere it is clear that we also have

$$Y_2 = y_2\left(1 + \frac{a^3}{2r^3}\right), \quad Y_3 = y_3\left(1 + \frac{a^3}{2r^3}\right), \quad r = |\mathbf{y}|.$$

Thus far-field Green's function (2.16.9) for the sphere is

$$G(\mathbf{x}, \mathbf{y}) = \frac{-1}{4\pi |\mathbf{x}|}\left[1 + \frac{x_j y_j}{|\mathbf{x}|^2}\left(1 + \frac{a^3}{2|\mathbf{y}|^3}\right)\right], \quad \mathbf{y} \sim O(a), \ |\mathbf{x}| \to \infty. \tag{2.16.11}$$

This represents the far-field velocity potential of the motion produced by a point source at $\mathbf{y}$ close to the sphere. Because $|\mathbf{y}|/|\mathbf{x}|$ is small the second ('dipole') term in the brackets is always *small* compared with 1. This appears to suggest that, after all, the sphere has a relatively small effect on the motion at large distances! This is certainly true for monopole sources, but most sources of interest in applications are dipoles or quadrupoles, and in these circumstances we shall see that it is the small, second term that dominates the far field.

EXAMPLE 1. RADIAL DIPOLE ADJACENT TO A SPHERE Let us determine the far field produced by a radially orientated dipole point source at distance L from the centre of a rigid sphere of radius a (Figure 2.16.3). We have to solve

$$\nabla^2 \varphi = f_1 \frac{\partial}{\partial x_1}[\delta(x_1 - L)\delta(x_2)\delta(x_3)], \quad \text{where} \quad \frac{\partial \varphi}{\partial x_n} = 0 \ \text{on} \ |\mathbf{x}| = a.$$

The dipole is orientated in the x_1 direction and lies on the x_1 axis at $(L, 0, 0)$. The solution is given by

$$\varphi(\mathbf{x}) = \int f_1 \frac{\partial}{\partial y_1}[\delta(y_1 - L)\delta(y_2)\delta(y_3)] \, G(\mathbf{x}, \mathbf{y})d^3\mathbf{y},$$

where the integration is over the fluid and $\partial G/\partial x_n = 0$ on the sphere. The source term is zero everywhere except at $(L, 0, 0)$. To evaluate the integral we write

$$\varphi(\mathbf{x}) = f_1 \int \frac{\partial}{\partial y_1}[G(\mathbf{x}, \mathbf{y})\delta(y_1 - L)\delta(y_2)\delta(y_3)] \, d^3\mathbf{y}$$

$$- f_1 \int \delta(y_1 - L)\delta(y_2)\delta(y_3)\frac{\partial G}{\partial y_1}(\mathbf{x}, \mathbf{y})d^3\mathbf{y}.$$

The first integral is zero because $\delta(y_1 - L) = 0$ on the boundaries of the region of integration,

$$\therefore \quad \varphi(\mathbf{x}) = -f_1 \left[\frac{\partial G}{\partial y_1}(\mathbf{x}, \mathbf{y}) \right]_{\mathbf{y}=(L,0,0)}. \tag{2.16.12}$$

Thus far the calculation is exact. To determine the solution in the far field we use far-field approximation (2.16.11) for $G(\mathbf{x}, \mathbf{y})$. We see immediately that the differentiation with respect to y_1 will be applied only to the 'small' second term in the brackets of (2.16.11), giving

$$\begin{aligned}
\varphi(\mathbf{x}) &= \frac{f_1 x_j}{4\pi |\mathbf{x}|^3} \left\{ \frac{\partial}{\partial y_1} \left[y_j \left(1 + \frac{a^3}{2|\mathbf{y}|^3} \right) \right] \right\}_{\mathbf{y}=(L,0,0)} \\
&= \frac{f_1 x_1}{4\pi |\mathbf{x}|^3} \left(1 - \frac{a^3}{L^3} \right) \\
&= \frac{f_1 \cos\theta}{4\pi |\mathbf{x}|^2} \left(1 - \frac{a^3}{L^3} \right), \quad |\mathbf{x}| \to \infty,
\end{aligned}$$

where θ is the angle between the x_1 axis and the $\mathbf{x}$ direction (so that $x_1 = |\mathbf{x}| \cos\theta$).

By setting $a = 0$ in this formula we recover solution (2.7.3) for a dipole in the absence of the sphere. The sphere accordingly *reduces* the far-field velocity relative to that produced by a free-field dipole, and it vanishes when $L \to a$. In this limit the surface of the sphere is effectively plane in the vicinity of the dipole and an equal and opposite 'image dipole' is formed in the sphere. The motion is then equivalent to that produced by a *quadrupole*, and to calculate the motion in this case it would be necessary to use a more accurate approximation to $G(\mathbf{x}, \mathbf{y})$. This conclusion applies only to *radially* oriented dipoles (see Problems 2), but it remains true for *any* rigid surface when a dipole orientated in the direction of the local surface normal approaches the surface.

EXAMPLE 2. TRANSLATIONAL MOTION OF A RIGID SPHERE Let $v_n(\mathbf{x}, t)$ denote the normal velocity on the surface of a sphere S of radius a. The velocity potential of the motion produced at time t in the fluid is the same as that generated by a distribution of monopoles of strength $v_n(\mathbf{x}, t)$ per unit area of S when S is assumed to be stationary (*rigid*). The corresponding source strength $q(\mathbf{x}, t)$ in Laplace equation (2.15.1) is

$$q(\mathbf{x}, t) = v_n(\mathbf{x}, t)\delta(s_\perp - \epsilon), \quad (\epsilon \to +0),$$

where $s_\perp$ is distance measured in the normal direction from S into the fluid and $\epsilon > 0$ places the sources just within the fluid adjacent to S. The velocity potential $\varphi(\mathbf{x}, t)$ is therefore

$$\begin{aligned}
\varphi(\mathbf{x}, t) &= \int_{\text{fluid}} v_n(\mathbf{y}, t)\delta(s_\perp - \epsilon)G(\mathbf{x}, \mathbf{y})d^3\mathbf{y} \quad (\epsilon \to +0) \\
&= \oint_S v_n(\mathbf{y}, t)G(\mathbf{x}, \mathbf{y})dS(\mathbf{y}), \quad \text{where} \quad \frac{\partial G}{\partial x_n}(\mathbf{x}, \mathbf{y}) = 0 \text{ on S.} \tag{2.16.13}
\end{aligned}$$

Consider the case in which the sphere translates as a rigid body at velocity $U(t)$ in the x_1 direction with its centre on the x_1 axis. Green's function $G(\mathbf{x}, \mathbf{y})$ is strictly also a function of time, because of the need to satisfy $\partial G/\partial x_n = 0$ on the moving surface. However, we can avoid explicit representation of this dependence by using moving axes, taking the coordinate origin at the centre of the sphere, as in Figure 2.16.2.

Then

$$v_n(\mathbf{y}, t) = U(t) \cos \vartheta,$$

and when $|\mathbf{x}| \gg a$, we can evaluate the integral in (2.16.13) by using far-field approximation (2.16.11) for $G(\mathbf{x}, \mathbf{y})$, i.e.,

$$\varphi(\mathbf{x}, t) \approx \frac{-1}{4\pi |\mathbf{x}|} \left[\oint_S v_n(\mathbf{y}, t) dS(\mathbf{y}) + \frac{x_j}{|\mathbf{x}|^2} \oint_S y_j \left(1 + \frac{a^3}{2|\mathbf{y}|^3} \right) v_n(\mathbf{y}, t) dS(\mathbf{y}) \right].$$

The first integral represents the net 'volume flux' through S and vanishes identically for rigid-body translational motion. The second integral is non-zero only for $j = 1$, when $y_1 = a \cos \vartheta$ and $|\mathbf{y}| = a$ on S, and we can take $dS = 2\pi a^2 \sin \vartheta \, d\vartheta$ [so that the surface integral becomes $3\pi a^3 U(t) \int_0^\pi \cos^2 \vartheta \sin \vartheta \, d\vartheta = 2\pi a^3 U(t)$]. Hence

$$\varphi(\mathbf{x}, t) \approx \frac{-U(t)a^3 x_1}{2|\mathbf{x}|^3} \equiv \frac{-U(t)a^3 \cos \theta}{2|\mathbf{x}|^2}, \quad |\mathbf{x}| \to \infty,$$

where θ is the angle between the x_1 axis and the far-field direction $\mathbf{x}$ (see Figure 2.7.1).

This result is actually identical to exact solution (2.11.9) for any $|\mathbf{x}| > a$ because, as we have already seen (§2.7), a moving sphere is precisely equivalent to a dipole located at its centre, and use of the far-field Green's function leads to the exact predictions of the monopole and dipole motions generated by a source field.

2.17 Far-field Green's function for cylindrical bodies

The reciprocal calculation of the Green's function described in §2.15 for the bounded three-dimensional bodies is easily modified to deal with cylindrical bodies.

Figure 2.17.1 illustrates the case of an infinite circular cylinder of radius a whose axis lies along the y_3 axis. The source point $\mathbf{y}$ is adjacent to the cylinder and is temporarily assumed to be within an axial distance $|y_3| \sim \ell$ from the coordinate origin O (this condition will be relaxed in §2.18). In this region expansion (2.16.6) remains valid with $\varphi_3^*(\mathbf{y}) \equiv 0$, because the impinging flow described by velocity potential (2.16.4) for $j = 3$ is unaffected by the cylinder. Hence we can take

$$G(\mathbf{x}, \mathbf{y}) = \frac{-1}{4\pi |\mathbf{x}|} \left(1 + \frac{x_j Y_j}{|\mathbf{x}|^2} \right), \quad \mathbf{y} \sim O(\ell), \ |\mathbf{x}| \to \infty, \tag{2.17.1}$$

where the Kirchhoff vector $\mathbf{Y}$ has the components

$$Y_1 = y_1 - \varphi_1^*(\mathbf{y}), \quad Y_2 = y_2 - \varphi_2^*(\mathbf{y}), \quad Y_3 = y_3. \tag{2.17.2}$$

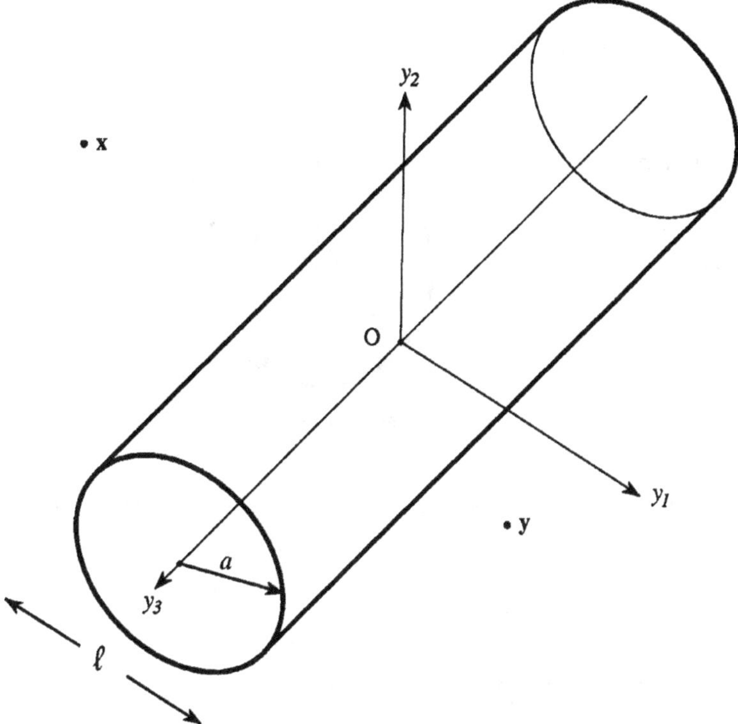

Figure 2.17.1

2.17.1 The circular cylinder

The potentials $\varphi_1^*(\mathbf{y})$, $\varphi_2^*(\mathbf{y})$ for the circular cylinder of radius a (Figure 2.17.2) can be found by the method of §2.16.2.

For $j = 1$ the flow is symmetric about the y_1 axis and is independent of the 'spanwise' coordinate y_3. With polar coordinates $(y_1, y_2) = r(\cos\vartheta, \sin\vartheta)$, condition (2.16.8) to be satisfied on the cylinder is

$$\frac{\partial\varphi_1^*}{\partial r} = \cos\vartheta \quad \text{at} \quad r = a. \tag{2.17.3}$$

As in the case of the sphere, we try a solution of the form

$$\varphi_1^* = \Psi(r)\cos\vartheta,$$

which satisfies the polar form of Laplace's equation

$$\left[\frac{1}{r}\frac{\partial}{\partial r}\left(r\frac{\partial}{\partial r}\right) + \frac{1}{r^2}\frac{\partial^2}{\partial\vartheta^2}\right]\Psi(r)\cos\vartheta = 0,$$

provided that

$$r^2\frac{d^2\Psi}{dr^2} + r\frac{d\Psi}{dr} - \Psi = 0.$$

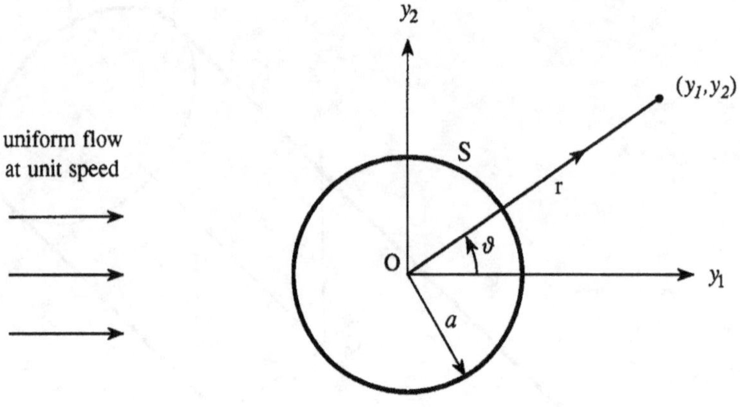

Figure 2.17.2

The general solution is $\Psi = Ar + B/r$. The component Ar must be rejected because it does not decay as $r \to \infty$;

$$\therefore \quad Y_1 \equiv y_1 - \varphi_1^* = r \cos \vartheta - \frac{B}{r} \cos \vartheta,$$

and condition (2.17.3) yields $B = -a^2$. Therefore,

$$Y_1 = r \cos \vartheta + \frac{a^2}{r} \cos \vartheta \equiv y_1 \left(1 + \frac{a^2}{r^2} \right).$$

Similarly,

$$Y_2 = y_2 \left(1 + \frac{a^2}{r^2} \right).$$

Hence the far-field Green's function for a circular cylinder, with source near the origin, is

$$G(\mathbf{x}, \mathbf{y}) = \frac{-1}{4\pi |\mathbf{x}|} \left(1 + \frac{x_j Y_j}{|\mathbf{x}|^2} \right), \quad \mathbf{y} \sim O(\ell), \ |\mathbf{x}| \to \infty, \tag{2.17.4}$$

where

$$Y_j = y_j \left(1 + \frac{a^2}{y_1^2 + y_2^2} \right), \quad j = 1, 2, \quad Y_3 = y_3. \tag{2.17.5}$$

2.17.2 The rigid strip

The rigid strip of 'chord' $2a$ and infinite span provides a simple model of a sharp-edged airfoil. In Figure 2.17.3 the airfoil occupies $-a < y_1 < a$, $y_2 = 0$, $-\infty < y_3 < \infty$. The airfoil has no influence on a uniform mean flow in the y_1 direction nor on one in the y_3 direction, so that potential functions $\varphi_1^*(\mathbf{y}) \equiv 0$ and $\varphi_3^*(\mathbf{y}) \equiv 0$.

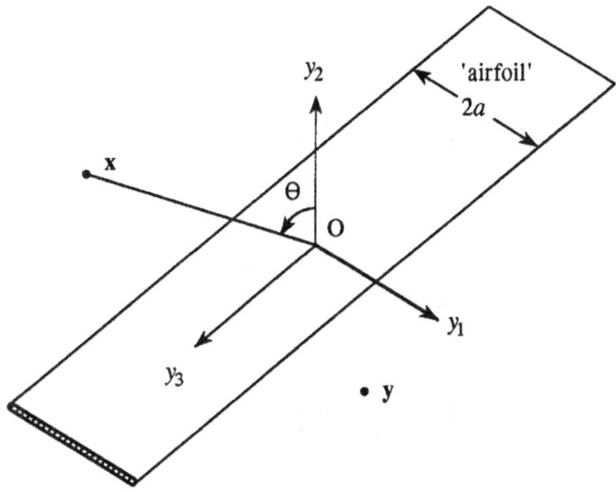

Figure 2.17.3

The potential $\varphi_2^*(\mathbf{y}) \equiv \varphi_2^*(y_1, y_2)$ can be determined by the method of conformal transformation (§3.8). If $z = y_1 + iy_2$, the cross section of the airfoil in the z plane is mapped onto the circular cylinder $|Z| = a$ in the Z plane by the transformation

$$Z = z + \sqrt{z^2 - a^2}.$$

Because $Z \sim 2z$ as $|z| \to \infty$ a uniform flow at unit speed in the y_2 direction in the z plane at large distances from the airfoil corresponds to a uniform flow at speed $\frac{1}{2}$ in the direction of the imaginary Z axis at large distances from the cylinder. This flow can be found by the method previously discussed for the circular cylinder (or see §3.8) and determines $Y_2 = y_2 - \varphi_2^*(y_1, y_2) = \operatorname{Re} w(z)$, where w is the complex potential

$$
\begin{aligned}
w(z) &= -\frac{i}{2}\left(Z - \frac{a^2}{Z}\right) \\
&= -\frac{i}{2}\left(z + \sqrt{z^2 - a^2} - \frac{a^2}{z + \sqrt{z^2 - a^2}}\right) \\
&= -i\sqrt{z^2 - a^2}.
\end{aligned}
$$

Thus the far-field Green's function for a strip, with source near the origin, is

$$G(\mathbf{x}, \mathbf{y}) = \frac{-1}{4\pi|\mathbf{x}|}\left(1 + \frac{x_j Y_j}{|\mathbf{x}|^2}\right), \quad \mathbf{y} \sim O(a), \ |\mathbf{x}| \to \infty, \tag{2.17.6}$$

where the components of the Kirchhoff vector are

$$Y_1 = y_1, \quad Y_2 = \operatorname{Re}\left(-i\sqrt{z^2 - a^2}\right), \quad Y_3 = y_3, \quad z = y_1 + iy_2. \tag{2.17.7}$$

Figure 2.17.4 shows the streamline pattern of the flow past the strip defined by velocity potential $Y_2(\mathbf{y})$. The streamlines crowd together and change very rapidly near the sharp edges. This is an indication that the interaction with an edge of a source of dipole or quadrupole (or higher-order) type can make an important contribution to the velocity potential in the far field because ∇Y_2 is very large near an edge.

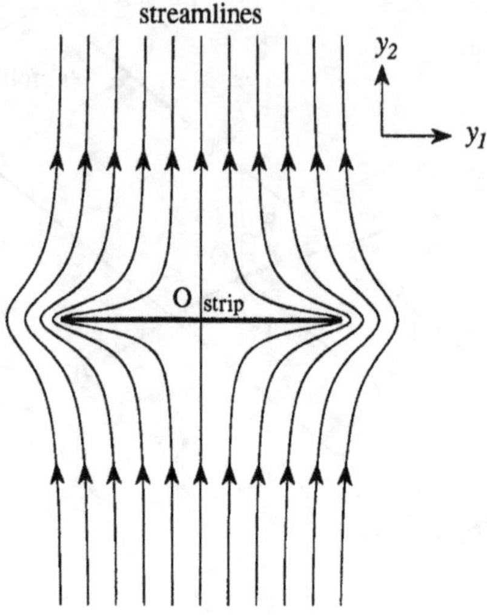

Figure 2.17.4

EXAMPLE 1. Calculate the far-field velocity potential when

$$\nabla^2\varphi = f_2\frac{\partial}{\partial x_2}\left[\delta(x_1 - L)\delta(x_2)\delta(x_3)\right], \quad L > a,$$

where $\dfrac{\partial\varphi}{\partial x_2} = 0$ on the airfoil $-a < x_1 < a,\ x_2 = 0,\ -\infty < x_3 < \infty.$

The dipole source is orientated in the x_2 direction and is positioned just to the right of the edge at $y_1 = a$ in Figure 2.17.3. The solution is given by the following form of Equation (2.16.12):

$$\varphi(\mathbf{x}) = -f_2\left[\frac{\partial G}{\partial y_2}(\mathbf{x}, \mathbf{y})\right]_{\mathbf{y}=(L,0,0)} \approx \frac{f_2 x_2}{4\pi|\mathbf{x}|^3}\left(\frac{\partial Y_2}{\partial y_2}\right)_{\mathbf{y}=(L,0,0)} \quad \text{as} \quad |\mathbf{x}| \to \infty,$$

where, from (2.17.7),

$$\frac{\partial Y_2}{\partial y_2} = \mathrm{Re}\left(-i\frac{\partial}{\partial y_2}\sqrt{z^2 - a^2}\right), \quad z = y_1 + iy_2$$

$$= \mathrm{Re}\left(\frac{z}{\sqrt{z^2 - a^2}}\right),$$

$$\therefore \quad \varphi(\mathbf{x}) \approx \frac{f_2 x_2}{4\pi|\mathbf{x}|^3}\frac{L}{\sqrt{L^2 - a^2}} = \frac{f_2 L\cos\Theta}{4\pi|\mathbf{x}|^2\sqrt{L^2 - a^2}}, \quad |\mathbf{x}| \to \infty,$$

where $\Theta = \cos^{-1}(x_2/|\mathbf{x}|)$ is the angle between the normal to the strip and the radius vector $\mathbf{x}$ shown in Figure 2.17.3.

The velocity potential is increased by a factor $L/\sqrt{L^2 - a^2}$ relative to that produced by the same dipole in free space and is unbounded as $L \to a$, when the dipole approaches the edge. It reduces to its value in the absence of the airfoil when $L \to \infty$.

2.18 Symmetric far-field Green's function

Definition (2.16.9) of the far-field Green's function can be recast to exhibit the reciprocal character of source and observer positions $\mathbf{y}$ and $\mathbf{x}$. To do this note, first, that for a body of characteristic diameter ℓ,

$$Y_j(\mathbf{y}) = y_j - \varphi_j^*(\mathbf{y}) \sim O(\ell),$$

and therefore that $Y_j/|\mathbf{x}| \ll 1$. Hence

$$G(\mathbf{x}, \mathbf{y}) \approx \frac{-1}{4\pi|\mathbf{x}|}\left(1 + \frac{x_j Y_j}{|\mathbf{x}|^2}\right) \equiv \frac{-1}{4\pi|\mathbf{x}|}\left(1 + \frac{\mathbf{x}\cdot\mathbf{Y}}{|\mathbf{x}|^2}\right)$$

$$\approx \frac{-1}{4\pi\left(|\mathbf{x}| - \frac{\mathbf{x}\cdot\mathbf{Y}}{|\mathbf{x}|}\right)},$$

$$\approx \frac{-1}{4\pi|\mathbf{x} - \mathbf{Y}|}, \quad \mathbf{Y} \sim O(\ell), \quad |\mathbf{x}| \to \infty, \tag{2.18.1}$$

where on the last line we have used far-field approximation (2.7.14):

$$|\mathbf{x} - \mathbf{Y}| \approx |\mathbf{x}| - \frac{\mathbf{x}\cdot\mathbf{Y}}{|\mathbf{x}|}, \quad |\mathbf{x}| \to \infty.$$

Now let $\mathbf{X}(\mathbf{x})$ denote the Kirchhoff vector for the body expressed in terms of $\mathbf{x}$, i.e. let

$$X_j(\mathbf{x}) = x_j - \varphi_j^*(\mathbf{x}). \tag{2.18.2}$$

Then, because $\varphi_j^*(\mathbf{x}) \to 0$ as $|\mathbf{x}| \to \infty$, we also have $|\mathbf{X}| \sim |\mathbf{x}|$ as $|\mathbf{x}| \to \infty$, and therefore

$$|\mathbf{x} - \mathbf{Y}| \approx |\mathbf{X} - \mathbf{Y}|, \quad \text{when } |\mathbf{x}| \to \infty.$$

Thus, to the same approximation, (2.18.1) can be written as

$$G(\mathbf{x}, \mathbf{y}) \approx \frac{-1}{4\pi|\mathbf{X} - \mathbf{Y}|}, \quad \mathbf{Y} \sim O(\ell), \quad |\mathbf{x}| \to \infty.$$

This result is the basis of our revised definition of the far-field Green's function;

$$G(\mathbf{x}, \mathbf{y}) = \frac{-1}{4\pi|\mathbf{X} - \mathbf{Y}|}, \tag{2.18.3}$$

where $\mathbf{X} = \mathbf{x} - \boldsymbol{\varphi}^*(\mathbf{x})$, $\mathbf{Y} = \mathbf{y} - \boldsymbol{\varphi}^*(\mathbf{y})$ are the Kirchhoff vectors for the body expressed, respectively, in terms of $\mathbf{x}$ and $\mathbf{y}$. The components X_j and Y_j are the velocity potential of incompressible flow past the body having unit speed in the j direction at large distances from the body; φ_j^* is the velocity potential of the incompressible flow that would be produced by rigid-body motion of S at unit speed in the j direction.

Our new definition exhibits complete accord with reciprocity. Also, because of the symmetrical way in which $\mathbf{x}$ and $\mathbf{y}$ enter (2.18.3), we may now remove any restriction on the position of the coordinate origin. The approximation is valid for arbitrary source and observer locations provided that *at least one of them* lies in the far field of the body.

When *both* $\mathbf{x}$ and $\mathbf{y}$ are in the far field (so that $\mathbf{X} \sim \mathbf{x}$ and $\mathbf{Y} \sim \mathbf{y}$) predictions made with the far-field Green's function will be the same as when the body is *absent*. When $\mathbf{x}$ is close to the body the source must be in the far field; $G(\mathbf{x}, \mathbf{y})$ then determines the modification by the body of flow near the body produced by the distant source.

Definition (2.18.3) is easily recalled because it is an obvious generalization of the free-space Green's function (2.15.3). In applications it is necessary to remember that it is valid for determining only the far fields of the surface monopole and dipole sources induced on the body by sources in the fluid. When used in calculations $G(\mathbf{x}, \mathbf{y})$ must normally be expanded to *first order only* in the Kirchhoff source vector $\mathbf{Y}(\mathbf{y})$.

2.18.1 Far field of an arbitrarily moving body

To illustrate calculations with the far-field Green's function (2.18.3) we consider again the problem of determining the asymptotic form of the velocity potential at large distances from the arbitrarily moving body of Figure 2.9.1. The far-field Green's function determines the monopole and dipole components of the far field, and in what follows it is demonstrated how G should be expanded only to first order in the Kirchhoff vector $\mathbf{Y}$.

The far field is calculated from formula (2.16.13) with $G(\mathbf{x}, \mathbf{y})$ given by (2.18.3):

$$\varphi(\mathbf{x}, t) \approx - \oint_S \frac{v_n(\mathbf{y}, t)}{4\pi |\mathbf{X} - \mathbf{Y}|} dS(\mathbf{y}). \qquad (2.18.4)$$

Take the coordinate origin at the centre of volume $\mathbf{x}_o(t)$ of the body and expand $G(\mathbf{x}, \mathbf{y})$ to *first order* in $\mathbf{Y}$:

$$\frac{1}{|\mathbf{X} - \mathbf{Y}|} \approx \frac{1}{|\mathbf{x}|} \left(1 + \frac{\mathbf{x} \cdot \mathbf{Y}}{|\mathbf{x}|^2} \right) \quad \left(\mathbf{X} \sim \mathbf{x} \text{ as } |\mathbf{x}| \to \infty \right).$$

Then

$$\varphi(\mathbf{x}, t) \approx - \frac{1}{4\pi |\mathbf{x}|} \oint_S v_n(\mathbf{y}, t) \left(1 + \frac{\mathbf{x} \cdot \mathbf{Y}}{|\mathbf{x}|^2} \right) dS(\mathbf{y})$$

$$= - \frac{1}{4\pi |\mathbf{x}|} \oint_S v_n(\mathbf{y}, t) \, dS(\mathbf{y}) - \frac{x_i}{4\pi |\mathbf{x}|^3} \oint_S v_n(\mathbf{y}, t) \, Y_i(\mathbf{y}) dS(\mathbf{y}).$$

The first integral represents a volume *monopole* source flow and is non-zero only if the volume enclosed by S changes with time, that is, only for a *pulsating* body. It would then be the most important component of the far field – the second integral is the dipole contribution and is smaller by a factor of $O(\ell/|\mathbf{x}|) \ll 1$.

The monopole vanishes for a *rigid body*. To evaluate the dipole we use (2.13.7) to express $v_n(\mathbf{y}, t)$ in the form [the coordinate origin being at $\mathbf{x}_o(t)$]

$$v_n(\mathbf{y}, t) = \left(\mathbf{U}_o + \mathbf{\Omega} \wedge \mathbf{y} \right) \cdot \mathbf{n},$$

where $\mathbf{n}$ is the surface normal directed into the fluid. Making the substitution $Y_j = y_j - \varphi_j^*(\mathbf{y})$ in the second integral, we obtain

$$\varphi(\mathbf{x}, t) \approx -\frac{x_i}{4\pi |\mathbf{x}|^3} \oint_S (y_i - \varphi_i^*)\left(\mathbf{U}_o \cdot \mathbf{n} + \Omega \wedge \mathbf{y} \cdot \mathbf{n}\right) dS(\mathbf{y})$$

$$= -\frac{x_i}{4\pi |\mathbf{x}|^3} \oint_S (y_i - \varphi_i^*)\left(U_{oj} n_j + \Omega_j \{\mathbf{y} \wedge \mathbf{n}\}_j\right) dS(\mathbf{y}), \quad |\mathbf{x}| \to \infty. \qquad (2.18.5)$$

The divergence theorem applied to the interior volume bounded by S yields

$$\oint_S y_i \left(U_{oj} n_j + \Omega_j \{\mathbf{y} \wedge \mathbf{n}\}_j\right) dS(\mathbf{y}) = \Delta U_{oi},$$

where Δ is the volume of the fluid displaced by the body. The integral involving φ_i^* is evaluated by use of definitions (2.13.12) of the inertia coefficients of the body:

$$-\oint_S \varphi_i^* \left(U_{oj} n_j + \Omega_j \{\mathbf{y} \wedge \mathbf{n}\}_j\right) dS(\mathbf{y}) = -\oint_S \left(U_{oj} n_j \varphi_i^* + \Omega_j \varphi_i^* \frac{\partial \chi_j^*}{\partial y_n}\right) dS(\mathbf{y})$$

$$\equiv U_{oj} \hat{M}_{ij} + \Omega_j \hat{C}_{ij}.$$

Hence

$$\varphi(\mathbf{x}, t) \approx \frac{\partial}{\partial x_i} \left[\frac{(\Delta \delta_{ij} + \hat{M}_{ij}) U_{oj} + \hat{C}_{ij} \Omega_j}{4\pi |\mathbf{x}|}\right], \quad |\mathbf{x}| \to \infty, \qquad (2.18.6)$$

which coincides with (2.13.14).

2.19 Far-field Green's function summary and special cases

2.19.1 General form

$$G(\mathbf{x}, \mathbf{y}) = \frac{-1}{4\pi |\mathbf{X} - \mathbf{Y}|} \qquad (2.19.1)$$

where

$$\left.\begin{array}{l} \mathbf{X} = \mathbf{x} - \varphi^*(\mathbf{x}) \\ \mathbf{Y} = \mathbf{y} - \varphi^*(\mathbf{y}) \end{array}\right\} \text{ Kirchhoff vectors for the body.}$$

The vector components $X_j(\mathbf{x})$ and $Y_j(\mathbf{y})$ are the velocity potentials of incompressible flow past the body having unit speed in the j direction at large distances from the body; φ_j^* is the velocity potential of the incompressible flow that would be produced by rigid-body motion of S at unit speed in the j direction. For a cylindrical body parallel to the x_3 direction, we take

$$X_3 = x_3, \quad Y_3 = y_3.$$

Standard special cases are listed in Table 2.19.1.

Table 2.19.1 *Standard special cases*

Body	X_1	X_2	X_3
Sphere of radius a center at origin	$x_1\left(1 + \frac{a^3}{2\|\mathbf{x}\|^3}\right)$	$x_2\left(1 + \frac{a^3}{2\|\mathbf{x}\|^3}\right)$	$x_3\left(1 + \frac{a^3}{2\|\mathbf{x}\|^3}\right)$
Circular cylinder of radius a coaxial with the x_3 axis	$x_1\left(1 + \frac{a^2}{x_1^2 + x_2^2}\right)$	$x_2\left(1 + \frac{a^2}{x_1^2 + x_2^2}\right)$	x_3
Strip airfoil $-a < x_1 < a,\ x_2 = 0,\ -\infty < x_3 < \infty$	x_1	$\mathrm{Re}(-i\sqrt{z^2 - a^2})$ $z = x_1 + ix_2$	x_3

2.19.2 Airfoil of variable chord

The far-field Green's function defined by (2.17.6) and (2.17.7) for a rigid strip can be generalized to include the finite-span, variable chord airfoil illustrated in Figure 2.19.1. The coordinate axes are orientated as in Figure 2.17.3 for the strip airfoil, with y_2 normal to the plane of the airfoil and y_3 in the spanwise direction. The airfoil span is assumed to be large, and the chord $2a \equiv 2a(y_3)$ is a 'slowly' varying function of y_3. The potential Y_2 of flow past the airfoil in the y_2 direction may then be approximated locally by the formula for an airfoil of uniform chord $2a(y_3)$. Therefore we obtain a first approximation to the far-field Green's function (2.19.1) for an airfoil of span L occupying the interval $-\frac{1}{2}L < y_3 < \frac{1}{2}L$ by taking

$$Y_1 = y_1, \quad Y_2 = \begin{cases} \mathrm{Re}\left[-i\sqrt{z^2 - a(y_3)^2}\right], & |y_3| < \frac{1}{2}L \\ y_2, & |y_3| > \frac{1}{2}L \end{cases}, \quad Y_3 = y_3, \quad z = y_1 + iy_2. \quad (2.19.2)$$

This model has been found to give predictions within a few per cent of those based on the exact value of $Y_2(\mathbf{y})$ in the case of an airfoil of an *elliptic* planform whose

$$\text{aspect ratio} = \frac{\text{airfoil span}}{\text{midchord}} > 5.$$

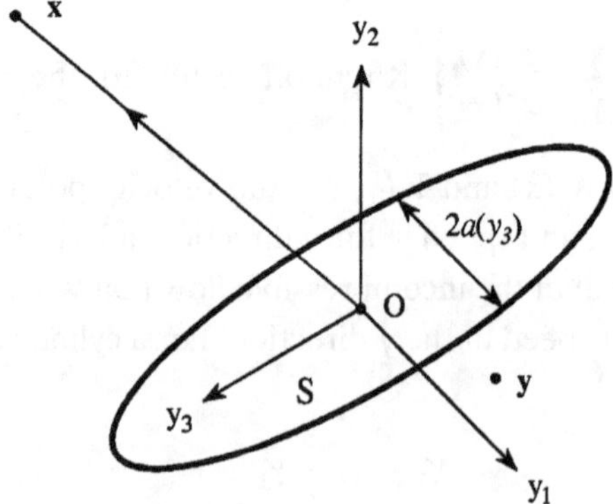

Figure 2.19.1

Figure 2.19.2

2.19.3 Projection or cavity on a plane wall

Let the plane wall be rigid and coincide with $x_2 = 0$ (Figure 2.19.2). When the projection or cavity is absent, the Green's function with vanishing normal derivative on the wall is

$$G_o(\mathbf{x}, \mathbf{y}) = -\frac{1}{4\pi\,|\mathbf{x} - \mathbf{y}|} - \frac{1}{4\pi\,|\bar{\mathbf{x}} - \mathbf{y}|},$$

where $\bar{\mathbf{x}} = (x_1, -x_2, x_3)$ is the image of the observer position $\mathbf{x}$ in the plane wall.

The figure illustrates the case for a projection, but the following discussion applies without change to wall cavities. Assume first that the origin is close to the projection. Let $|\mathbf{x}| \to \infty$ (noting that $|\bar{\mathbf{x}}| = |\mathbf{x}|$) and expand G_o near the projection to first order in $\mathbf{y}$ (i.e., correct to 'dipole' order):

$$G_o(\mathbf{x}, \mathbf{y}) \approx \frac{-1}{4\pi\,|\mathbf{x}|}\left[\left(1 + \frac{\mathbf{x} \cdot \mathbf{y}}{|\mathbf{x}|^2}\right) + \left(1 + \frac{\bar{\mathbf{x}} \cdot \mathbf{y}}{|\mathbf{x}|^2}\right)\right]$$

$$\approx \frac{-1}{4\pi\,|\mathbf{x}|}\left[2 + \frac{2(x_1 y_1 + x_3 y_3)}{|\mathbf{x}|^2}\right].$$

We require a corrected expression that has vanishing normal derivative (as a function of $\mathbf{y}$) on the wall and on the projection. By inspection, we obtain this simply by replacing the factor

$$\frac{2(x_1 y_1 + x_3 y_3)}{|\mathbf{x}|^2} \quad \text{with} \quad \frac{2(x_1 Y_1 + x_3 Y_3)}{|\mathbf{x}|^2},$$

where $Y_1 = y_1 - \varphi_1^*(\mathbf{y})$, $Y_3 = y_3 - \varphi_3^*(\mathbf{y})$ are the velocity potentials of 'horizontal' flows past the projection that are parallel to the wall and have unit speeds respectively in the y_1 and y_3 directions as $|\mathbf{y}| \to \infty$.

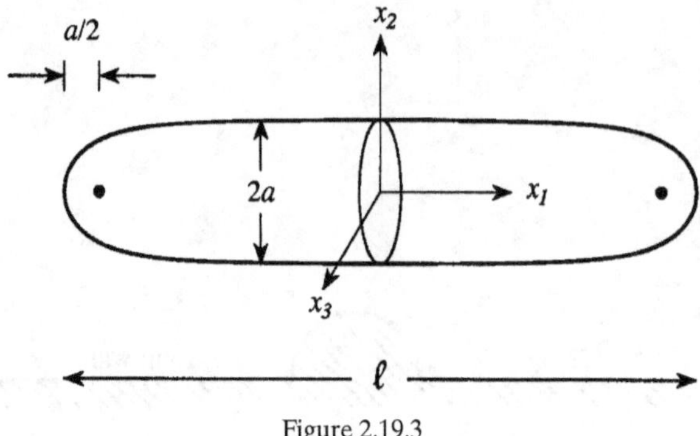

Figure 2.19.3

It may now be verified that (in the usual notation) the required far-field Green's function is

$$G(\mathbf{x}, \mathbf{y}) = -\frac{1}{4\pi |\mathbf{X} - \mathbf{Y}|} - \frac{1}{4\pi |\bar{\mathbf{X}} - \mathbf{Y}|},$$

(2.19.3)

where

$$Y_1 = y_1 - \varphi_1^*(\mathbf{y}), \quad Y_2 = y_2, \quad Y_3 = y_3 - \varphi_3^*(\mathbf{y}),$$
$$X_1 = x_1 - \varphi_1^*(\mathbf{x}), \quad X_2 = x_2, \quad X_3 = x_3 - \varphi_3^*(\mathbf{x}),$$

(2.19.4)

and $\bar{\mathbf{X}} = (X_1, -X_2, X_3)$.

We can use these formulae also for a two-dimensional projection or cavity that is uniform, say, in the x_3 direction simply by setting $Y_3 = y_3$, $X_3 = x_3$.

To complete this summary of far-field Green's function a selection of useful examples, without proofs, is now given.

2.19.4 Rankine ovoid

Let the ovoid (Lighthill 1986) have overall length ℓ (Figure 2.19.3) and let a be the uniform radius of the cylindrical midsection. Take the origin at the geometrical centre; the radius ϖ of a circular cross section is then given by

$$\varpi^2(x_1) = \frac{a^2 - (h - |x_1|)^2}{2} + \frac{(h - |x_1|)}{2}\sqrt{(h - |x_1|)^2 + 2a^2}, \quad \text{provided} \quad \frac{2a}{\ell} < 0.3,$$

(2.19.5)

where

$$h = \frac{\ell - a}{2}.$$

The far-field Green's function is given by (2.19.1) with

$$Y_1 = y_1 + \frac{a^2}{4}\left(\frac{1}{|\mathbf{y} - i h|} - \frac{1}{|\mathbf{y} + i h|}\right), \quad \mathbf{i} = (1, 0, 0);$$

$$Y_j = \begin{cases} y_j\left(1 + \frac{\varpi^2(y_1)}{y_2^2 + y_3^2}\right), & |y_1| < \frac{\ell}{2} \\ y_j, & |y_1| > \frac{\ell}{2} \end{cases}, \quad j = 2, 3.$$

(2.19.6)

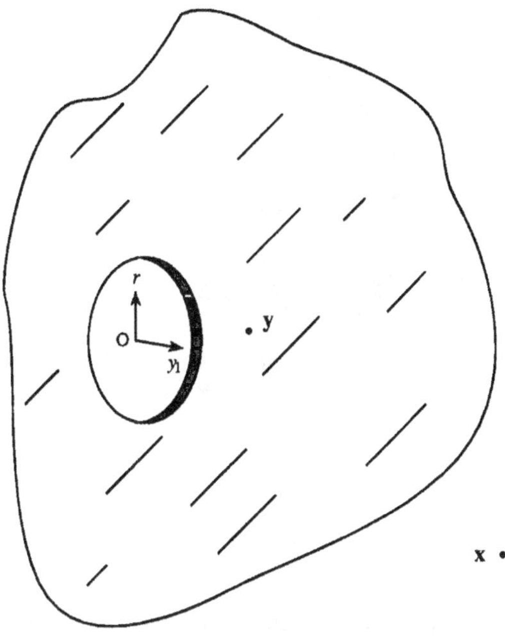

Figure 2.19.4

Only the diagonal elements of the added-mass tensor M_{ij} are non-zero for the coordinate system in the figure, and

$$M_{11} = \frac{2\pi a^3 \rho_o}{3}, \quad M_{22} = M_{33} \approx m_o, \quad \text{where } m_o = \rho_o \pi a^2 \left(\ell - \frac{5a}{3}\right)$$

$$= \text{mass of displaced fluid,}$$

and the volume of the ovoid $= \pi a^2 \, (\ell - 5a/3)$. Note that Y_1, the potential of flow past the ovoid in the y_1 direction, is formed by the combination of the unit mean flow potential y_1 with that for a point source at $\mathbf{y} = -\mathbf{i}h$ and a sink of equal strength at $\mathbf{y} = +\mathbf{i}h$ (at the points indicated by heavy filled circles in the figure).

2.19.5 Circular aperture

Consider an aperture of radius a with centre at the origin in a thin rigid wall coinciding with the plane $x_1 = 0$ (Figure 2.19.4). Let the source point $\mathbf{y}$ be close to the aperture on *either side* of the wall, and let $\mathbf{x}$ be in the far field at distance $\gg a$ from the aperture in $x_1 > 0$.

The far-field Green's function now has the special form

$$\left. \begin{aligned} G(\mathbf{x}, \mathbf{y}) &= \frac{-[1 - \Phi(\mathbf{y})]}{2\pi |\mathbf{x}|} \quad \text{for} \quad y_1 \geq 0 \\[2mm] &= \frac{-\Phi(\mathbf{y})}{2\pi |\mathbf{x}|} \quad \text{for} \quad y_1 \leq 0 \end{aligned} \right\} \quad |\mathbf{x}| \gg a, \qquad (2.19.7)$$

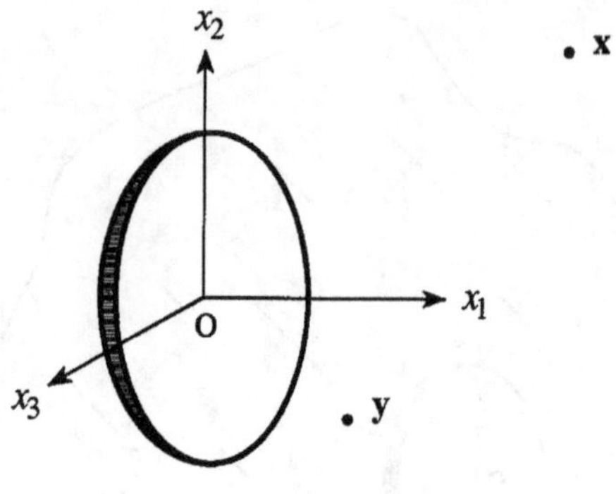

Figure 2.19.5

in which

$$\Phi(\mathbf{y}) = \frac{1}{\pi} \int_0^\infty \frac{\sin(\xi a)}{\xi} e^{-\xi|y_1|} J_0(\xi r) d\xi, \quad r = \sqrt{y_2^2 + y_3^2},$$

where J_0 is the Bessel function of the order of 0. Note that $G(\mathbf{x}, \mathbf{y})$ and its derivatives with respect to $\mathbf{y}$ are continuous as $\mathbf{y}$ passes through the aperture, and that

$$\Phi(\mathbf{y}) \sim \frac{a}{\pi|\mathbf{y}|} \quad \text{when} \quad |\mathbf{y}| \gg a.$$

2.19.6 Circular disc

Let the disc have radius a with centre at the origin and be orientated normal to the x_1 axis (Figure 2.19.5). The far-field Green's function has the standard form of (2.19.1), where

$$Y_1 = \frac{2a\,\mathrm{sgn}(y_1)}{\pi} \int_0^\infty \left[\frac{\sin(\xi a)}{\xi a} - \cos(\xi a) \right] \frac{e^{-\xi|y_1|} J_0(\xi r) d\xi}{\xi}, \quad r = \sqrt{y_2^2 + y_3^2},$$

$$Y_2 = y_2, \quad Y_3 = y_3. \tag{2.19.8}$$

The added mass is $M_{11} = \frac{8}{3}\rho_o a^3$.

PROBLEMS 2

1. Deduce Bernoulli's equation $\frac{\partial \varphi}{\partial t} + \frac{p}{\rho_o} + \frac{1}{2}v^2 - \Phi = f(t)$ from Equation (2.2.4).

2. Find the quadrupole far-field approximation for $\varphi = -\int \frac{q(\mathbf{y},t)d^3\mathbf{y}}{4\pi|\mathbf{x}-\mathbf{y}|}$, given that $\int q(\mathbf{y}, t)d^3\mathbf{y} = \int y_j q(\mathbf{y}, t)d^3\mathbf{y} = 0$.

3. A *collapsing* spherical cavity in a large body of water has radius $a(t)$ at time t. If the pressure p vanishes within the cavity and $p \to P > 0$ at large distances from

the cavity, show that when body forces are ignored the pressure at radial distance r from the centre of the cavity is given by

$$p = P + \frac{\rho_o a}{r}\left[a\frac{d^2 a}{dt^2} + 2\left(\frac{da}{dt}\right)^2\right] - \frac{\rho_o a^4}{2r^4}\left(\frac{da}{dt}\right)^2,$$

where ρ_o is the density of the water. By applying this equation at $r = a$, show that the pressure at distance r can also be written

$$p = P\left(1 - \frac{a}{r}\right) + \frac{\rho_o}{2}\left(\frac{da}{dt}\right)^2\left(\frac{a}{r} - \frac{a^4}{r^4}\right)$$

and that $a(t)$ satisfies the differential equation

$$\frac{da}{dt} = -\sqrt{\frac{2P}{3\rho_o}}\left(\frac{a_o^3}{a^3} - 1\right)^{\frac{1}{2}}$$

where $a(t) = a_o$ is the initial radius of the cavity at $t = 0$ (when $da/dt = 0$).

Show by numerical integration of this equation that the cavity disappears at time

$$t \approx 0.915 t_o \quad \text{where} \quad t_o = \sqrt{\frac{\rho_o a_o^2}{P}}.$$

Use the numerical solution to determine the maximum pressure ratio p/P in the fluid as a function of t/t_o. Deduce that the maximum pressure exceeds P when $t > 0.73 t_o$, and find the final maximum value of p/P.

4. A rigid sphere of radius a is immersed in an ideal, incompressible fluid of uniform density ρ_o that is at rest at infinity. The sphere translates parallel to the x_1 axis such that at time t its centre is at $\mathbf{X} = [X_1(t), 0, 0]$. The velocity potential of the fluid motion is

$$\varphi(\mathbf{x}, t) = \frac{-a^3\left[x_1 - X_1(t)\right]U(t)}{2|\mathbf{x} - \mathbf{X}(t)|^3}, \quad \text{where} \quad U(t) = \frac{dX_1}{dt}.$$

Use Bernoulli's equation to calculate the pressure at any point in the fluid. Hence show that the sphere experiences a drag force equal to

$$\frac{m_o}{2}\frac{dU}{dt}(t),$$

where m_o is the mass of fluid displaced by the sphere.

5. The surface of a sphere of radius a with centre at the origin vibrates in an ideal incompressible fluid with small-amplitude normal velocity $v_n = u(t)P_2(\cos\theta)$, where θ is the angle between the radius vector and the positive x_1 axis. Verify that the vibrations occur without changing the volume of the sphere, and that it can therefore be used to model the vibrations of a bell. Find the pressure fluctuations at large distances from the sphere.

6. A rigid sphere of radius a rotates in an ideal incompressible fluid at constant angular velocity Ω about a tangent line that coincides with the z axis. If the sphere touches the z axis at the origin, show that the velocity potential at large distances is given approximately by

$$\varphi = \frac{-a^4 \Omega \sin\theta \sin(\phi - \Omega t)}{2r^2},$$

where the spherical polar coordinates (r, θ, ϕ) are defined by

$$x = r\sin\theta\cos\phi, \quad y = r\sin\theta\sin\phi, \quad z = r\cos\theta.$$

7. Find the velocity potential of the flow generated in an ideal incompressible fluid by two equal and opposite point sources of strengths $\pm q(t)$ situated at opposite ends of a diameter of a rigid sphere of radius a with centre at the origin.

8. Establish the uniqueness of the monopole and dipole formulae (2.9.6) and (2.9.7). Use the facts that for these two respective cases $[\varphi] = 0$ and $[\partial\varphi/\partial x_n] = 0$ across S.

9. Deduce from Green's formula (2.8.7) that the formula remains valid in the limit as $\mathbf{x}$ approaches S from the fluid region $f > 0$ provided H is replaced with $\frac{1}{2}$.

10. Verify the monopole and dipole representations

$$\varphi(\mathbf{x}) = -\frac{3}{8\pi} \oint_S \frac{\mathbf{n} \cdot \mathbf{U} dS(\mathbf{y})}{|\mathbf{x} - \mathbf{y}|} = \frac{3}{8\pi} \frac{\partial}{\partial x_j} \oint_S \frac{n_j \mathbf{y} \cdot \mathbf{U} dS(\mathbf{y})}{|\mathbf{x} - \mathbf{y}|}$$

for the velocity potential $\varphi = -Ua^3\cos\theta/r^2$, $r = |\mathbf{x}|$ of motion produced by a rigid sphere of radius a moving at velocity $\mathbf{U}$ along the x axis, where the origin is taken at the centre of the sphere (see Figure 2.10.2).

11. A rigid, spherical ball is projected horizontally from the origin in the x direction at speed U_o in an ideal, incompressible fluid of density ρ_o. If the acceleration that is due to gravity g acts in the direction of the negative y axis, show that the path of the ball is

$$y = -\frac{(\rho_s - \rho_o)}{(2\rho_s + \rho_o)} \frac{gx^2}{U_o^2},$$

where ρ_s is the mass density of the material of the ball.

12. Show that the image in a rigid sphere of radius a of a radially orientated dipole source of strength μ situated at distance $r = h$ from the centre of the sphere is a radial dipole of strength $-\mu a^3/h^3$ at the inverse point $r = a^2/h$.

13. Show that a point source of strength q situated a distance h from the centre of a rigid sphere of radius $a \ll h$ in an ideal fluid of density ρ_o exerts a suction force on the sphere equal to $\rho_o q^2 a^3/4\pi h^5$.

14. Use the result of §2.11, Example 5, to write an expression for the stream function for the motion produced by a point source adjacent to a rigid sphere.

15. A rigid sphere of mass M and radius a executes small translational *oscillations* at velocity $\mathbf{V}(t)$ in an ideal, unbounded incompressible fluid of density ρ_o. If the

motion is produced by uniform oscillations of the fluid at velocity $\mathbf{U}(t)$ parallel to the x direction, show that

$$\mathbf{V}(t) = \frac{3m_o}{2M + m_o}\,\mathbf{U}(t),$$

where $m_o = \frac{4}{3}\pi a^3 \rho_o$ is the mass of fluid displaced by the sphere.

16. Show that, in a reference frame in which the Rankine solid in Figure 2.12.4 translates at constant speed U in the negative x direction, the kinetic energy T_o of the fluid is determined by the relations

$$T_o = -\frac{\rho_o}{2}\oint_S \varphi\frac{\partial\varphi}{\partial x_n}\,dS, \qquad \varphi = \frac{-Ua^2}{4(\varpi^2 + x^2)^{\frac{1}{2}}},$$

where S is the solid surface and x_n a local coordinate normal to S. Deduce that

$$T_o = \frac{\rho_o U^2 \pi a}{2}\int_0^a \varpi\sqrt{1 - \frac{\varpi^2}{a^2}}\,d\varpi = \frac{\pi a^3 \rho_o U^2}{6}.$$

17. Let $\varphi(x, \varpi)$ be the velocity potential of axisymmetric, irrotational flow from the thin-walled, circular cylindrical duct of Figure 2.12.8, where the coordinate origin is at the centre of the duct exit plane. Deduce from (2.12.16) that the axial component of velocity in $\varpi < R\,(= \text{cylinder radius})$ is given by

$$\frac{\partial\varphi}{\partial x}(x, \varpi) = \frac{U}{2} - \frac{U}{2\pi}\int_0^\infty I_0\left(\lambda\frac{\varpi}{R}\right)\sqrt{\frac{2K_1(\lambda)}{I_1(\lambda)}}\sin\left\{\lambda\left[\frac{x}{R} + \mathcal{F}(\lambda)\right]\right\}\,d\lambda,$$

where I_0 is a zero-order Bessel function.

    The velocity potential satisfies

$$\varphi(x, \varpi) \to U(x - \ell) \qquad \text{for } |x| \gg R \text{ inside the duct,}$$
$$\to -AU/4\pi|\mathbf{x}| \quad \text{for } |\mathbf{x}| = (x^2 + \varpi^2)^{\frac{1}{2}} \gg R \text{ outside,}$$

where ℓ is called the 'end correction' of the duct.

    (i) Show that

$$\ell = \int_{-\infty}^0 \left[\frac{\partial\varphi^*}{\partial x}(x, 0, 0) - 1\right]dx + \int_0^\infty \frac{\partial\varphi^*}{\partial x}(x, 0, 0)\,dx, \qquad \text{where } \varphi^* = \frac{\varphi}{U};$$

    (ii) that

$$\ell = \frac{1}{2\pi}\int_{-\infty}^\infty dx \int_0^\infty \left(\frac{2\sin(\lambda x/R)}{\lambda} - \sqrt{\frac{2K_1(\lambda)}{I_1(\lambda)}}\sin\left\{\lambda\left[\frac{x}{R} + \mathcal{F}(\lambda)\right]\right\}\right)d\lambda;$$

    (iii) and therefore, by means of the formula

$$\int_{-\infty}^\infty \sin\left\{\lambda\left[\frac{x}{R} + \mathcal{F}(\lambda)\right]\right\}dx = \text{Im}\left\{2\pi R\delta(\lambda)e^{i[\lambda\mathcal{F}(\lambda)]}\right\}$$
$$= 2\pi R\delta(\lambda)\sin\left[\lambda\mathcal{F}(\lambda)\right],$$

that

$$\ell = -R\mathcal{F}(0) = -\frac{R}{\pi} \int_0^\infty \frac{\ln[2\mathrm{K}_1(\mu)\mathrm{I}_1(\mu)]}{\mu^2}\, d\mu.$$

(iv) Use numerical integration to deduce that $\ell \approx 0.6127R$.

18. Use the far-field Green's function to solve

$$\nabla^2 \varphi = f_2 \frac{\partial}{\partial x_2} [\delta(x_1 - L)\delta(x_2)\delta(x_3)], \quad \text{where} \quad \frac{\partial \varphi}{\partial x_n} = 0 \text{ on } |\mathbf{x}| = a,$$

for the velocity potential as $|\mathbf{x}| \to \infty$ of an azimuthally orientated dipole adjacent to a rigid sphere.

19. Use the far-field Green's function to solve

$$\nabla^2 \varphi = f_1 \frac{\partial}{\partial x_1} [\delta(x_1 - L)\delta(x_2)\delta(x_3)], \quad \text{where} \quad \frac{\partial \varphi}{\partial x_n} = 0 \text{ on } (x_1^2 + x_2^2)^{\frac{1}{2}} = a,$$

for the velocity potential as $|\mathbf{x}| \to \infty$ of a radially orientated dipole adjacent to a rigid circular cylinder.

20. Repeat Question 19 for the dipoles

$$f_2 \frac{\partial}{\partial x_2} [\delta(x_1 - L)\delta(x_2)\delta(x_3)], \quad f_3 \frac{\partial}{\partial x_3} [\delta(x_1 - L)\delta(x_2)\delta(x_3)].$$

21. Verify [correct to $O(\epsilon^2)$] that when the far-field Green's function (2.16.11) for the sphere is used in Equation (2.16.12), the far-field velocity potential produced by the spinning prolate spheroid in Figure 2.11.3 is null. Derive the correct quadrupole behaviour by making use of the next order of approximation to the far-field Green's function $G(\mathbf{x}, \mathbf{y})$. Do this by first showing that the next term in the brackets of expansion (2.16.3) for $G_o(\mathbf{x}, \mathbf{y})$ is

$$\frac{|\mathbf{y}|^2}{|\mathbf{x}|^2} P_2 \left(\frac{\mathbf{x} \cdot \mathbf{y}}{|\mathbf{x}||\mathbf{y}|} \right).$$

22. A solid body translates without rotation at velocity $\mathbf{U}$. Show that the kinetic energy T_o of the fluid is given by

$$T_o = \frac{1}{2}\rho_o \mathbf{U} \cdot \mathbf{I} - \frac{1}{2} m_o U^2$$

where $\mathbf{I}$ is the specific total impulse and m_o is the mass of the displaced fluid.

23. For the Rankine ovoid of Figure 2.19.3 show that

$$m_o = \rho_o \pi a^2 \left(\ell - \frac{5a}{3} \right), \quad \mathrm{M}_{11} = \frac{2\pi a^3 \rho_o}{3}, \quad \mathrm{M}_{22} = \mathrm{M}_{33} \approx m_o \ (\ell \gg a),$$

where m_o is the mass of the displaced fluid. Show similarly (for $\ell \gg a$) that the inertia tensor $\mathrm{C}_{ij} \equiv 0$ and that the only non-zero components of B_{ij} are

$$\mathrm{B}_{22} = \mathrm{B}_{33} \approx \frac{m_o \ell^2}{12}.$$

Explain why M_{11}, which is just equal to the added mass of a sphere of radius a, is very much smaller than M_{22} and M_{33}.

24. The Rankine ovoid of Question 23 has mass m_s and is free to rotate about an axis through its centre coinciding with the x_3 axis. It is placed in a *steady*, uniform irrotational stream of velocity $(0, -U, 0)$, as illustrated in the figure. If the axis of the ovoid makes a small angle θ with the x_1 direction, show that moment M_3 exerted on the fluid is given by

$$M_3 = B_{33}\frac{d^2\theta}{dt^2} + (M_{22} - M_{11})U^2\theta,$$

and that the ovoid can execute simple harmonic oscillations governed by

$$(\mathcal{I}_3 + B_{33})\frac{d^2\theta}{dt^2} + (M_{22} - M_{11})U^2\theta = 0,$$

where $\mathcal{I}_3 \approx m_s\ell^2/12$ is the moment of inertia of the ovoid about the x_3 axis. Deduce that the period of the oscillations is approximately equal to

$$\frac{\pi\ell}{U}\sqrt{\frac{m_s + m_o}{3m_o}}, \quad \ell \gg a.$$

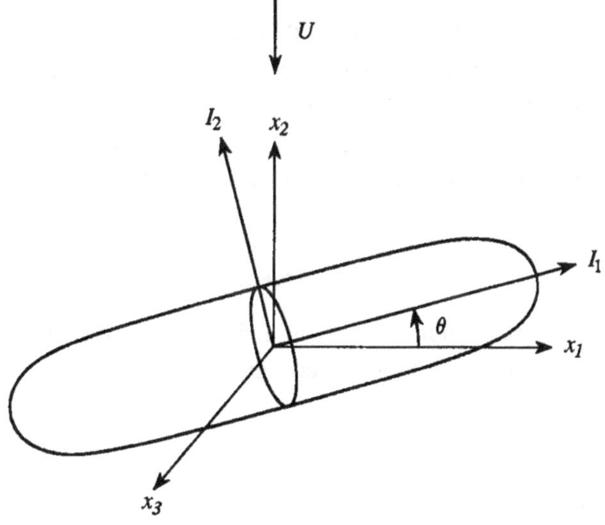

25. Deduce from (2.14.6) that the moment about the i direction exerted on a rigid body moving in irrotational flow at rest at infinity is given by

$$M_i = \rho_o\frac{d}{dt}\oint_S \left\{U_{oj}\varphi_j^*\{(\mathbf{x} - \mathbf{x}_o) \wedge \mathbf{n}\}_i + \Omega_j\chi_j^*\{(\mathbf{x} - \mathbf{x}_o) \wedge \mathbf{n}\}_i\right\} dS + \rho_o(\mathbf{I} \wedge \mathbf{U}_o)_i$$

$$= -\frac{d}{dt}\left(U_{oj}C_{ji} + \Omega_j B_{ji}\right) + \rho_o(\mathbf{I} \wedge \mathbf{U}_o)_i,$$

where $\mathbf{I}$ is given in terms of the inertia coefficients by (2.13.13).

3

Ideal Flow in Two Dimensions

3.1 Complex representation of fluid motion

Fluid motion that is the same in all planes parallel to the xy plane, say, with no velocity in the z direction, is said to be two dimensional, with all flow quantities dependent on only x, y, and possibly the time t. Fully three-dimensional flow problems can often be regarded as locally two dimensional in small, confined neighbourhoods, and the two-dimensional solutions can sometimes be pieced together to provide a useful representation of the three-dimensional flow. Furthermore, considerable insight into properties of more general flows can frequently be obtained by a study of the solutions of analogous two-dimensional flows. Many of these solutions can be obtained in analytic closed form by methods based on the theory of complex variables; these methods are studied in this chapter for an ideal, incompressible fluid.

3.1.1 The stream function

Let the velocity $\mathbf{v}$ have components u, v in the x and y directions, respectively. For incompressible motion, the equation of continuity

$$\frac{\partial u}{\partial x} + \frac{\partial v}{\partial y} = 0$$

can be satisfied by introduction of the *stream function* $\psi(x, y)$, where

$$u = \frac{\partial \psi}{\partial y}, \quad v = -\frac{\partial \psi}{\partial x}. \tag{3.1.1}$$

Consider two points, 1 at (x_1, y_1) and 2 at (x_2, y_2), lying on curve C of Figure 3.1.1. The volume flux from left to right through C (per unit 'span', i.e. per unit length normal to the plane of the paper) is $\int_C \mathbf{v} \cdot \mathbf{n} ds$, where s is arc length measured along C from 1 to 2 and $\mathbf{n} = (dy/ds, -dx/ds)$ is the unit normal on C. Hence, using (3.1.1), we obtain

$$\text{Volume flux} = \int_C \mathbf{v} \cdot \mathbf{n} ds = \int_C \left(\frac{\partial \psi}{\partial y} dy + \frac{\partial \psi}{\partial x} dx \right) = \int_C d\psi = \psi_2 - \psi_1.$$

102

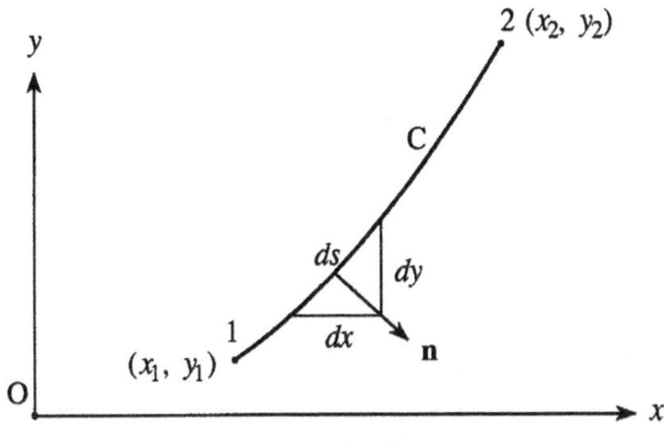

Figure 3.1.1

Thus the change in the value of ψ between any two points is equal to the volume flux per unit span across *any* curve joining those points. It is evident that ψ is defined only to within an arbitrary constant, so that if point 1 lies on the x axis and we define $\psi_1 = 0$, then

$$\psi(x, y) = \text{flux between the } x \text{ axis and } (x, y).$$

Also, on C

$$\mathbf{v} \cdot \mathbf{n} = \frac{\partial \psi}{\partial y} \frac{dy}{ds} + \frac{\partial \psi}{\partial x} \frac{dx}{ds} = \frac{\partial \psi}{\partial s},$$

i.e., we obtain the velocity in any given direction by differentiating at right angles *to the left*.

Similarly, the streamlines of the flow are defined by

$$\frac{dx}{u} = \frac{dy}{v}, \quad \text{i.e.,} \quad \frac{dx}{\partial \psi / \partial y} = \frac{-dy}{\partial \psi / \partial x},$$

$$\therefore \quad \frac{\partial \psi}{\partial x} dx + \frac{\partial \psi}{\partial y} dy \equiv d\psi = 0,$$

$$\text{i.e.,} \quad \psi = \text{constant}.$$

All of these consequences of definitions (3.1.1) of the stream function are the two-dimensional analogues of those in §2.12 for the Stokes stream function.

The vorticity $\boldsymbol{\omega}$ in two-dimensional flow can have only one component $\boldsymbol{\omega} = \omega\mathbf{k}$, where $\mathbf{k}$ is the unit vector perpendicular to the plane of the flow (out of the plane of the paper in Figure 3.1.1), and

$$\omega = \frac{\partial v}{\partial x} - \frac{\partial u}{\partial y}.$$

Therefore, in terms of ψ,

$$\frac{\partial^2 \psi}{\partial x^2} + \frac{\partial^2 \psi}{\partial y^2} = -\omega. \tag{3.1.2}$$

3.1.2 The complex potential

When the motion is irrotational $\omega = 0$, and there exists a velocity potential φ such that $\mathbf{v} = \nabla\varphi \equiv (\partial\varphi/\partial x,\ \partial\varphi/\partial y)$. In that case Equations (3.1.1) show that

$$\frac{\partial\varphi}{\partial x} = \frac{\partial\psi}{\partial y}, \quad \frac{\partial\varphi}{\partial y} = -\frac{\partial\psi}{\partial x}.$$

These are the Cauchy–Riemann equations that imply that $\varphi(x, y)$, $\psi(x, y)$ are the real and imaginary parts of a *regular* function,

$$w(z) = \varphi(x, y) + i\psi(x, y),$$

of the complex variable $z = x + iy$. $w(z)$ is called the *complex potential*; its real and imaginary parts both satisfy Laplace's equation:

$$\frac{\partial^2\varphi}{\partial x^2} + \frac{\partial^2\varphi}{\partial y^2} = 0, \quad \frac{\partial^2\psi}{\partial x^2} + \frac{\partial^2\psi}{\partial y^2} = 0.$$

Also, $\nabla\varphi \cdot \nabla\psi = 0$, which means that the streamlines $\psi = $ constant intersect the 'equipotentials' $\varphi = $ constant at right angles.

When $w(z)$ is known the flow velocity can be calculated from the *complex velocity*,

$$w'(z) \equiv \frac{dw}{dz} = \frac{\partial\varphi}{\partial x} - i\frac{\partial\varphi}{\partial y} \equiv u - iv,$$

which is also a regular function of z.

SUMMARY In two dimensions the conditions that the flow is both irrotational and incompressible are contained in the statement that $dw/dz = u - iv$ is a regular function. Indeed, for any closed contour C in the z plane, which can be collapsed to a point without meeting any boundaries:

$$\oint_C \left(\frac{dw}{dz}\right) dz = \oint_C \mathbf{v} \cdot d\mathbf{x} + i \oint_C \mathbf{v} \cdot \mathbf{n}\, ds = \text{(circulation around C)} + i\,\text{(flux out of C)} \equiv 0,$$

where the notation is the same as in Figure 3.1.1.

3.1.3 Uniform flow

The complex potential for flow at uniform speed U is as follows:

Figure 3.1.2(a), in the x direction,

$$w = Uz, \quad \varphi = Ux, \quad \psi = Uy;$$

Figure 3.1.2(b), in a direction at angle α to the x axis,

$$w = Uze^{-i\alpha}, \quad \varphi = Ur\cos(\theta - \alpha), \quad \psi = Ur\sin(\theta - \alpha),$$

where $z = re^{i\theta}$.

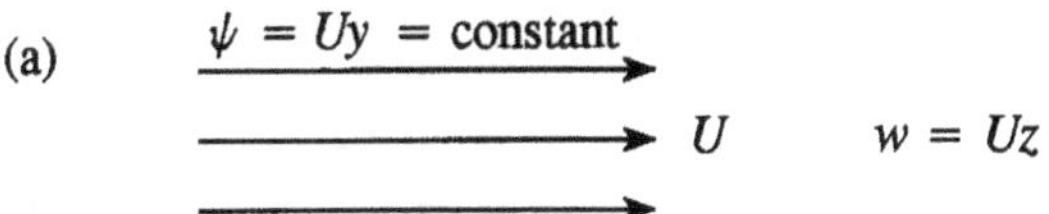

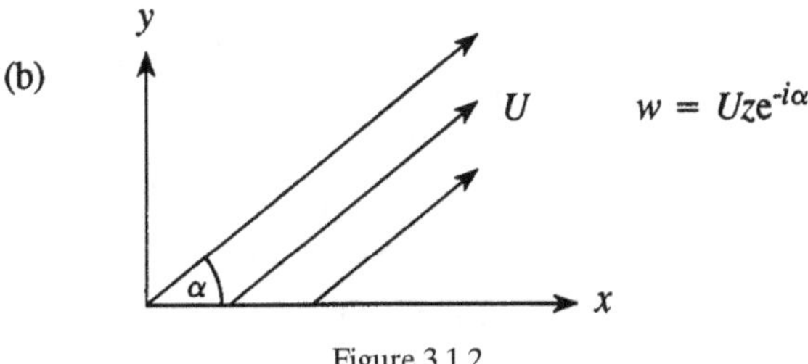

Figure 3.1.2

3.1.4 Flow past a cylindrical surface

A fundamental problem of the two-dimensional theory is the determination of the complex potential of flow past a solid, cylindrical surface C (Figure 3.1.3).

The surface must coincide with a streamline of the flow, so that in the case of a uniform impinging flow at angle α to the x axis, it is required to determine

$$w(z) = \varphi + i\psi,$$

where
$$\left.\begin{array}{c} \mathbf{n} \cdot \nabla\varphi = 0 \\ \psi = \text{constant} \end{array}\right\} \quad \text{on C,}$$

and
$$w'(z) \sim Ue^{-i\alpha} \quad \text{as} \quad |z| \to \infty.$$

The two conditions on C are, of course, equivalent, as $\mathbf{n} \cdot \nabla\varphi = \partial\psi/\partial s = 0$ implies that $\psi = \text{constant}$ on C, where s is distance measured along C.

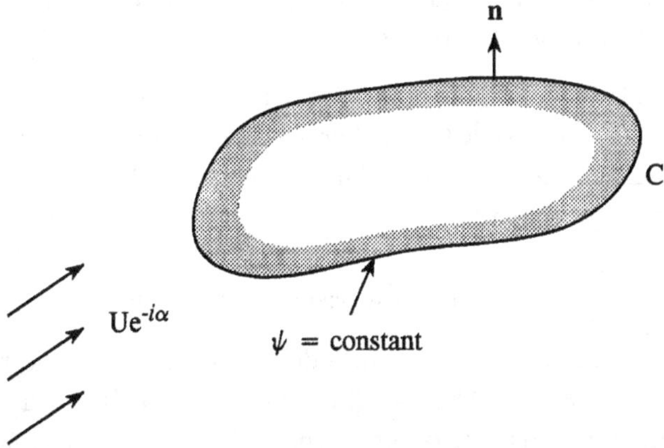

Figure 3.1.3

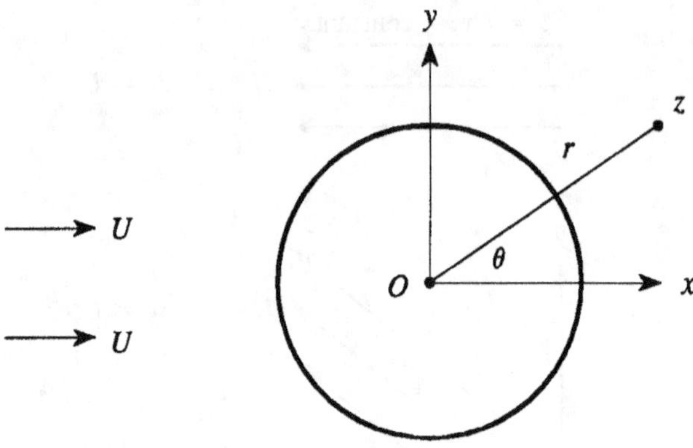

Figure 3.2.1

3.2 The circular cylinder

Problems involving irrotational flow in the vicinity of a rigid, circular cylinder are particularly important and can be treated by use of the circle theorem (Milne-Thomson 1940).

3.2.1 Circle theorem

(1) Let $f(z)$ be the velocity potential of a possible flow *in the absence of rigid boundaries* for which any singularities of $f(z)$ lie in the region $|z| > a > 0$;
(2) let a rigid cylinder of radius a and boundary C be inserted into the flow with its centre at $z = 0$.

Then the complex potential of the flow in the presence of the cylinder is

$$w(z) = f(z) + f^* \left(\frac{a^2}{z} \right), \tag{3.2.1}$$

where the asterisk denotes the complex conjugate.

PROOF Because $z^* = a^2/z$ on the circle, the complex function $w(z) = \varphi + i\psi$ is purely real on $|z| = a$ and therefore $\psi = 0$ on C.

If z lies in the fluid (outside C) the point a^2/z is inside C, and vice versa. By hypothesis, all singularities of $f(z)$ are outside C, and therefore all singularities of $f^*(a^2/z)$ lie within C. Thus $w(z)$ has exactly the same singularities as $f(z)$ within the flow and therefore satisfies all conditions of the problem.

3.2.2 Uniform flow past a circular cylinder

Let the flow at large distances from the cylinder be at speed U in the x direction, with velocity potential $f(z) = Uz$. When a cylinder of radius a is inserted into this flow (Figure 3.2.1) with its centre at the origin, we have $f^*(a^2/z) = Ua^2/z$, and by the circle theorem the complex potential is

$$w = U \left(z + \frac{a^2}{z} \right), \quad |z| > a. \tag{3.2.2}$$

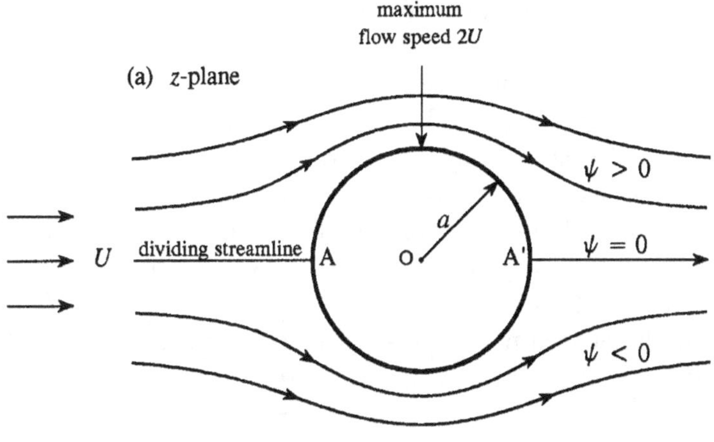

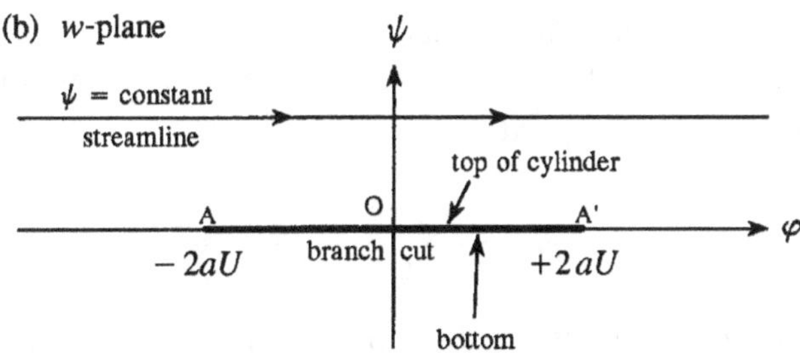

Figure 3.2.2

We can express this in polar form by putting $z = r e^{i\theta}$:

$$w = \varphi + i\psi, \quad \text{where} \quad \varphi = U \cos\theta \left(r + \frac{a^2}{r} \right), \quad \psi = U \sin\theta \left(r - \frac{a^2}{r} \right).$$

We see immediately that $\psi = 0$ on the surface $r = a$ of the cylinder, and also that

$$\mathbf{n} \cdot \nabla\varphi = \frac{\partial\varphi}{\partial r} = U \cos\theta \left(1 - \frac{a^2}{r^2} \right)$$

vanishes on the cylinder.

The pattern of the streamlines is shown in Figure 3.2.2(a). The flow is obviously symmetric with respect to the x axis, which passes through the centre of the cylinder in the direction of the mean flow. The points A, A' are stagnation points ($z = \pm a$) where the complex velocity vanishes:

$$w'(z) = U \left(1 - \frac{a^2}{z^2} \right) = 0.$$

The streamline $\psi = 0$ consists of the x axis and the surface of the cylinder: It *divides* at A, passes symmetrically along the upper and lower halves of the cylinder, joins up at A', and proceeds along the positive x axis. The streamlines crowd together directly above and below the cylinder; in these regions continuity causes the flow speed $|w'(z)|$

to increase, and it attains a maximum value of $2U$ on the surface of the cylinder at $z = \pm ia$.

Equation (3.2.2) can be regarded as a transformation that maps the z plane of the flow onto the w plane [Figure 3.2.2(b)]. The transformation is singular at the stagnation points A, A$'$, where $dw/dz = 0$ and the flow divides. The interval $-2aU < \varphi < 2aU$ of the real axis in the w plane is a 'branch cut'. By putting $z = re^{i\theta}$ and letting $r \to a + 0$, we see that the upper half $0 < \theta < \pi$ of the cylinder maps onto the *upper* side of the branch cut, $\varphi = 2aU\cos\theta$, $\psi = +0$, and the lower side of the cylinder $-\pi < \theta < 0$ maps onto the lower side $\varphi = 2aU\cos\theta$, $\psi = -0$.

The nature of the singularities at the stagnation points is evident from the inverse transformation

$$ z = \frac{w}{2U} + \sqrt{\frac{w^2}{4U^2} - a^2}. \qquad (3.2.3) $$

In the w plane each horizontal straight line $\psi = \text{constant}$ is the image of a streamline of the flow past the cylinder. This provides a convenient means of plotting streamlines: First write (3.2.3) in the explicit form

$$ x + iy = \frac{\varphi + i\psi}{2U} + \sqrt{\frac{(\varphi + i\psi)^2}{4U^2} - a^2}. $$

For a fixed value of ψ, the real and imaginary parts of this expression define parametrically in terms of φ the coordinates (x, y) of a point on the streamline, which is mapped out by permitting φ to vary over the range $-\infty < \varphi < \infty$.

EXAMPLE 1. FLOW PAST A CYLINDER AT ANGLE α When the mean flow incident upon the cylinder of Figure 3.2.1 makes a direction α with the x direction, we have $f(z) = Uze^{-i\alpha}$ in the circle theorem, so that $f^*(a^2/z) = Ua^2e^{i\alpha}/z$, and the overall complex potential is therefore

$$ w(z) = U\left(ze^{-i\alpha} + \frac{a^2e^{i\alpha}}{z}\right). $$

EXAMPLE 2. PRESSURE DISTRIBUTION ON THE CYLINDER The pressure distribution for the steady flow past the cylinder of Figure 3.2.1 is given by Bernoulli's equation (in the absence of body forces):

$$ p = p_\infty + \frac{1}{2}\rho_o U^2 - \frac{1}{2}\rho_o v^2 = p_\infty + \frac{1}{2}\rho_o U^2 - \frac{1}{2}\rho_o \left|\frac{dw}{dz}\right|^2, $$

where p_∞ is the uniform pressure at large distances from the cylinder. On the cylinder $|z| = a$, $dw/dz = U\left(1 - e^{-2i\theta}\right)$, and $|dw/dz|^2 = 4U^2\sin^2\theta$;

$$ \therefore \quad p = p_\infty + \frac{\rho_o U^2}{2}\left(1 - 4\sin^2\theta\right) \quad \text{on the cylinder.} $$

The symmetry of the pressure distribution with respect to both its x and y directions implies that there is no net force on the cylinder (in accordance with d'Alembert's paradox). The pressure takes its minimum value $p = p_\infty - \frac{3}{2}\rho_0 U^2$ at $\theta = \pm\frac{\pi}{2}$; for large values of U the pressure becomes negative near these points, and this could produce 'cavitation', i.e., the formation of vacuous cavities, which might actually be filled with 'vapour'. This can occur in water, for example, at tips of rotor blades (and often at the tip of a rowing boat oar). The maximum pressure $p_\infty + \frac{1}{2}\rho_0 U^2$ occurs at the stagnation points ($\theta = 0,\ \pi$). The absence of mean drag is, of course, in conflict with experiment; vortex shedding into the wake produces a mean drag and a fluctuating lift force (§4.5).

3.2.3 The line vortex

The complex potential $w(z)$ for steady mean flow past a circular cylinder furnished by the circle theorem is regular and *single valued* at all points of the fluid. However, the flow is actually governed by the analytic properties of $w'(z)$, which determines $w(z)$ up to an arbitrary constant. This enables the range of possible solutions to be extended to include cases in which the flow *circulates* about the cylinder.

Consider the function

$$w = \frac{-i\Gamma}{2\pi}\ln z \quad \left(\varphi = \frac{\Gamma\theta}{2\pi},\ \psi = -\frac{\Gamma}{2\pi}\ln r,\ z = re^{i\theta}\right).$$

This is regular except at $z = 0$, but it is not single valued – it changes in value by $\pm\Gamma$ with each complete traversal of z around a closed contour enclosing the origin; $w'(z)$ is single valued with a simple pole at $z = 0$. The complex potential $w(z)$ describes the irrotational flow produced by a *line vortex* of 'strength' Γ concentrated at $z = 0$. To see this we formally extend the definition of the stream function ψ to include $r = 0$ by writing

$$\psi = \lim_{\epsilon \to 0} -\frac{\Gamma}{4\pi}\ln(r^2 + \epsilon^2). \tag{3.2.4}$$

Then

$$\nabla^2\psi \equiv \left(\frac{\partial^2}{\partial r^2} + \frac{1}{r}\frac{\partial}{\partial r}\right)\psi = \lim_{\epsilon\to 0}\frac{-\Gamma\epsilon^2}{\pi(r^2 + \epsilon^2)^2} = -\Gamma\delta(\mathbf{x}), \quad \text{where } \mathbf{x} = (x, y), \tag{3.2.5}$$

which, according to (3.1.2), implies the existence of a singular distribution of vorticity $\omega = \Gamma\delta(\mathbf{x})$ concentrated at the origin.

The streamlines are circles centred at $z = 0$, and the flow speed is $\partial\varphi/r\partial\theta = \Gamma/2\pi r$ in the anticlockwise direction (for $\Gamma > 0$). The circulation $\int_C \mathbf{v}\cdot d\mathbf{x} = \Gamma$ is called the strength of the vortex, where C is any contour encircling the vortex once, and the contour is traversed in the *positive* direction (with the *interior on the left*). When the vortex is at $z_0 = x_0 + iy_0$,

$$w = \frac{-i\Gamma}{2\pi}\ln(z - z_0).$$

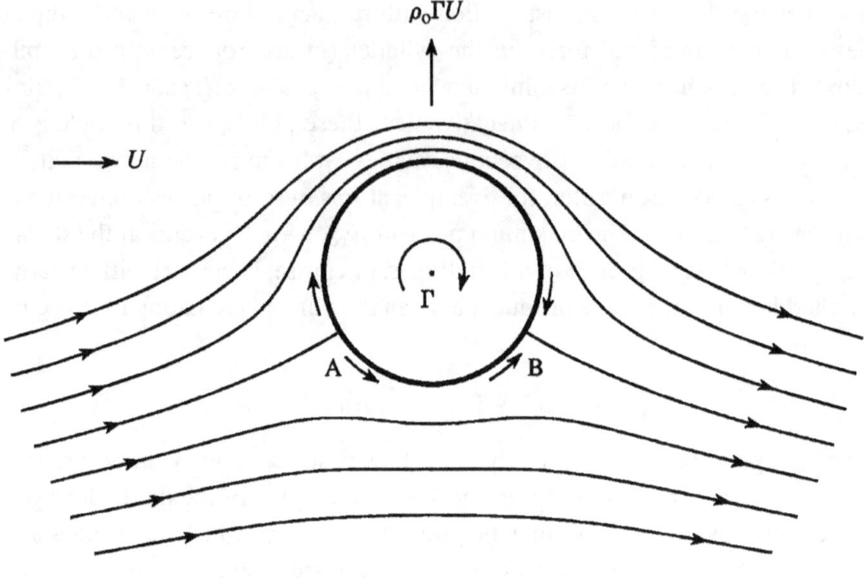

Figure 3.2.3

3.2.4 Circular cylinder with circulation

There is no force on the cylinder when the flow is steady, irrotational, and uniform at large distances from the body. This is no longer the case, however, in the presence of 'circulation' around the cylinder produced by a line vortex imagined to be placed on the cylinder axis, so that the condition that the normal velocity should vanish on the cylinder is still satisfied. Such a flow can be postulated for *any* value of the circulation, so that the solution appears to lack uniqueness without the imposition of further conditions. However, when the flow starts from rest, the circulation around any material circuit C must permanently vanish, provided viscous effects remain unimportant on C. This means that if circulation is subsequently found to exist around the cylinder then at some stage an equal and opposite amount of circulation must have been 'shed' into the flow from the cylinder surface (forming a vortex wake) in order that the net circulation around a large contour enclosing the cylinder and wake shall remain null.

With this understanding of the practical limitations, consider the situation illustrated in Figure 3.2.3, in which a cylinder of radius a (with centre at the origin) is placed in a uniform stream at speed U in the x direction. The circulation is assumed to be in the *negative* sense, the rotational motion being ascribed to a vortex of strength $-\Gamma$ lying along the axis of the cylinder, so that when the cylinder is required to coincide with the streamline $\psi = 0$, the complex potential can be taken in the form

$$w(z) = U\left(z + \frac{a^2}{z}\right) + \frac{i\Gamma}{2\pi}\ln\left(\frac{z}{a}\right). \tag{3.2.6}$$

On the cylinder $z \equiv re^{i\theta} = ae^{i\theta}$, and when $\Gamma = 0$ the stagnation points on the cylinder are at $\theta = 0,\ \pi$, with the incoming dividing streamline ($\psi = 0$) meeting the cylinder at $\theta = \pi$. When $\Gamma > 0$ the stagnation points occur at the points satisfying $w'(z) = 0$, i.e., at

$$z = a\left[-\frac{i\Gamma}{4\pi aU} \pm \sqrt{1 - \frac{\Gamma^2}{(4\pi aU)^2}}\right].$$

They are therefore shifted to angles at which

$$\sin\theta = \frac{\Gamma}{4\pi aU}, \quad \text{provided} \ \ \Gamma \le 4\pi aU.$$

The dividing streamline $\psi = 0$ intersects the cylinder at right angles at the point labelled A in Figure 3.2.3, the flows over the surface proceed around the upper and lower halves of the cylinder in opposite directions from A, joining up again and leaving the cylinder at the rear stagnation point B. Evidently we can adjust the locations of the stagnation points by changing the value of Γ, and this will later be seen to have important applications to the theory of airfoils.

The two stagnation points coincide at $\theta = -\frac{\pi}{2}$ when $\Gamma = 4\pi Ua$. There are no stagnation points on the cylinder when $\Gamma > 4\pi Ua$, but there is one within the fluid on the negative imaginary axis. The incident stream divides at this point, part breaking off to the right into the wake of the cylinder and part passing above and clockwise around the cylinder and returning to the stagnation point. Fluid trapped between the latter encircling streamline and the cylinder continually circulates around the cylinder and is not swept downstream by the main flow.

The clockwise circulation causes the streamlines to crowd together on the upper side of the cylinder and to diverge on the lower side. This asymmetry implies, from Bernoulli's equation, the existence of a net pressure force, or 'lift', on the cylinder in the y direction, at right angles to the direction of the impinging mean flow.

The surface pressure is

$$p = p_\infty + \frac{1}{2}\rho_o U^2 - \frac{1}{2}\rho_o v^2,$$

$$\text{where} \qquad v^2 = \left|\frac{dw}{dz}\right|^2 = \left(2U\sin\theta + \frac{\Gamma}{2\pi a}\right)^2 \quad \text{on the cylinder.}$$

Hence, if $\mathbf{n}$ is the unit normal on the cylinder, directed into the fluid, the force $\mathbf{F} = (F_1, F_2)$ on unit span of the cylinder is given by

$$\mathbf{F} = -a\int_0^{2\pi} \mathbf{n}p\,d\theta \equiv -a\int_0^{2\pi} (\cos\theta, \sin\theta)p\,d\theta,$$

$$\therefore \quad F_1 = 0, \quad F_2 = \rho_o\Gamma U.$$

The experimental realisation of the flow contemplated in Figure 3.2.3 can be achieved by rotation of the cylinder in the clockwise direction. The no-slip condition on the surface causes the production of *positive* vorticity on the lower half (or 'pressure' side) of the cylinder, the vorticity vector being orientated out of the plane of the paper in the figure. Either positive or a smaller amount of negative vorticity is produced on the upper ('suction') side of the cylinder, depending on the angular velocity of rotation. In either case a net amount of positive vorticity is produced and swept downstream by the mean flow. Kelvin's circulation theorem (§2.2) requires that an equal amount of negative (clockwise) circulation be left on the cylinder, leading ultimately to the steady state idealized in Figure 3.2.3.

However, if the cylinder rotation ceases, the ideal flow of Figure 3.2.3 would not persist. The irrotational flow over the surface would be retarded as it approaches on both sides the rear stagnation point B, where the pressure attains a maximum. In a real fluid this would cause surface 'boundary layers', through which the mean stream velocity rapidly decreases to zero on the surface, to thicken and separate from the surface shedding vorticity into the flow; the back reaction of this shed vorticity on the surface would rapidly lead to the formation of a broad region of separated flow, that is, to the formation of a 'wake' filled with energy bearing vorticity that is convected away in the mean stream. The resulting mean flow momentum deficit in the wake would then generate a substantial drag force and fluctuating lift forces.

The situation is different at the leading stagnation point. The impinging surface flow is accelerated away from the stagnation point A, causing a reduction in boundary-layer thickness, and there is a reduced tendency for separation. This suggests that in order to avoid large regions of separation in flow past a smooth body it will be advantageous to 'streamline' the rear section of the body to avoid the appearance of a rearward stagnation point. This is important in airfoil design, and we achieve this by having a sharp trailing edge; we avoid separation at the leading edge of the airfoil (for subsonic flow) by maintaining a rounded nose of finite radius of curvature.

3.2.5 Equation of motion of a cylinder with circulation

Consider a rigid circular cylinder of radius a and mass M per unit span moving in the z plane at velocity at time t $\mathbf{V}(t) = (U, V)$. Let the centre of the cylinder be at $z_o(t)$, so that

$$\dot{z}_o = U + iV,$$

and let Γ be the circulation around the cylinder (see Figure 3.2.4).

Relative to the moving cylinder the ambient fluid has the translational complex potential $-\dot{z}_o^*(z - z_o)$, and therefore, by the circle theorem, the potential w_{rel} of the whole flow (relative to the cylinder) is

$$w_{\text{rel}} = -\left[\dot{z}_o^*(z - z_o) + \frac{a^2 \dot{z}_o}{(z - z_o)}\right] - \frac{i\Gamma}{2\pi} \ln(z - z_o).$$

Thus, relative to fixed axes in the z plane, the complex potential is

$$w = \frac{-a^2 \dot{z}_o}{(z - z_o)} - \frac{i\Gamma}{2\pi} \ln(z - z_o). \tag{3.2.7}$$

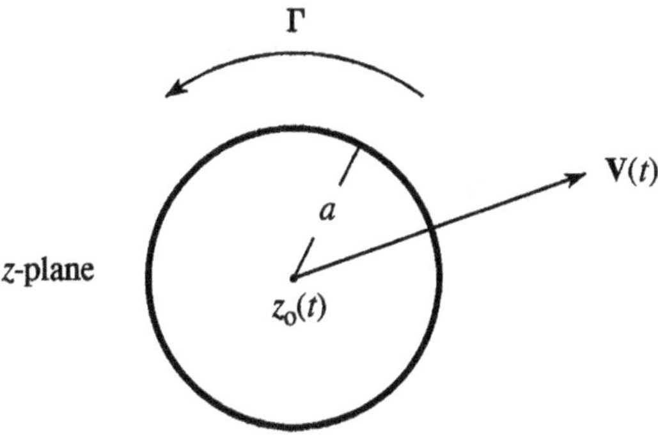

Figure 3.2.4

If $z - z_o = a e^{i\theta}$ on the cylinder, the force $\mathbf{F} = (F_1, F_2)$ exerted on the cylinder by the flow (per unit span) is

$$\mathbf{F} = -a \int_0^{2\pi} (\cos\theta, \sin\theta) p \, d\theta, \qquad (3.2.8)$$

where the pressure p is calculated from Bernoulli's equation

$$p = -\rho_o \frac{\partial \varphi}{\partial t} - \frac{1}{2}\rho_o |w'(z)|^2 + \text{constant}.$$

On the cylinder we obtain, from (3.2.7),

(1) $$\frac{\partial \varphi}{\partial t} = \operatorname{Re}\left(\frac{\partial w}{\partial t}\right) = \operatorname{Re}\left(-a\ddot{z}_o e^{-i\theta} - \dot{z}_o^2 e^{-2i\theta} + \frac{i\Gamma \dot{z}_o}{2\pi a} e^{-i\theta}\right),$$

(2) $$|w'(z)|^2 = \left|\dot{z}_o - \frac{i\Gamma}{2\pi a} e^{i\theta}\right|^2 = |\dot{z}_o|^2 + \frac{\Gamma^2}{(2\pi a)^2} + \frac{i\Gamma}{2\pi a}\left(\dot{z}_o e^{-i\theta} - \dot{z}_o^* e^{i\theta}\right).$$

Using these results to evaluate the pressure and substituting into (3.2.8), we find

$$\left. \begin{aligned} F_1 &= -m_o \ddot{x}_o - \rho_o \Gamma \dot{y}_o, \\ F_2 &= -m_o \ddot{y}_o + \rho_o \Gamma \dot{x}_o, \\ \text{where} \quad z_o &= x_o + i y_o, \end{aligned} \right\} \qquad (3.2.9)$$

and $m_o = \rho_o \pi a^2$ is the added mass (per unit span) of the cylinder, which in the present case is also the mass of fluid displaced by the cylinder. In the notation of §2.14, the (two-dimensional) added-mass tensor for a circular cylinder is simply $M_{ij} = m_o \delta_{ij}$.

We can also write

$$\mathbf{F} = -m_o \frac{d\mathbf{V}}{dt} + \rho_o \mathbf{\Gamma} \wedge \mathbf{V}, \quad \text{where} \quad \mathbf{\Gamma} = \Gamma \mathbf{k}.$$

Hence, if $\mathcal{F}$ denotes an external force applied to the cylinder (for example, because of gravity), the equation of motion becomes

$$(M + m_o)\frac{d\mathbf{V}}{dt} = \rho_o \mathbf{\Gamma} \wedge \mathbf{V} + \mathcal{F}.$$

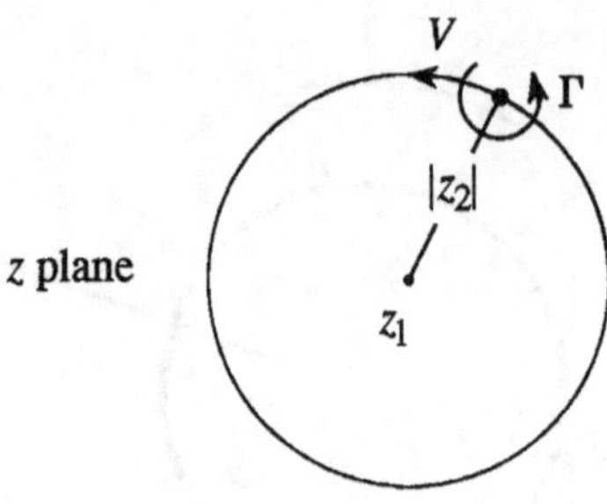

Figure 3.2.5

This can also be cast in the complex form

$$(M + m_o)\frac{d^2 z_o}{dt^2} = i\rho_o\Gamma\frac{dz_o}{dt} + (\mathcal{F}_1 + i\mathcal{F}_2). \qquad (3.2.10)$$

EXAMPLE 3. MOTION UNDER NO FORCES When $\mathcal{F} = 0$ the solution of (3.2.10) can be written as (see Figure 3.2.5)

$$z_o(t) = z_1 + z_2 e^{i\Omega t}, \quad \Omega = \frac{\rho_o\Gamma}{M + m_o}, \qquad (3.2.11)$$

where z_1 and z_2 are complex constants. This describes motion in a circle with centre z_1 and radius $|z_2|$ at angular velocity Ω.

Thus the translational speed of the cylinder axis $V = \Omega|z_2|$;

$$\therefore \quad \text{radius of path} \quad = \frac{V(M + m_o)}{\rho_o\Gamma}.$$

EXAMPLE 4. Let the cylinder be projected parallel to the x axis subject to the initial conditions

$$z_o = 0, \quad \dot{z}_o = U_o \ \text{(real)} \ \text{ at } t = 0.$$

Then

$$z_o = \frac{U_o}{i\Omega}\left(e^{i\Omega t} - 1\right),$$

the path being a circle with centre $z_1 = U_o/i\Omega$.

For small times

$$z_o \approx U_o t + \frac{i U_o \Omega t^2}{2} + \cdots +,$$

which shows how 'backspin' causes the cylinder to move 'to the left'.

EXAMPLE 5. MOTION UNDER GRAVITY The net gravitational force on the cylinder is

$$\mathcal{F}_2 = m_o g - Mg = (\text{Archimedean upthrust}) - (\text{weight of cylinder}).$$

Therefore equation of motion (3.2.10) becomes

$$(M + m_o)\frac{d^2 z_o}{dt^2} = i\rho_o\Gamma\frac{dz_o}{dt} - i(M - m_o)g,$$

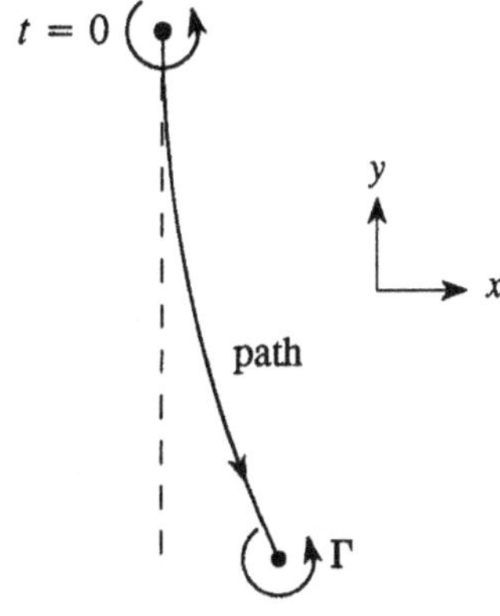

Figure 3.2.6

with the general solution

$$z_o = z_1 + z_2 e^{i\Omega t} + \frac{(M - m_o)gt}{\rho_o \Gamma}.$$

If the motion starts from rest at $z = 0$,

$$0 = z_1 + z_2, \quad 0 = i\Omega z_2 + \frac{(M - m_o)g}{\rho_o \Gamma};$$

$$\therefore \quad z_o = \frac{i(M - m_o)g}{\rho_o \Gamma \Omega} \left(e^{i\Omega t} - 1 - i\Omega t\right).$$

For small times this becomes

$$z_o = \frac{M - m_o}{M + m_o} \left(-\frac{igt^2}{2} + \frac{\rho_o g \Gamma t^3}{6(M + m_o)} + \cdots +\right),$$

again illustrating how the cylinder 'swings to the left' as it falls, the radius of curvature of the path being $(M^2 - m_o^2)g/\rho_o^2 g^2$ (Figure 3.2.6 shows the situation when $M > m_o$).

3.3 The Blasius force and moment formulae

The force and moment exerted (per unit span) on a solid body in two-dimensional irrotational flow can be expressed in terms of complex contour integrals in the z plane around the boundary C of the body. The most important applications are to cases in which the body is *at rest*. Localized regions of vorticity (e.g., the presence of one or more line vortices) within the flow usually make the motion unsteady, however, so that the pressure p acting on C must be determined from the general Bernoulli equation

$$p = -\rho_o \frac{\partial \varphi}{\partial t} - \frac{1}{2}\rho_o v^2 + \text{constant}.$$

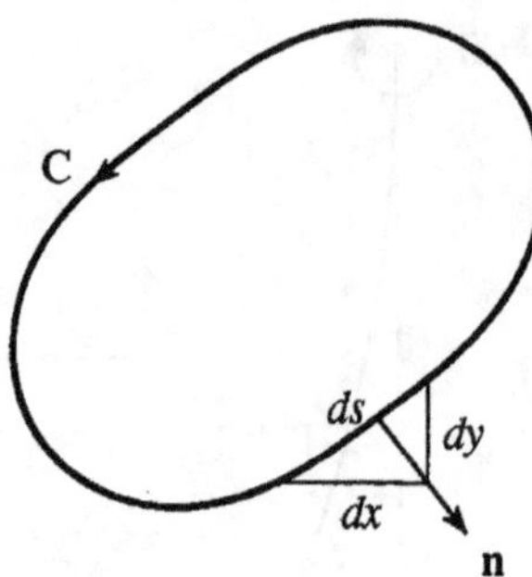

Figure 3.3.1

The force $\mathbf{F} = (F_1, F_2)$ on C in an ideal fluid is given by

$$\mathbf{F} = -\oint_C \mathbf{n} p \, ds, \quad \text{where} \quad \mathbf{n} = \left(\frac{dy}{ds}, -\frac{dx}{ds} \right) \tag{3.3.1}$$

is the unit normal directed into the fluid and ds is the element of arc length on C (Figure 3.3.1).

3.3.1 Blasius' force formula for a stationary rigid body

Blasius' force formula for a stationary rigid body is

$$F_1 - i F_2 = \left(-i\rho_o \oint_C \frac{\partial w}{\partial t} dz \right)^* + \frac{i\rho_o}{2} \oint_C \left(\frac{dw}{dz} \right)^2 dz. \tag{3.3.2}$$

PROOF From (3.3.1)

$$F_1 - i F_2 = -\oint_C p(dy + i dx) = -i \oint_C p \, dz^*.$$

Now $w = \varphi + i\psi$, where $\psi \equiv \psi(t)$ on C, because the surface coincides with an instantaneous streamline of the steady or unsteady motion. Hence $dw = dw^*$ on C, on which we can write

$$v^2 dz^* = \frac{dw}{dz} \frac{dw^*}{dz^*} dz^* = \left(\frac{dw}{dz} \right)^2 dz.$$

Also $\oint_C f \, dz^* \equiv 0$, where f is constant or a function of the time alone. Therefore

$$\oint_C \frac{\partial \varphi}{\partial t} dz^* = \oint_C \frac{\partial w^*}{\partial t} dz^*.$$

Collecting together these results, we now see from Bernoulli's equation that, for a stationary body,

$$F_1 - i F_2 = i\rho_o \oint_C \frac{\partial w^*}{\partial t} dz^* + \frac{i\rho_o}{2} \oint_C \left(\frac{dw}{dz} \right)^2 dz,$$

which is equivalent to (3.3.2).

3.3.2 Blasius' moment formula for a stationary rigid body

For a stationary rigid body a similar formula can be derived for the moment M_3 of the surface forces on C (about an axis in the $\mathbf{k}$ direction, out of the plane of the paper). For *steady* motion the reader can easily show by the same procedure as that of the previous subsection that the moment on C (per unit span) about an arbitrary point z_o is given by

$$M_3 = -\mathrm{Re}\left[\frac{\rho_o}{2}\oint_{\mathrm{S}}(z - z_o)\left(\frac{dw}{dz}\right)^2 dz\right].\tag{3.3.3}$$

3.3.3 Kutta–Joukowski lift force

Circulation Γ around an arbitrary cylindrical body C placed in a uniform mean stream of velocity U produces a lift on C equal to $\rho_o \Gamma U$ per unit span. Indeed, let the mean flow be in the x direction and suppose the circulation Γ is *clockwise*, as in Figure 3.3.2. At large distances the disturbance to the flow produced by the body becomes negligible, and the complex potential can be expanded in the form

$$w = Uz + \frac{i\Gamma}{2\pi}\ln z + \frac{a_1}{z} + \frac{a_2}{z^2} + \cdots +,\tag{3.3.4}$$

where a_1, a_2, $\ldots$, are constants.

The force on C is given by the steady form of (3.3.2). However, $w'(z)$ is regular within the fluid; therefore the integration contour C can be expanded onto a circle C_∞ 'at infinity'. Hence

$$F_1 - i F_2 = \frac{i\rho_o}{2}\oint_{\mathrm{C}}\left(\frac{dw}{dz}\right)^2 dz = \frac{i\rho_o}{2}\oint_{C_\infty}\left(U + \frac{i\Gamma}{2\pi z} - \frac{a_1}{z^2} + \cdots +\right)^2 dz.$$

The only nontrivial contribution from the z-dependent terms in the integrand is from $i\Gamma U/\pi z$, which supplies the desired result

$$F_1 - i F_2 = -i\rho_o \Gamma U.\tag{3.3.5}$$

EXAMPLE 1. CIRCULAR CYLINDER WITH CIRCULATION Complex potential (3.2.6) behaves as in (3.3.4) as $|z| \to \infty$. When the exact formula is used in the steady form of Blasius formula (3.3.2) we have

$$F_1 - i F_2 = \frac{i\rho_o}{2}\oint_{\mathrm{S}}\left[U\left(1 - \frac{a^2}{z^2}\right) + \frac{i\Gamma}{2\pi z}\right]^2 dz,$$

which can be evaluated by residues to supply the net force per unit span:

$$F_1 = 0, \quad F_2 = \rho_o \Gamma U.$$

The force is at right angles to the undisturbed direction of the mean stream.

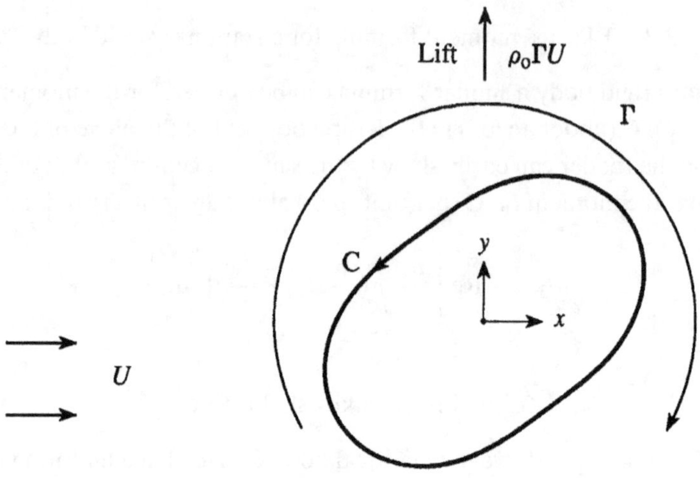

Figure 3.3.2

3.3.4 Leading-edge suction

Irrotational flow around a knife-edge was shown in §2.12 to be maintained by a singular
suction force. The representation of this flow in terms of the locally two-dimensional
velocity potential and stream function of (2.12.17) actually corresponds to a complex
potential that has a square-root singularity at the edge (see §3.5). Thus the complex
potential of flow around the edge $z = -a$, say, of a finite or semi-infinite rigid plate that
extends in the positive direction along the real axis from $z = -a$ (Figure 3.3.3) is given
in the neighbourhood of the edge by

$$w = C\sqrt{z+a} + w_o, \tag{3.3.6}$$

where C is a *real* constant and w_o is constant (so that $\psi = \text{Im}\, w$ has the same constant
value on the upper and lower surfaces of the plate). Formal application of Blasius
formula (3.3.2) shows that, for both steady and unsteady flows, the force *on the edge* is
given by

$$F_1 - i F_2 = \frac{i\rho_o}{2} \oint_C \frac{C^2}{4(z+a)}\, dz,$$

where C is the surface of the plate traversed in the anticlockwise sense around the edge.
Hence $F_2 = 0$ and $F_1 = -\frac{\pi}{4}\rho_o C^2 < 0$. This conclusion, that a suction force is produced
parallel to the plane of the plate, is obviously independent of the orientation of the plate;
in the general case of a plate of arbitrary orientation, C is complex, and the magnitude

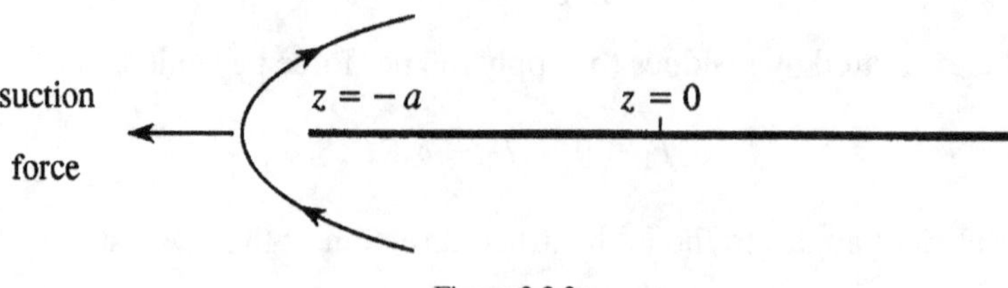

Figure 3.3.3

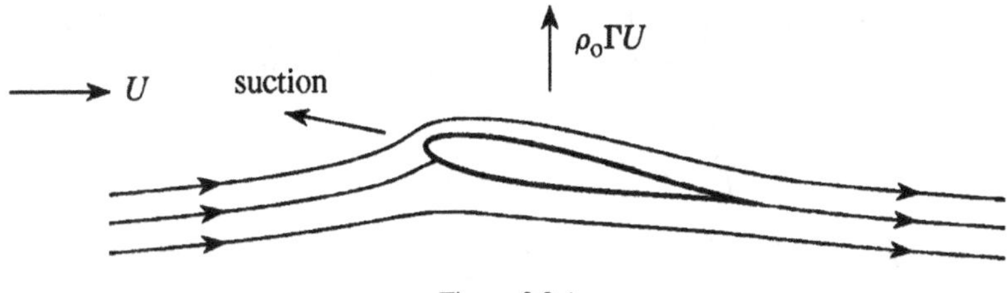

Figure 3.3.4

of the suction force is found merely by the replacement of C with $|C|$ in the suction force formula.

Irrotational flow theory predicts that the velocity at the edge becomes infinite like $1/|z + a|^{\frac{1}{2}}$ because the radius of curvature $\mathcal{R}$ vanishes at the edge; Equations (2.3.9) then show that an infinite centripetal pressure gradient is required for turning the flow around the edge. This produces a net 'suction' force parallel to the plate, concentrated at the edge, such that

$$\text{at an edge where} \quad w \sim C\sqrt{z + a} \quad \text{the suction force} \ = \frac{\pi \rho_0 |C|^2}{4}. \tag{3.3.7}$$

In practice very low pressures of this kind in which a flow negotiates an edge can be avoided if the edge is well 'rounded', as at the leading edge of the airfoil in Figure 3.3.4. In this case clockwise circulation Γ about the airfoil has removed the singular flow at the trailing edge (an example of the Kutta–Joukowski hypothesis of §3.8.1) and therefore removed suction at the sharp trailing edge where the dividing streamline flows over the upper and lower sides of the airfoil reunite smoothly. However, the relatively high-speed 'attached' flow around the nose of the airfoil still produces a finite suction force – just enough to ensure that the lift is exactly at right angles to the direction of the mean flow. At a knife-edge, fluid inertia causes the edge flow to *separate*, the very high-velocity gradients at the edge promoting viscous diffusion of vorticity from the surface into the flow, where it is rapidly swept away by convection in the outer flow, although (in accordance with the Kutta–Joukowski hypothesis) the amount of shed vorticity is often insensitive to the magnitude of the viscosity coefficient.

3.4 Sources and line vortices

The point source in two-dimensional flow produces a symmetrical radial flow in all directions. A source of strength q at the origin is represented by the complex potential

$$w = \frac{q}{2\pi} \ln z,$$

which is regular everywhere except at the source point $z = 0$. The real and imaginary parts,

$$\varphi = \frac{q}{2\pi} \ln r, \quad \psi = \frac{q\theta}{2\pi}, \quad (z = re^{i\theta}),$$

define the radially outward flow at speed $\partial\varphi/\partial r = q/2\pi r$ along streamlines $\theta = $ constant. The origin is a singularity at which fluid volume is *created* at a rate equal to

$\oint_C \nabla\varphi \cdot \mathbf{n}\,ds$, where C is any simple closed curve enclosing the origin with outward normal $\mathbf{n}$, and ds is the element of arc length on C. Taking C to be a circle of radius r, we obtain

$$\oint_C \nabla\varphi \cdot \mathbf{n}\,ds = \int_0^{2\pi} \frac{\partial\varphi}{\partial r} r\,d\theta = q.$$

The source strength q is therefore the rate of production of fluid volume (per unit span).

For a source at $z_0 = x_0 + iy_0$,

$$w = \frac{q}{2\pi}\ln(z - z_0) \quad \text{and} \quad \varphi = \frac{q}{2\pi}\ln[(x - x_0)^2 + (y - y_0)^2]^{\frac{1}{2}}.$$

The velocity potential φ satisfies the Laplace equation $\nabla^2\varphi = 0$ except at the source point. However, by 'regularising' the definition of φ by setting (for a source at $z = 0$)

$$\varphi \equiv \frac{q}{2\pi}\ln r = \lim_{\epsilon \to 0} \frac{q}{4\pi}\ln(r^2 + \epsilon^2),$$

we find, as in (3.2.5), that

$$\nabla^2\left(\frac{q}{2\pi}\ln r\right) = q\delta(\mathbf{x}), \quad \mathbf{x} = (x, y). \tag{3.4.1}$$

Hence the potential

$$G(\mathbf{x}, \mathbf{y}) = \frac{1}{2\pi}\ln|\mathbf{x} - \mathbf{y}|, \quad \big[\mathbf{x} = (x_1, x_2), \quad \mathbf{y} = (y_1, y_2)\big], \tag{3.4.2}$$

produced by a *unit* source at $\mathbf{x} = \mathbf{y}$ satisfies

$$\nabla^2 G = \delta(\mathbf{x} - \mathbf{y}), \tag{3.4.3}$$

and G may be taken to define the 'free- space' Green's function in two dimensions.

EXAMPLE 1. SOURCE ADJACENT TO A WALL The function

$$w = \frac{q}{2\pi}\left[\ln(z - z_0) + \ln(z - z_0^*)\right], \quad \left[\varphi = \frac{1}{2\pi}(\ln r_1 + \ln r_2)\right],$$

represents the flow produced by two equal point sources of strength q at $z_0 = x_0 + iy_0$ and $z_0^* = x_0 - iy_0$ (Figure 3.4.1). The motion is symmetric with respect to the x axis, and $\partial\varphi/\partial y = 0$ on $y = 0$. Therefore, in the region $y > 0$, the potential also describes the flow produced by a point source at z_0 *adjacent to a rigid wall* at $y = 0$ (the presence of the wall is said to be accounted for by an 'image' source).

The source exerts a *suction* force on the wall on which, taking $z_o = ih$, $w'(z) \equiv w'(x) = qx/\pi(x^2 + h^2)$. Using the steady form of Bernoulli's equation, we have

$$F_2 = -\int_\infty^\infty p(x, 0)\,dx = \frac{1}{2}\rho_o \int_{-\infty}^\infty |w'(x)|^2\,dx = \frac{\rho_o q^2}{2\pi^2}\int_{-\infty}^\infty \frac{x^2\,dx}{(x^2 + h^2)^2} = \frac{\rho_o q^2}{4\pi h}.$$

EXAMPLE 2. SOURCE ADJACENT TO A RIGID CIRCULAR CYLINDER Let the cylinder have radius a with centre at the origin, and consider a source q at $x = h > a$ on the x axis.

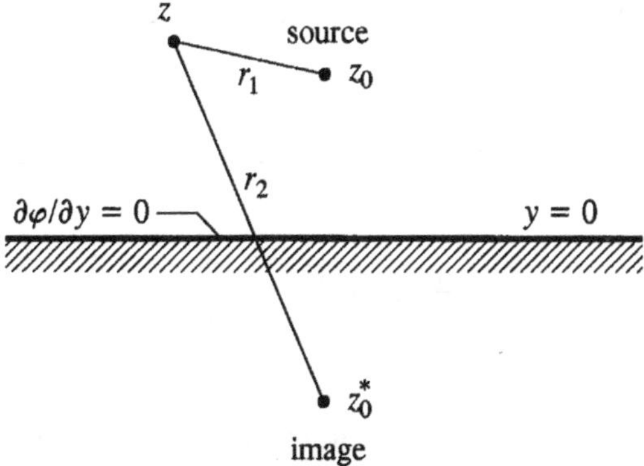

Figure 3.4.1

The velocity potential in the absence of the cylinder is $f(z) = (q/2\pi)\ln(z - h)$. By the circle theorem (§3.2), when the cylinder is present the complex potential is

$$w(z) = \frac{q}{2\pi}\ln(z - h) + \frac{q}{2\pi}\ln\left(\frac{a^2}{z} - h\right).$$

This can be put in the form

$$w(z) = \frac{q}{2\pi}\ln(z - h) + \frac{q}{2\pi}\ln\left(z - \frac{a^2}{h}\right) - q\ln z + q\ln(-h).$$

Discarding the irrelevant complex constant $q\ln(-h)$, we see that the effect of the cylinder is to augment the complex potential of the original source at $z = h$ by a source q at the *inverse point* $z = a^2/h$ and a 'sink' (or 'negative' source) q at the centre of the cylinder (see Figure 3.4.2). The two 'image' sources have equal and opposite strengths to ensure that there is no net flux of fluid through the surface of the cylinder.

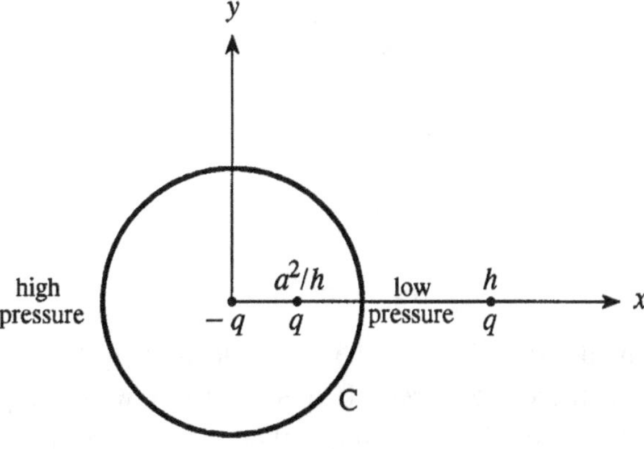

Figure 3.4.2

The force exerted on the cylinder is calculated by use of the Blasius formula (3.3.2) for steady flow:

$$w'(z) = \frac{q}{2\pi}\left(\frac{1}{z-h} + \frac{1}{z-a^2/h} - \frac{1}{z}\right),$$

$$\therefore \quad F_1 - iF_2 = \frac{i\rho_0 q^2}{8\pi^2}\oint_C\left[\frac{1}{(z-h)^2} + \frac{2}{(z-h)}\left(\frac{1}{z-a^2/h} - \frac{1}{z}\right)\right.$$

$$\left. + \frac{1}{(z-a^2/h)^2} + \frac{1}{z^2} - \frac{2}{z(z-a^2/h)}\right]dz$$

i.e., by residues, $\qquad F_1 = \dfrac{\rho_0 q^2 a^2}{2\pi h(h^2-a^2)} > 0, \quad F_2 = 0.$

The cylinder is therefore subject to a suction force towards the source. The result is independent of the sign of q, so the cylinder is also attracted by a sink.

3.4.1 Line vortices

Line vortices have already been introduced in §3.2.3 to model circulatory flow about cylindrical bodies. The singular vorticity distribution

$$\operatorname{curl}\mathbf{v} = \omega = \Gamma\mathbf{k}\delta(\mathbf{x})$$

is associated with a two-dimensional flow that is irrotational for $r = |\mathbf{x}| > 0$ with complex potential $w = (-i\Gamma/2\pi)\ln z$.

The line vortex can be regarded as an idealisation of a rectilinear vortex of the same total circulation but with the vorticity $\omega = \Gamma/\pi\epsilon^2$ spread uniformly over a central circular 'core' of radius ϵ, say. The fluid in $r < \epsilon$ is therefore in solid-body rotation at angular velocity $\frac{1}{2}\omega$ (§1.4.1). At a radial distance r from the centre the flow velocity is v_θ tangential to the circle of radius r, where

$$v_\theta = \begin{cases} \dfrac{\omega r}{2} \equiv \dfrac{\Gamma r}{2\pi\epsilon^2}, & \text{for } r < \epsilon \\[2ex] \dfrac{\omega\epsilon^2}{2r} \equiv \dfrac{\Gamma}{2\pi r}, & \text{for } r \geq \epsilon \end{cases}.$$

3.4.2 Motion of a line vortex

According to this model (see Figure 3.4.3) the velocity vanishes at the centre of the vortex core, which is therefore at rest. It therefore follows more generally that the vortex core will remain fixed in space unless there exists a local flow velocity $\mathbf{v}_o$, say, in addition to the rotational velocity field of the vortex. In other words, the vortex is convected by the local flow in which it is immersed. This result can also be expressed

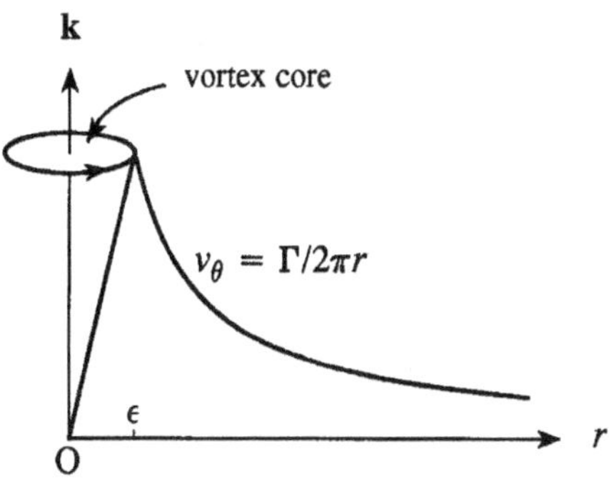

Figure 3.4.3

analytically as follows. Use identity (1.5.5) to write momentum equations (2.1.1) (in the absence of body forces) in the form

$$\frac{\partial \mathbf{v}}{\partial t} + \nabla\left(\frac{1}{2}v^2\right) + \omega \wedge \mathbf{v} = -\frac{1}{\rho_o}\nabla p. \tag{3.4.4}$$

Take the curl of this equation, and note that

$$\mathrm{curl}(\omega \wedge \mathbf{v}) = (\mathbf{v} \cdot \nabla)\omega + (\mathrm{div}\,\mathbf{v})\omega - (\omega \cdot \nabla)\mathbf{v} - (\mathrm{div}\,\omega)\mathbf{v}$$
$$= (\mathbf{v} \cdot \nabla)\omega - (\omega \cdot \nabla)\mathbf{v} \tag{3.4.5}$$

because $\mathrm{div}\,\mathbf{v} = \mathrm{div}\,\omega = 0$. Furthermore, in two-dimensional flow, $(\omega \cdot \nabla)\mathbf{v} = \omega(\mathbf{k} \cdot \nabla)\mathbf{v} \equiv 0$. Therefore, the curl of (3.4.4) reduces to

$$\frac{\partial \omega}{\partial t} + \mathbf{v} \cdot \nabla\omega \equiv \frac{D\omega}{Dt} = 0, \tag{3.4.6}$$

which shows that vorticity is convected without change by the flow, i.e., it is always attached to the same fluid particles.

We can obtain a more explicit deduction of this dynamical result by considering the net pressure force applied to the vortex core, which must vanish as the core radius $\epsilon \to 0$ in order to avoid infinite accelerations. Thus, at time t, suppose a vortex of circulation Γ is at $z = z_0(t)$, and let the complex potential be expressed in the form

$$w(z) = w_0(z) - \frac{i\Gamma}{2\pi}\ln(z - z_0),$$

where $w_0(z)$ is the complex potential with the 'self-potential' of the vortex removed. Then it is required to show that

$$\frac{dz_o^*}{dt} \equiv \dot{z}_o^* = w'(z_o). \tag{3.4.7}$$

The net pressure force on Γ is given by

$$\mathbf{F} = \lim_{r \to 0}\oint_C -p\mathbf{n}\,ds = \lim_{r \to 0}r\rho_o\int_0^{2\pi}\left(\frac{\partial \varphi}{\partial t} + \frac{1}{2}v^2\right)\mathbf{n}\,d\theta,$$

where the integration is around the circular contour $|z - z_o| = re^{i\theta}$ of Figure 3.4.4.

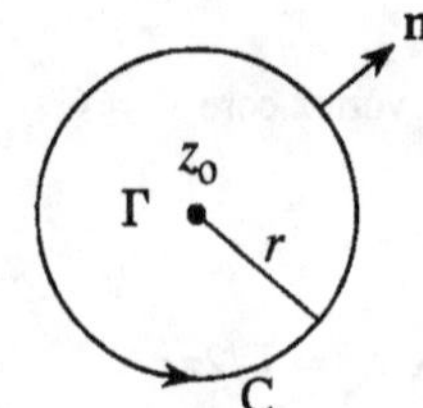

Figure 3.4.4

Near the vortex

$$\frac{\partial w}{\partial t} \approx \dot{w}_o(z_o) + \frac{i\Gamma \dot{z}_o}{2\pi(z - z_o)}, \quad w'(z) \approx w'_o(z_o) - \frac{i\Gamma}{2\pi(z - z_o)}.$$

Therefore, if we put

$$\dot{z}_o = |\dot{z}_o|e^{i\theta_o} \quad \text{and} \quad w'(z_o) = |w'(z_o)|e^{i\theta_w},$$

then

$$\mathbf{F} = \lim_{r \to 0} r\rho_o \int_0^{2\pi} \left[\operatorname{Re}(\dot{w}_o) - \frac{\Gamma|\dot{z}_o|\sin(\theta_o - \theta)}{2\pi r} \right.$$

$$\left. + \frac{1}{2}|w'(z_o)|^2 - \frac{\Gamma|w'(z_o)|\sin(\theta_w + \theta)}{2\pi r} + \frac{\Gamma^2}{2(2\pi r)^2} \right] (\cos\theta, \ \sin\theta)d\theta$$

$$= -\frac{\rho_o\Gamma}{2} \left[|\dot{z}_o|\sin\theta_o + |w'(z_o)|\sin\theta_w, \ -|\dot{z}_o|\cos\theta_o + |w'(z_o)|\cos\theta_w \right].$$

Hence, the requirement that $\mathbf{F} = 0$ becomes

$$|\dot{z}_o|\sin\theta_o = -|w'(z_o)|\sin\theta_w, \quad |\dot{z}_o|\cos\theta_o = |w'(z_o)|\cos\theta_w,$$

$$\therefore \quad \dot{z}_o^* = |\dot{z}_o|(\cos\theta_o - i\sin\theta_o) = |w'(z_o)|(\cos\theta_w + i\sin\theta_w) = w'(z_o). \quad \text{Q. E. D.}$$

EXAMPLE 3. The complex potential

$$w = -\frac{i\Gamma}{2\pi}\ln(z - z_0) + \frac{i\Gamma}{2\pi}\ln(z - z_0^*)$$

describes the flow produced by two line vortices of circulations $\pm\Gamma$ respectively at $z_0 = x_0 + iy_0$, $z_0^* = x_0 - iy_0$ (Figure 3.4.5). The stream function $\psi = \operatorname{Im} w$ vanishes on the x axis, which is a streamline on which $-\partial\psi/\partial x = \partial\varphi/\partial y = 0$. In the region $y > 0$ the potential therefore represents the motion that is due to a *single* vortex of strength Γ at z_0 *adjacent to a rigid wall* at $y = 0$, whose presence is accounted for by the equal and opposite 'image' vortex. Each vortex translates parallel to the wall at the local flow speed $u = \Gamma/4\pi y_o$ produced by the velocity potential of its image.

EXAMPLE 4. SPINNING VORTICES Two vortices at $z = \pm z_o(t)$, each of circulation Γ and distance 2ℓ apart, rotate about the origin (midway between them) at angular velocity $\Omega = \Gamma/4\pi\ell^2$ (see Figure 3.4.6). The velocity potential is

$$w = -\frac{i\Gamma}{2\pi}\ln(z - z_o) - \frac{i\Gamma}{2\pi}\ln(z + z_o).$$

Each vortex moves under the velocity potential of the other.

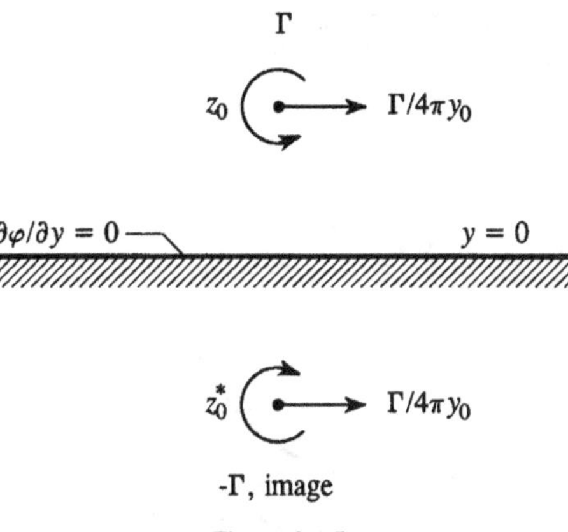

Figure 3.4.5

EXAMPLE 5. VORTEX MOTION OUTSIDE A CYLINDER A vortex Γ at $z = z_o$ is distance h from the centre $z = 0$ of a rigid cylinder of radius a ($< h$) (Figure 3.4.7). If there is no net circulation around the cylinder, the circle theorem gives a complex potential in the form

$$w(z) = w_o(z) - \frac{i\Gamma}{2\pi} \ln(z - z_o),$$

where

$$w_o(z) = \frac{i\Gamma}{2\pi} \ln\left(z - \frac{a^2}{z_o^*}\right) - \frac{i\Gamma}{2\pi} \ln z + \frac{i\Gamma}{2\pi} \ln(-z_o^*).$$

There are image vortices $-\Gamma$ at the centre of the cylinder and $+\Gamma$ at the inverse point $z = a^2/z_o^*$ on the ray from the centre to the vortex at z_o. The velocities induced by the images are at right angles to this line, so that the vortex follows a circular path around the cylinder (in the clockwise sense in the figure) at speed

$$v_o = \frac{\Gamma}{2\pi(h - a^2/h)} - \frac{\Gamma}{2\pi h} = \frac{\Gamma a^2}{2\pi h(h^2 - a^2)}. \tag{3.4.8}$$

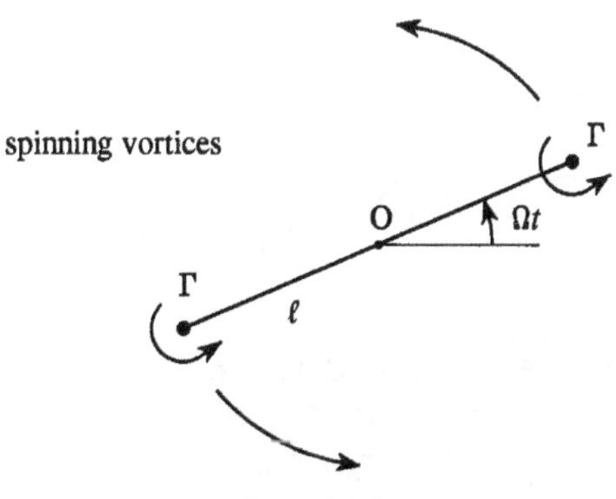

Figure 3.4.6

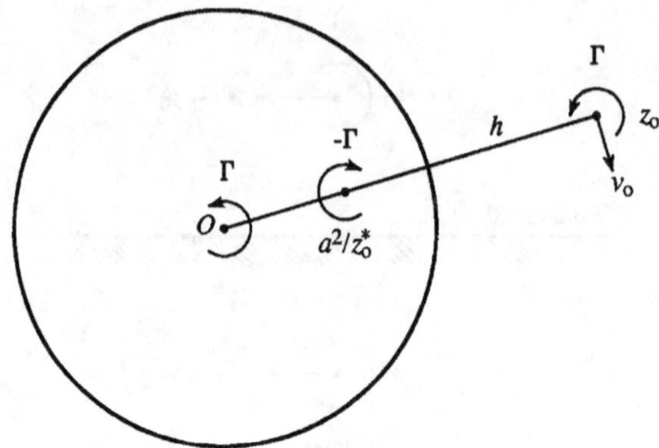

Figure 3.4.7

EXAMPLE 6. INFINITE ROW OF VORTICES Line vortices of equal strength Γ are placed on the x axis at $x = 0,\ \pm a,\ \pm 2a,\ \pm 3a,\ \dots$ [Figure 3.4.8(a)]. The complex potential is

$$w = -\frac{i\Gamma}{2\pi} \ln\left[\prod_{-\infty}^{\infty} (z - na)\right] = -\frac{i\Gamma}{2\pi} \ln\left[\frac{\pi z}{a} \prod_{n=1}^{\infty}\left(1 - \frac{z^2}{n^2 a^2}\right)\right] + \text{constant}$$

$$= -\frac{i\Gamma}{2\pi} \ln \sin\left(\frac{\pi z}{a}\right) + \text{constant}. \tag{3.4.9}$$

The vortices remain at rest, and the array constitutes a discrete model of a 'shear layer' or vortex sheet. At distances exceeding a above and below from the array, $w \sim \mp \Gamma z/2a$, so that the flow velocity $\sim \mp \Gamma/2a$; the configuration is, however, unstable.

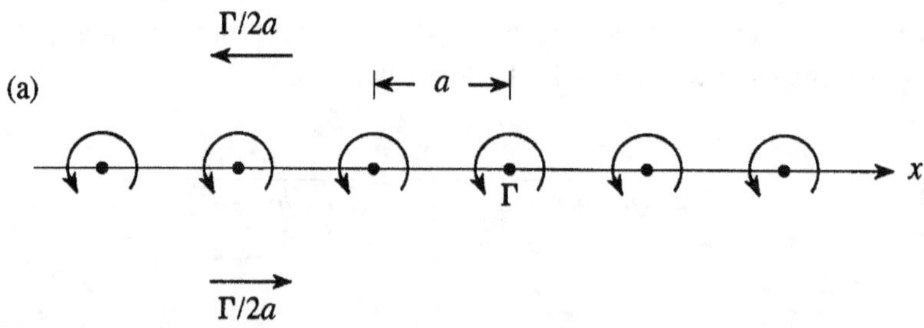

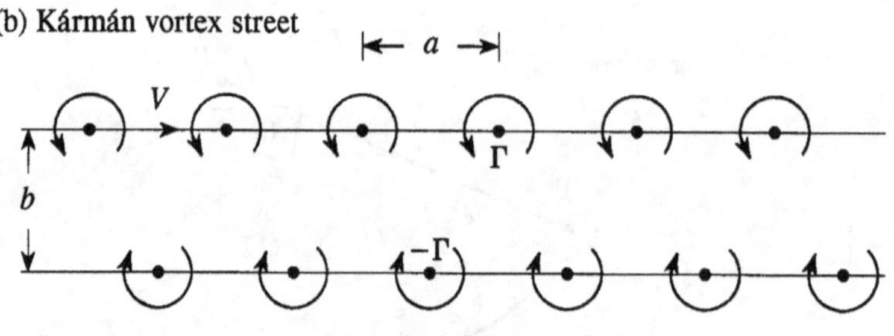

Figure 3.4.8

3.4.3 Kármán vortex street

This consists of a double row of vortices with respective circulations $\pm\Gamma$ a distance b apart, with each vortex placed opposite the midpoint between two vortices in the opposite row. The overall velocity potential can be evaluated from (3.4.9). For the arrangement of Figure 3.4.8(b) the system translates to the right at velocity

$$V = \frac{\Gamma}{2a}\tanh\left(\frac{\pi b}{a}\right). \tag{3.4.10}$$

The configuration was proposed by von Kármán as a simple two-dimensional model of the vortex wake of a cylindrical body at intermediate values of the Reynolds number (§4.5). It is stable to small disturbances when

$$\frac{b}{a} \approx 0.281.$$

3.4.4 Kinetic energy of a system of rectilinear vortices

Consider a line vortex with vorticity $\omega = \Gamma/\pi\epsilon^2$ distributed over a circular core of radius ϵ, as in Figure 3.4.3. Then we can take

$$\psi = \begin{cases} \dfrac{\Gamma}{4\pi\epsilon^2}(\epsilon^2 - r^2), & \text{for } r < \epsilon \\[2ex] -\dfrac{\Gamma}{2\pi}\ln\left(\dfrac{r}{\epsilon}\right), & \text{for } r \geq \epsilon \end{cases} \tag{3.4.11}$$

The kinetic energy of the motion (per unit span) is therefore

$$T = \frac{1}{2}\rho_o \int (\nabla\psi)^2 dxdy = \frac{\rho_o\Gamma^2}{4\pi\epsilon^4}\int_0^\epsilon r^3 dr + \frac{\rho_o\Gamma^2}{4\pi}\int_\epsilon^\infty \frac{dr}{r}.$$

The first integral yields a finite 'core kinetic energy' $\mathcal{E}_o = \rho_o\Gamma^2/16\pi$, but the kinetic energy of the exterior, irrotational flow is infinite.

The kinetic energy is finite for a system of vortices of strengths Γ_1, $\Gamma_2, \ldots$, respectively, at z_1, $z_2, \ldots$, provided $\sum_j \Gamma_j = 0$. In that case,

$$\psi = -\sum_j \frac{\Gamma_j}{2\pi}\ln\left(\frac{|\mathbf{x} - \mathbf{x}_j|}{\epsilon_j}\right) \sim \sum_j \frac{\Gamma_j}{2\pi}\frac{\mathbf{x}\cdot\mathbf{x}_j}{|\mathbf{x}|^2} + \sum_j \frac{\Gamma_j}{2\pi}\ln\epsilon_j, \quad \text{as } |\mathbf{x}| \to \infty, \tag{3.4.12}$$

where $\mathbf{x}_j = (x_j, y_j)$ and ϵ_j is the core radius of the jth vortex, which is assumed to be small compared with the shortest distance between the vortices. Then, for the kinetic energy we have

$$T = \frac{1}{2}\rho_o \int (\nabla\psi)^2 dxdy = \frac{1}{2}\rho_o \int \left[\text{div}(\psi\nabla\psi) - \psi\nabla^2\psi\right] dxdy.$$

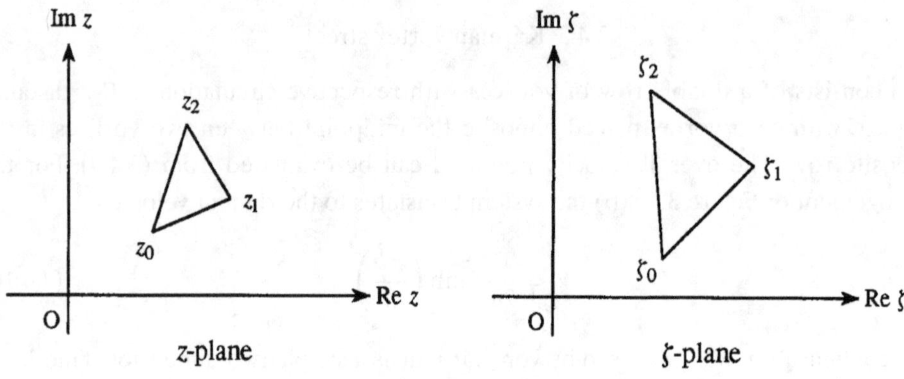

Figure 3.5.1

The divergence term can be replaced with an integral over a 'contour at infinity', which is null because of (3.4.12). The remaining integration is evaluated with the aid of Equation (3.1.2) and a local representation of the type (3.4.11) for each vortex, yielding

$$T = \frac{\rho_o}{2} \int \psi \omega \, dx dy = \sum_j \frac{\rho_o \Gamma_j^2}{16\pi} - \frac{\rho_o}{4\pi} \sum_{i \neq j} \Gamma_i \Gamma_j \ln \left(\frac{|\mathbf{x}_i - \mathbf{x}_j|}{\epsilon_j} \right). \qquad (3.4.13)$$

EXAMPLE 8. KINETIC ENERGY OF A VORTEX PAIR For the vortex pair of Figure 3.4.5 of circulations $\pm\Gamma$, each of radius ϵ and separated by a distance $2y_o$, the total kinetic energy is

$$T = \frac{\rho_o \Gamma^2}{8\pi} + \frac{\rho_o \Gamma^2}{2\pi} \ln \left(\frac{2y_o}{\epsilon} \right). \qquad (3.4.14)$$

3.5 Conformal transformations

A complex function $\zeta = f(z)$ defines a transformation between points in the z plane and points $\zeta \equiv \xi + i\eta$ in the ζ plane (see Figure 3.5.1). Suppose $f(z)$ is regular in a region $\mathcal{D}$ of the z plane and consider three neighbouring points z_0, z_1, z_2 in $\mathcal{D}$ and their corresponding *images* ζ_0, ζ_1, ζ_2 in the ζ plane. When z_1, z_2 are *very close* to z_0 we can write

$$\zeta_1 - \zeta_0 = f'(z_0)(z_1 - z_0), \quad \zeta_2 - \zeta_0 = f'(z_0)(z_2 - z_0), \quad \text{provided} \ f'(z_0) \neq 0.$$

Therefore

$$|\zeta_1 - \zeta_0| = |f'(z_0)||z_1 - z_0|, \quad |\zeta_2 - \zeta_0| = |f'(z_0)||z_2 - z_0|,$$

so that *small* distances between points in the z plane in the vicinity of z_0 are *magnified* in the ζ plane by the factor $|f'(z_0)|$. Also,

$$\arg(\zeta_1 - \zeta_0) = \arg[f'(z_0)(z_1 - z_0)] = \arg[f'(z_0)] + \arg(z_1 - z_0)$$

$$\arg(\zeta_2 - \zeta_0) = \arg[f'(z_0)(z_2 - z_0)] = \arg[f'(z_0)] + \arg(z_2 - z_0),$$

which means that the angle between the rays $\zeta_1 - \zeta_0$ and $\zeta_2 - \zeta_0$ has the *same magnitude and sense* as the angle between the rays $z_1 - z_0$ and $z_2 - z_0$.

Thus the effect of the transformation is to *rotate* straight-line elements near z_0 through the same angle $\arg[f'(z_0)]$ and increase their lengths by a factor of $|f'(z_0)|$. Small-*area* elements are increased by a factor of $|f'(z_0)|^2$. Similarly, when two curves intersect in the z plane their images in the ζ plane will *intersect at the same angle*. A transformation with these properties is said to be 'conformal'.

The transformation is *not* conformal at a *critical point* $z = z_0$ where $f'(z_0) = 0$.

3.5.1 Transformation of Laplace's equation

If

$$w(z) = \varphi(x, y) + i\psi(x, y), \quad z = x + iy,$$

is *regular* in $\mathcal{D}$, both $\varphi(x, y)$ and $\psi(x, y)$ satisfy

$$\frac{\partial^2 \varphi}{\partial x^2} + \frac{\partial^2 \varphi}{\partial y^2} = 0, \quad \frac{\partial^2 \psi}{\partial x^2} + \frac{\partial^2 \psi}{\partial y^2} = 0 \ \text{ in } \mathcal{D}.$$

Let $\zeta = f(z)$ define a conformal transformation of $\mathcal{D}$ into a region $\mathcal{D}'$ in the ζ plane. Let $W(\zeta)$ be regular in $\mathcal{D}'$ with real and imaginary parts $\Phi(\xi, \eta)$, $\Psi(\xi, \eta)$. Then

$$\frac{\partial^2 \Phi}{\partial \xi^2} + \frac{\partial^2 \Phi}{\partial \eta^2} = 0, \quad \frac{\partial^2 \Psi}{\partial \xi^2} + \frac{\partial^2 \Psi}{\partial \eta^2} = 0, \ \text{ in } \mathcal{D}'.$$

The transformation $\zeta = f(z)$ permits us to define a corresponding function $w(z) \equiv \varphi(x, y) + i\psi(x, y) = W[f(z)]$, which is regular in $\mathcal{D}$, with derivative $w'(z) = f'(z)W'[f(z)]$. For corresponding points in $\mathcal{D}$ and $\mathcal{D}'$ we have

$$\varphi(x, y) = \Phi\left[\xi(x, y), \eta(x, y)\right], \quad \psi(x, y) = \Psi\left[\xi(x, y), \eta(x, y)\right].$$

In other words: The solutions Φ and Ψ of Laplace's equation in $\mathcal{D}'$ are also solutions of Laplace's equation in $\mathcal{D}$.

These results have the following significance: The solution of Laplace's equation within a given two-dimensional bounded region $\mathcal{D}$ is equivalent to the solution of Laplace's equation within the transformed region $\mathcal{D}'$, where the boundaries of $\mathcal{D}$ are mapped onto corresponding boundaries of $\mathcal{D}'$. Therefore the flow past a system of rigid boundaries in the z plane is represented by the transformation $\zeta = f(z)$ by an equivalent flow in the ζ plane. If it is possible to solve the problem in the ζ plane, the solution to the original problem in $\mathcal{D}$ can be found by use of the transformation.

We can treat problems involving sources and line vortex singularities by observing that sources and vortices map into *identical* sources and vortices when the

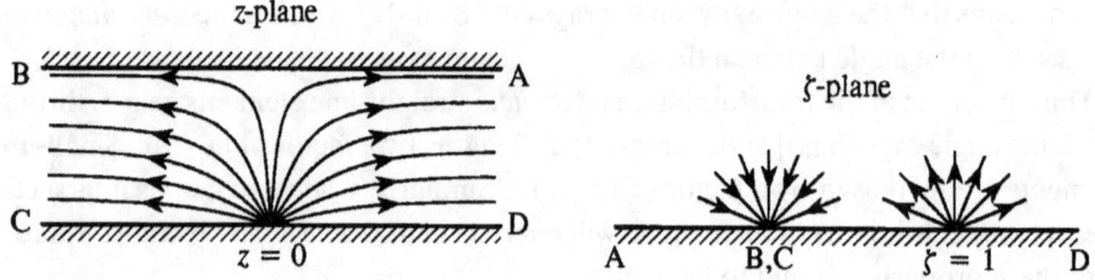

Figure 3.5.2

transformation is conformal. Indeed, if $\zeta = \zeta_0$ is the image of, say, a vortex Γ at $z = z_0$, the complex potential in the neighbourhood of ζ_0 [where $\zeta - \zeta_0 \approx f'(z_0)(z - z_0)$] is

$$W(\zeta) = w(z) = -\frac{i\Gamma}{2\pi} \ln(z - z_0) + \text{terms finite at } z_0$$

$$= -\frac{i\Gamma}{2\pi} \ln\left[\frac{\zeta - \zeta_0}{f'(z_0)}\right] + \text{terms finite at } \zeta_0$$

$$= -\frac{i\Gamma}{2\pi} \ln(\zeta - \zeta_0) + \text{terms finite at } \zeta_0.$$

The vortex in the z plane therefore maps into an equal vortex at the image point in the ζ plane.

EXAMPLE 1. SOURCE FLOW INTO AN INFINITE DUCT Let the source of strength q be at the origin in the z plane, with the real axis as the lower side of the duct ABCD of width h. The fluid flows equally to the left and right in the duct, so that the ends AD and BC are equivalent to sinks each of strength $q/2$. We solve the problem by transforming the interior of the duct onto the upper half of the $\zeta = \xi + i\eta$ plane.

To find the transformation we seek a solution of

$$\frac{\partial^2 \eta}{\partial x^2} + \frac{\partial^2 \eta}{\partial y^2} = 0$$

such that $\eta = 0$ on the walls of the duct and $\eta > 0$ within the duct. The only separable solution satisfying these conditions is $\eta = f(x)\sin(\pi y/h)$, in which case

$$f(x) = Ae^{\pi x/h} + Be^{-\pi x/h}.$$

The upper and lower sides of the duct will map onto $\eta = 0$, and the sink at the end BC will lie on $\eta = 0$ provided $A > 0$ and $B = 0$. Thus,

$$\eta = Ae^{\pi x/h} \sin \frac{\pi y}{h}.$$

This implies that $\xi = Ae^{\pi x/h} \cos \frac{\pi y}{h}$. If $A = 1$ is taken, it follows that the source at $z = 0$ lies at $\zeta = 1$, as indicated in Figure 3.5.2. The required transformation is therefore

$$\zeta = e^{\pi z/h}. \tag{3.5.1}$$

Half of the flux q from the source at $\zeta = 1$ is absorbed by the sink of strength $q/2$ at $\zeta = 0$, and the remainder by an equal sink at $|\zeta| = \infty$. Thus, observing that the total flux from the source flows into an angle π in front of the wall and that the potential of a sink situated at $|\zeta| \to \infty$ is a large negative constant C, we can take for the complex potential

$$w(\zeta) = \frac{q}{\pi} \ln(\zeta - 1) - \frac{q}{2\pi} \ln \zeta + C \equiv \frac{q}{\pi} \ln\left(\zeta^{\frac{1}{2}} - \zeta^{-\frac{1}{2}}\right) + C,$$

$$\therefore \quad w(z) = \frac{q}{\pi} \ln\left(\sinh \frac{\pi z}{2h}\right),$$

where constants have been absorbed into w.

The inverse relation

$$\frac{\pi z}{2h} = \ln\left(e^{\frac{\pi w}{q}} + \sqrt{1 + e^{\frac{2\pi w}{q}}}\right), \quad w = \varphi + i\psi,$$

provides a parametric representation of the streamlines in terms of φ for fixed values of ψ.

Observe that the solution is also applicable to the flows produced by a point source placed midway between two parallel planes at a distance $2h$ apart and by a source at the corner of a semi-infinite rectangular duct.

EXAMPLE 2. FLOW AROUND THE EDGE OF A RIGID HALF-PLANE The transformation $\zeta = i\sqrt{z}$ maps the z plane cut along the negative real axis (i.e., the region $-\pi < \arg z < \pi$) onto the upper ζ plane. The uniform flow in the ζ plane specified by $w = U\zeta$ (for real $U > 0$) in the ξ direction (parallel to the boundary $\eta = 0$) maps into a clockwise flow around the edge of the half-plane (see Figure 3.5.3). In the z plane, with $z = re^{i\theta}$,

$$w = iU\sqrt{z} \equiv -U\sqrt{r} \sin\left(\frac{\theta}{2}\right) + iU\sqrt{r} \cos\left(\frac{\theta}{2}\right), \quad -\pi < \theta < \pi.$$

The polar components of velocity are therefore

$$\mathbf{v} = (v_r, v_\theta) = \left(\frac{\partial \varphi}{\partial r}, \frac{1}{r}\frac{\partial \varphi}{\partial \theta}\right) = \frac{-U}{2\sqrt{r}}\left(\sin\frac{\theta}{2}, \cos\frac{\theta}{2}\right),$$

where $v_\theta = 0$ on the upper and lower surfaces of the half-plane at $\theta = \pm\pi$. The streamlines $\psi = $ constant are the parabolas $\sqrt{r}\cos\left(\frac{\theta}{2}\right) = $ constant, i.e.,

$$y = \pm 2\beta\sqrt{1 - \frac{x}{\beta}}, \quad \text{where } x < \beta = \text{positive constant.}$$

The streamline $\beta = +0$ corresponds to the upper and lower surfaces of the half-plane, which maps into the streamline $\eta = 0$ in the ζ plane. The velocity is infinite like $1/\sqrt{r}$ as $r \to 0$ at the sharp edge.

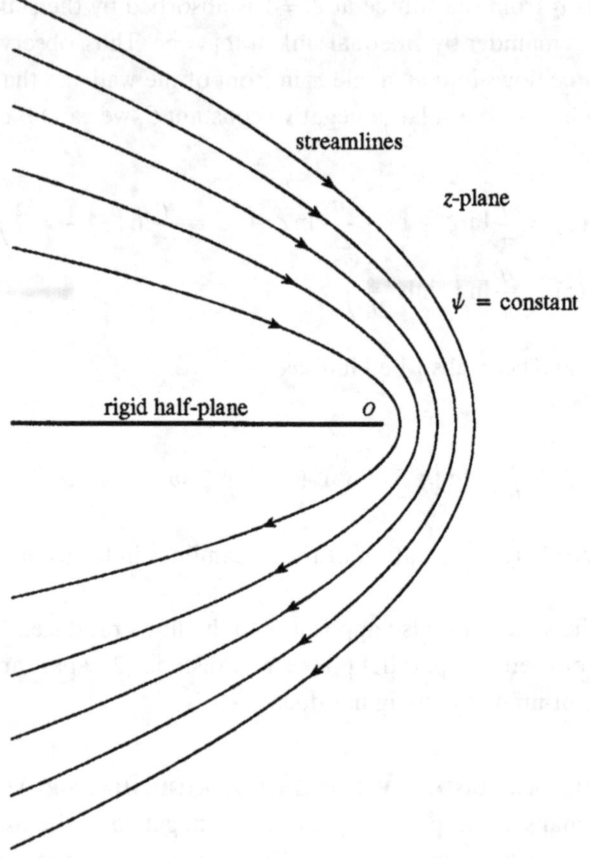

Figure 3.5.3

3.5.2 Equation of motion of a line vortex

The equation of motion of a line vortex of strength Γ at $z = z_o(t)$ in flow governed by the complex potential $w(z)$ is (§3.4.2)

$$\frac{dz_o^*}{dt} = w'(z_o), \quad \text{where} \quad w_o(z) = w(z) + \frac{i\Gamma}{2\pi}\ln(z - z_o). \qquad (3.5.2)$$

Some care is required in evaluating $w_o'(z_o)$ in cases in which $w(z)$ is defined by the conformal transformation of a solution from the ζ plane. In these circumstances, if $\zeta = \zeta(z_o)$ is the image of the vortex in the ζ plane, we can usually write

$$w(z) = -\frac{i\Gamma}{2\pi}\ln\left[\zeta(z) - \zeta(z_o)\right] + F(z), \qquad (3.5.3)$$

where $\zeta(z)$, $F(z)$ are regular functions of z in the neighbourhood of the vortex at $z = z_o$. When $|z - z_o|$ is small,

$$\zeta(z) = \zeta(z_o) + (z - z_o)\zeta'(z_o) + \frac{(z - z_o)^2}{2}\zeta''(z_o) + \cdots +,$$

where the primes denote differentiation with respect to z. Thus, near the vortex,

$$w_o(z) = w(z) + \frac{i\Gamma}{2\pi} \ln(z - z_o)$$

$$= -\frac{i\Gamma}{2\pi} \ln\left[\zeta(z) - \zeta(z_o)\right] + \frac{i\Gamma}{2\pi} \ln(z - z_o) + F(z)$$

$$\approx -\frac{i\Gamma}{2\pi} \ln\left[\zeta'(z_o) + \frac{1}{2}\zeta''(z_o)(z - z_o)\right] + F(z). \tag{3.5.4}$$

Hence, by differentiation and substitution into (3.5.2), the equation of motion becomes

$$\frac{dz_o^*}{dt} = -\frac{i\Gamma\zeta''(z_o)}{4\pi\,\zeta'(z_o)} + F'(z_o). \tag{3.5.5}$$

The real and imaginary parts of this equation supply two nonlinear first-order ordinary differential equations for the position $[x_o(t),\, y_o(t)]$ of the vortex.

3.5.3 Numerical integration of the vortex path equation

It is frequently necessary to integrate Equation (3.5.5) numerically. The time and space variables are first nondimensionalised with respect to convenient time and length scales of the problem. The integration is started from a prescribed point on the trajectory through which the vortex is required to pass.

Let us consider integration by means of a fourth-order *Runge–Kutta* algorithm. Write (3.5.5) in the form

$$\frac{dz_o}{dt} = [f(z_o)]^*, \quad \text{where} \quad f(z_o) = -\frac{i\Gamma\zeta''(z_o)}{4\pi\,\zeta'(z_o)} + F'(z_o),$$

and let h be a suitably small integration time step (that need not be constant). Assume that at $t = t_n$ the vortex is at $z_o(t_n) = z_o^n$. To determine the position z_o^{n+1} at time $t_{n+1} = t_n + h$ we evaluate

$$k_1 = h[f(z_o^n)]^*, \quad k_2 = h[f(z_o^n + k_1/2)]^*, \quad k_3 = h[f(z_o^n + k_2/2)]^*, \quad k_4 = h[f(z_o^n + k_3)]^*,$$

and then we find

$$z_o^{n+1} = z_o^n + \frac{1}{6}(k_1 + 2k_2 + 2k_3 + k_4).$$

EXAMPLE 3. MOTION OF A LINE VORTEX OF ADJACENT TO A RIGID HALF-PLANE
Let the half-plane correspond to the negative real axis $[x < 0,\ y = 0$; Figure 3.5.4(a)]. The transformation $\zeta = i\sqrt{z}$, $-\pi < \arg z < \pi$, maps the flow onto the upper half of the ζ plane [Figure 3.5.4(b)]. Let the vortex at $z_o(t)$ map into a vortex at $\zeta = \zeta_o(t)$. The velocity potential $w(\zeta)$ of the motion in the ζ plane is found by the introduction of an image vortex of strength $-\Gamma$ at $\zeta = \zeta_o^*(t)$ (as in Example 3 of §3.4), in which case

$$w = \frac{-i\Gamma}{2\pi} \ln(\zeta - \zeta_o) + \frac{i\Gamma}{2\pi} \ln(\zeta - \zeta_o^*).$$

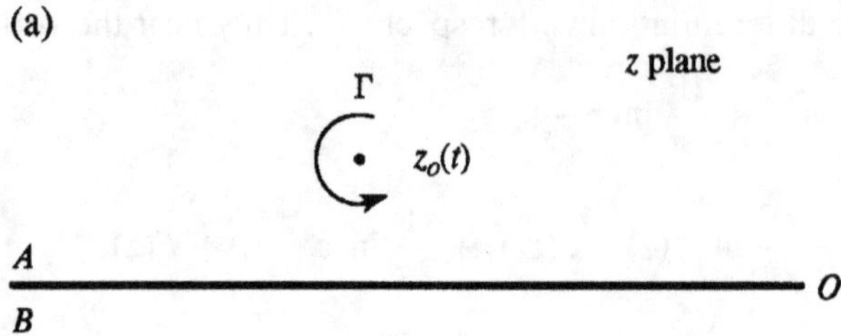

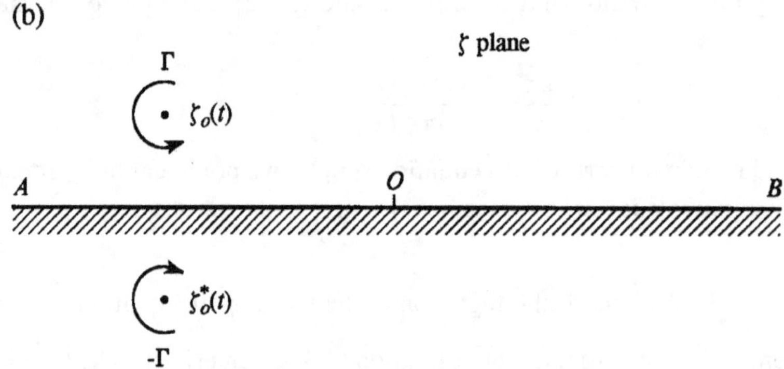

Figure 3.5.4

In the z plane this becomes

$$w(z) = -\frac{i\Gamma}{2\pi}\ln\left[\zeta(z) - \zeta(z_o)\right] + \frac{i\Gamma}{2\pi}\ln\left[\zeta(z) - \zeta^*(z_o)\right],$$

which is of the form (3.5.3). Hence equation of motion (3.5.5) becomes

$$\frac{dx_o}{dt} - i\frac{dy_o}{dt} = \frac{i\Gamma}{8\pi z_o} + \frac{i\Gamma}{4\pi\sqrt{z_o}[\sqrt{z_o} + (\sqrt{z_o})^*]}.$$

This can be integrated in closed form. Let $z_o = re^{i\theta}$; then the real and imaginary parts of the equation are

$$\frac{dx_o}{dt} \equiv \cos\theta\frac{dr}{dt} - r\sin\theta\frac{d\theta}{dt} = \frac{\Gamma}{8\pi r}\left(\sin\theta + \tan\frac{\theta}{2}\right),$$

$$\frac{dy_o}{dt} \equiv \sin\theta\frac{dr}{dt} + r\cos\theta\frac{d\theta}{dt} = -\frac{\Gamma}{8\pi r}(\cos\theta + 1).$$

Therefore

$$\frac{dr}{dt} = -\frac{\Gamma}{8\pi r}\tan\frac{\theta}{2}, \quad \frac{d\theta}{dt} = -\frac{\Gamma}{4\pi r^2}, \tag{3.5.6}$$

i.e.,
$$r\frac{d\theta}{dr} = 2\cot\frac{1}{2}\theta,$$

$$\therefore \quad r = \ell\sec\frac{1}{2}\theta, \quad \ell = \text{constant}. \tag{3.5.7}$$

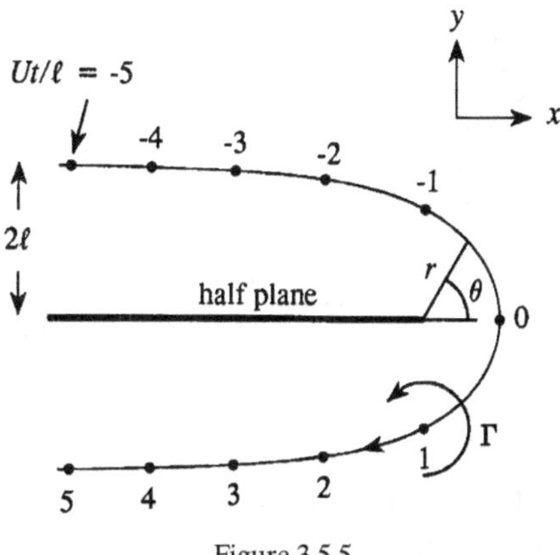

Figure 3.5.5

This is the polar equation of the trajectory plotted in Figure 3.5.5. The constant length ℓ is the distance of closest approach of the vortex to the edge of the half-plane, which occurs at $\theta = 0$. Substituting for r in the second of equations (3.5.6), we find

$$\sec^2\left(\frac{1}{2}\theta\right)\frac{d\theta}{dt} = -\frac{\Gamma}{4\pi\ell^2}, \quad \therefore \quad \theta = 2\tan^{-1}\left(-\frac{\Gamma t}{8\pi\ell^2}\right),$$

where time is measured from the instant at which $\theta = 0$. The dependence of r on t is now obtained by substitution into (3.5.7).

Collecting together these results we have

$$r = \ell\sqrt{1+\left(\frac{Ut}{\ell}\right)^2}, \quad \theta = 2\tan^{-1}\left(-\frac{Ut}{\ell}\right);$$

$$\frac{x_o}{\ell} = \frac{1-(Ut/\ell)^2}{\sqrt{1+(Ut/\ell)^2}}, \quad \frac{y_o}{\ell} = \frac{-2Ut/\ell}{\sqrt{1+(Ut/\ell)^2}}, \quad \text{where } U = \frac{\Gamma}{8\pi\ell}. \quad (3.5.8)$$

When $\Gamma > 0$ the vortex starts above the half-plane at $t = -\infty$ at $x_o = -\infty$, $y_o = 2\ell$ and translates towards the edge, initially at speed U parallel to the plane. It crosses the x axis at $t = 0$ at $x = \ell$ and proceeds along a symmetrical path below the half-plane.

3.6 The Schwarz–Christoffel transformation

The interior of a straight-sided polygon in the z plane can be mapped onto the upper half of the ζ plane, such that the sides of the polygon are transformed into the real ζ axis, by means of the formula

$$\frac{dz}{d\zeta} = K(\zeta-\xi_1)^{\frac{\alpha_1}{\pi}-1}(\zeta-\xi_2)^{\frac{\alpha_2}{\pi}-1}\ldots(\zeta-\xi_n)^{\frac{\alpha_n}{\pi}-1}, \quad (3.6.1)$$

where $\alpha_1, \alpha_2, \ldots, \alpha_n$ are the internal angles of the polygon, $\xi_1, \xi_2, \ldots, \xi_n$ ($\xi_j < \xi_{j+1}$) are the points on the real ζ axis corresponding to the ordered angular points $z_1, z_2, \ldots, z_n$ in the z plane, and K is a complex constant.

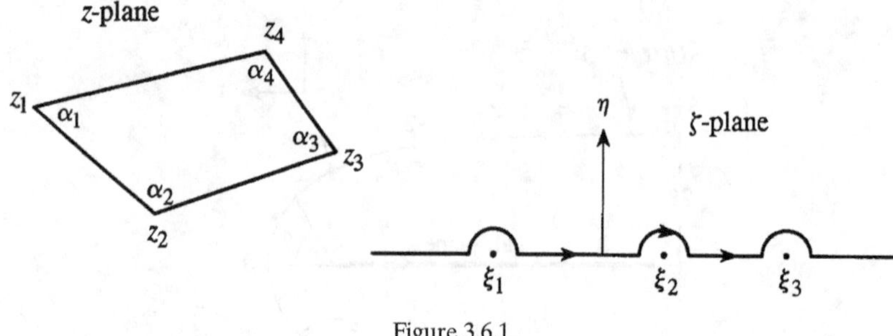

Figure 3.6.1

To understand this result, note that $\arg(dz/d\zeta)$ is constant when ζ is real and does not coincide with any of the ξ_j. Therefore $\arg dz = $ constant when ζ lies on the real axis between any two ξ_j, ξ_{j+1}, so that all points z that correspond to points between ξ_j and ξ_{j+1} on the real axis lie on a straight line in the z plane.

Next consider the change in $\arg(dz/d\zeta)$ as ζ moves along the real axis and passes through the point ξ_2, say, in Figure 3.6.1. The only factor in (3.6.1) that produces a change is $(\zeta - \xi_2)^{\frac{\alpha_2}{\pi}-1}$. Let the path near ξ_2 be deformed into a small semicircle of radius r in the upper half-plane, on which $\zeta - \xi_2 = re^{i\theta}$, so that

$$(\zeta - \xi_2)^{\frac{\alpha_2}{\pi}-1} = r^{\frac{\alpha_2}{\pi}-1}e^{i\left(\frac{\alpha_2}{\pi}-1\right)\theta}.$$

In passing from left to right around, the semi-circle θ decreases from π to 0, and therefore $\arg(dz/d\zeta)$ increases by $\pi - \alpha_2$. Therefore the change in direction in the z plane is equal to $\pi - \alpha_2$, so that the lines in the z plane corresponding to $\xi_1\xi_2$ and $\xi_2\xi_3$ are inclined at angle $\pi - \alpha_2$ to one another, making the internal angle of the polygon at the corner corresponding to ξ_2 equal to α_2.

The perimeter of the polygon maps onto the entire real axis, so that one or more of the points ξ_j must be at $\xi = \infty$. If $\xi_k = \infty$ the factor $(\zeta - \xi_k)^{\frac{\alpha_k}{\pi}-1}$ must be omitted from the right-hand side of (3.6.1).

The integrated form of (3.6.1),

$$z = K\int^{\zeta} (\zeta - \xi_1)^{\frac{\alpha_1}{\pi}-1}(\zeta - \xi_2)^{\frac{\alpha_2}{\pi}-1} \cdots (\zeta - \xi_n)^{\frac{\alpha_n}{\pi}-1}\, d\zeta + L,$$

introduces an additional arbitrary constant L. The constant K determines the orientation and size of the polygon; L determines its position in the z plane. Thus in the transformation there are $2n + 3$ independent parameters:

(1) the n real numbers ξ_1, ξ_2, ..., ξ_n;
(2) $n - 1$ of the interior angles α_1, α_2, ..., α_n, which must satisfy

$$\alpha_1 + \alpha_2 + \cdots + \alpha_n = (n - 2)\pi;$$

(3) four constants from the real and imaginary parts of K, L.

However, the polygon is fully defined in the z plane by the $2n$ real and imaginary parts of the vertices z_1, z_2, ..., z_n. Hence three of the parameters in the transformation may be prescribed arbitrarily.

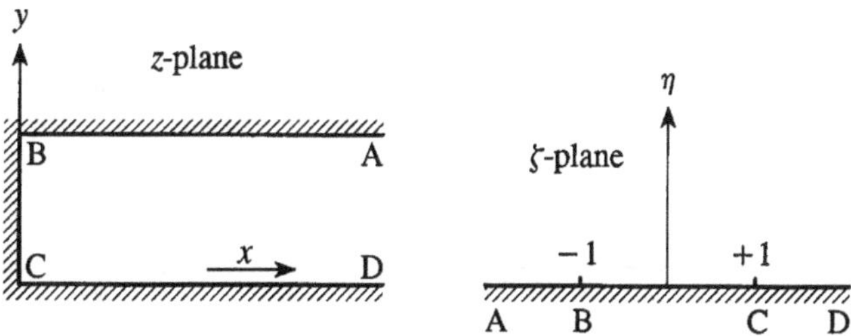

Figure 3.6.2

EXAMPLE 1. THE SEMI-INFINITE RECTANGULAR CHANNEL Consider the rectangular domain ABCD of width h in the z plane (Figure 3.6.2), with respective corners $z = +\infty + ih$, ih, $0, +\infty$. For the purpose of mapping onto the upper half of the ζ plane, the infinite rectangle may be replaced with a *triangle* with vertices AD, B, and C and corresponding interior angles 0, $\frac{\pi}{2}$, and $\frac{\pi}{2}$. If points B and C are taken to map into $\zeta = -1$ and $\zeta = +1$, the transformation may be imagined to 'open out' the triangle onto the real ζ axis so that the point AD maps onto $\zeta = \infty$, with A at $\xi = $ '$-\infty$' and D at '$+\infty$'. There is no explicit reference to AD in transformation formula (3.6.1), which becomes

$$\frac{dz}{d\zeta} = K(\zeta + 1)^{-\frac{1}{2}}(\zeta - 1)^{-\frac{1}{2}} = \frac{K}{(\zeta^2 - 1)^{\frac{1}{2}}},$$

$$\therefore \quad z = K \cosh^{-1} \zeta + L \equiv K \ln\left(\zeta + \sqrt{\zeta^2 - 1}\right) + L. \tag{3.6.2}$$

Because we have already specified the images in the ζ plane of the three vertices of the triangle, the values of K and L cannot be specified arbitrarily. First, the point $\zeta = 1$ is the image of $z = 0$; therefore $L = 0$. To evaluate the transformation at other values of ζ care must be taken to use the same *branch* of $\sqrt{\zeta^2 - 1}$ and of the logarithm $\ln\left(\zeta + \sqrt{\zeta^2 - 1}\right)$. We shall define these to be real and positive on the real axis for $\xi > 1$. When ζ moves to other points in the upper ζ plane it can do so only by passing *above* singularities on the real axis. Thus, passing above the branch point at $\zeta = +1$ onto the real axis in the interval $-1 < \xi < +1$, we must take

$$\sqrt{\zeta^2 - 1} = +i\sqrt{1 - \xi^2} \quad \text{and} \quad \ln\left(\zeta + \sqrt{\zeta^2 - 1}\right) = \ln\left(\xi + i\sqrt{1 - \xi^2}\right),$$

where $\sqrt{1 - \xi^2} > 0$. Thus $\ln\left(\zeta + \sqrt{\zeta^2 - 1}\right) \to i\pi$ when $\zeta \to -1$, and because this corresponds to $z = ih$, it follows that $K = h/\pi$. Hence the transformation becomes

$$z = \frac{h}{\pi} \cosh^{-1} \zeta \equiv \frac{h}{\pi} \ln\left(\zeta + \sqrt{\zeta^2 - 1}\right). \tag{3.6.3}$$

EXAMPLE 2. SOURCE AT AN INTERIOR CORNER OF A SEMI-INFINITE RECTANGULAR CHANNEL Let a source of strength q be placed at the corner $z = 0$ (C) of the rectangular channel of Figure 3.6.2. In the ζ plane the source is at $\zeta = +1$, from which the volume

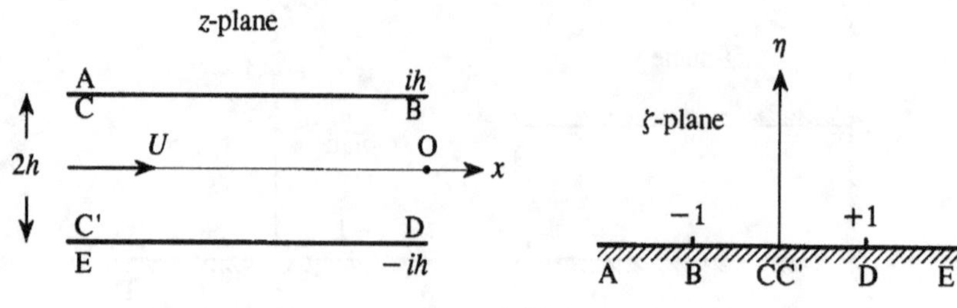

Figure 3.6.3

flux q flows radially into the upper half-plane into a circular sector of angle π, so that $w(\zeta) = \frac{q}{\pi} \ln(\zeta - 1)$. From (3.6.3) we can take $\zeta = \cosh \frac{\pi z}{h}$; therefore

$$w = \frac{q}{\pi} \ln \left(\cosh \frac{\pi z}{h} - 1 \right) \equiv \frac{2q}{\pi} \ln \sinh \frac{\pi z}{2h} + \text{constant}.$$

Apart from a constant factor this is equivalent to the solution in §3.5.1, Example 1, for flow into an infinite duct of width h.

3.6.1 Irrotational flow from an infinite duct

Uniform irrotational flow from the semi-infinite channel ABCC'DE of Figure 3.6.3 is the two-dimensional analogue of the irrotational nozzle flow problem considered in §2.12.6. Let the channel have width $2h$ and take the origin at O at the centre of the exit plane BD, so that the lips B, D correspond respectively to $z = \pm ih$. The points C, C' are respectively at $-\infty \pm ih$ within the duct, and A and E are the corresponding points at infinity just outside the duct.

The fluid region in the z plane is mapped onto the upper ζ plane with the correspondence of points shown in Figure 3.6.3. The points C, C' are regarded as one point that maps into the origin $\zeta = 0$ with interior polygonal angle equal to zero. At B and D the interior angles are both equal to 2π. The points A and E may be assumed to coincide at infinity (where DE and BA meet at interior angle -2π), but do not appear explicitly in the transformation formula:

$$\frac{dz}{d\zeta} = K\zeta^{-1}(\zeta + 1)(\zeta - 1) = K\left(\zeta - \frac{1}{\zeta} \right),$$

$$\therefore \quad z = K\left(\frac{\zeta^2}{2} - \ln \zeta \right) + L.$$

The correspondences $z = \pm ih$ and $\zeta = \mp 1$ yield $K = -2h/\pi$ and $L = h/\pi - ih$, so that

$$z = \frac{2h}{\pi}\left(\ln \zeta - \frac{\zeta^2}{2} \right) + \frac{h}{\pi} - ih. \tag{3.6.4}$$

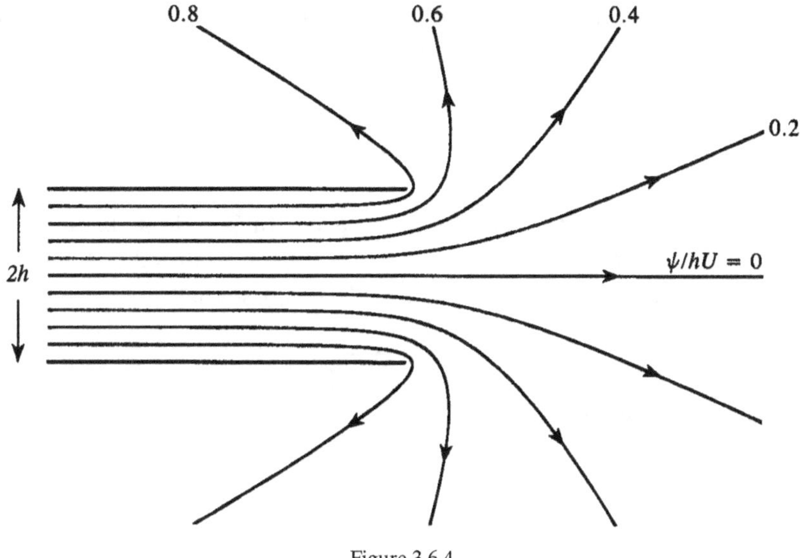

Figure 3.6.4

The uniform flow at speed U from $x = -\infty$ within the duct is generated by a source of strength $q = 2hU$ at CC′. This source flows into an angular interval π in the ζ plane, with complex potential

$$w = \frac{2hU}{\pi} \ln \zeta - ihU. \tag{3.6.5}$$

By symmetry the duct axis (the x axis) maps onto the imaginary axis in the ζ plane, and the constant $-ihU$ has been inserted on the right-hand side of (3.6.5) to make this correspond to the streamline $\psi = 0$.

Equations (3.6.4) and (3.6.5) provide a parametric relation between w and z that determines the flow from the duct. To plot the streamlines, however, it is more convenient to take $w = \varphi + i\psi$ as the parameter by first rewriting (3.6.5) in the form

$$\zeta = i e^{\pi w/2hU}$$

and then substituting into (3.6.4) to obtain

$$z = \frac{h}{\pi} + \frac{w}{U} + \frac{h}{\pi} e^{\pi w/hU},$$

i.e.,

$$\frac{x}{h} = \frac{1}{\pi} + \frac{\varphi}{hU} + \frac{1}{\pi} \cos\left(\frac{\pi\psi}{hU}\right) e^{\pi\varphi/hU},$$

$$\tag{3.6.6}$$

$$\frac{y}{h} = \frac{\psi}{hU} + \frac{1}{\pi} \sin\left(\frac{\pi\psi}{hU}\right) e^{\pi\varphi/hU}.$$

The streamlines may now be plotted by allowing φ/hU to vary over the range $(-\infty, \infty)$ for fixed values of ψ in the range $-hU < \psi < hU$, as illustrated in Figure 3.6.4.

EXAMPLE 3. SUCTION FORCE ON DUCT LIP Suppose p_o is the far-field free-space pressure for the duct flow of Figures 3.6.3 and 3.6.4. Upstream within the duct, Bernoulli's equation gives

$$p + \frac{1}{2}\rho_o U^2 = p_o.$$

Let F denote the suction force per unit span on each of the duct lips. The rate of change of mean flow momentum per unit span is $-2\rho_o h U^2$ and is produced by the net suction force $-2F$ plus a net pressure force equal to $2h(p - p_o)$, so that

$$-2F + 2h(p - p_o) = -2\rho_o h U^2.$$

Using Bernoulli's equation to evaluate $p - p_o$, we find

$$F \equiv \text{suction force per unit span} = \frac{1}{2}\rho_o h U^2.$$

3.6.2 Irrotational flow through a wall aperture

Fluid escaping from a large vessel through a wall aperture usually emerges in the form of a jet, so that the flow cannot be irrotational. The two-dimensional version of this problem for an ideal fluid is discussed in §3.7, but we shall first consider the case in which the flow is entirely irrotational. The problem is not completely devoid of practical interest because, for example, it plays an important part in the approximate treatment of the *linear theory* of the diffraction of sound by a perforated wall.

Consider a two-dimensional aperture of width $2h$ that occupies the interval $-h < x < h$ of a thin, rigid wall coinciding with the real axis in the z plane. In Figure 3.6.5 the edges of the aperture are at B and E where $z = \mp h$. Let the flow be from the upper to the lower halves of the z plane, and let U denote the mean normal velocity (in the negative y direction) in the plane of the aperture, so that the net volume flux through the aperture is $2hU$ per unit span. We find the velocity potential of the flow by mapping the fluid region in the z plane (i.e., the z plane 'cut' along the sections $-\infty < x < -h$ and $h < x < +\infty$ of the real axis) onto the upper ζ plane, with the correspondence of points indicated in Figure 3.6.5(c).

To obtain the transformation consider first the simpler problem of transforming the interior of the 'diamond-shaped' domain in Figure 3.6.5(b) onto the upper ζ plane. We map the point CD onto $\zeta = 0$ and B and E onto $\zeta = \mp 1$; the point AF is mapped onto the point at infinity, and this vertex does not appear explicitly in transformation formula (3.6.1). The points AF and CD in Figure 3.6.5(b) are imagined to recede to infinity, so that the interior polygonal angles BAFE and BCDE ultimately vanish. The aperture geometry of Figure 3.6.5(a) is recovered by rotation of the rays BA, BC, EF, and ED onto the real axis in the directions indicated in the figure. After rotation the interior angles BAFE and BCDE are both equal to $-\pi$; the angles at B and E are equal to 2π. The 'polygon' has four sides, and the sum of the interior angles is just equal to 2π, as required.

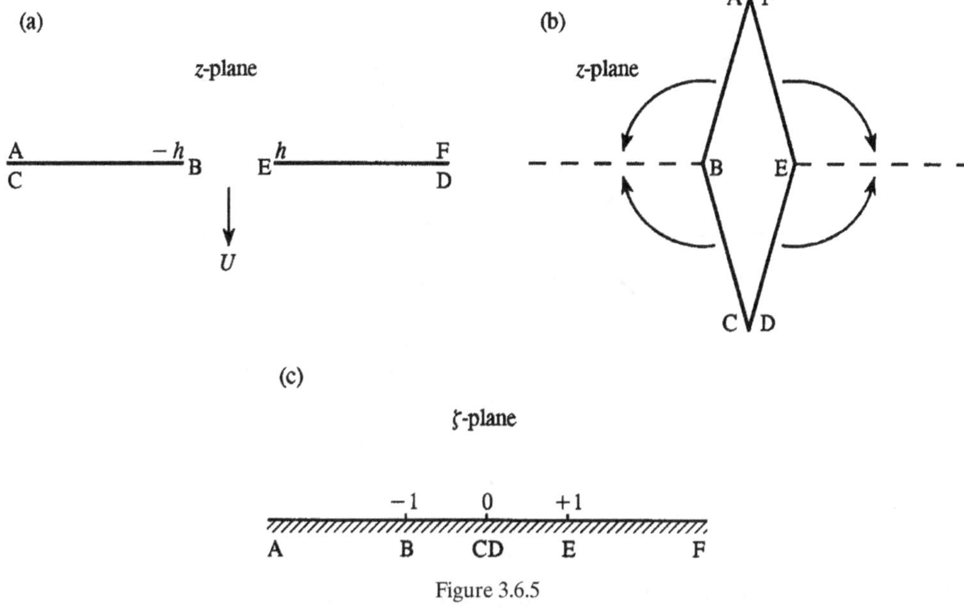

Figure 3.6.5

We therefore accomplish the mapping from Figure 3.6.5(a) to 3.6.5(c) by writing

$$\frac{dz}{d\zeta} = K\zeta^{-2}(\zeta + 1)(\zeta - 1), \quad \therefore \quad z = K\left(\zeta + \frac{1}{\zeta}\right) + L.$$

The conditions that $z = \pm h$ at $\zeta = \pm 1$ yield $K = h/2$, $L = 0$, so that

$$z = \frac{h}{2}\left(\zeta + \frac{1}{\zeta}\right), \quad \text{or} \quad \zeta = \frac{z}{h} + \sqrt{\frac{z^2}{h^2} - 1}. \tag{3.6.7}$$

The branch cuts for $\sqrt{z^2/h^2 - 1}$ are taken to coincide with the solid sections of the wall, such that $\sqrt{z^2/h^2 - 1} \sim \pm z/h$ respectively as $|z| \to \infty$ in the upper and lower half-planes. Thus, as $|z| \to \infty$ in the lower half-plane $\zeta \sim h/2z \to 0$.

The volume flux $2hU$ through the aperture flows into a sink at $|z| = \infty$ in the lower z plane. The complex potential in the ζ plane therefore corresponds to a sink of strength $2hU$ at $\zeta = 0$. Because this sink flow has angular width π, we must have

$$w = -\frac{2hU}{\pi}\ln\zeta = -\frac{2hU}{\pi}\ln\left(\frac{z}{h} + \sqrt{\frac{z^2}{h^2} - 1}\right). \tag{3.6.8}$$

To plot the streamlines we write $\zeta = e^{-\pi w/2hU}$; then (3.6.7) gives

$$z \equiv x + iy = h\cosh\left(\frac{\pi w}{2hU}\right) \equiv h\cosh\left(\frac{\pi\varphi}{2hU}\right)\cos\left(\frac{\pi\psi}{2hU}\right) + ih\sinh\left(\frac{\pi\varphi}{2hU}\right)\sin\left(\frac{\pi\psi}{2hU}\right).$$

Therefore the streamlines are the family of hyperbolas

$$\frac{x^2}{\cos^2\left(\pi\psi/2hU\right)} - \frac{y^2}{\sin^2\left(\pi\psi/2hU\right)} = h^2, \quad -2hU < \psi < 0,$$

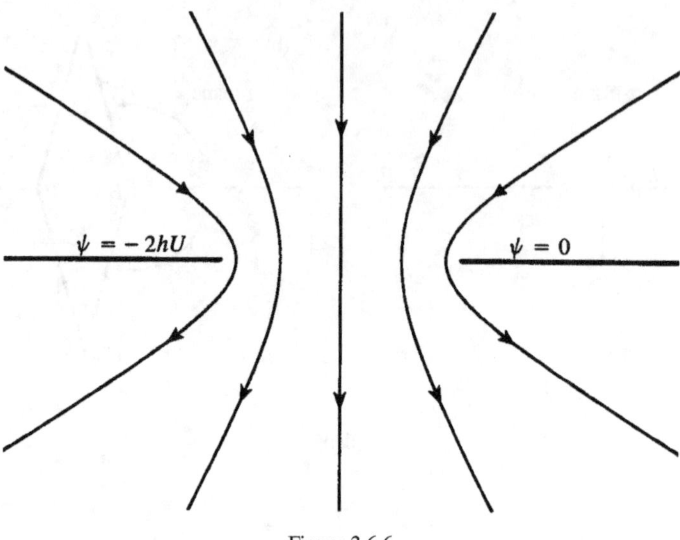

Figure 3.6.6

where $\psi = -2hU$ on the wall ABC and $\psi = 0$ on DEF (see Figure 3.6.6). However, the streamline pattern can be interpreted differently: Solution (3.6.8) also represents the potential of flow through a 'hyperbolic' aperture whose walls correspond to two streamlines $\psi_1 = \text{constant}$ and $\psi_2 = \text{constant}$, where $-2hU \le \psi_1 < \psi_2 \le 0$.

Near the edges B and E

$$w(z) \sim -2ihU + \frac{2i\sqrt{2hU}}{\pi}\sqrt{z+h} \quad \text{and} \quad -\frac{2\sqrt{2hU}}{\pi}\sqrt{z-h}.$$

Therefore, by (3.3.7), each edge experiences a suction force equal to $2\rho_o hU^2/\pi$ per unit span.

3.7 Free-streamline theory

The axisymmetric nozzle flow of Figure 2.12.7(a) and the analogous jet problems of Figure 2.12.9 involve curvilinear vortex sheets separating the mean flow in the jet from nominally stationary fluid outside the jet. For the cylindrical jet flow shown in Figure 2.12.7(a) the boundary of the jet is defined trivially by extension of the duct. For Borda's mouthpiece and the aperture flow problems of Figure 2.12.9, however, the 'free-streamline problem' of determining the position and shape of the vortex sheet boundary forms an integral part of the solution of the equations of motion for the whole flow. Many two-dimensional problems of this kind can be treated by the method of conformal transformations for an ideal fluid. The basic approach is outlined in this section.

3.7.1 Coanda edge flow

The 'Coanda effect' is a phenomenon of common occurrence that involves the *failure* of water flowing under gravity from the spout of a kettle, for example, to form a jet.

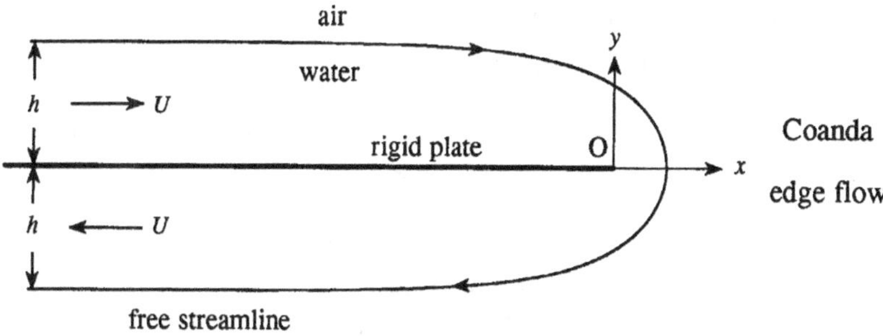

Figure 3.7.1

Instead, the flow appears to 'cling' to the surface and passes around the lip of the spout in much the manner of flow emerging irrotationally. A gross simplification of a flow of this kind is illustrated in Figure 3.7.1 (Keller 1957). A thin layer of water flows two dimensionally and symmetrically around the edge O of the half-plane, $x < 0$, $y = 0$. Far from the edge above and below the half-plane, the water layer has thickness h and the motion is uniform at speed U.

If gravity is assumed to act in the $-y$ direction, Bernoulli's equation for steady flow is

$$\frac{p}{\rho_o} + \frac{v^2}{2} + gy = \text{constant.}$$

In a first approximation, free-streamline theory neglects the influence of gravity. Because y varies over distances of the order of h within the water, this is expected to be valid provided the flow is rapid enough for the effects of gravity to be small, i.e., provided the 'Froude number' $U^2/gh \gg 1$. Similarly, the Reynolds number should be large so that viscosity is negligible except in thin surface boundary layers. Pressure fluctuations in the ambient air can be ignored because the mass density of air ρ', say, is negligible compared with the density ρ_o of the water; this need not imply that the air is at rest, but it does mean that the speed v of the water is constant on the free streamline. On the other hand, in cases in which the density change across a free streamline is small (so that $\rho' \sim \rho_o$), for example, for gas jets in air or liquid layers flowing in liquid, the 'vortex sheet' interface is unstable and predictions of free-streamline theory become unreliable or of only qualitative relevance (§4.1, Example 5).

In summary, the main requirements of the theory are

$$\frac{U^2}{gh} \gg 1, \quad \text{and} \quad \frac{\rho'}{\rho_o} \ll 1, \tag{3.7.1}$$

and the fundamental objective is the determination of a velocity potential φ such that

(1) $\partial\varphi/\partial x_n = 0$ on fixed boundaries of the flow,
(2) $|\nabla\varphi| = \text{constant}$ on the free boundaries.

The dynamical condition that ambient pressure changes can be ignored determines the shape of the free streamlines by means of Bernoulli's equation, which then implies that the flow speed just inside the free streamline is constant (equal to U in

Figure 3.7.1). This is an essential ingredient of the Helmholtz–Kirchhoff–Joukowski method of solution, which in its simplest form involves the following steps:

(1) Find a representation (i.e. transformation) that expresses the complex potential w of the water flow in terms of a complex valued parameter t.
(2) Find a representation of

$$U\frac{dz}{dw} \quad \text{in terms of } t.$$

(3) Solve for the details of the flow by manipulation of the results of (1) and (2).

The complex parameter t is generally required to lie in the upper half-plane, and this typically permits steps (1) and (2) to be performed by means of the Schwarz–Christoffel transformation.

To effect the transformations (1) and (2) it is first necessary to examine the w plane and the plane of the *hodograph* variable

$$\Omega = \ln\left(U\frac{dz}{dw}\right) = \ln\left(\frac{U}{q}\right) + i\theta, \tag{3.7.2}$$

where q (> 0) and θ are defined in terms of the complex velocity by

$$w'(z) = u - iv = qe^{-i\theta}. \tag{3.7.3}$$

Thus q is the magnitude of the flow velocity and θ is the direction of flow measured from the positive x direction. On solid boundaries θ must be in the tangential direction to the surface; on free streamlines $q = \text{constant}$.

Introduce the labelling of the inner and outer boundaries of the flow as in Figure 3.7.2(a). The total volume flux around edge B is hU per unit span. We can therefore set $\psi = 0$ on the surface streamline ABC and $\psi = hU$ on the free streamline A′B′C′. As one passes along these streamlines in the clockwise direction, the potential φ varies monotonically over the interval $(-\infty, +\infty)$. The flow therefore corresponds to the region within the infinite strip in the w plane bounded by $0 < \psi < hU$ [Figure 3.7.2(b)].

We find the relevant region in the Ω plane [Figure 3.7.2(c)] by first noting that $q = U$ at all points on the free streamline, and that $\theta = 0$, $-\pi$ respectively on the upper and lower surfaces of the half-plane. Similarly $\theta = 0$, $-\frac{\pi}{2}$, $-\pi$ at A′, B′, and C′, and $q \to +\infty$ at B, in particular as the edge is approached from the upper and lower sides at B⁻ and B⁺. We can therefore construct the following table of correspondences that defines the semi-infinite rectangular region of Figure 3.7.2(c).

z:	AA′	B′	CC′	B⁻	B⁺
Ω:	0	$-i\pi/2$	$-i\pi$	$-\infty + i0$	$-\infty - i\pi$

The Schwarz–Christoffel transformation formula (3.6.1) may now be used to map the strip of Figure 3.7.2(b) and the rectangle of Figure 3.7.2(c) onto the upper t plane of

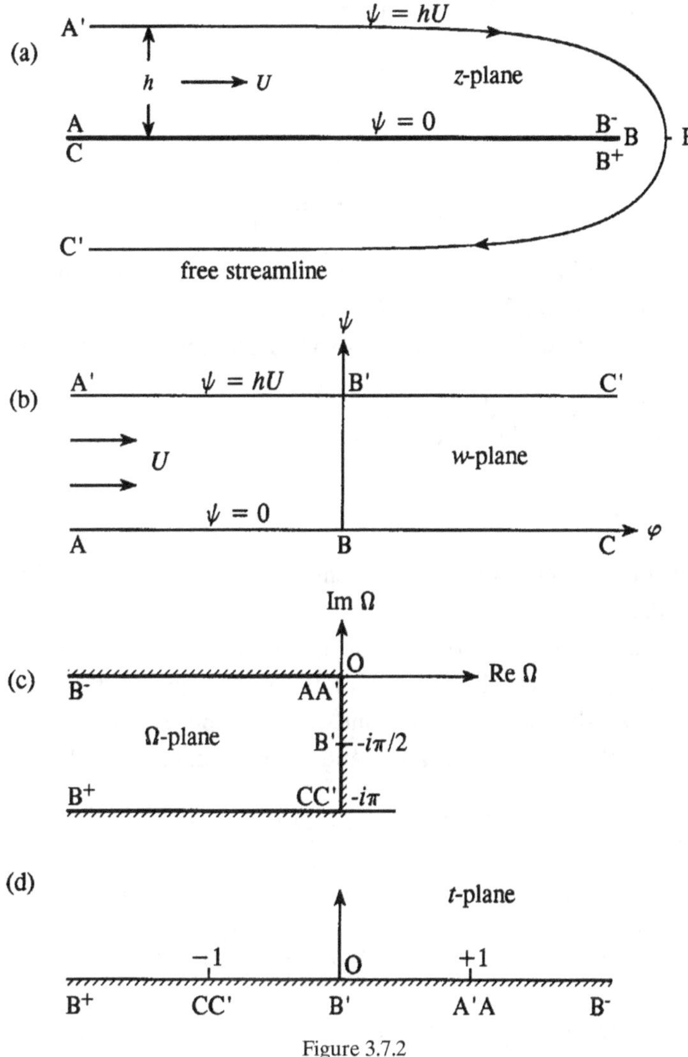

Figure 3.7.2

Figure 3.7.2(d) with the indicated correspondence of common points. Thus, for the w–t transformation we have

$$\frac{dw}{dt} = \frac{K}{t^2 - 1}, \quad \therefore \quad w = \frac{K}{2}\left[\ln(t-1) - \ln(t+1)\right] + L.$$

The conditions $w = 0$ as $\mathrm{Re}\,t \to \pm\infty$ and $w = ihU$ at $t = 0$ yield $K = 2hU/\pi$, $L = 0$. Hence the required parametric representation becomes

$$w = \frac{hU}{\pi}\left[\ln(t-1) - \ln(t+1)\right] \quad \text{and} \quad \frac{dw}{dt} = \frac{2hU}{\pi(t^2-1)}. \tag{3.7.4}$$

Next, for the Ω–t transformation,

$$\frac{d\Omega}{dt} = \frac{K}{\sqrt{t^2-1}}, \quad \therefore \quad \Omega = K\ln\left(t + \sqrt{t^2-1}\right) + L,$$

where we now find $K = -1,\ L = 0$, so that

$$\Omega = -\ln\left(t + \sqrt{t^2 - 1}\right).$$

Hence, using definition (3.7.2) we have

$$U\frac{dz}{dw} = \frac{1}{t + \sqrt{t^2 - 1}}. \tag{3.7.5}$$

Equations (3.7.4) and (3.7.5) are the desired parametric representations from which all details of the irrotational flow can be found.

Combining the second of Equations (3.7.4) and Equation (3.7.5) we find

$$\frac{dz}{dt} = \frac{2h}{\pi(t^2 - 1)[t + \sqrt{t^2 - 1}]},$$

$$\therefore\quad z = \frac{h}{\pi}\left[\ln(t - 1) + \ln(t + 1) - 2\ln(t + \sqrt{t^2 - 1}) + 2\ln 2\right], \tag{3.7.6}$$

where the constant of integration has been chosen to ensure that $z = 0$ at the edge of the half-plane where $t \to \infty$.

This formula can be used to obtain the equation of the free streamline, which corresponds to the interval $-1 < t < +1$ of the real t axis. In this interval set $t = \sin\vartheta$ $\left(-\frac{\pi}{2} < \vartheta < \frac{\pi}{2}\right)$; then $\vartheta > 0$ corresponds to the upper branch ($y > 0$) of the free streamline and $\vartheta < 0$ to the lower branch, and on the streamline (3.7.6) gives

$$z = x + iy = \frac{2h}{\pi}\left[\ln(2\cos\vartheta) + i\vartheta\right], \quad -\frac{\pi}{2} < \vartheta < \frac{\pi}{2},$$

i.e.,

$$x = \frac{2h}{\pi}\ln(2\cos\vartheta), \quad y = \frac{2h\vartheta}{\pi} \quad\text{or}\quad x = \frac{2h}{\pi}\ln\left[2\cos\left(\frac{\pi y}{2h}\right)\right]. \tag{3.7.7}$$

Thus the width of the stream of fluid at its narrowest point (where the free streamline crosses the x axis) is $2h\ln 2/\pi$, where the mean flow velocity is increased to $\pi U/2\ln 2 \approx 2.27U$. The velocity at the edge $z = 0$ is infinite, however. This singularity is necessary because suction must be applied to the fluid at the edge to 'pull' the stream around the half-plane. Because the rate of *decrease* of x momentum is just equal to $2\rho_o hU^2$ per unit span, this must also be the magnitude of the suction force. We can verify the consistency of our solution by evaluating the suction force explicitly from the behaviour of $w(z)$ close to the edge.

Near $z = 0$ (where $t \to \infty$) Equations (3.7.4) and (3.7.6) imply that

$$w \sim -\frac{2hU}{\pi t} \quad\text{and}\quad z \sim -\frac{h}{2\pi t^2},$$

$$\therefore\quad w \sim 2iU\sqrt{\frac{2hz}{\pi}} \quad\text{for}\quad |z| \ll h.$$

According to (3.3.7) the corresponding suction force is precisely $2\rho_o hU^2$, in agreement with our conclusion above.

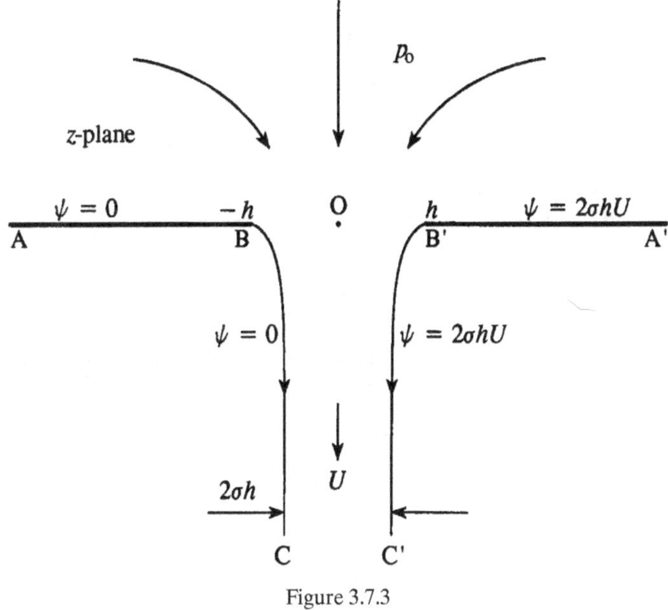

Figure 3.7.3

3.7.2 Mapping from the w plane to the t plane

In solving most free-streamline problems it is usually sufficient to have an expression for dw/dt in terms of t. This can often be found by inspection from the following observation: The function $w(t)$ is regular in the upper t plane with isolated singularities on the real t axis, and $\psi = \mathrm{Im}\, w$ is constant on all sections of the real t axis between the singularities. Therefore $w(t)$ may be regarded as the complex potential of a flow in the t plane produced by sources, sinks, etc., at the singular points. Thus the flow in Figure 3.7.2(b) is produced by a source of strength hU at AA$'$ and a sink of equal strength at CC$'$. Mapping (3.7.4) can therefore be found immediately to within an additive arbitrary constant.

3.7.3 Separated flow through an aperture

Consider the problem of flow through a two-dimensional aperture in a thin wall, previously discussed for entirely irrotational flow in §3.6.2. Assume the fluid flows through an aperture of width $2h$ from the upper to the lower halves of the z plane, as in Figure 3.7.3, and let p_o denote the mean excess pressure drop across the wall. The flow separates at the edges B and B$'$ forming a jet, whose ultimate uniform velocity U is determined by Bernoulli's equation to be $U = \sqrt{2p_o/\rho_o}$. This is also the velocity on the free streamlines BC, B$'$C$'$, where the excess pressure vanishes. The ultimate width of the jet is $2\sigma h$, where σ is the *contraction ratio*.

The total volume flux is $2\sigma hU$, and we can therefore set $\psi = 0$ on the streamline ABC and $\psi = 2\sigma hU$ on A$'$B$'$C$'$. The potential φ increases over the range $(-\infty, +\infty)$ along these streamlines, so that the flow corresponds to the region within the infinite strip $0 < \psi < 2\sigma hU$ in the w plane [Figure 3.7.4(a)].

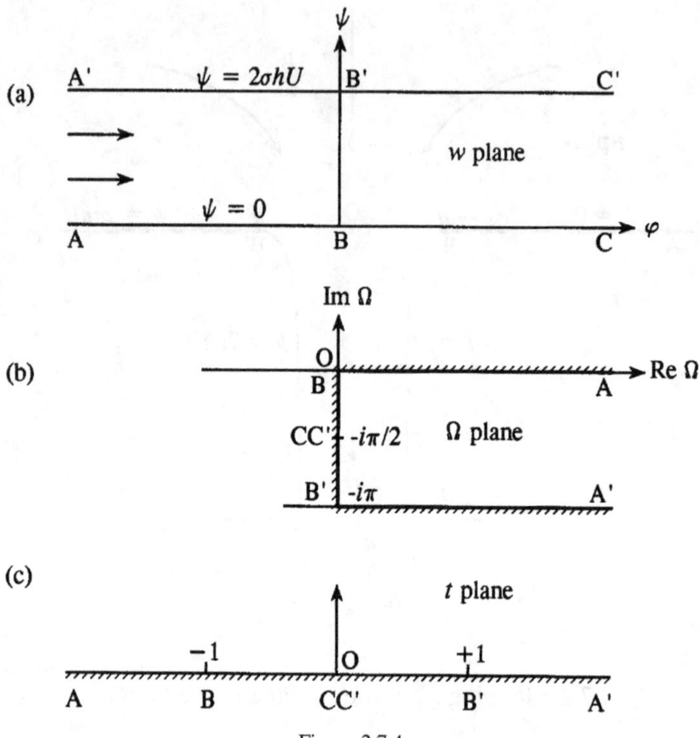

Figure 3.7.4

Inspection of Figure 3.7.3 reveals that the region of variation in the plane of $\Omega = \ln(U\,dz/dw)$ is the semi-infinite rectangle of Figure 3.7.4(b) determined by the following table of correspondences:

z:	A	B	CC'	B'	A'
Ω:	$\infty + i0$	0	$-i\pi/2$	$-i\pi$	$\infty - i\pi$

The Schwarz–Christoffel transformation (3.6.1) is now applied to the map strip of Figure 3.7.4(a) and the rectangle of Figure 3.7.4(b) onto the upper t plane of Figure 3.7.4(c) with the indicated correspondence of common points.

For the w–t transformation,

$$\frac{dw}{dt} = \frac{K}{t} \quad \text{and} \quad w = K\ln t + L.$$

Therefore, because $w = 0$ at $t = -1$ and $w = 2i\sigma hU$ at $t = 1$, we have

$$w = -\frac{2\sigma hU}{\pi}\ln t + 2i\sigma hU \quad \text{and} \quad \frac{dw}{dt} = -\frac{2\sigma hU}{\pi t}. \tag{3.7.8}$$

Similarly,

$$\frac{d\Omega}{dt} = \frac{K}{\sqrt{t^2 - 1}}, \quad \therefore \quad \Omega = K\ln\left(t + \sqrt{t^2 - 1}\right) + L,$$

where $\Omega = -i\pi$ at $t = 1$ and $\Omega = 0$ at $t = -1$, and therefore

$$\Omega = \ln\left(t + \sqrt{t^2 - 1}\right) - i\pi,$$

and

$$U\frac{dz}{dw} = -\left(t + \sqrt{t^2 - 1}\right). \tag{3.7.9}$$

Equations (3.7.8) and (3.7.9) determine the flow.

THE CONTRACTION RATIO From the second of Equations (3.7.8) and Equation (3.7.9),

$$\frac{dz}{dt} = \frac{2\sigma h}{\pi}\left(1 + \frac{\sqrt{t^2 - 1}}{t}\right). \tag{3.7.10}$$

Now the free streamline BC corresponds to the interval $-1 < t < 0$ of the real t axis. Therefore, on this streamline

$$z + h = \frac{2\sigma h}{\pi}\int_{-1}^{t}\left(1 + i\frac{\sqrt{1 - \tau^2}}{\tau}\right)d\tau.$$

At $t = 0$ we have $x = -\sigma h$; therefore, taking the real part of the integral, we find

$$-\sigma h + h = \frac{2\sigma h}{\pi}\int_{-1}^{0} d\tau, \quad \therefore \quad \sigma = \frac{\pi}{2 + \pi} \approx 0.611. \tag{3.7.11}$$

This is larger than the approximation $\sigma = \frac{1}{2}$ we obtain by assuming that the surface pressure on the upper wall near the aperture is equal to p_o (§2.12.7), but marginally smaller than the experimental value $\sigma \sim 0.63$.

EXAMPLE 1. JET FLOW FROM A FUNNEL Consider the general case of a jet exhausting from a two-dimensional funnel formed by two plates inclined at angle 2α. Let the plates be arranged symmetrically with respect to the x axis as in Figure 3.7.5, with the jet flowing to the right from a gap of width $2h$ between the plates. The asymptotic jet width and velocity are respectively $2\sigma h$ and U, where σ is the contraction ratio.

It is sufficient to determine the motion in the upper half-plane bounded by the streamlines AB (the real axis), on which we take $\psi = 0$, and DCB', where $\psi = \sigma hU$. The w, Ω, and t planes are shown in Figure 3.7.6, and we have the relations

$$w = -\frac{\sigma hU}{\pi}\ln t + i\sigma hU,$$

$$\Omega = \frac{2\alpha}{\pi}\ln\left(\sqrt{t} + \sqrt{t - 1}\right) - i\alpha,$$

$$\frac{dz}{dt} = -\frac{\sigma he^{-i\alpha}}{\pi t}\left(\sqrt{t} + \sqrt{t - 1}\right)^{\frac{2\alpha}{\pi}}.$$

The free streamline CB' corresponds to the interval $0 < t < 1$ of the real t axis. By setting $t = \cos^2\vartheta$ $(0 < \vartheta < \pi/2)$ in this interval, we can cast the equations defining the jet boundary in the form

$$\frac{dx}{d\vartheta} = \frac{2\sigma h\tan\vartheta}{\pi}\cos\left[\alpha\left(\frac{2\vartheta}{\pi} - 1\right)\right], \quad \frac{dy}{d\vartheta} = \frac{2\sigma h\tan\vartheta}{\pi}\sin\left[\alpha\left(\frac{2\vartheta}{\pi} - 1\right)\right].$$

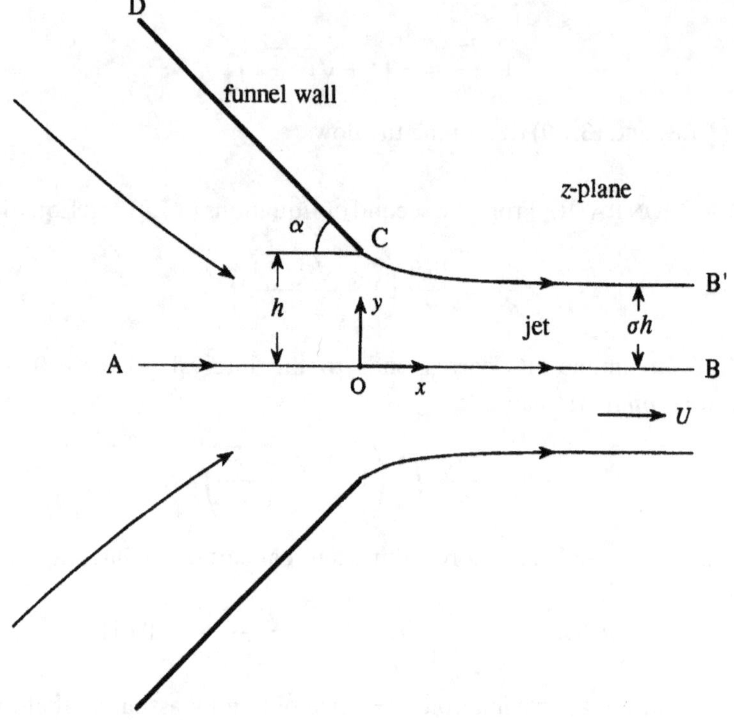

Figure 3.7.5

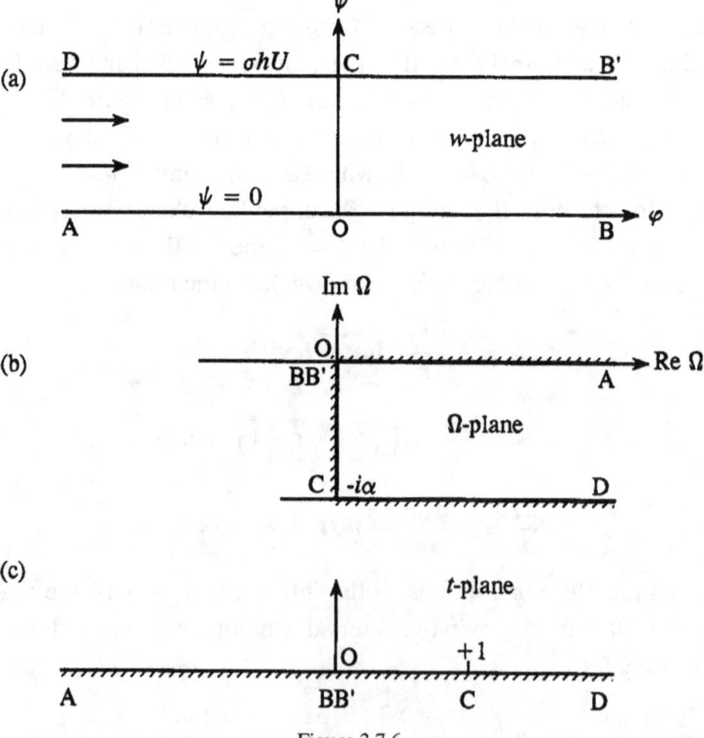

Figure 3.7.6

Table 3.7.1 *After Birkhoff & Zarantonello 1957*

α	0	$\pi/8$	$\pi/4$	$3\pi/8$	$\pi/2$	$5\pi/8$	$3\pi/4$	$7\pi/8$	π
σ, theory	1	0.853	0.747	0.669	0.611	0.568	0.537	0.515	0.500
σ, experiment	1	0.882	0.753	0.684	0.632	0.606	0.577	0.546	0.541

At the edge C, $x = 0$, $y = h$, and $\vartheta = 0$; at B$'$, $\vartheta \to \frac{\pi}{2}$ and $y = \sigma h$. Hence, the second of these equations yields

$$\sigma = 1 \left/ \left\{ 1 - \frac{2}{\pi} \int_0^{\frac{\pi}{2}} \tan \vartheta \, \sin\left[\alpha \left(\frac{2\vartheta}{\pi} - 1 \right) \right] d\vartheta \right\} \right. . \tag{3.7.12}$$

Table 3.7.1 compares values of the contraction coefficient calculated from this formula with experiment for an *axisymmetric* funnel (i.e. a *circular* jet).

3.7.4 The wake of a flat plate

Separated flow of an ideal fluid normally incident at speed U upon a finite two-dimensional plate (Figure 3.7.7) can also be treated by free-streamline theory and was originally examined to throw light on the question of drag produced by a wake. A 'dead-water' wake is formed in which the fluid is at rest and the pressure uniform. In the simplest approximation the theory assumes that the wake pressure is equal to the pressure in the flow at large distances from the plate, so that the velocity along the free streamlines is also equal to U.

However, the pressure is high on the front of the plate, where the flow stagnates, and this produces a drag (per unit span) predicted by free-streamline theory to be

$$\text{drag} = \frac{\rho_0 U^2 \pi \ell}{\pi + 4},$$

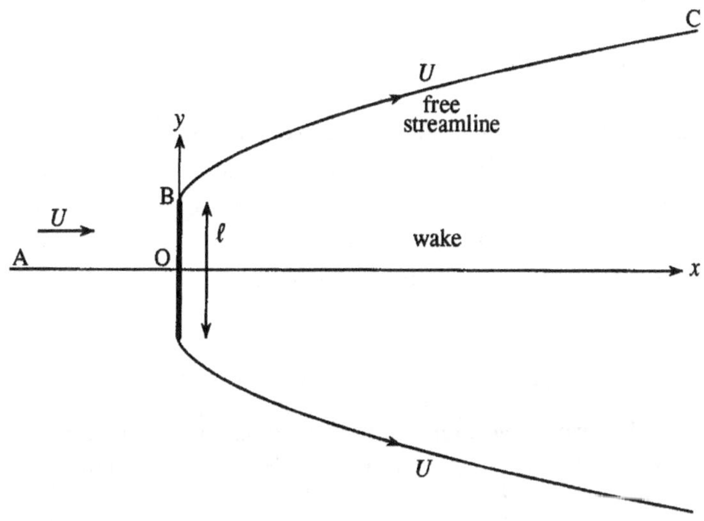

Figure 3.7.7

where ℓ is the width of the plate. It is convenient to express results of this kind in terms of a 'drag coefficient':

$$C_D = \frac{\text{drag}}{\frac{1}{2}\rho_o U^2 \times \text{frontal area}} = \frac{\text{drag}}{\frac{1}{2}\rho_o U^2 \times \ell} \equiv \frac{2\pi}{\pi + 4} \approx 0.88. \qquad (3.7.13)$$

Measured values of the drag coefficient in a real high-Reynolds-number flow for elongated plates of this type are typically $C_D \sim 2$. Of course, in a real fluid the wake is unstable, and any free streamlines would rapidly break down into streams of eddies; the motion within the wake region becomes turbulent, and the pressure just behind the plate is smaller than p_o. The ideal theory predicts that the width of the wake bounded by the free streamlines increases indefinitely with distance downstream (as the square root of the distance); this unrealistic behaviour gives a fair representation of the mean shape of the wake close to the plate, but is probably responsible for the failure to predict a reasonable estimate for the drag. However the 'shape' of the pressure distribution on the front of the plate is well predicted by the present theory.

More reasonable predictions are given by free-streamline theory for the drag on a two-dimensional plate inserted normally into a uniform jet of *finite* width. Figure 3.7.8 shows a 'blade' tip penetrating a distance d into a jet of speed U and initially of uniform width h. The flow bifurcates on either side of the stagnation streamline that meets the blade at C, forming a downward moving jet of width h_1 with downward momentum flux $\rho_o U^2 h_1$; the upper part of the jet leaves the blade at D, forming a jet of width $h_2 = h - h_1$ bounded on its lower side by the free streamline DE. This jet moves in a direction making an angle χ relative to the undisturbed jet with momentum flux $\rho_o U^2 h_2$.

Conservation of vertical momentum implies that $h_1 = h_2 \sin \chi$ and that

$$h_1 = \frac{h \sin \chi}{1 + \sin \chi}, \quad h_2 = \frac{h}{1 + \sin \chi}.$$

The drag force on the plate (per unit span) can therefore be expressed in terms of the deflection angle χ alone:

$$\text{drag} = \rho_o U^2 (h - h_2 \cos \chi) \equiv \rho_o U^2 h \left(1 - \frac{\cos \chi}{1 + \sin \chi}\right). \qquad (3.7.14)$$

All of the horizontal momentum of the jet is destroyed by the blade when $\chi \to \frac{\pi}{2}$, when the drag becomes equal to $\rho_o U^2 h$. This limit is attained only after the blade tip has sliced through the jet a distance $\sim h$ (the jet width) beyond the opposite side of the jet ($d > 2h$); then $h_1 = h_2$ and the drag coefficient $C_D \equiv 2$. The precise variation of the drag with penetration depth d is shown in the figure.

3.7.5 Flow past a curved boundary

Steady irrotational flow of a two-dimensional jet from a long duct with a profiled nozzle contraction [Figure 3.7.9(a)] can be treated approximately by modification of the above method.

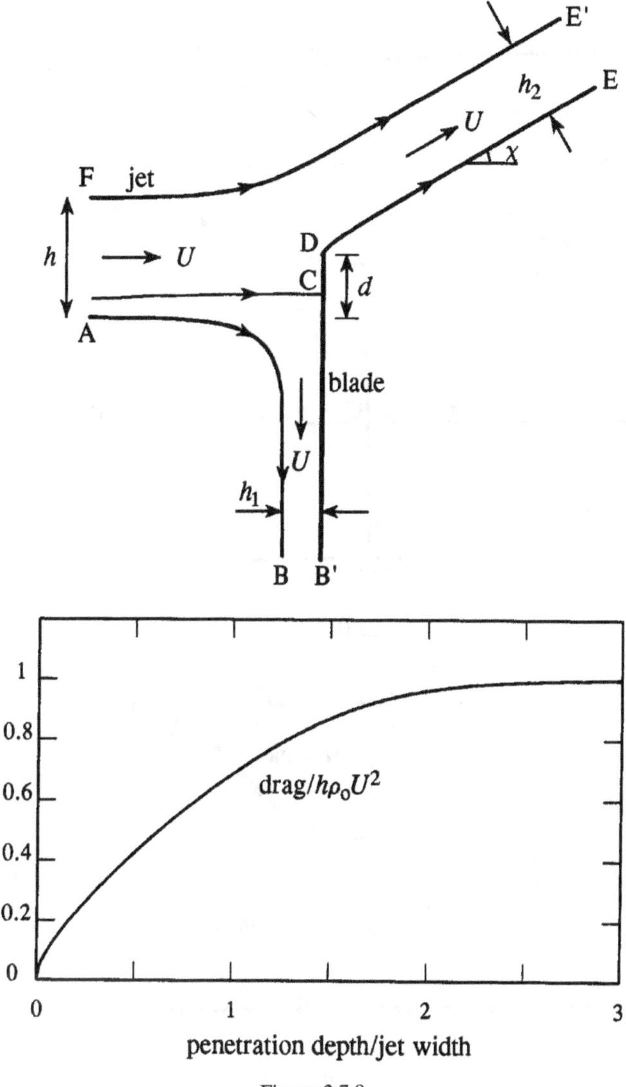

Figure 3.7.8

Let the asymptotic free jet speed be U, and consider the motion in the upper z plane bounded by the streamlines AB on the real axis, where $\psi = 0$, and DCB' on which $\psi = \sigma h U$, giving the usual picture [Figure 3.7.9(b)] in the w plane. The duct width at AD far upstream is $2h$, the nozzle width at O is $2\mu h$, and the ultimate width of the jet is denoted by $2\sigma h$ where $\sigma \le \mu$.

Suppose the tangent to the nozzle profile at the exit C makes an angle $-\alpha$ with the positive x direction. The region of variation in the plane of $\Omega = \ln(U dz/dw)$ is then determined by the following correspondences:

z:	A	B	B'	C	D
Ω:	$\ln\left(\frac{1}{\sigma}\right)$	0	0	$-i\alpha$	$\ln\left(\frac{1}{\sigma}\right)$

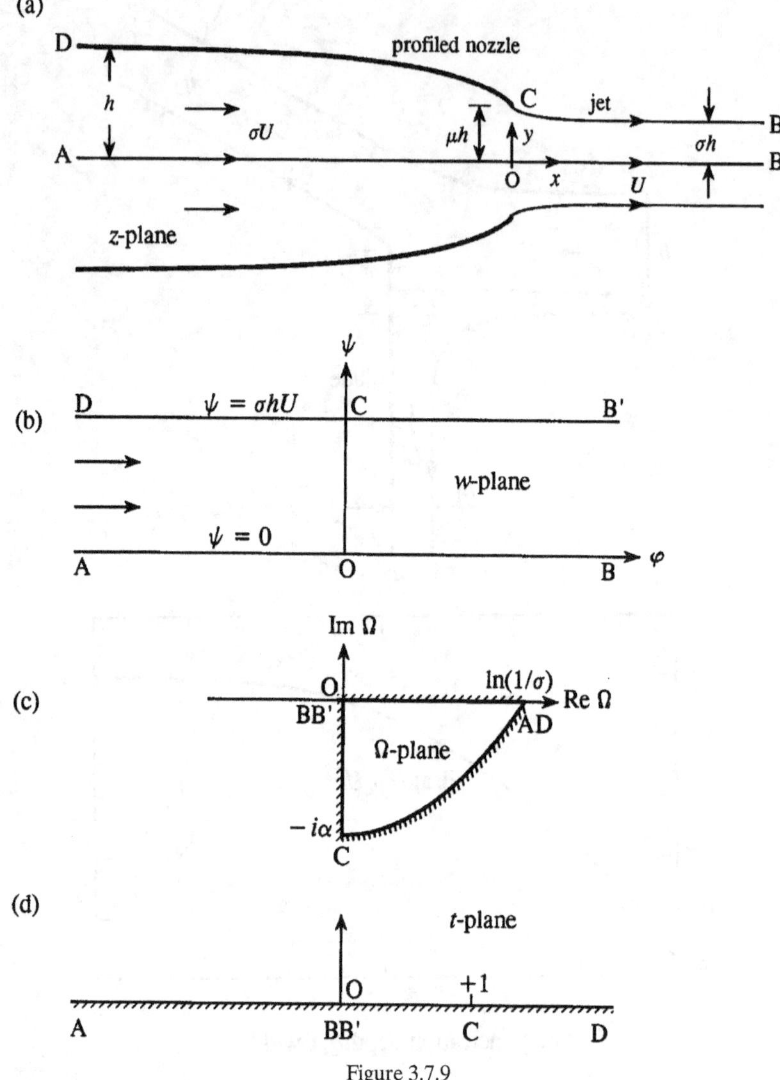

Figure 3.7.9

The relevant region of the Ω plane is therefore bounded, with vertices at BB′, C, and AD. The curvilinear, triangular domain drawn in Figure 3.7.9(c) represents just one of any number of possibilities. Only the edges AB and B′C (corresponding respectively to the central streamline AB and the free streamline CB′) are definite, whereas the path followed by the curvilinear edge between C and D depends on the precise *shape* of the profiled nozzle.

When the points A, BB′, C, and D are identified on the real t axis, as indicated in Figure 3.7.9(d),

$$w = -\frac{\sigma hU}{\pi}\ln t + i\sigma hU, \qquad \frac{dw}{dt} = -\frac{\sigma hU}{\pi t}, \tag{3.7.15}$$

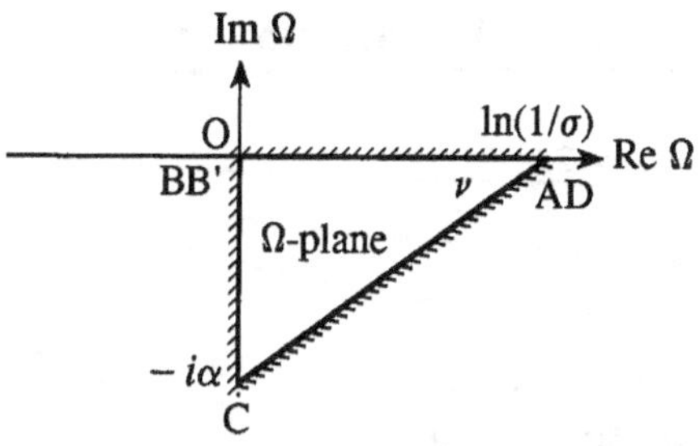

Figure 3.7.10

and the definition $\Omega(t) = \ln(U\,dz/dw) \equiv \ln(Ue^{i\theta}/q)$ permits us to write

$$dz = e^{\Omega(t)}\frac{dw}{U} = -\frac{\sigma h}{\pi}\frac{e^{\Omega(t)}dt}{t} \equiv -\frac{\sigma hU}{\pi}\frac{e^{i\theta(t)}}{q(t)t}\,dt.$$

Therefore, integrating along the *positive* section of the real t axis from $t = 1$, we obtain

$$z(t) \equiv x(t) + iy(t) = i\mu h - \frac{\sigma h}{\pi}\int_1^t \frac{e^{\Omega(\tau)}\,d\tau}{\tau}. \tag{3.7.16}$$

When $\Omega(t)$ is known and $0 < t < 1$, this equation constitutes a parametric representation of the free streamline CB'. The nozzle profile is defined by values of $t > 1$; $\tan\theta(t) = dy/dx$ is the profile tangent provided the flow remains *attached* along the curvilinear boundary of the nozzle all the way to the nozzle exit. This condition is usually satisfied when the nozzle profile has no inflexion points.

By assigning different functional forms to $\Omega(t)$ we can generate solutions for different nozzle profiles determined by (3.7.16). The simplest possible choice corresponds to the case in which the edge CD in the Ω plane is straight (Figure 3.7.10). The Schwarz–Christoffel formula (3.6.1) then supplies

$$\frac{d\Omega}{dt} = \frac{K}{t^{\frac{1}{2}}(t-1)^{\frac{\nu}{\pi}+\frac{1}{2}}}, \quad \nu = \tan^{-1}\left[\frac{\alpha}{\ln(1/\sigma)}\right] < \frac{\pi}{2},$$

where ν is the angle in Figure 3.7.10 between CD and the real Ω axis. Integration and application of the conditions $\Omega(0) = 0$, $\Omega(1) = -i\alpha$, then yield

$$\Omega(t) = \frac{\alpha e^{i\nu}}{B(\frac{1}{2}, \frac{1}{2} - \frac{\nu}{\pi})}\int_0^t \frac{d\tau}{\tau^{\frac{1}{2}}(\tau-1)^{\frac{\nu}{\pi}+\frac{1}{2}}}, \quad \operatorname{Im} t > 0, \tag{3.7.17}$$

where

$$B(x, y) \equiv \frac{\Gamma(x)\Gamma(y)}{\Gamma(x+y)} = \int_0^1 \tau^{x-1}(1-\tau)^{y-1}d\tau$$

is the beta function (Γ is the usual gamma function).

Equation (3.7.17) can be used in (3.7.16) to determine the geometry of the free streamline and nozzle for the choice of CD in Figure 3.7.10. First, $y(t) \to \sigma h$ when

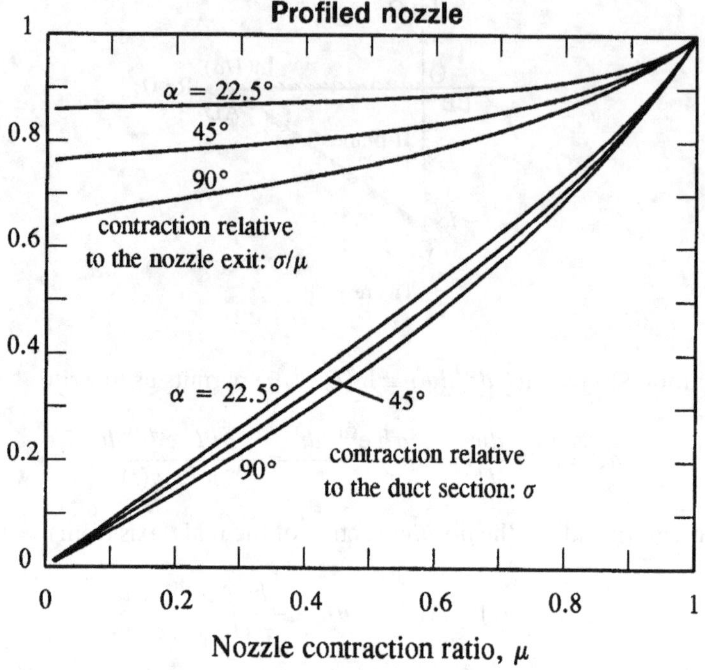

Figure 3.7.11

$t \to +0$, so that the imaginary part of (3.7.16) then reduces to the following equation for the jet contraction ratio σ:

$$\mu = \sigma + \frac{\sigma}{\pi} \int_0^1 \sin\left[\frac{\alpha}{B\left(\frac{1}{2}, \frac{1}{2} - \frac{\nu}{\pi}\right)} \int_0^\tau \frac{d\lambda}{\lambda^{\frac{1}{2}}(1-\lambda)^{\frac{\nu}{\pi}+\frac{1}{2}}} \right] \frac{d\tau}{\tau}. \tag{3.7.18}$$

The solution $\sigma = \sigma(\mu)$ of this equation is the jet contraction ratio as a function of the contraction ratio μ of the nozzle for different values of the angle α defining the slope $dy/dx = -\tan\alpha$ of the nozzle profile at the exit $(x = 0)$. Because $\nu = \tan^{-1}[\alpha/\ln(1/\sigma)]$, we solve the equation by inverse interpolation first by computing $\mu = \mu(\sigma)$ for $0 < \sigma < 1$. The results in Figure 3.7.11 also include plots of $\sigma(\mu)/\mu$, the contraction ratio relative to the nozzle exit width $2\mu h$; the values of σ/μ are similar in magnitude to the contraction ratio for a wall aperture and a funnel (§3.7.3).

Second, equation (3.7.16) gives the following parametric representation of the nozzle profile:

$$x(t) + iy(t) = i\mu h - \frac{\sigma h e^{-i\alpha}}{\pi} \int_1^t \exp\left[\frac{\alpha e^{i\nu}}{B(\frac{1}{2}, \frac{1}{2} - \frac{\nu}{\pi})} \int_1^\tau \frac{d\lambda}{\lambda^{\frac{1}{2}}(\lambda-1)^{\frac{\nu}{\pi}+\frac{1}{2}}} \right] \frac{d\tau}{\tau}, \quad t > 1.$$

This formula was used to draw the nozzle profile illustrated in Figure 3.7.9(a) for $\alpha = 90°$, $\mu = 0.5$, in terms of which $\nu \approx 58°$, $\sigma = 0.379$, and $\sigma/\mu = 0.757$.

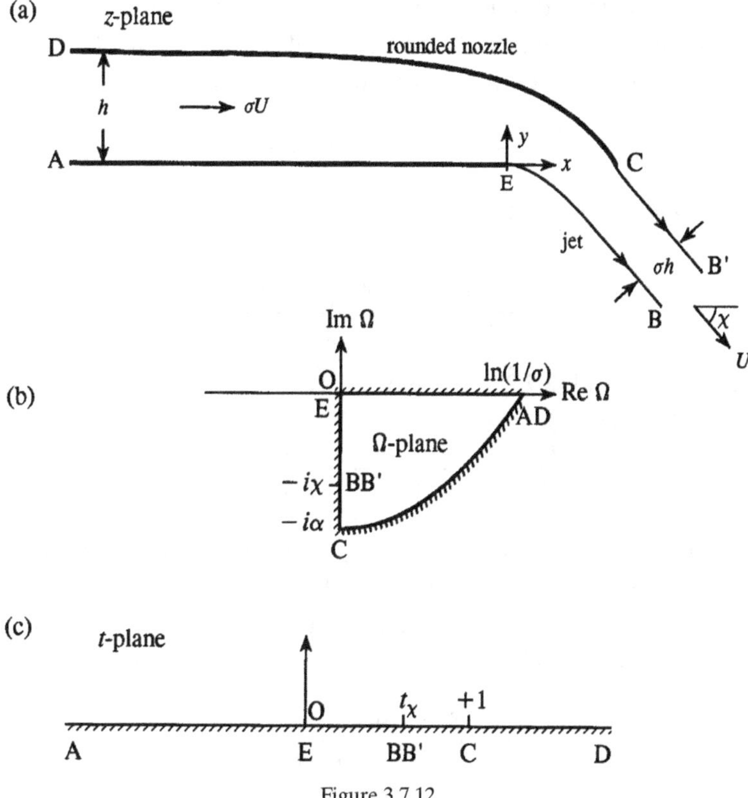

Figure 3.7.12

EXAMPLE 2. FLOW FROM A CURVED, ASYMMETRIC NOZZLE Consider steady, free-streamline flow from the nozzle of Figure 3.7.12(a). Let the uniform section of the duct at AD and the exit EC have the same width h. Take the origin in the z plane at E, so that C is at $z = h$. The upper wall of the duct is rounded, and its tangent at C makes an angle $-\alpha$ with the positive direction of the real axis. The asymptotic direction of the jet makes an angle χ with the real axis, and the jet width and velocity are respectively σh and U, where σ is the contraction ratio.

The curvilinear triangular region of variation in the $\Omega = \ln(U\,dz/dw)$ plane [Figure 3.7.12(b)] is determined partly by the correspondences:

z:	A	E	B	B′	C	D
Ω:	$\ln(1/\sigma)$	0	$-i\chi$	$-i\chi$	$-i\alpha$	$\ln(1/\sigma)$

The curved section CD will be replaced with a straight line, as in Figure 3.7.10 inclined at angle $v = \tan^{-1}[\alpha/\ln(1/\sigma)]$ to the real Ω-axis. Then if E, C, AD map into the points $0, 1, \infty$ in the t plane [Figure 3.7.12(c)] we find

$$\Omega \equiv \ln\left(U\frac{dz}{dw}\right) = \frac{\alpha e^{iv}}{B\left(\frac{1}{2}, \frac{1}{2} - \frac{v}{\pi}\right)} \int_0^t \frac{d\tau}{\tau^{\frac{1}{2}}(\tau - 1)^{\frac{v}{\pi} + \frac{1}{2}}}, \quad \text{Im}\, t > 0. \quad (3.7.19)$$

The w plane is configured much as in Figure 3.7.9. The volume flow rate σhU is generated by a source of strength σhU at DA and flows into a sink of equal strength at BB'. If $t = t_\chi$ is the image of BB' in the t plane, it follows that

$$w = -\frac{\sigma hU}{\pi} \ln(t - t_\chi) + \text{constant}. \qquad (3.7.20)$$

We obtain the equation relating t_χ to the angle χ and the contraction ratio σ by setting $\Omega = -i\chi$ in (3.7.19):

$$\chi = \frac{\alpha}{B\left(\frac{1}{2}, \frac{1}{2} - \frac{\nu}{\pi}\right)} \int_0^{t_\chi} \frac{d\lambda}{\lambda^{\frac{1}{2}}(1 - \lambda)^{\frac{\nu}{\pi}+\frac{1}{2}}}. \qquad (3.7.21)$$

Combining (3.7.19) and (3.7.20), we find

$$\frac{dz}{dt} = \frac{-\sigma h}{\pi(t - t_\chi)} \exp\left[\frac{\alpha e^{i\nu}}{B\left(\frac{1}{2}, \frac{1}{2} - \frac{\nu}{\pi}\right)} \int_0^t \frac{d\lambda}{\lambda^{\frac{1}{2}}(\lambda - 1)^{\frac{\nu}{\pi}+\frac{1}{2}}}\right]. \qquad (3.7.22)$$

This equation defines $z(t)$ in terms of the two parameters t_χ and σ (or, equivalently, in terms of χ, σ). Recall that $z(0) = 0$, $z(1) = h$; thus the real and imaginary parts of the equation obtained by integration of (3.7.22) from $t = 0$ to $t = 1$ (from E to C) along a contour just above the real axis yield

$$\frac{\pi \cos \chi}{\sigma} + \ln\left|\frac{1 - t_\chi}{t_\chi}\right| + \int_0^1 \frac{\{\cos[\Theta(\mu) - \chi] - 1\}}{\mu - t_\chi} d\mu = 0,$$

$$\frac{\pi \sin \chi}{\sigma} - \pi - \int_0^1 \frac{\sin[\Theta(\mu) - \chi]}{\mu - t_\chi} d\mu = 0, \qquad (3.7.23)$$

where

$$\Theta(\mu) = \frac{\alpha}{B\left(\frac{1}{2}, \frac{1}{2} - \frac{\nu}{\pi}\right)} \int_0^\mu \frac{d\lambda}{\lambda^{\frac{1}{2}}(1 - \lambda)^{\frac{\nu}{\pi}+\frac{1}{2}}}, \quad 0 < \mu < 1.$$

The solution of simultaneous equations (3.7.23) supplies the values of σ, χ, and t_χ. When the upper wall of the duct meets the real axis at $\alpha = 90°$ we find

$$\sigma = 0.494, \quad \chi = 48.3°, \quad t_\chi = 0.989. \qquad (3.7.24)$$

This is the case shown in Figure 3.7.12(a); the free streamlines of the jet, and the rounded shape of the upper wall of the duct are found by integration of (3.7.22) in turn along the sections EB, CB', and CD of the real t axis.

3.7.6 The hodograph transformation formula

Free-streamline problems are tractable by the methods already described provided the moving fluid occupies a *simply connected* domain. This ensures the existence of a mapping onto the upper t plane, with the flow boundary mapping onto the real axis. When the rigid boundaries are sectionally straight, Im Ω is constant on these boundaries and $\mathrm{Re}\,\Omega$ is constant on the free streamlines. Joukowski pointed out that the mapping from the t plane onto the region of variation of Ω is then given by a particular instance of the formula

$$\frac{d\Omega}{dt} = \frac{m\,f(t)}{\sqrt{(t-a_1)(t-a_2)(t-a_3)\ldots}}\,, \tag{3.7.25}$$

where a_1, a_2, a_3, $\ldots$, are real constants, m is a constant equal to 1 or i, and $f(t)$ is a rational function with real coefficients that can have *simple* poles with residues b_1, b_2, b_3, $\ldots$, at points $t = c_1$, c_2, c_3, $\ldots$, on the real axis, i.e.,

$$f(t) = \hat{f}(t) + \frac{b_1}{t-c_1} + \frac{b_2}{t-c_2} + \frac{b_3}{t-c_3}, \ldots,$$

where $\hat{f}(t)$ is a polynomial with real coefficients.

The points $t = a_1$, a_2, a_3, $\ldots$, define intervals of the real axis within which $d\Omega/dt$ is either real or pure imaginary, and therefore respectively the intervals where $\mathrm{Im}\,\Omega$ and $\mathrm{Re}\,\Omega$ are constant. The points $t = c_1$, c_2, c_3, $\ldots$, are the images of those points on the rigid boundaries where the flow experiences a discontinuous change in direction (where $\mathrm{Im}\,\Omega$ changes discontinuously); these correspond to rigid corners and stagnation points. Also, because Ω can have infinite singularities only of *logarithmic* order, it follows that $d\Omega/dt$ decreases at least as fast as $1/t$ as $|t| \to \infty$.

In any particular problem (3.7.25) is used with the corresponding formula for dw/dt that gives the mapping onto the w plane (derived according to the general principles of §3.7.2). The coefficients a_k, b_k, c_k are determined by the flow geometry. This is often difficult, and we can sometimes make more progress by considering the inverse problem, in which the values of the coefficients are prescribed and the flow geometry subsequently deduced from the solution.

3.7.7 Chaplygin's singular point method

The hodograph transformation formula (3.7.25) for free-streamline flows with piecewise straight boundaries was derived for the problems discussed in §§3.7.1 – 3.7.3 by consideration of the geometry of the region of variation of Ω and application of the Schwarz–Christoffel formula. Chaplygin observed that it can also be deduced from a knowledge of the singularities of $d\Omega/dt$. His method has the advantage of being applicable also to flows containing point sources and vortices, which cause $d\Omega/dt$ to have poles in $\mathrm{Im}\,t \geq 0$; such problems are difficult or impossible to treat by simple geometrical arguments.

The procedure requires that the singularities of $dw(t)/dt$ and $dz(t)/dt$ should first be found. We facilitate this by noting that singularities on curvilinear boundaries in the z plane (e.g. at the intersection of two parallel free streamlines at infinity) are the same as for straight boundaries. Knowledge of the singularities determines the behaviour of $d\Omega/dt$ in the neighbourhood of its poles c_1, c_2, $c_3, \ldots$. Liouville's theorem and the

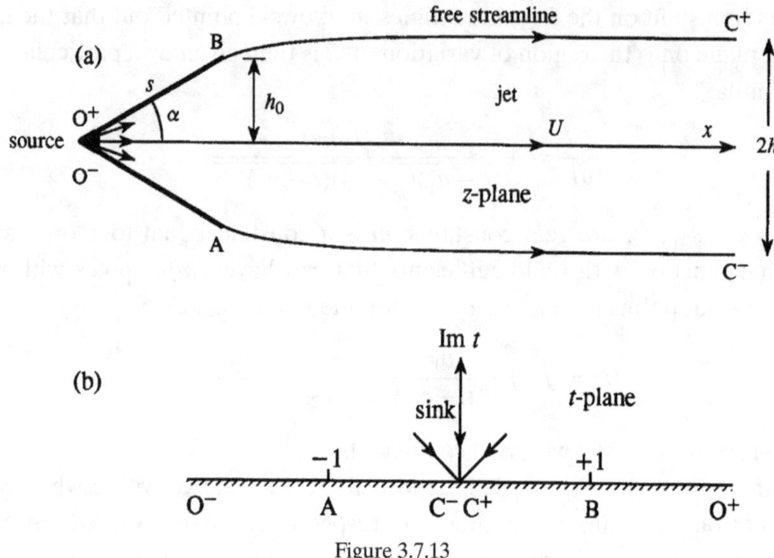

Figure 3.7.13

properties of the general formula (3.7.25) can then be invoked to deduce the functional form of $d\Omega/dt$ for all t. The following examples illustrate the basic ideas.

3.7.8 Jet produced by a point source

A uniform jet of speed U and width $2h$ is produced by a source of strength $2hU$ situated at the angle between two equal plates [Figure 3.7.13(a)]. The plates have length s and meet at the source ($z = 0$) with angle 2α, and the flow is assumed to separate smoothly from their free ends A and B ($z = se^{\mp i\alpha}$).

Let the t plane be configured as in Figure 3.7.13(b) with a sink of strength $2hU$ absorbing the jet flow at $C^- C^+$ ($t = 0$). Then

$$\frac{dw}{dt} = \frac{-2hU}{\pi t}. \tag{3.7.26}$$

To construct the corresponding equation for Ω we write

$$\frac{d\Omega}{dt} \equiv \frac{d}{dt} \ln\left(U \frac{dz}{dw}\right) = \frac{d}{dt} \ln\left(\frac{dz}{dt}\right) - \frac{d}{dt} \ln\left(\frac{dw}{dt}\right). \tag{3.7.27}$$

The transformation from the z plane to the t plane is conformal, so that dz/dt is regular in $\mathrm{Im}\, t > 0$. However, dz/dt has simple poles on the real axis at the images of the rigid corner O^+O^- ($t = \infty$) and the intersection C^+C^- ($t = 0$) of the free streamlines. According to Joukowski's formula (3.7.25) only those poles produced by corners on *rigid* boundaries contribute to $d\Omega/dt$ (in the absence of sources and vortices within the flow). Thus, because the pole at $t = \infty$ cannot make a finite contribution to $d\Omega/dt$, it follows for this problem that $d\Omega/dt$ has no poles. To verify this, observe that the mapping of the corner C^+C^- onto $t = 0$ implies that $dz/dt \sim O(1/t)$ near $t = 0$ (cf. §3.6.1). Therefore, using this and (3.7.26), we find

$$\frac{d\Omega}{dt} = f^+(t) - \frac{1}{t} + \frac{1}{t} = f^+(t),$$

where $f^+(t)$ is regular in $\mathrm{Im}\, t > 0$ and $|f^+(t)| \leq O(1/t)$ as $|t| \to \infty$.

Now $d\Omega/dt$ is real on the solid boundaries O^-A, O^+B (where $|t| > 1$) and pure imaginary on the free streamlines AC^-, BC^+ ($|t| < 1$). Thus,

$$\sqrt{t^2 - 1}\frac{d\Omega}{dt} \equiv \sqrt{t^2 - 1}\, f^+(t)$$

is real on the whole of the real t axis, and regular in $\mathrm{Im}\, t > 0$. By the *reflection principle* it is therefore regular throughout the whole of the t plane. Because it is also bounded it follows from Liouville's theorem that it is equal to a constant K, say, and therefore (in the usual notation) that

$$\frac{d\Omega}{dt} = \frac{K}{\sqrt{t^2 - 1}},$$

i.e. $\quad \Omega \equiv \ln\left(U\frac{dz}{dw}\right) = K\ln\left(t + \sqrt{t^2 - 1}\right) + L.$

The values of K, L are determined from the conditions $\Omega = 0$ at $t = 0$, and $\Omega = \pm i\alpha$ at $t = \pm 1$.

Hence,

$$\Omega = i\alpha - \frac{2\alpha}{\pi}\ln\left(t + \sqrt{t^2 - 1}\right),$$

and the solution now proceeds in the usual way. In particular, because $z = 0$ at the source O^-O^+ ($t = \infty$) and $z = se^{\pm i\alpha}$ at $t = \pm 1$, we have

$$\frac{h}{s} = \frac{\pi}{\tilde{\psi}\left(\frac{\alpha}{2\pi} + \frac{1}{2}\right) - \tilde{\psi}\left(\frac{\alpha}{2\pi}\right) - \frac{\pi}{\alpha}},$$

where $\tilde{\psi}(x) = d\ln[\Gamma(x)]/dx$. The upper free streamline in Figure 3.7.13(a) is defined parametrically by

$$z = se^{i\alpha} + \frac{he^{i\alpha}}{\pi}\int_0^\theta \tan\left(\frac{\lambda}{2}\right)e^{-i\alpha\lambda/\pi}\, d\lambda, \quad 0 < \theta < \pi.$$

The case illustrated in the figure obtains for $\alpha = 30°$, when $h/s = 0.648$; the ratio of the asymptotic jet width to the initial width $2h_0$ at AB is $h/h_0 = h/(s\sin\alpha) = 1.297$.

3.7.9 Deflection of trailing-edge flow by a source

Consider steady parallel flow at speed U in $\mathrm{Im}\, z > 0$ towards the edge $z = 0$ of the semi-infinite rigid plate AO of Figure 3.7.14(a). The plate coincides with the negative real axis, and the free streamline OB from the edge is assumed to be disturbed by flow from a point source of strength $q > 0$ on the plate near the edge. The source flow produces a

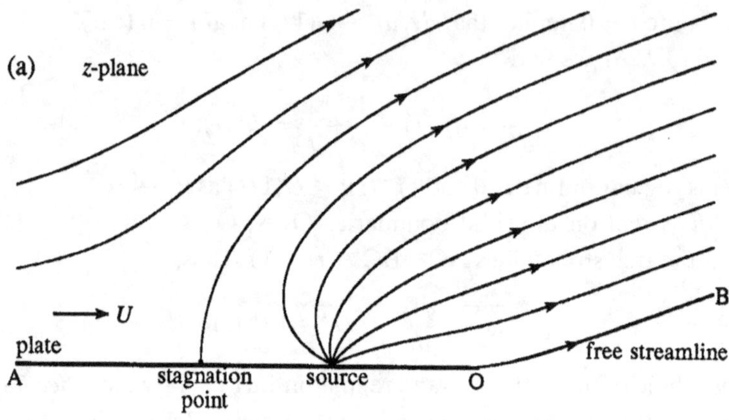

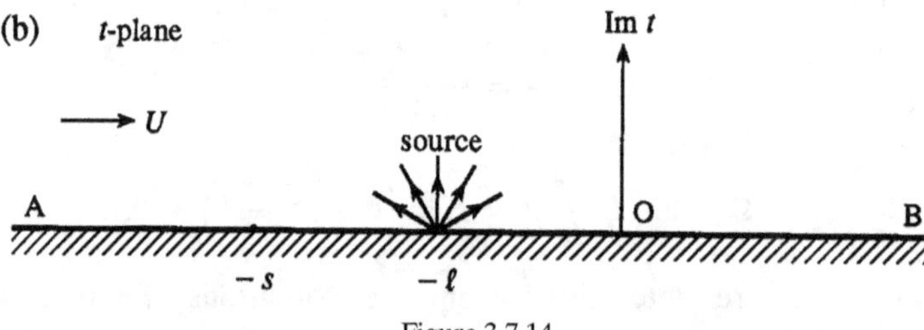

Figure 3.7.14

stagnation point a short distance upstream, which corresponds to a pole in Joukowski's formula (3.7.25).

Let the flow be mapped onto the upper t plane as in Figure 3.7.14(b), with the source at $t = -\ell$ and stagnation point at $t = -s$. Then

$$\frac{dw}{dt} = U + \frac{q}{\pi(t+\ell)}$$

$$= U\left(\frac{t+s}{t+\ell}\right), \quad \text{where} \quad s = \ell + \beta, \quad \beta = \frac{q}{\pi U}. \tag{3.7.28}$$

Hence [cf. (3.7.27)]

$$\frac{d\Omega}{dt} = f^+(t) - \frac{1}{t+s} + \frac{1}{t+\ell},$$

where $f^+(t)$ is regular in $\operatorname{Im} t > 0$ and $|f^+(t)| \le O(1/|t|)$ as $|t| \to \infty$.

Now $d\Omega/dt$ is pure imaginary on OB and real on AO. Therefore

$$i\sqrt{t}\,\frac{d\Omega}{dt} \quad \text{is real on the whole of the real } t \text{ axis.}$$

This means that

$$i\sqrt{t}\,\frac{d\Omega}{dt} + \frac{i\sqrt{-s}}{t+s} - \frac{i\sqrt{-\ell}}{t+\ell}$$

is real for all real values of t and regular in $\mathrm{Im}\, t > 0$. By the reflection principle it is regular throughout the whole of the t plane, and, because it vanishes (at least as fast as $1/\sqrt{t}$) as $|t| \to \infty$, it must actually be identically zero. Thus,

$$\frac{d\Omega}{dt} = \frac{-i\sqrt{s}}{\sqrt{t}(t+s)} + \frac{i\sqrt{\ell}}{\sqrt{t}(t+\ell)}. \tag{3.7.29}$$

Integrating and noting that $\Omega \to 0$ as $|t| \to \infty$, we find

$$U\frac{dz}{dw} = \left(\frac{\sqrt{t}+i\sqrt{s}}{\sqrt{t}-i\sqrt{s}}\right)\left(\frac{\sqrt{t}-i\sqrt{\ell}}{\sqrt{t}+i\sqrt{\ell}}\right).$$

We now find the mapping between the z and the t planes by combining this result with (3.7.28) to obtain

$$\frac{dz}{dt} = \left(\frac{\sqrt{t}+i\sqrt{s}}{\sqrt{t}+i\sqrt{\ell}}\right)^2,$$

i.e. $\quad z = t + \left(\sqrt{s} - \sqrt{\ell}\right)$

$$\times \left\{4i\sqrt{t} + 2\left(3\sqrt{\ell} - \sqrt{s}\right)\left[\ln(\sqrt{t/\ell}+i) - i\pi/2\right] + \frac{2\sqrt{t}(\sqrt{s} - \sqrt{\ell})}{\sqrt{t}+i\sqrt{\ell}}\right\}, \tag{3.7.30}$$

provided that $z = 0$ at $t = 0$.

This formula is used to plot the free streamline by taking t to vary over the interval $(0, +\infty)$ of the real axis. The case shown in Figure 3.7.14(a) corresponds to $\beta = q/\pi U = \ell$. The streamlines are plotted with (3.7.30) and the formula

$$t_i = (t_r + \ell)\tan\left(\frac{\psi/U - t_i}{\beta}\right), \quad t = t_r + it_i,$$

where the stream function $\psi = 0$ on the free streamline and $\psi = \mathrm{constant} > 0$ on a streamline of the flow.

EXAMPLE 3. SOURCE NEAR THE EDGE OF A JET Fluid occupying the region $y > 0$ is in uniform flow at speed U in the x direction. Find the perturbed mean streamline boundary of the flow when a source of strength q is inserted in the stream close to the boundary [Figure 3.7.15(a)].

Let the flow be mapped onto the upper t plane with the source at $t = ih$. Then

$$\frac{dw}{dt} = U + \frac{q}{2\pi}\left(\frac{1}{t-ih} + \frac{1}{t+ih}\right) \equiv \frac{U(t-s)(t-s^*)}{t^2+h^2}, \tag{3.7.31}$$

where

$$s = \beta + i\sqrt{h^2 - \beta^2}, \quad \beta = \frac{q}{2\pi U},$$

is the image of the *stagnation point*. The stagnation point lies within the flow ($\mathrm{Im}\, s > 0$) when $\beta < h$, i.e. provided the volume flow from the source is not too large.

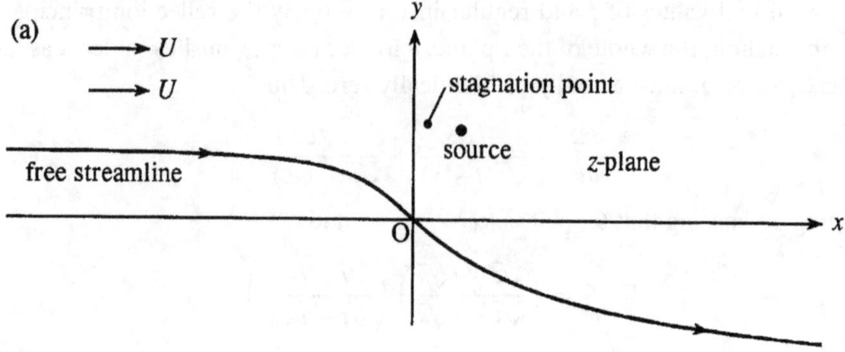

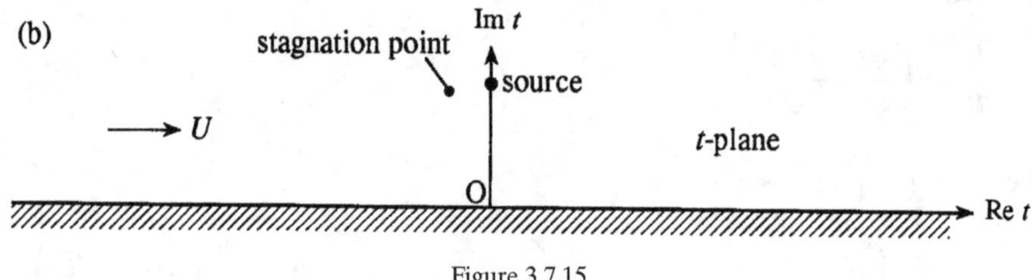

Figure 3.7.15

The transformation $z = z(t)$ is evidently regular in $\operatorname{Im} t \geq 0$, and therefore

$$\frac{d\Omega}{dt} = f^+(t) + \frac{1}{t - ih} - \frac{1}{t - s}, \quad \operatorname{Im} t \geq 0,$$

where $f^+(t)$ is regular in the upper half-plane. Thus $d\Omega/dt$ has poles at $t = ih$ and $t = s$ in the upper half of the t plane and must be pure imaginary along the whole of the real t axis.

The function

$$f(t) = \frac{1}{t - ih} - \frac{1}{t + ih} - \frac{1}{t - s} + \frac{1}{t - s^*}$$

also satisfies these conditions. Moreover, $d\Omega/dt - f(t)$ is regular in $\operatorname{Im} t > 0$, is pure imaginary on the real axis, and vanishes as $|t| \to \infty$. Liouville's theorem therefore implies that it vanishes everywhere, so that $d\Omega/dt = f(t)$, and, by integration,

$$U\frac{dz}{dw} = \left(\frac{t - ih}{t + ih}\right)\left(\frac{t - s^*}{t - s}\right), \tag{3.7.32}$$

where the constant of integration is found from the condition that $\Omega = 0$ at $t = \infty$.

By combining (3.7.31) and (3.7.32) we find

$$\frac{dz}{dt} = \left(\frac{t - s^*}{t + ih}\right)^2 \equiv 1 - \frac{2(s^* + ih)}{t + ih} + \left(\frac{s^* + ih}{t + ih}\right)^2,$$

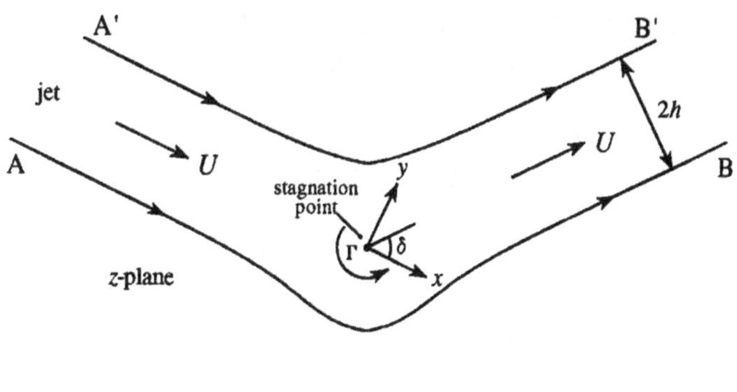

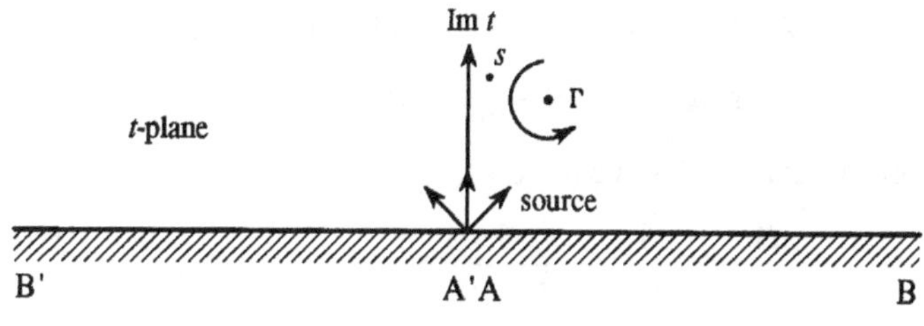

Figure 3.7.16

and, by taking $z = 0$ at $t = 0$,

$$z = t - (s^* + ih)\left[2\ln\left(\frac{t}{h} + i\right) - i\pi + \frac{it(s^* + ih)}{h(t + ih)}\right].$$

This equation and the interval $-\infty < t < \infty$ of the real axis define the free-streamline boundary of the disturbed flow. In Figure 3.7.15 we have taken $\beta/h \equiv q/2\pi Uh = 0.35$. When β/h exceeds about 0.96, the free streamline becomes self-intersecting, indicating that the source has forced its way out of the jet and that the postulated steady flow is actually unstable. By considering the limiting values of Im z as $t \to \pm\infty$ we find that the source causes the overall width of the flow to increase by $\delta y = 2\pi\beta = q/U$, i.e. by precisely the amount necessary to accommodate the additional volume flow (at an asymptotic speed U) from the source.

EXAMPLE 4. JET DEFLECTION BY A VORTEX (Simmons 1939) The force exerted by a steady flow on a *stationary* line vortex produces a mean deflection of the flow that is similar to that produced by an airfoil or by a cylinder with circulation. Consider the case of a uniform jet of width $2h$ whose centreline coincides with the x axis and that initially moves at speed U in the positive x direction. Suppose the jet impinges on a vortex of strength Γ situated on the y axis at $y = h_\Gamma$. Figure 3.7.16 illustrates the situation for $h_\Gamma = 0$. It is required to determine the dependence of the angular deflection δ of the jet on Γ and h_Γ.

Let the mapping onto $\operatorname{Im} t > 0$ be as indicated in the figure. The jet flow is created by a point source of strength $2hU$ at $t = 0$. Thus, if the vortex maps onto $t = \gamma \equiv \ell e^{i\psi}$, then

$$\frac{dw}{dt} = \frac{2hU}{\pi t} - \frac{i\Gamma}{2\pi}\left(\frac{1}{t-\gamma} - \frac{1}{t-\gamma^*}\right),$$

$$\therefore \quad \frac{\pi}{2hU}\frac{dw}{dt} = \frac{t^2 - t[\gamma + \gamma^* + i\beta(\gamma - \gamma^*)] + \ell^2}{t(t-\gamma)(t-\gamma^*)}, \quad \text{where} \quad \beta = \frac{\Gamma}{4hU}$$

$$\equiv \frac{(t-s)(t-s^*)}{t(t-\gamma)(t-\gamma^*)}, \tag{3.7.33}$$

where $t = s$ is the image of the stagnation point, given by

$$s \equiv \ell e^{i\chi}, \quad \cos\chi = \cos\psi - \beta\sin\psi. \tag{3.7.34}$$

By the method of Example 3, we have

$$\frac{d\Omega}{dt} = \frac{1}{t-\gamma} - \frac{1}{t-\gamma^*} - \frac{1}{t-s} + \frac{1}{t-s^*},$$

which is pure imaginary on the real axis. Integrating and using the condition that $dw/dz \to Ue^{-i\delta}$ as $t \to \infty$, we find

$$U\frac{dz}{dw} = e^{i\delta}\left(\frac{t-\gamma}{t-\gamma^*}\right)\left(\frac{t-s^*}{t-s}\right). \tag{3.7.35}$$

Because $dw/dz \to U$ as $t \to 0$, this equation implies that $e^{i(\delta+2\psi-2\chi)} = 1$ and therefore that

$$\chi = \psi + \frac{\delta}{2}. \tag{3.7.36}$$

Hence, from (3.7.33), (3.7.35), and the relation $s/\gamma = e^{i\delta/2}$, we find

$$\frac{\pi}{2h}\frac{dz}{dt} = \frac{e^{i\delta}}{t}\left(\frac{t-s^*}{t-\gamma^*}\right)^2 \equiv \frac{1}{t} + \frac{e^{i\delta}-1}{t-\gamma^*} + \gamma^*\left(\frac{e^{i\delta/2}-1}{t-\gamma^*}\right)^2,$$

$$\therefore \quad z = z_0 + \frac{2h}{\pi}\left[\ln t' + (e^{i\delta}-1)\ln(t'-e^{-i\psi}) - \frac{e^{-i\psi}(e^{i\delta}-1)^2}{t'-e^{-i\psi}}\right], \tag{3.7.37}$$

where $t' = t/\ell$. The conditions that $z = ih_\Gamma$ at $t' = e^{i\psi}$ and $\operatorname{Im} z \to \pm h$ as $t' \to \mp 0$ yield the relations

$$\frac{z_0}{h} = \sin\delta + \frac{4}{\pi}\left[\sin^2(\delta/2)\ln(2\sin\psi) - \frac{\sin^2(\delta/4)\sin(\delta/2-\psi)}{\sin\psi}\right]$$

$$+ i\left\{\frac{4}{\pi}[\pi - \psi + \tan(\delta/4)]\sin^2(\delta/2) - 1\right\},$$

and

$$\frac{h_\Gamma}{h} = \left(\frac{2\psi}{\pi} - 1\right)\cos\delta + \frac{2}{\pi}\left[\sin\delta\,\ln(2\sin\psi) - \frac{2\sin^2(\delta/4)\cos(\delta/2+\psi)}{\sin\psi}\right]. \tag{3.7.38}$$

To use these results we first combine (3.7.34) and (3.7.36) to obtain

$$\tan\psi = \frac{1-\cos(\delta/2)}{\beta-\sin(\delta/2)}, \quad \beta = \frac{\Gamma}{4hU}. \tag{3.7.39}$$

Thus, when the vortex strength β and offset h_Γ are prescribed, Equations (3.7.38) and (3.7.39) can be solved for the deflection angle δ and the vortex image angle ψ. The shape of the jet can then be plotted by use of (3.7.37), the upper and lower streamlines in the figure corresponding respectively to the intervals $(-\infty, 0)$, $(0, \infty)$ of the real t axis.

For the case shown in Figure 3.7.16,

$$h_\Gamma = 0, \quad \beta \equiv \frac{\Gamma}{4hU} = 0.5, \quad \delta = 51.3°, \quad \psi = 55.7°.$$

3.8 The Joukowski transformation

The Joukowski transformation

$$\zeta = \frac{z}{a} + \sqrt{\frac{z^2}{a^2} - 1}, \quad a > 0, \quad z = x + iy, \tag{3.8.1}$$

and its inverse $\quad \dfrac{z}{a} = \dfrac{1}{2}\left(\zeta + \dfrac{1}{\zeta}\right), \quad \zeta = \xi + i\eta,$

have important applications to the theory of airfoils. It is conformal everywhere except at $z = \pm a$, where $d\zeta/dz = \infty$ (i.e. at $\zeta = \pm 1$, where $dz/d\zeta = 0$). In particular, this means that smooth contours in the ζ plane in $|\zeta| > 1$ map into smooth contours in the z plane. Also, $z \sim a\zeta/2$ when $|z|$ and $|\zeta|$ are large, so that uniform flow at large distances from the origin in either plane maps into a uniform flow *in the same direction* in the other plane.

Consider steady flow in the ζ plane at speed V in a direction inclined at angle α ($0 < \alpha < \frac{\pi}{2}$) to the positive ξ axis, with complex potential $V\zeta e^{-i\alpha}$. According to the circle theorem (§3.2), when a circular cylinder $|\zeta| = \beta$ with centre at the origin and radius β is inserted into the flow the complex potential becomes

$$w = V\left(\zeta e^{-i\alpha} + \frac{\beta^2 e^{i\alpha}}{\zeta}\right). \tag{3.8.2}$$

The streamlines of this flow are illustrated in Figure 3.8.1. We can plot them by first inverting (3.8.2) to obtain for $\zeta = \xi + i\eta$

$$\xi + i\eta = \beta e^{i\alpha}\left(\frac{w}{2V\beta} + \sqrt{\frac{w^2}{(2V\beta)^2} - 1}\right), \quad \text{where} \quad w = \varphi + i\psi; \tag{3.8.3}$$

the principal value of the square root is taken to ensure that $w \sim V\zeta e^{-i\alpha}$ as $|\zeta| \to \infty$. This provides a parametric representation of the streamlines (on which $\psi = $ constant) in terms of φ ($-\infty < \varphi < +\infty$).

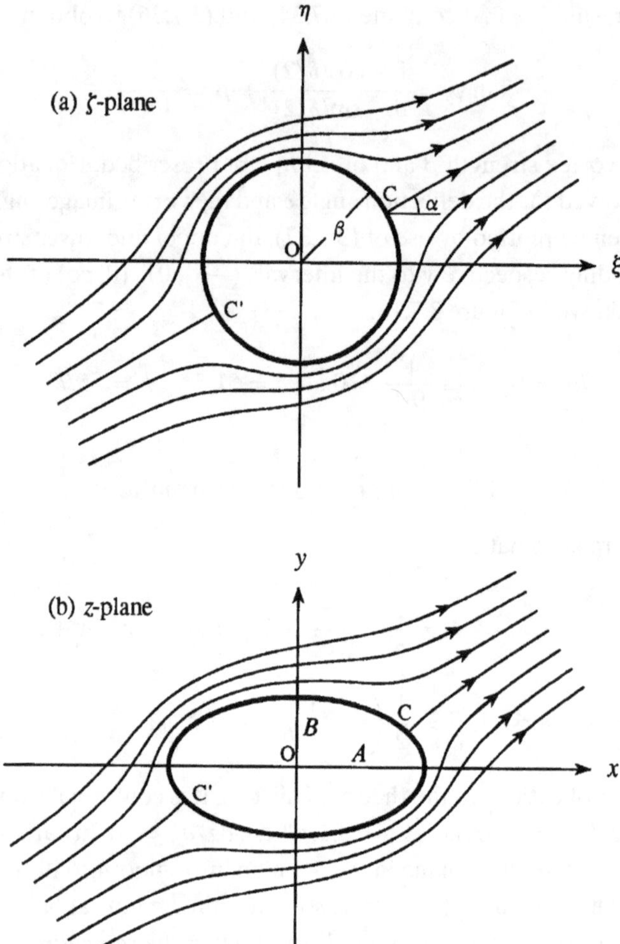

Figure 3.8.1

When the radius $\beta > 1$ the circular cylinder maps into a cylinder of elliptic cross section in the z plane [Figure 3.8.1(b)]. If $\zeta = \beta e^{i\Theta}$ on the circular cylinder, the second of (3.8.1) gives

$$\frac{z}{a} = \frac{1}{2}\left(\beta e^{i\Theta} + \frac{1}{\beta}e^{-i\Theta}\right),$$

so that the elliptic cylinder is given parametrically by

$$x = \frac{a}{2}\left(\beta + \frac{1}{\beta}\right)\cos\Theta, \quad y = \frac{a}{2}\left(\beta - \frac{1}{\beta}\right)\sin\Theta,$$

i.e.,
$$\frac{x^2}{A^2} + \frac{y^2}{B^2} = 1, \quad \text{where} \quad A = \frac{a}{2}\left(\beta + \frac{1}{\beta}\right), \quad B = \frac{a}{2}\left(\beta - \frac{1}{\beta}\right), \quad (3.8.4)$$

$$a = \sqrt{A^2 - B^2}, \quad \beta = \sqrt{\frac{A+B}{A-B}} > 1.$$

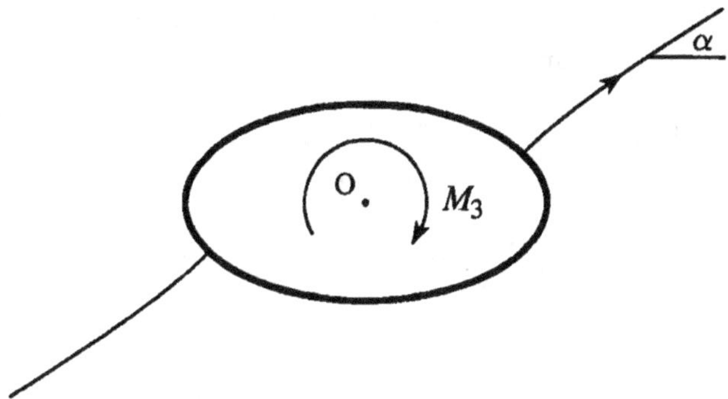

Figure 3.8.2

Thus in the z plane the mean flow in the α direction is around the elliptic cylinder. Now as $|\zeta|,\ |z| \to \infty$,

$$w \sim \zeta V e^{-i\alpha} \sim \frac{2zV}{a} e^{-i\alpha}$$

because $\zeta \sim 2z/a$. Hence, if the speed in the z plane at large distances from the elliptic cylinder is U, we must set $U = 2V/a$, and the corresponding complex potential is then given by (3.8.2) and the first of (3.8.1) in the form

$$\begin{aligned}
w &= \frac{U}{2}\left[\left(z+\sqrt{z^2-a^2}\right)e^{-i\alpha} + \frac{\beta^2 a^2 e^{i\alpha}}{\left(z+\sqrt{z^2-a^2}\right)}\right] \\
&= \frac{U}{2}\left[\left(z+\sqrt{z^2-a^2}\right)e^{-i\alpha} + \beta^2\left(z-\sqrt{z^2-a^2}\right)e^{i\alpha}\right]
\end{aligned} \tag{3.8.5}$$

We plot the streamline pattern around the elliptic cylinder [Figure 3.8.1(b)] by using (3.8.3) to calculate $\zeta = \xi + i\eta$ in terms of $w = \varphi + i\psi$ and then substituting into the second of (3.8.1) to obtain the parametric representation of a point $z = x + iy$ in terms of φ $(-\infty < \varphi < \infty)$ on each streamline $\psi = \text{constant}$.

It is evident from the symmetry of both of the ideal flows in Figure 3.8.1 that the fore–aft surface pressure forces are in equilibrium, and that there is no net force on either body (D'Alembert's paradox). However, Bernoulli's equation,

$$p = \frac{1}{2}\rho_o U^2 - \frac{1}{2}\rho_o(\nabla\varphi)^2,$$

shows that the excess pressure p is a maximum on the surface at the stagnation points, labelled C, C′ in the figure. For the elliptic cylinder this means that there must be a net moment on the cylinder, tending to turn it in the clockwise direction, so that its major axis is facing the oncoming mean stream. To verify this the moment M_3 (per unit span) about the origin is calculated by means of Blasius formula (3.3.3) (see Figure 3.8.2):

$$M_3 = -\mathrm{Re}\left[\frac{\rho_o}{2}\oint z\left(\frac{dw}{dz}\right)^2 dz\right],$$

where the integration can be taken around any contour enclosing the cylinder. In particular, we can integrate around a circle of radius $|z| \to \infty$, where

$$\frac{dw}{dz} = \frac{U}{2} \left\{ e^{-i\alpha} \left[1 + \frac{z}{(z^2 - a^2)^{\frac{1}{2}}} \right] + \beta^2 e^{i\alpha} \left[1 - \frac{z}{(z^2 - a^2)^{\frac{1}{2}}} \right] \right\}$$

$$\sim U \left[e^{-i\alpha} + \frac{a^2}{4z^2} \left(e^{-i\alpha} - \beta^2 e^{i\alpha} \right) + \cdots + \right].$$

Then, by residues,

$$\text{moment} = \text{Re} \left[-\frac{1}{2} \rho_o \pi i U^2 a^2 \left(e^{-2i\alpha} - \beta^2 \right) \right]$$

$$= -\frac{\pi a^2}{2} \rho_o U^2 \sin 2\alpha = -\frac{1}{2} \pi (A^2 - B^2) \rho_o U^2 \sin 2\alpha. \tag{3.8.6}$$

Thus the mean flow exerts a clockwise couple on the cylinder, tending to turn it broadside to the stream, a conclusion that is applicable for any elongated body (such as a boat in a stream).

The couple vanishes when $\alpha = 0$, when the major axis of the cylinder is parallel to the stream. However, this configuration is clearly unstable. On the other hand, the other equilibrium position $\alpha = \frac{\pi}{2}$ is stable.

3.8.1 The flat-plate airfoil

When $\beta \to 1$ elliptic image (3.8.4) in the z plane of the circular cylinder of Figure 3.8.1(a) collapses onto the strip $|x| < a$, $y = 0$, which can be regarded as a 'flat-plate' approximation to an airfoil (of infinite span). The points $\zeta = \pm 1$ on the cylinder correspond respectively to the trailing edge $z = -a$ and the leading edge $z = +a$ of the airfoil. These are singular points at which $d\zeta/dz = \infty$, where the Joukowski transformation (3.8.1) ceases to be conformal.

The generation of *lift* by an airfoil requires it to be inclined at a small *angle of attack* α to the mean flow direction. This is, of course, just the problem considered previously for the elliptic cylinder. We find that the velocity potential of the mean flow is given by setting $V = aU/2$ and $\beta = 1$ in (3.8.2):

$$w = \frac{aU}{2} \left(\zeta e^{-i\alpha} + \frac{e^{i\alpha}}{\zeta} \right). \tag{3.8.7}$$

In the z plane we have, setting $\beta = 1$ in (3.8.5),

$$w \equiv \varphi + i\psi = U \left(z \cos\alpha - i \sin\alpha \sqrt{z^2 - a^2} \right). \tag{3.8.8}$$

The streamline pattern ($\psi = $ constant) is plotted in Figure 3.8.3. The flow exhibits fore–aft asymmetry, with leading and trailing stagnation points A and B where the excess pressure is a maximum. The complex velocity

$$\frac{dw}{dz} = U \left(\cos\alpha - \frac{iz \sin\alpha}{\sqrt{z^2 - a^2}} \right)$$

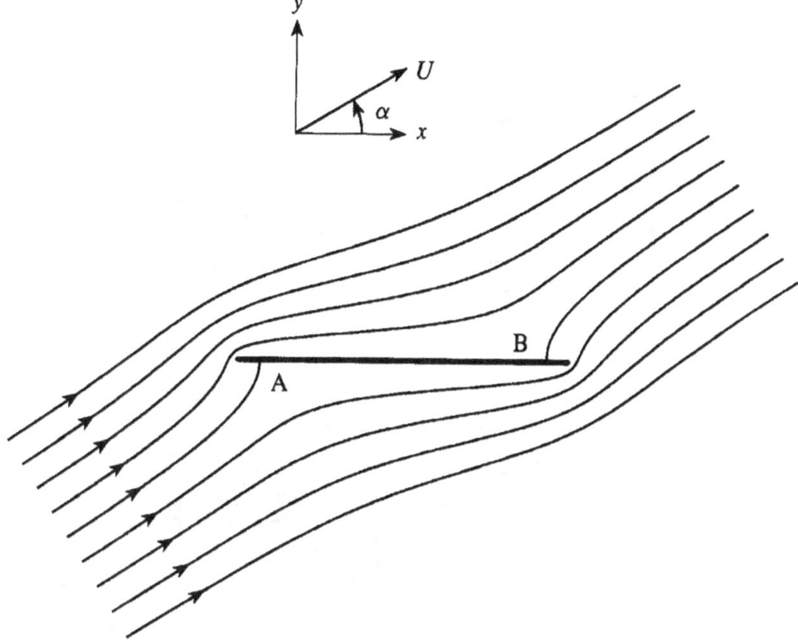

Figure 3.8.3

is singular at the edges $z = \pm a$, where streamlines are required to turn through $180°$. There is no net force on the airfoil, but there is a couple of magnitude $-\frac{1}{2}\pi a^2 \rho_o U^2 \sin 2\alpha$ tending to rotate it in the clockwise direction.

The flow past an airfoil in practice is subject to relatively strong viscous action in the neighbourhood of the trailing edge. When the motion starts from rest the flow everywhere is initially irrotational. The high-speed flow around the trailing edge must necessarily be rapidly slowed as it approaches the stagnation point B. In reality, however, the flow at the edge immediately separates resulting in the formation of a 'starting vortex' of positive circulation Γ that is convected away by the mean flow. According to Kelvin's circulation theorem the progressive shedding of circulation from the trailing edge must be compensated by the growth in an equal amount of *negative* circulation around the airfoil, the effect of which is to progressively shift the stagnation point B to the trailing edge where it ultimately cancels the irrotational flow edge singularity. Thus vorticity continues to be shed until the flow at the trailing edge becomes continuous, that is, until the edge flow is smooth and leaves the edge tangentially. The process is very rapid in practice, the shed vorticity coagulating into the starting vortex that is quickly carried away in the mean flow, so that it eventually ceases to affect the motion at the airfoil. However, the influence of the circulation induced around the airfoil by shedding is to maintain the smooth flow from the trailing edge.

The requirement that the flow be smooth at the trailing edge determines the magnitude of Γ. In the ζ plane we obtain the velocity potential in the presence of the *negative* circulation about the airfoil by adding $(i\Gamma/2\pi)\ln\zeta$ to the right-hand side of (3.8.7):

$$w = \frac{aU}{2}\left(\zeta e^{-i\alpha} + \frac{e^{i\alpha}}{\zeta}\right) + \frac{i\Gamma}{2\pi}\ln\zeta. \tag{3.8.9}$$

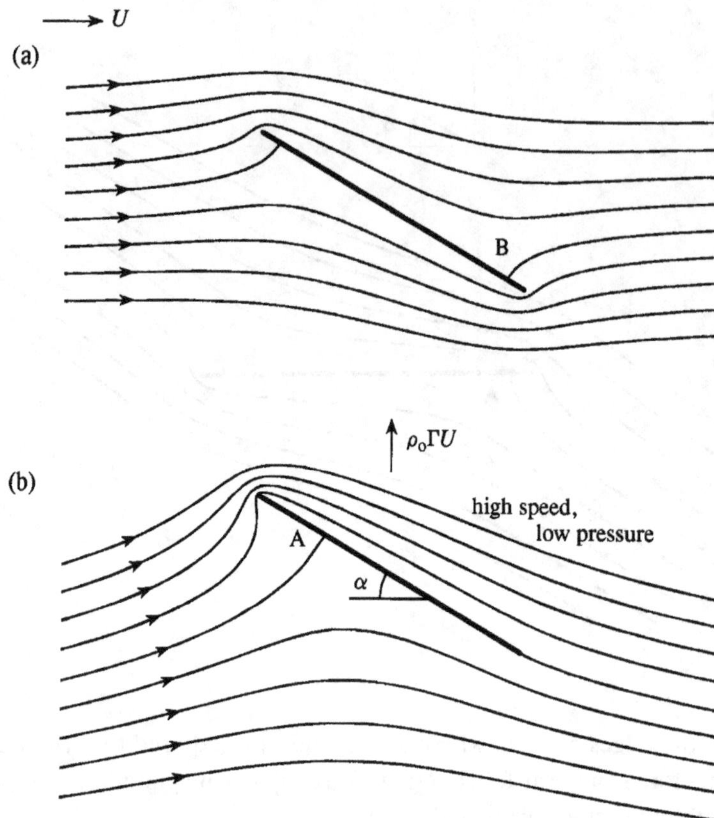

Figure 3.8.4

The complex velocity in the z plane then becomes

$$\frac{dw}{dz} = U\left(\cos\alpha - \frac{iz\sin\alpha}{\sqrt{z^2 - a^2}}\right) + \frac{i\Gamma}{2\pi\sqrt{z^2 - a^2}}. \tag{3.8.10}$$

The value of Γ is found by application of the *Kutta–Joukowski hypothesis* that the velocity should remain finite at $z = a$, which yields

$$\Gamma = 2\pi Ua\sin\alpha, \tag{3.8.11}$$

in which case

$$\frac{dw}{dz} = U\cos\alpha - iU\sin\alpha\sqrt{\frac{z-a}{z+a}}. \tag{3.8.12}$$

The flow therefore leaves the trailing edge tangentially with velocity $U\cos\alpha$ on both sides of the airfoil. The situation is illustrated in Figure 3.8.4. This compares the stream-line patterns (a) without circulation, and (b) with circulation (3.8.11) around the airfoil, when the airfoil and flow are rotated clockwise through the angle of attack α, so that the incident mean flow is horizontal.

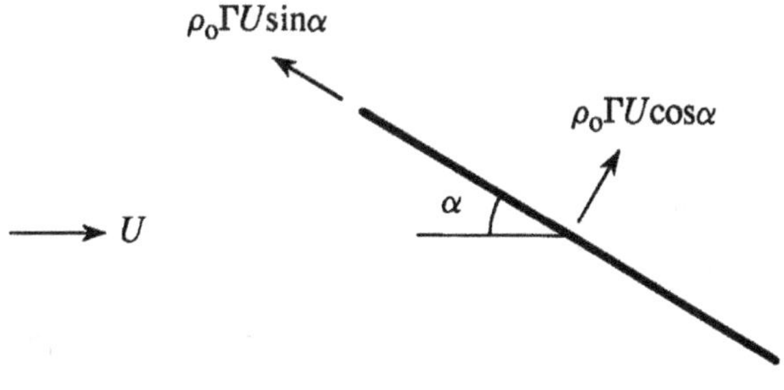

Figure 3.8.5

3.8.2 Calculation of the lift

There is no net force between the airfoil and fluid except in the presence of circulation. In this case the force is determined in two dimensions by use of the time-independent form of Blasius formula (3.3.2). For steady flow dw/dz is given by (3.8.10) (with respect to coordinate axes orientated as in Figure 3.8.3), so that

$$F_1 - i F_2 = \frac{i\rho_o}{2} \oint_S \left(\frac{dw}{dz}\right)^2 dz = -\frac{i\rho_o}{2} \oint_S \left(U\cos\alpha - \frac{izU\sin\alpha}{\sqrt{z^2 - a^2}} + \frac{i\Gamma}{2\pi\sqrt{z^2 - a^2}}\right)^2 dz.$$

The integrand is regular throughout the region occupied by the fluid, and the value of the integral is therefore the same as that evaluated on a large circle $|z| = R \to \infty$. On this circle only the term in the integrand that behaves like C/z ($C = \text{constant}$) can make a nontrivial contribution to the integral, by an amount equal precisely to $2\pi i C$. Hence

$$F_1 - i F_2 = -\rho_o \Gamma U \sin\alpha - i\rho_o \Gamma U \cos\alpha, \quad \text{i.e.,} \quad (F_1, F_2) = \rho_o \Gamma U(-\sin\alpha, \cos\alpha).$$

The net force **F** on the airfoil (orientated as in Figure 3.8.3) accordingly consists of a component $\rho_o \Gamma U \sin\alpha$ in the negative x direction and a component $\rho_o \Gamma U \cos\alpha$ in the y direction. The overall force is therefore a lift force of magnitude

$$\text{lift} = \rho_o \Gamma U \equiv 2\pi \rho_o U^2 a \sin\alpha \quad \text{per unit span,} \tag{3.8.13}$$

directed as indicated in Figure 3.8.4(b) at right angles to the impinging mean flow. The lift is the resultant of a component of magnitude $\rho_o \Gamma U \cos\alpha$ normal to the plane of the airfoil and the component $\rho_o \Gamma U \sin\alpha$ parallel to the airfoil (Figure 3.8.5); the latter is produced by leading-edge suction. There is no suction at the trailing edge because the flow velocity remains finite there.

3.8.3 Lift calculated from the Kirchhoff vector force formula

The lift can also be calculated in a very convenient fashion by use of formula (4.5.15) of Chapter 4. For steady motion in the x direction and when viscous forces are neglected, the lift (in the y direction) is given by

$$\text{lift} = \rho_o \int_V \nabla X_2 \cdot \boldsymbol{\omega} \wedge \mathbf{v}_{\text{rel}} d^3\mathbf{x}, \tag{3.8.14}$$

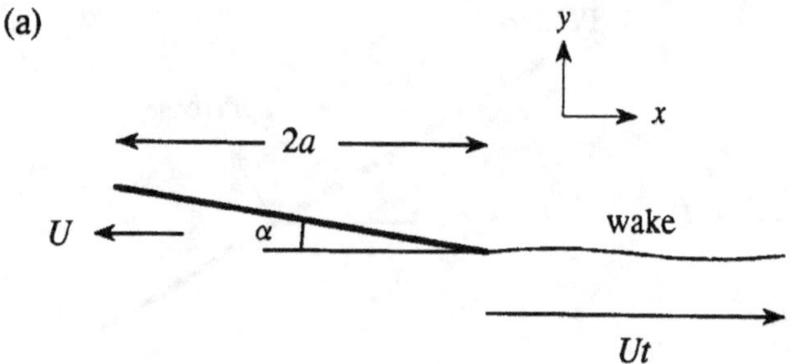

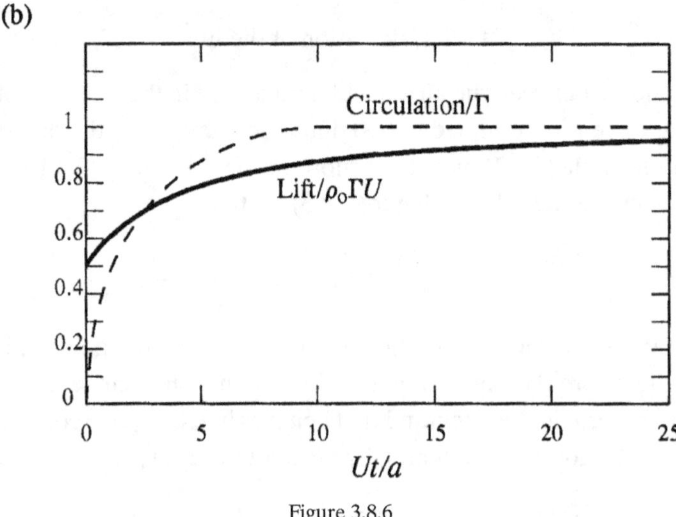

Figure 3.8.6

where $X_2(\mathbf{x})$ is the y component of the Kirchhoff vector (the velocity potential of flow past the airfoil having unit speed in the y direction at large distances from the airfoil) and $\mathbf{v}_{\mathrm{rel}}$ is the convection velocity of the shed vorticity relative to the airfoil.

When the motion is steady, all of the wake vorticity shed from the trailing edge has been swept away towards $x = +\infty$. At such points $X_2 \sim y$, where it may be assumed that the convection velocity $\mathbf{v}_{\mathrm{rel}} = (U, 0)$:

$$\therefore \quad \text{lift per unit span} = \rho_o \int_V \omega(x - Ut, y)U\,dx\,dy \equiv \rho_o \Gamma U,$$

where ω is the shed vorticity distribution of total circulation Γ.

3.8.4 Lift developed by a starting airfoil

When the angle of attack α is small it is possible to obtain an analytical representation of the growth of the lift for an airfoil that starts impulsively from rest at $t = 0$ and proceeds to translate at constant speed U (Wagner 1925). Let the motion be in the negative x direction in an otherwise stationary fluid [Figure 3.8.6(a)], and let the coordinate origin translate with the airfoil at the midchord position. In a linearised approximation (to first order in α) it may be assumed that the vorticity ω shed from the trailing edge is confined to a *vortex sheet* lying along the x axis between $x = a$ and $x = a + Ut$ [that is, that $\mathbf{v}_{\mathrm{rel}} = (U, 0)$], so that

$$\omega(\mathbf{x}, t) = \gamma_o(x, t)\delta(y), \tag{3.8.15}$$

where $\gamma_o(x, t)$ is the circulation per unit length of the sheet.

It may then be shown (by the method discussed below in §3.12) that

$$\gamma_o(x, t) = \frac{2\alpha U}{\pi} \int_{-\infty}^{\infty} \frac{e^{-ik(Ut-x)}dk}{(k + i0)[\mathrm{H}_0^{(1)}(ka) + i\mathrm{H}_1^{(1)}(ka)]}, \quad a < x < a + Ut, \tag{3.8.16}$$

where $\mathrm{H}_0^{(1)}$ and $\mathrm{H}_1^{(1)}$ are Hankel functions, and the integration path passes just above any singularities on the real k axis. This integral is easily evaluated numerically and can be used with formula (3.8.14) to calculate the lift force as a function of time.

The circulation in the wake at time t is

$$\int_a^{a+Ut} \gamma_o(x, t)dx,$$

which is effectively equal to the total shed circulation $\Gamma \approx 2\pi\alpha a U$ when Ut/a exceeds about 10; it is plotted as the dashed curve in Figure 3.8.6(b).

Because $\omega \sim O(\alpha)$, the Kirchhoff vector X_2 in (3.8.14) can be approximated by its value for an airfoil at zero angle of attack (Table 2.19.1), namely

$$X_2 = \mathrm{Re}\left(-i\sqrt{z^2 - a^2}\right),$$

in terms of which

$$\text{lift per unit span} = \rho_o U \int_V \gamma_o(x, t)\delta(y)\frac{\partial X_2}{\partial y}dxdy = \rho_o U \int_a^{a+Ut} \frac{\gamma_o(x, t)xdx}{\sqrt{x^2 - a^2}}. \tag{3.8.17}$$

When $Ut/a \gg 1$ the shed vorticity is far downstream of the airfoil, where $\partial X_2/\partial y \to 1$, and the lift tends to the Kutta–Joukowski value $2\pi\alpha a\rho_o U^2 \equiv \rho_o\Gamma U$. The approach to this limit is plotted as the solid curve in Figure 3.8.6(b), which also indicates that the lift immediately jumps to half its final value as soon as the airfoil begins to move.

3.9 The Joukowski airfoil

The generation and release of the starting vortex removes the singular velocity at the trailing edge and eliminates the associated suction force. The corresponding pressure distribution on the airfoil – a combination of leading-edge suction and a sideways pressure force – produces a lift force in a direction precisely at right angles to that of the impinging mean flow. For the thin-plate airfoil, however, there remains a leading-edge flow with a nominally infinite maximum velocity, which is not realisable in practice and at which a real flow would immediately separate; this would create a region above the airfoil of relative low velocity and high pressure, filled with vorticity, that would ultimately cause the airfoil to 'stall'. This is avoided by proper airfoil design. The leading edge must be sufficiently thickened and *rounded* and thereby furnished with a nonzero

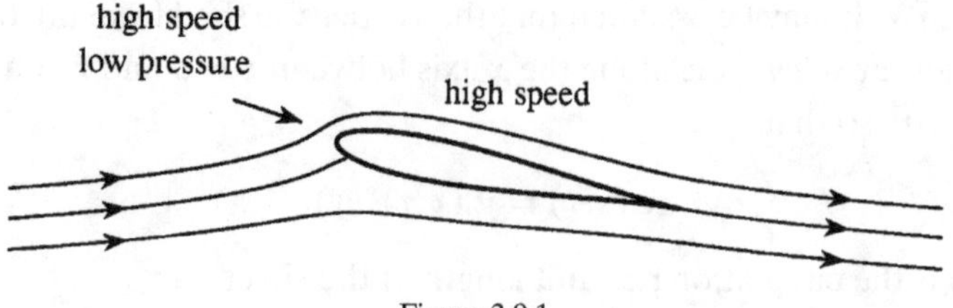

Figure 3.9.1

radius of curvature (see Figure 3.9.1). The effect is that separation does not occur provided the angle of attack α is sufficiently small (say, less than about 8°–10°). Indeed, the convergence of the streamlines in Figure 3.8.4(b) at the leading edge shows that flow over the airfoil forward of the stagnation point A is *accelerated*. This is also true for a rounded edge: The flow is accelerated into the low-pressure region above the airfoil and therefore tends to remain attached to the surface provided the radius of curvature at the leading edge is not too small. This would not happen at the trailing edge of Figure 3.8.4(a), however, even when rounded, because the edge flow is directed towards the high-pressure stagnation point B, which retards the fluid and inevitably leads to separation.

3.9.1 Streamline flow past an airfoil

The Joukowski transformation (3.8.1) can be used to study airfoils with rounded nose profiles. Recall that the edges of the thin-plate airfoil $z = \pm a$ correspond to the singular points of the transformation where $d\zeta/dz = \infty$. These points are at $\zeta = \pm 1$ in the ζ plane. They lie inside a cylinder $\zeta = \beta e^{i\Theta}$ $(-\pi < \Theta \le \pi)$ when $\beta > 1$, which must therefore be mapped conformally into a smooth profile (without edges) in the z plane. According to the second of Equations (3.8.1) the profile in the z plane is the elliptic cylinder (Figure 3.9.2)

$$\frac{z}{a} = \frac{x}{a} + i\frac{y}{a} = \frac{1}{2}\left(\beta + \frac{1}{\beta}\right)\cos\Theta + \frac{i}{2}\left(\beta - \frac{1}{\beta}\right)\sin\Theta, \quad -\pi < \Theta \le \pi.$$

Suppose the circle $|\zeta| = \beta$ in Figure 3.9.2 is shifted a distance δ to the left along the real axis until it just touches the circle $|\zeta| = 1$ at $\zeta = 1$. Then $\beta = 1 + \delta$, and the new circle (in Figure 3.9.3) is

$$\zeta + \delta = (1 + \delta)e^{i\Theta}, \quad -\pi < \Theta \le \pi, \ \delta > 0. \tag{3.9.1}$$

Mapping (3.8.1) of this circle is nonconformal at $\zeta = 1$, and its image in the z plane has a sharp edge at $z = a$. The leading edge remains smooth, however, because the other singular point at $\zeta = -1$ is still an interior point.

The profile in the z plane is the symmetric *Joukowski airfoil* illustrated in Figure 3.9.3 for $\delta = 0.1$. The airfoil extends along the x axis over the interval

$$-\frac{a}{2}\left(1 + 2\delta + \frac{1}{1 + 2\delta}\right) < x < a$$

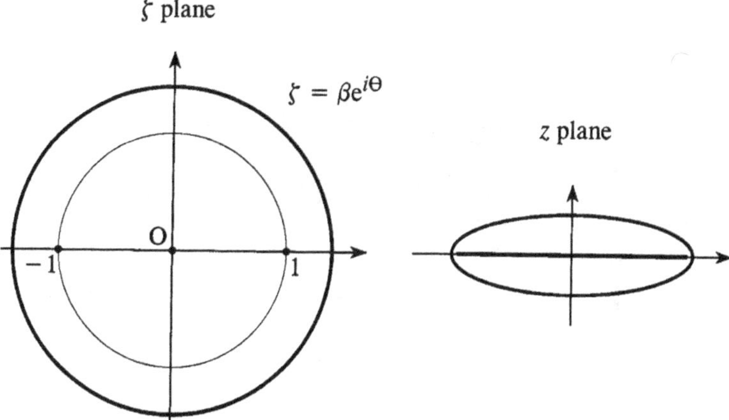

Figure 3.9.2

with chord $2a(1+\delta)^2/(1+2\delta)$, and the radius of curvature of the nose is

$$\mathcal{R} = \frac{8a\delta^2(1+\delta)^3}{(1+2\delta)[(1+2\delta)^3 - 2\delta]} \sim 8a\delta^2.$$

In the ζ plane the complex potential of flow at speed U at angle α to the real axis is given by the following modification of (3.8.9):

$$w = \frac{Ua}{2}\left[(\zeta + \delta)e^{-i\alpha} + \frac{(1+\delta)^2 e^{i\alpha}}{(\zeta + \delta)}\right] + \frac{i\Gamma}{2\pi}\ln(\zeta + \delta). \tag{3.9.2}$$

The trailing edge of the airfoil corresponds to the point $\zeta = 1$, and the velocity will remain finite provided $dw/d\zeta = 0$ there, which yields

$$\Gamma = 2\pi Ua(1+\delta)\sin\alpha.$$

When this condition is satisfied the velocity is finite everywhere on the airfoil. The airfoil experiences a lift force equal to $\rho_o \Gamma U$ per unit span, as in the case of the thin-plate airfoil.

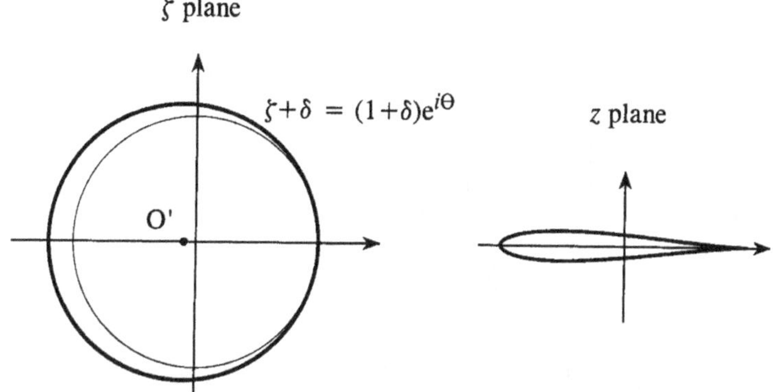

Figure 3.9.3

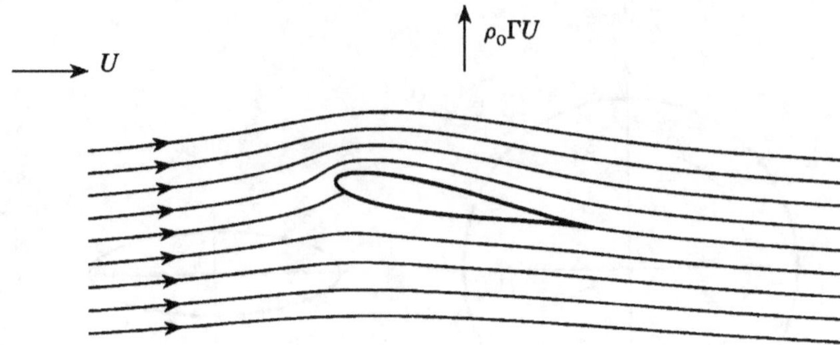

Figure 3.9.4

Figure 3.9.4 shows the typical streamline pattern when the angle of attack $\alpha = 10°$ (a large angle for practical airfoils, chosen for the purpose of illustration), when the airfoil and flow are rotated through this angle so that the incident flow is horizontal.

Symmetric airfoils of this kind are used for tailplanes and rudders when there is no preference in the desired direction of lift. For aircraft wings, however, improved lift characteristics at smaller angles of attack are obtained by use of cambered profiles, where the lower side of the airfoil is either flat or concave. We can generate such profiles by considering flow around a circle in the ζ plane that again passes through the singular point $\zeta = 1$, but with its centre shifted to $\zeta = -\delta + i\delta'$ (O′ in Figure 3.9.5), that is the circle

$$\zeta = -\delta + i\delta' + \sqrt{(1+\delta)^2 + \delta'^2}\,e^{i\Theta}, \quad -\pi < \Theta \leq \pi.$$

The leading edge of the airfoil is rounded because $\zeta = -1$ lies within the circle. The case shown in the figure corresponds to $\delta = 0.1$, $\delta' = 0.2$. The flow pattern can be calculated as before, but such details are left as an exercise for the reader.

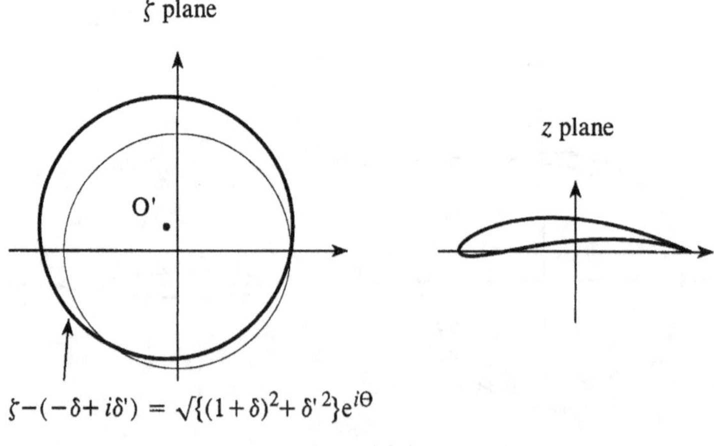

$$\zeta - (-\delta + i\delta') = \sqrt{\{(1+\delta)^2 + \delta'^2\}}\,e^{i\Theta}$$

Figure 3.9.5

3.10 Separation and stall

The presence of the viscous boundary layers on an airfoil, whether laminar or turbulent, can under certain conditions cause the flow to *separate*, forming a free shear layer or velocity discontinuity near the surface. The mean pressure gradient outside the boundary layer has a strong influence on boundary-layer development. If the external flow is accelerated by a fall in pressure in the direction of motion fluid particles in the boundary layer (which of necessity are travelling more slowly) also experience an acceleration in the direction of motion and continue to move over the surface of the body within an *attached* boundary layer. If the pressure increases in the flow direction, however, tending to retard the external flow, the slower-moving fluid in the boundary layer is yet more strongly retarded and ultimately, when the kinetic energy of boundary-layer particles has been lost to the increasing pressure, the direction of flow in the boundary layer may be reversed. Thus flow far from the wall tends to continue in the same direction, but near the wall, where the initial kinetic energy of the fluid is smaller, the flow can come to rest or be reversed. This stationary or slowly moving fluid close to the wall gradually forms a thickened region between the surface and the higher-speed external flow; any such thickening region of backward-moving fluid must push the external flow away from the surface and cause the flow to 'separate'. A free shear layer formed in this way is unstable and rapidly evolves into a succession of vortices, which feed back on the flow over the airfoil, modifying the pressure distribution and leading to the formation of a mean state for the separated flow.

Separation always occurs when the angle of attack α is large enough (typically greater than about $8°$–$10°$), starting close to the trailing edge on the 'upper' or suction side, and moving progressively forward towards the nose as the angle of attack increases. The outer flow is accelerated around the nose but slows down towards the trailing edge, where the pressure increases towards that in the free stream. A schematic representation of separation is shown in Figure 3.10.1 for a thin, symmetric airfoil of chord c. Separation occurs at a distance x_s along the airfoil centreline from the nose. The lower curve in Figure 3.10.2 shows how the separation point progressively moves forward from the trailing edge with increasing angle of attack when α exceeds about $10°$.

At small angles of attack the flow remains attached and the lift is well approximated by irrotational flow formula (3.8.13). Setting $c = 2a$ in that formula, we can write

$$\text{lift per unit span} = c\pi\alpha\rho_o U^2 = \frac{1}{2}\rho_o U^2 c\, C_L, \quad \alpha \ll 1,$$

where the *lift coefficient*

$$C_L = 2\pi\alpha \quad (\alpha \text{ in radians}). \tag{3.10.1}$$

This linear dependence on α for small angles of attack corresponds to the straight-line section of the empirical solid curve labelled C_L in Figure 3.10.2. For this particular airfoil the lift peaks at $\alpha \sim 14°$, when separation occurs at about one fifth of a chord length ($0.2c$) ahead of the trailing edge. The lift decreases quickly beyond this angle, as the

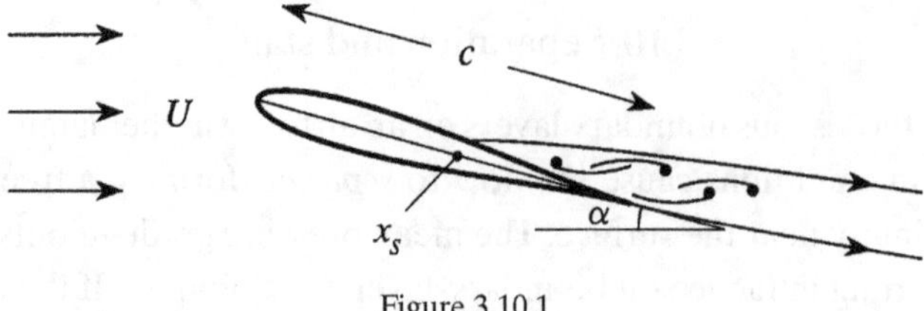

Figure 3.10.1

separation point moves rapidly forward and the flow above the airfoil becomes 'stalled' and turbulent with a mean pressure within the turbulent region approaching that of the free stream.

3.10.1 Linear theory of separation

A simplified form of free-streamline theory (§3.7) can be used to extend the small-angle approximation (3.10.1) for C_L into the region where stall occurs (Khrabrov & Ol 2004). To do this the airfoil is first imagined to be replaced with a thin strip of chord c inclined at angle $\alpha \ll 1$ to the direction of the oncoming stream. Take a reference frame in which the airfoil is at rest with the origin in the z plane at the leading edge [O in Figure 3.10.3(a)], so that the section $0 < x < c$ of the real axis coincides with the airfoil and the complex potential of the impinging mean flow can be taken as $w_o = Uze^{-i\alpha}$ ($U =$ constant). The lower side $y = -0$ of the interval $0 < x < c$ of the real axis corresponds to the lower or 'pressure' side of the airfoil. On the upper ('suction') side only the section $0 < x < x_s$, $y = +0$ of the airfoil is in 'contact' with the mean flow. Referring to Figure 3.10.1, when the flow is steady the region of separation can be regarded as bounded below by a free streamline springing from the sharp trailing edge and above by one emanating from the separation point x_s. In the simplest approximation the pressure within the separated zone is assumed to be equal to the undisturbed pressure at infinity

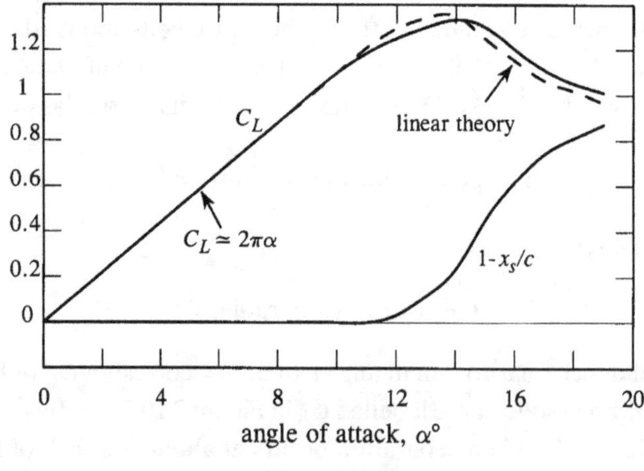

Figure 3.10.2

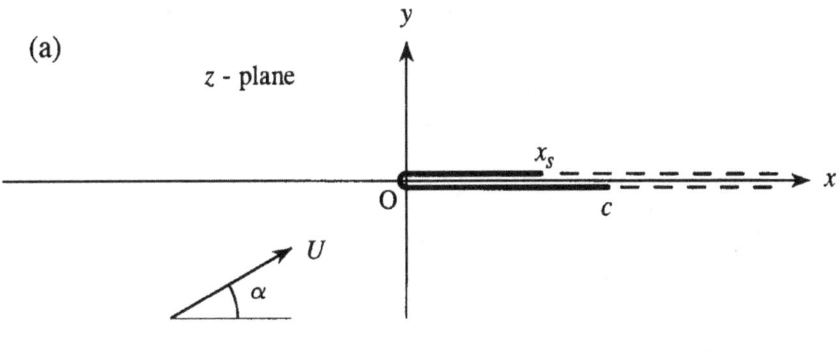

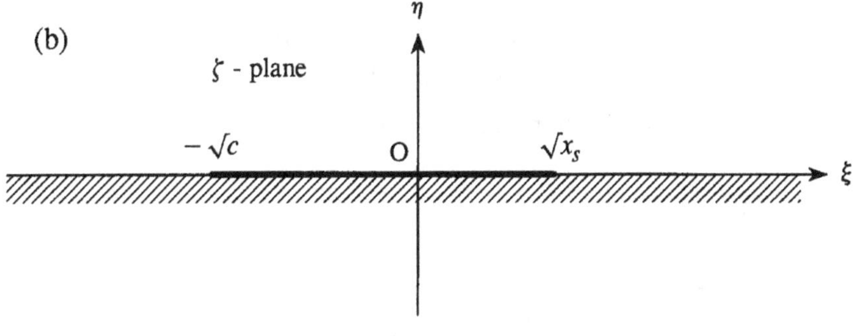

Figure 3.10.3

p_o, say, and therefore the flow speed in the main flow on the free streamlines is just equal to U.

Linear theory is applicable for sufficiently small values of α that the free streamlines may be regarded as coincident with the upper and lower sides of the real axis, as indicated by the dashed lines in Figure 3.10.3(a). Then, if the complex potential is

$$w = w_o(z) + \hat{w}(z) \equiv Uze^{-i\alpha} + \hat{w}(z),$$

where $\hat{w}(z)$ is the $O(\alpha)$ perturbation produced by the airfoil, the perturbation velocity $\hat{w}'(z) = \hat{u} - i\hat{v}$ must satisfy $\hat{u} = 0$ on the free streamlines. Similarly, the normal (y-) component of velocity $\sim U\alpha + \hat{v}$ must vanish on those sections of the airfoil where the mean flow is attached.

Hence a complex velocity $\hat{w}'(z) = \hat{u} - i\hat{v}$ is required that satisfies

$$\hat{u} = 0 \text{ for } \begin{cases} x > x_s, & y = +0 \\ x > c, & y = -0 \end{cases} \quad \text{and} \quad \hat{v} = -\alpha U \text{ for } \begin{cases} 0 < x < x_s, & y = +0 \\ 0 < x < c, & y = -0 \end{cases}. \quad (3.10.2)$$

At large distances from the airfoil $\hat{w}'(z) \sim i\Gamma/2\pi z$, where Γ is equal to the circulation around a contour C that just encloses the airfoil:

$$\Gamma = -\oint_C \hat{u}\,dx \equiv \int_0^{x_s} \hat{u}(x, +0)dx - \int_0^c \hat{u}(x, -0)dx, \quad (3.10.3)$$

and the lift per unit span is $\rho_o \Gamma U$. The Kutta condition requires that $\hat{w}'(z)$ should remain finite at the trailing edge $z = c$ and also at the separation point $x = x_s$, $y = +0$.

To determine $\hat{u}$ for use in (3.10.3) the fluid region is mapped into the upper half of the ζ plane [$\zeta = \xi + i\eta$, Figure 3.10.3(b)] by means of the transformation

$$\zeta = \sqrt{z}.$$

Then the lower side of the airfoil corresponds to the section $-\sqrt{c} < \xi < 0$ of the real axis in the ζ plane, and the 'wetted' (or attached) section of the suction side of the airfoil maps into the interval $0 < \xi < \sqrt{x_s}$; the upper and lower free streamlines in Figure 3.10.3(a) correspond respectively to the intervals $\sqrt{x_s} < \xi < +\infty$ and $-\infty < \xi < -\sqrt{c}$ of the real axis.

Also

$$\hat{u} - i\hat{v} = \frac{d\hat{w}}{dz} = \frac{d\zeta}{dz}\frac{d\hat{w}}{d\zeta} = \frac{1}{2\zeta}\frac{d\hat{w}}{d\zeta}, \tag{3.10.4}$$

so that on the real ζ axis, conditions (3.10.2) imply that

$$\mathrm{Re}\left(\frac{d\hat{w}}{d\zeta}\right) = 0 \ \ \text{for} \ \ -\infty < \xi < -\sqrt{c} \ \ \text{and} \ \ \sqrt{x_s} < \xi < +\infty,$$

$$\mathrm{Im}\left(\frac{d\hat{w}}{d\zeta}\right) = 2\alpha U\xi \ \ \text{for} \ \ -\sqrt{c} < \xi < \sqrt{x_s}. \tag{3.10.5}$$

Now $d\hat{w}/d\zeta$ is regular in $\mathrm{Im}\,\zeta > 0$ and decreases like $1/\zeta$ as $|\zeta| \to \infty$. Thus, if the reciprocal $1/g(\zeta)$ of a function $g(\zeta)$ is also regular and bounded in the upper half-plane, Cauchy's integral theorem implies that

$$\frac{d\hat{w}}{d\zeta} = \frac{g(\zeta)}{2\pi i}\int_{-\infty}^{\infty}\frac{d\hat{w}/d\zeta'\,d\zeta'}{(\zeta' - \zeta)g(\zeta')}, \ \ \mathrm{Im}\,\zeta > 0, \tag{3.10.6}$$

where the integration contour lies just above the real axis. For any well-behaved function $f(\zeta)$,

$$\int_{-\infty}^{\infty}\frac{f(\zeta')d\zeta'}{\zeta' - \zeta} \ \to \ \fint_{-\infty}^{\infty}\frac{f(\xi')d\xi'}{\xi' - \xi} + \pi i f(\xi)$$

as $\zeta = \xi + i\eta$ approaches the real axis from above ($\eta \to +0$), where the integral on the right-hand side (now taken along the real axis) is a principal value. Hence (3.10.6) yields

$$\frac{d\hat{w}}{d\xi} = \frac{g(\xi)}{\pi i}\fint_{-\infty}^{\infty}\frac{d\hat{w}/d\xi'\,d\xi'}{(\xi' - \xi)g(\xi')}, \ \ \mathrm{Im}\,\zeta \equiv \eta = +0. \tag{3.10.7}$$

Let us now take for $g(\zeta)$ the function

$$g(\zeta) = i\sqrt{(\zeta + \sqrt{c})(\zeta - \sqrt{x_s})}, \ \ \mathrm{Im}\,\zeta \geq 0, \tag{3.10.8}$$

where the positive square root is taken on the real axis for $\xi > \sqrt{x_s}$. This choice ensures that the Kutta condition is satisfied at $\xi = -\sqrt{c}$ and $\sqrt{x_s}$, and also that, *on the real axis*,

$$g(\xi) \text{ is } \begin{cases} \text{real on the airfoil, } -\sqrt{c} < \xi < \sqrt{x_s} \\ \text{pure imaginary elsewhere} \end{cases}.$$

Hence, by substituting into (3.10.7) and taking account of conditions (3.10.5) we find, by taking the real part when ξ lies on the airfoil $(-\sqrt{c} < \xi < \sqrt{x_s})$,

$$\mathrm{Re}\left(\frac{d\hat{w}}{d\xi}\right) = \frac{2\alpha U}{\pi}\sqrt{(\xi + \sqrt{c})(\sqrt{x_s} - \xi)} \int_{-\sqrt{c}}^{\sqrt{x_s}} \frac{\xi' d\xi'}{(\xi' - \xi)\sqrt{(\xi' + \sqrt{c})(\sqrt{x_s} - \xi')}}. \quad (3.10.9)$$

We evaluate the integral by putting

$$\xi' = \tfrac{1}{2}(\sqrt{x_s} + \sqrt{c})\cos\vartheta + \tfrac{1}{2}(\sqrt{x_s} - \sqrt{c}), \quad \xi = \tfrac{1}{2}(\sqrt{x_s} + \sqrt{c})\cos\Theta + \tfrac{1}{2}(\sqrt{x_s} - \sqrt{c})$$

and using the formula

$$\int_0^\pi \frac{\cos n\vartheta \, d\vartheta}{\cos\vartheta - \cos\Theta} = \frac{\pi \sin n\Theta}{\sin\Theta}.$$

Therefore

$$\int_{-\sqrt{c}}^{\sqrt{x_s}} \frac{\xi' d\xi'}{(\xi' - \xi)\sqrt{(\xi' + \sqrt{c})(\sqrt{x_s} - \xi')}} = \int_0^\pi \frac{\left[\cos\vartheta + \left(\frac{\sqrt{x_s} - \sqrt{c}}{\sqrt{x_s} + \sqrt{c}}\right)\right] d\vartheta}{\cos\vartheta - \cos\Theta} = \pi.$$

Hence, the tangential component of velocity $\hat{u}$ on the upper and lower sides of the airfoil becomes [as a function of ξ, from (3.10.4)]

$$\hat{u} = \frac{\alpha U}{\xi}\sqrt{(\xi + \sqrt{c})(\sqrt{x_s} - \xi)}, \quad -\sqrt{c} < \xi < \sqrt{x_s},$$

and for the circulation Γ Equation (3.10.3) yields (with $dx = 2\xi\, d\xi$)

$$\Gamma = 2\alpha U \int_{-\sqrt{c}}^{\sqrt{x_s}} \sqrt{(\xi + \sqrt{c})(\sqrt{x_s} - \xi)}\, d\xi = \frac{c\pi\alpha U}{4}\left(1 + \sqrt{\frac{x_s}{c}}\right)^2. \quad (3.10.10)$$

The lift per unit span and the lift coefficient for separated airfoil flow are therefore

$$\mathrm{lift} = c\pi\alpha\rho_o U^2 \frac{1}{4}\left(1 + \sqrt{\frac{x_s}{c}}\right)^2, \quad C_L = \frac{\alpha\pi}{2}\left(1 + \sqrt{\frac{x_s}{c}}\right)^2. \quad (3.10.11)$$

These are the required representations correct to first order in the angle of attack α. However, they involve the position $x = x_s$ of the separation point that also depends on α. To illustrate the predicted behaviour of the lift it is accordingly necessary to use a separate source of data that gives the dependence of x_s on α; for example, the lower curve in Figure 3.10.2 can be used to calculate x_s/c as a function of α. When this is done approximation (3.10.11) for the lift coefficient C_L is represented by the dashed curve in Figure 3.10.1 – a conclusion that agrees remarkably well with the exact variation.

3.11 Sedov's method

Representation (3.10.6) of the complex velocity in separated flow is a particular case of Sedov's (1965) treatment of potential flow past a set of coplanar thin-plate airfoils. Consider the flat-plate airfoil of Figure 3.11.1 occupying the interval $|x| < a$ of the real axis with an impinging uniform flow $w_o = Uze^{-i\alpha}$. Only the component of the

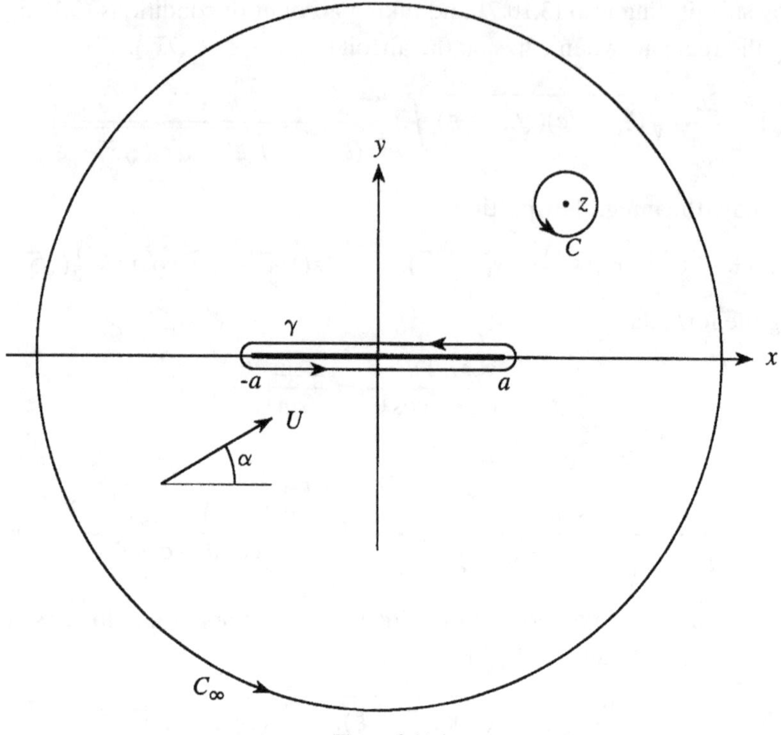

Figure 3.11.1

incident flow in the y direction normal to the plate interacts with the plate, producing a perturbation potential $\hat{w}(z)$ such that the complex potential is

$$w = Uze^{-i\alpha} + \hat{w}(z)$$

where the perturbation complex velocity $\hat{w}' = \hat{u}(x, y) - i\hat{v}(x, y)$ vanishes at least as fast as $1/z$ at infinity.

3.11.1 Boundary conditions

The net normal component of velocity must vanish on both sides of the airfoil, so that

$$\hat{v}(x, \pm 0) = -U \sin \alpha \quad \text{for } |x| < a. \tag{3.11.1}$$

In addition, however, Sedov's method depends for its success on the fact that

$$\hat{u}(x, +0) = -\hat{u}(x, -0) \quad \text{for all } x, \tag{3.11.2}$$

which we can establish by first considering Cauchy's integral theorem applied to the contour C enclosing the point z:

$$\hat{w}'(z) = \frac{1}{2\pi i} \oint_C \frac{\hat{w}'(\zeta)d\zeta}{(\zeta - z)}.$$

Let C be deformed onto the circle C_∞ at infinity and the contour γ that just encloses the airfoil. There is no contribution from C_∞, so that when z lies on the real axis in $|x| > a$ we can write, by equating real and imaginary parts,

$$\hat{u}(x,0) = \frac{1}{2\pi} \oint_\gamma \frac{\hat{v}(\zeta)d\zeta}{\zeta - x}, \quad \hat{v}(x,0) = -\frac{1}{2\pi} \oint_\gamma \frac{\hat{u}(\zeta)d\zeta}{\zeta - x},$$

where here and below all contours are assumed to be traversed in the anticlockwise or *positive* sense in Figure 3.11.1. Condition (3.11.1) implies that the first integral is null and therefore that, on the real axis,

$$\hat{u}(x,0) = 0 \quad \text{and} \quad \hat{w}'(z) = -i\hat{v}(x,0) \quad \text{for } |x| > a.$$

The first of these is equivalent to (3.11.2) for points of the real axis not on the airfoil.

However, $\hat{w}'(z)$ is regular for $y = \operatorname{Im} z \geqslant 0$ and can be analytically extended from its value in $|x| > a$ on the real axis as the function $\hat{w}'(z) = -i\hat{v}(x + iy, 0)$ (an example of the Schwarz reflection principle). Hence

$$\hat{u}(x, y) - i\hat{v}(x, y) = -i\hat{v}(x + iy, 0),$$

$$\hat{u}(x, -y) - i\hat{v}(x, -y) = -i\hat{v}(x - iy, 0).$$

Adding and subtracting these equations and noting that $\hat{v}(x + iy, 0)^* = \hat{v}(x - iy, 0)$, we find that

$$\hat{u}(x, y) = -\hat{u}(x, -y), \quad \hat{v}(x, y) = \hat{v}(x, -y), \tag{3.11.3}$$

which completes the proof of (3.11.2).

3.11.2 Sedov's formula

Let a function $g(z)$ be introduced such that $1/g(z)$ is regular in the fluid. We again apply Cauchy's integral theorem using the contour C of Figure 3.11.1, but this time using the function $\hat{w}'(z)/g(z)$, so that for z within C

$$\hat{w}'(z) = \frac{g(z)}{2\pi i} \oint_C \frac{\hat{w}'(\zeta)d\zeta}{(\zeta - z)g(\zeta)}.$$

Deform C onto C_∞ and the contour γ enclosing the airfoil to obtain

$$\hat{w}'(z) = \frac{g(z)}{2\pi i} \oint_{C_\infty} \frac{\hat{w}'(\zeta)d\zeta}{(\zeta - z)g(\zeta)} - \frac{g(z)}{2\pi i} \oint_\gamma \frac{\hat{w}'(\zeta)d\zeta}{(\zeta - z)g(\zeta)}, \tag{3.11.4}$$

where ζ is real on γ.

Now we choose $g(z)$ to satisfy $g(x + i0) = -g(x - i0)$ in the interval $-a < x < a$ occupied by the airfoil, and such that it exhibits appropriate potential flow velocity singularities at the edges $z = \pm a$. Suitable functions are given in Table 3.11.1, where the *positive* square roots are taken for z on the real axis to the right of the airfoil. These

Table 3.11.1

	No Kutta condition	Kutta condition applied at $z = a$,	$z = -a$
$g(z)$	$\dfrac{1}{\sqrt{z^2 - a^2}}$	$\sqrt{\dfrac{z-a}{z+a}}$	$\sqrt{\dfrac{z+a}{z-a}}$

functions are pure imaginary on the airfoil and assume equal and opposite values on the two sides. It follows from this and condition (3.11.2) that (3.11.4) reduces to

$$\hat{w}'(z) = \frac{g(z)}{2\pi i} \oint_{C_\infty} \frac{\hat{w}'(\zeta)d\zeta}{(\zeta - z)g(\zeta)} + \frac{g(z)}{2\pi} \oint_\gamma \frac{\hat{v}(\zeta)d\zeta}{(\zeta - z)g(\zeta)}. \qquad (3.11.5)$$

This is Sedov's formula determining the complex velocity induced by the airfoil in terms of conditions at infinity and the normal velocity on the airfoil.

EXAMPLE 1. AIRFOIL WITH CIRCULATION BUT NO MEAN FLOW When $U = 0$ (so that $\hat{v} = -U\sin\alpha = 0$ on γ) only the first integral on the right makes a contribution to (3.11.5), and then only if $\hat{w}' \sim 1/z$ on C_∞. This corresponds to an airfoil with circulation in the absence of mean flow. The motion must then be singular at both ends of the airfoil. Therefore we must take

$$g(z) = \frac{1}{\sqrt{z^2 - a^2}} \quad \text{and put} \quad \hat{w}' \sim \frac{i\Gamma}{2\pi z} \quad \text{on } C_\infty,$$

where Γ is the circulation (clockwise in Figure 3.11.1) about the airfoil. Then (3.11.5) yields

$$\hat{w}'(z) = \frac{i\Gamma}{2\pi\sqrt{z^2 - a^2}},$$

which corresponds to the result in §3.8 [second term on the right-hand side of (3.8.10)] calculated by means of the Joukowski transformation.

EXAMPLE 2. IRROTATIONAL FLOW WITH NO KUTTA CONDITION When there is no circulation around the airfoil but $U \neq 0$,

$$\hat{w}'(z) \sim \frac{1}{z^2} \quad \text{as} \quad |z| \to \infty.$$

The motion is singular at both ends of the airfoil and we must again take $g(z) = 1/\sqrt{z^2 - a^2}$. There is now no contribution from C_∞, so that, setting $\hat{v} = -U\sin\alpha$, we find that Equation (3.11.5) becomes

$$\hat{w}'(z) = \frac{-U\sin\alpha}{2\pi\sqrt{z^2 - a^2}} \oint_\gamma \frac{\sqrt{\zeta^2 - a^2}\,d\zeta}{\zeta - z},$$

where, by residues (deforming γ onto C_∞),

$$\frac{1}{2\pi i}\oint_\gamma \frac{\sqrt{\zeta^2-a^2}\,d\zeta}{\zeta-z} = -\sqrt{z^2-a^2} + \frac{1}{2\pi i}\oint_{C_\infty}\left(1+\frac{z}{\zeta}\right)d\zeta$$

$$= -\sqrt{z^2-a^2}+z.$$

Hence

$$w'(z) = w'_o(z) + \frac{iU\sin\alpha}{\sqrt{z^2-a^2}}\left(\sqrt{z^2-a^2}-z\right)$$

$$= U\left(\cos\alpha - \frac{iz\sin\alpha}{\sqrt{z^2-a^2}}\right),$$

which is just the first term on the right-hand side of (3.8.10) obtained by conformal transformation.

EXAMPLE 3. COMPLEX VELOCITY FOR A LIFTING AIRFOIL When the Kutta condition is imposed at the trailing edge $z=a$ of the airfoil in Figure 3.11.1 the perturbation velocity $\hat{w}'(z) \sim 1/z$ at infinity and we must take

$$g(z) = \sqrt{\frac{z-a}{z+a}}.$$

Only the second integral contributes to (3.11.5), where on the airfoil $\hat{v} = -U\sin\alpha$, so that

$$w'(z) = Ue^{-i\alpha} - \frac{U\sin\alpha}{2\pi}\sqrt{\frac{z-a}{z+a}}\oint_\gamma \sqrt{\frac{\zeta+a}{\zeta-a}}\frac{d\zeta}{(\zeta-z)}$$

$$= Ue^{-i\alpha} - \frac{U\sin\alpha}{2\pi}\sqrt{\frac{z-a}{z+a}}\times 2\pi i\left(1-\sqrt{\frac{z+a}{z-a}}\right)$$

$$= U\cos\alpha - iU\sin\alpha\sqrt{\frac{z-a}{z+a}},$$

which coincides with (3.8.12) obtained by conformal transformation.

3.11.3 Tandem airfoils

Sedov's method is applicable for any number of coplanar airfoils. For two 'tandem' airfoils (Figure 3.11.2) occupying the intervals (a_1, b_1), (a_2, b_2) of the real axis we have

$$w(z) = Uze^{-i\alpha} + \hat{w}(z),$$

$$\hat{w}'(z) = \frac{g(z)}{2\pi i}\oint_{C_\infty}\frac{\hat{w}'(\zeta)d\zeta}{(\zeta-z)g(\zeta)} + \frac{g(z)}{2\pi}\sum_{j=1}^{2}\oint_{\gamma_j}\frac{\hat{v}(\zeta)d\zeta}{(\zeta-z)g(\zeta)}, \qquad (3.11.6)$$

Table 3.11.2

	No Kutta condition	Kutta condition applied at $z = b_1,\ b_2$
$g(z)$	$\dfrac{1}{\sqrt{(z-a_1)(z-b_1)(z-a_2)(z-b_2)}}$	$\sqrt{\dfrac{(z-b_1)(z-b_2)}{(z-a_1)(z-a_2)}}$

where $g(z)$ is given in Table 3.11.2, the square roots being positive on the real axis to the right of the airfoils. The circulation Γ_j (clockwise) about the jth airfoil is given by

$$\Gamma_j = -\oint_{\gamma_j} \hat{w}'(\zeta)d\zeta. \tag{3.11.7}$$

EXAMPLE 4. TANDEM AIRFOILS WITH NO MEAN FLOW The motion is produced by circulation around the airfoils. Let the left and right airfoils of Figure 3.11.2 have the respective circulations Γ_1 and Γ_2 (in the clockwise sense). At large distances

$$\hat{w}'(z) \sim \frac{i\Gamma}{2\pi z} + \frac{c'}{z^2} + \cdots +,$$

where $\Gamma = \Gamma_1 + \Gamma_2$. If the net circulation $\Gamma = 0$, the velocity decreases like $1/z^2$ at large distances. We obtain the most general circulatory motion by taking $g(z)$ to be the

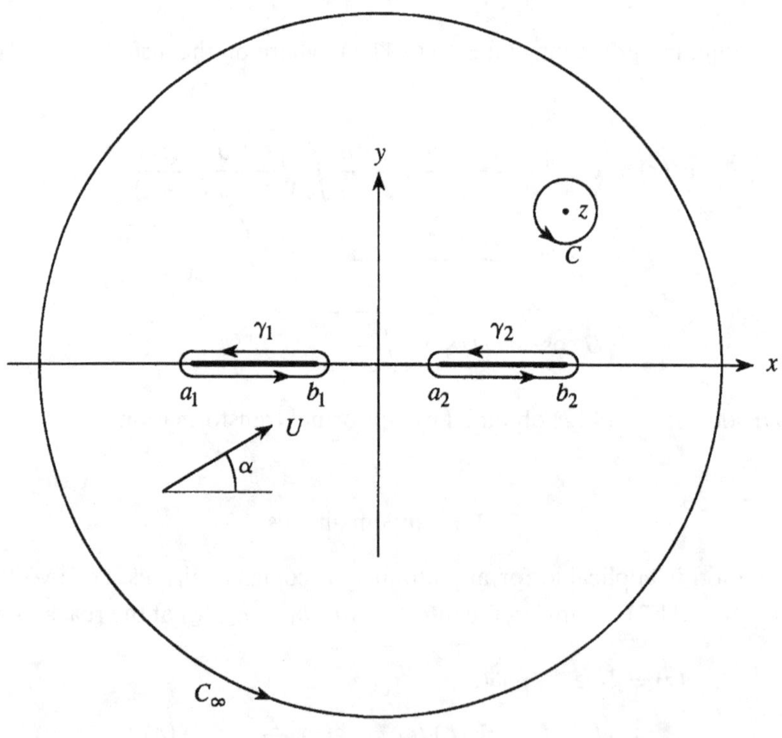

Figure 3.11.2

function in the 'No Kutta condition' column of Table 3.11.2. The final integral in (3.11.6) is null, and it then follows easily from the first that

$$\hat{w}'(z) = \frac{\Gamma z + c}{\sqrt{(z - a_1)(z - b_1)(z - a_2)(z - b_2)}} \quad \text{where} \quad c = c' - \frac{\Gamma(a_1 + a_2 + b_1 + b_2)}{2}.$$

$$(3.11.8)$$

If $\Gamma \neq 0$, the value of c can be chosen to make the velocity finite at one of the four edges.

EXAMPLE 5. LIFT PRODUCED BY TANDEM AIRFOILS Let the Kutta condition be applied at the trailing edges $x = b_1$, b_2, so that $g(z)$ is given by the third column of Table 3.11.2. Then $\hat{w}' \sim i\Gamma/2\pi z$ on C_∞, where Γ is the net circulation about the airfoils. Thus, there is no contribution from the first integral of (3.11.6), so that by taking $\hat{v} = -U \sin \alpha$ on the airfoils, we find

$$\hat{w}'(z) = \frac{-U \sin \alpha\, g(z)}{2\pi} \sum_{j=1}^{2} \oint_{\gamma_j} \left[\frac{(\zeta - a_1)(\zeta - a_2)}{(\zeta - b_1)(\zeta - b_2)}\right]^{\frac{1}{2}} \frac{d\zeta}{(\zeta - z)}$$

$$= \frac{-U \sin \alpha\, g(z)}{2\pi} \left\{ \frac{-2\pi i}{g(z)} + \oint_{C_\infty} \left[1 + \frac{z + (b_1 + b_2 - a_1 - a_2)/2}{\zeta} + \cdots + \right] \frac{d\zeta}{\zeta} \right\}$$

$$= iU \sin \alpha \left\{ 1 - \left[\frac{(z - b_1)(z - b_2)}{(z - a_1)(z - a_2)}\right]^{\frac{1}{2}} \right\}.$$

Hence,

$$\hat{w}'(z) \sim \frac{iU \sin \alpha}{2z} \left(c_1 + c_2\right) \quad \text{as} \quad |z| \to \infty,$$

where $c_1 = b_1 - a_1$, $c_2 = b_2 - a_2$ are the airfoil chords. The circulation about both airfoils is therefore $\Gamma = \pi U(c_1 + c_2) \sin \alpha$, and therefore the

$$\text{lift on the tandem airfoil} = \rho_o U \Gamma = (c_1 + c_2)\pi \sin \alpha\, \rho_o U^2. \qquad (3.11.9)$$

The lift is directed at right angles to the mean flow at speed U and is precisely equal to that predicted in ideal flow for a single airfoil whose chord is equal to the sum of the separate chords.

EXAMPLE 6. UNIFORM FLOW PAST TANDEM AIRFOILS WITH NO CIRCULATION Let the airfoils be of equal chord and be disposed symmetrically with respect to the origin (Figure 3.11.3), and consider uniform incident flow at speed V in the y direction. Then

$$w'(z) = -iV + \hat{w}'(z),$$

where $\hat{w}'(z) \sim c/z^2$ as $|z| \to \infty$ ($c = $ constant).

Take

$$g(z) = \frac{1}{\sqrt{(z^2 - a^2)(z^2 - b^2)}}$$

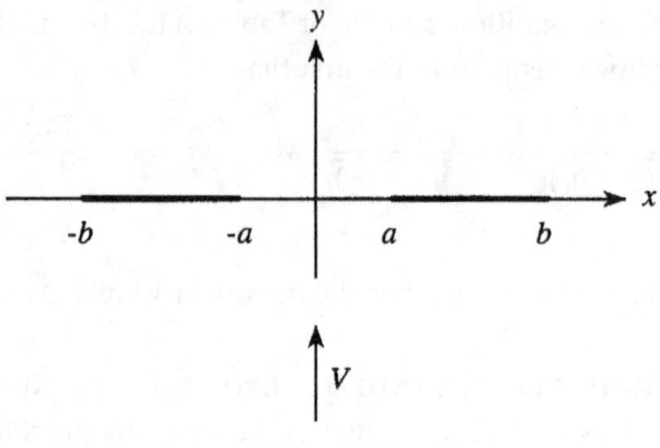

Figure 3.11.3

and put

$$\hat{w}'(z) = cg(z) - \frac{Vg(z)}{2\pi} \sum_{j=1}^{2} \oint_{\gamma_j} \frac{d\zeta}{(z-\zeta)g(\zeta)}$$

$$= iV + \frac{c - iV[z^2 - (a^2 + b^2)/2]}{\sqrt{(z^2 - a^2)(z^2 - b^2)}}.$$

We now choose c to ensure that $\Gamma_1 = -\Gamma_2 \equiv 0$, i.e. such that

$$\oint_{\gamma_2} \frac{c - iV[\zeta^2 - (a^2 + b^2)/2]\,d\zeta}{\sqrt{(\zeta^2 - a^2)(\zeta^2 - b^2)}} = 0,$$

which gives

$$c = iV\left[\frac{b^2 E(k)}{K(k)} + \frac{b^2 - a^2}{2}\right], \quad k = \frac{\sqrt{b^2 - a^2}}{b},$$

where $\qquad E(k) = \int_0^{\frac{\pi}{2}} \sqrt{1 - k^2 \sin^2 \vartheta}\,d\vartheta, \quad K(k) = \int_0^{\frac{\pi}{2}} d\vartheta / \sqrt{1 - k^2 \sin^2 \vartheta}$

are complete elliptic integrals.

Hence the complex velocity becomes

$$w'(z) = \frac{iV\{b^2[1 + E(k)/K(k)] - z^2\}}{\sqrt{(z^2 - a^2)(z^2 - b^2)}}. \tag{3.11.10}$$

3.11.4 High-lift devices

Improved lift and motion control of an aircraft during take-off and landing are achieved by use of retractable 'flaps'. Simple properties of flaps can be investigated by Sedov's method in a linearised approximation when the angle of attack α and the flap deflection angle ϑ are both small (Figure 3.11.4; cf. §3.10).

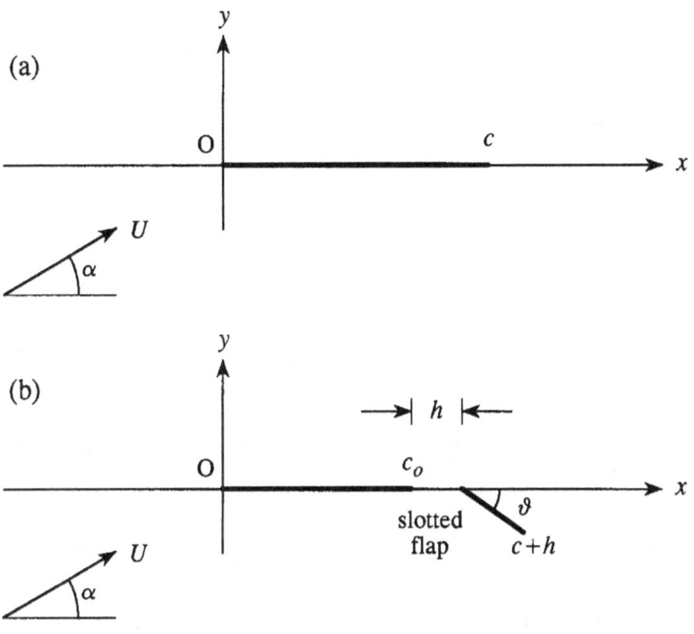

Figure 3.11.4

Let the airfoil have chord c and occupy the interval $0 < x < c$ of the real axis. At small angles of attack the lift coefficient C_{L_o} when the flap is not deployed is

$$C_{L_o} = \frac{\text{lift per unit span}}{c\frac{1}{2}\rho_o U^2} = 2\pi\alpha. \tag{3.11.11}$$

In the deployed state of Figure 3.11.4(b), for a 'slotted flap', the main section of the airfoil has chord c_o, the slot width is h and the flap has chord $c_f = c - c_o$ and is deflected at a small angle ϑ. The linearised complex velocity is

$$w'(z) = U(1 - i\alpha) + \hat{w}'(z), \quad \hat{w}' = \hat{u} - i\hat{v},$$

where on the main airfoil section and on the flap

$$\hat{v}(x, 0) = -U\alpha \text{ for } 0 < x < c_o; \quad \hat{v}(x, 0) = -U(\alpha + \vartheta) \quad \text{for } c_o + h < x < c + h. \tag{3.11.12}$$

Using the result of Example 5, we can write

$$w'(z) = U - i\alpha U g(z) - \frac{\vartheta U g(z)}{2\pi} \oint_{\gamma_f} \frac{d\zeta}{(\zeta - z)g(\zeta)}$$

where

$$g(z) = \left[\frac{(z - c_o)(z - c - h)}{z(z - c_o - h)}\right]^{\frac{1}{2}}$$

and the contour γ_f just encloses the interval $c_o + h < x < c + h$ of the real axis occupied by the flap, and is traversed in the anticlockwise sense.

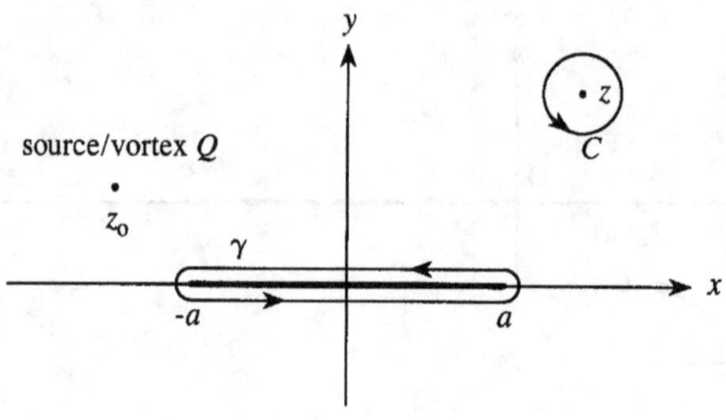

Figure 3.11.5

The lift per unit span $= \rho_o \Gamma U$, where Γ is the circulation around the airfoil and flap, and

$$w'(z) \sim U + \frac{i\Gamma}{2\pi z} = U + \frac{i\alpha Uc}{2z} + \frac{\vartheta Uc}{2\pi z} \oint_{\gamma_f} \frac{d\zeta}{g(\zeta)}, \quad |z| \to \infty.$$

Hence, recalling definition (3.11.11), we find that the lift coefficient C_L in the deployed state is given by

$$\frac{C_L}{C_{L_o}} = 1 + \frac{4\vartheta}{\alpha\pi} \int_0^{\sqrt{1-c_0/c}} \sqrt{\frac{\left(1 + \frac{h}{c} - \xi^2\right)\left(1 - \frac{c_o}{c} - \xi^2\right)}{\left(1 + \frac{h-c_o}{c} - \xi^2\right)}}\, d\xi. \tag{3.11.13}$$

3.11.5 Plain flap or aileron

This is a hinged trailing-edge flap with no slot ($h = 0$). Downward deflection of the flap ($\vartheta > 0$) effectively modifies the lift by simply increasing the camber of the airfoil. In this case (3.11.13) becomes

$$\frac{C_L}{C_{L_o}} = 1 + \frac{2\vartheta}{\alpha\pi}\left[\sin^{-1}\left(\sqrt{\frac{c_f}{c}}\right) + \left(1 - \frac{c_f}{c}\right)\sqrt{\frac{c_f}{c}}\right].$$

The following table shows how the lift coefficient increases with ϑ/α when $c_f = 0.2c$:

ϑ/α	0.5	1	1.5	2
C_L/C_{L_o}	1.26	1.52	1.78	2.04

3.11.6 Point sources and vortices

The solution by Sedov's method in the presence of singularities that are due to sources and vortices can be handled by the superposing of separate solutions for the sources, vortices and the mean flow and adjustment of the values of the circulations about airfoils. Let us illustrate the simplest case of a singularity of strength $Q = q_o - i\Gamma_o$ at $z = z_o$ adjacent to the airfoil of Figure 3.11.5.

To fix ideas, consider the particular case in which there is no impinging mean flow and no circulation around the airfoil. Then

$$w'(z) = \frac{Q}{2\pi(z - z_o)} + \frac{g(z)}{2\pi} \oint_\gamma \frac{\hat{v}(\zeta)d\zeta}{(\zeta - z)g(\zeta)}, \quad \text{where} \quad g(z) = \frac{1}{\sqrt{z^2 - a^2}}.$$

On the airfoil $\hat{v}(z)$ is equal and opposite to the normal velocity produced by Q, i.e.,

$$\hat{v}(z) = \frac{1}{4\pi i}\left[\frac{Q}{(z - z_o)} - \frac{Q^*}{(z - z_o)^*}\right] \quad \text{on the airfoil.}$$

Hence,

$$\begin{aligned}
w'(z) &= \frac{Q}{2\pi(z - z_o)} + \frac{g(z)}{2\pi} \oint_\gamma \frac{1}{4\pi i}\left[\frac{Q}{(\zeta - z_o)} - \frac{Q^*}{(\zeta - z_o^*)}\right]\frac{d\zeta}{(\zeta - z)g(\zeta)} \\[2mm]
&= \frac{1}{4\pi}\left[\frac{Q}{z - z_o} + \frac{Q^*}{z - z_o^*}\right] + \frac{g(z)(Q - Q^*)}{4\pi} \\[2mm]
&\quad + \frac{g(z)}{4\pi}\left[\frac{Q}{(z - z_o)g(z_o)} - \frac{Q^*}{(z - z_o^*)g(z_o^*)}\right].
\end{aligned} \tag{3.11.14}$$

This result is regular in the fluid except at the singularity at $z = z_o$ and exhibits the correct behaviour $w'(z) \sim Q/2\pi z$ at infinity.

3.11.7 Flow through a cascade

Let us consider irrotational flow through a cascade of equal rigid parallel plates, evenly spaced along the y axis at distance b apart, as in Figure 3.11.6. This and similar problems involving periodically spaced blades can be treated by Sedov's method. For uniform flow at speed U incident from the left at angle α we write

$$w'(z) = Ue^{-i\alpha} + \frac{g(z)}{2ib} \oint \frac{\hat{w}'(\zeta)}{g(\zeta)}\left\{\coth\left[\frac{(\zeta - z)\pi}{b}\right] - 1\right\} d\zeta,$$

where $\hat{w}(z)$ is the complex potential representing the influence of the cascade on the incident stream and the integration is around the small circular contour enclosing z. The Kutta condition is imposed at the trailing edges at $x = +a$, and $\hat{w}(z)$ is required to satisfy

$$\hat{w}'(z) \sim \begin{cases} 0, & \text{as } x \to -\infty \\ \text{constant}, & \text{as } x \to +\infty \end{cases}.$$

The motion must be periodic in the y direction with period b, and we can therefore take

$$g(z) = \sqrt{\frac{\sinh\left[\frac{(z-a)\pi}{b}\right]}{\sinh\left[\frac{(z+a)\pi}{b}\right]}},$$

where the positive square root is taken on the real axis to the right of the cascade.

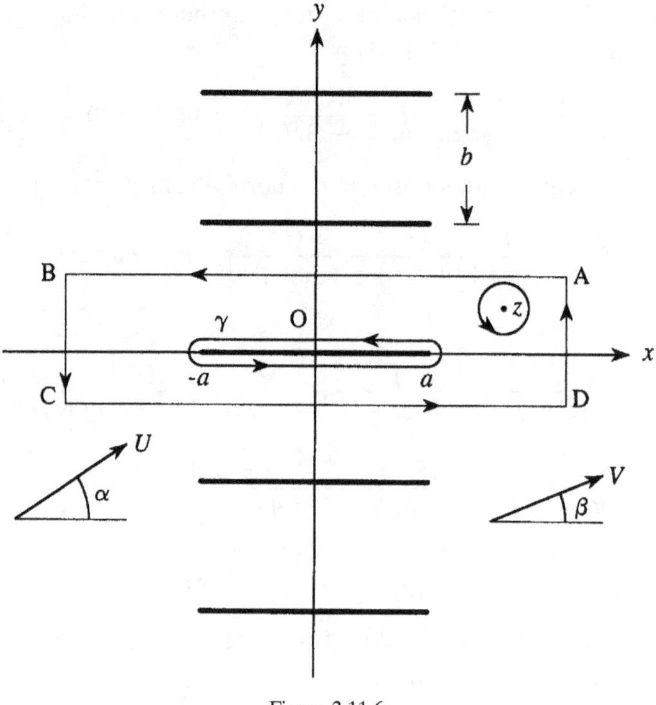

Figure 3.11.6

The integration contour is deformed onto the rectangle ABCD, where the sides AD, BC are of length b, and onto a contour γ that just encloses the blade on the real axis. The integrals along AB and CD are equal and opposite; the contributions from DA and BC vanish when the ends of the rectangle respectively recede to $x = \pm\infty$, because $\hat{w}'(z)$ ultimately vanishes on BC, and the term in the braces of the integrand vanishes on DA. In the usual way we then find (putting $\hat{v} = -U\sin\alpha$ on γ)

$$w'(z) = Ue^{-i\alpha} - \frac{g(z)U\sin\alpha}{2b}\oint_\gamma \left\{\coth\left[\frac{(\zeta - z)\pi}{b}\right] - 1\right\}\frac{d\zeta}{g(\zeta)},$$

where γ is traversed in the anticlockwise sense in Figure 3.11.6.

We can evaluate the integral by deforming the contour onto the infinite rectangle ABCD, yielding

$$w'(z) = U[\cos\alpha - ig(z)e^{-\frac{\pi a}{b}}\sin\alpha].$$

Thus, far to the right of the cascade the flow is at speed V in the direction β to the x axis, where

$$V = U\sqrt{\cos^2\alpha + e^{-\frac{4\pi a}{b}}\sin^2\alpha}, \quad \tan\beta = e^{-\frac{2\pi a}{b}}\tan\alpha.$$

3.12 Unsteady thin-airfoil theory

A change in the flight speed or in angle of attack of an airfoil must be accompanied by the shedding of additional vorticity from the trailing edge to preserve the smooth tangential edge flow. The resulting forces experienced by the airfoil can be determined analytically for a two-dimensional strip airfoil in cases in which the unsteady motions are small enough for the equations of motion to be linearised. We do this by confining attention to manoeuvres involving only small changes in angle of attack and small-amplitude oscillations. This also permits the wake occupied by the continuously shed vorticity to be treated as a vortex sheet.

3.12.1 The vortex sheet wake

Take the coordinate origin at the midchord of the undisturbed position of the airfoil, which occupies the interval $-a < x < a$ of the x axis. The mean flow relative to the airfoil is at speed U in the positive x direction. The vorticity released from the trailing edge because of the unsteady airfoil motions is assumed to lie in a vortex sheet stretching along the x axis between $a < x < \infty$, as indicated in Figure 3.12.1(a). The linearised version of two-dimensional vorticity equation (3.4.6), namely,

$$\left(\frac{\partial}{\partial t} + U\frac{\partial}{\partial x}\right)\omega = 0, \tag{3.12.1}$$

implies that ω is a function of $t - x/U$. Let Γ denote the net circulation of the wake at time t, and let $\gamma_{\mathrm{w}}(t - x/U)$ be the circulation *per unit length of the wake*. Then

$$\Gamma(t) = \int_a^\infty \gamma_{\mathrm{w}}\left(t - \frac{x'}{U}\right) dx', \tag{3.12.2}$$

and the wake vorticity ω is

$$\omega = \gamma_{\mathrm{w}}\left(t - \frac{x}{U}\right)\delta(y)\mathbf{k}, \quad x > a, \tag{3.12.3}$$

where $\mathbf{k}$ is a unit vector in the spanwise direction [out of the plane of the paper in Figure 3.12.1(a)].

We first determine the *linear theory* approximation to the complex potential produced by the wake vorticity. The vorticity is generated by small-amplitude airfoil motions, and in the first approximation we can calculate the velocity potential by assuming the airfoil occupies its undisturbed position on the x axis.

Consider a vortex element of circulation $d\Gamma = \gamma_{\mathrm{w}} dx'$ in the wake at $x = x'$ [Figure 3.12.1(b)]. The airfoil and $d\Gamma$ can be mapped into the ζ plane by Joukowski transformation (3.8.1). The image vortex is at $\zeta = \zeta'$ on the real ζ axis, and the airfoil maps into the unit circle with centre O at the origin [Figure 3.12.1(c)]. We obtain the complex potential dw_{w}, say, associated with $d\Gamma$ by placing an equal and opposite image vortex at the inverse point $\zeta = 1/\zeta'$ within the circle. However, in contrast to the analogous problem discussed in Example 5 of §3.4, there is now no need to place a vortex of strength $d\Gamma$

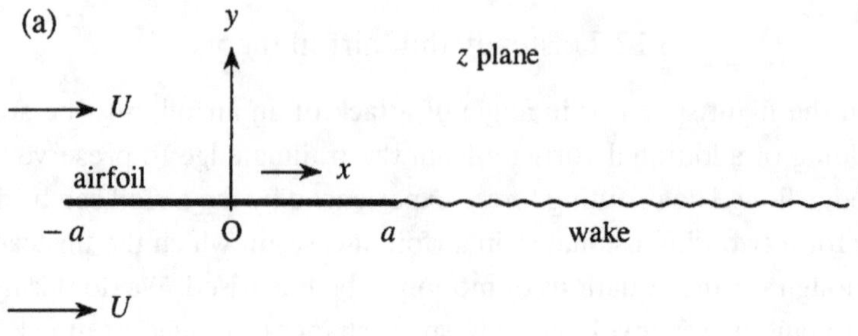

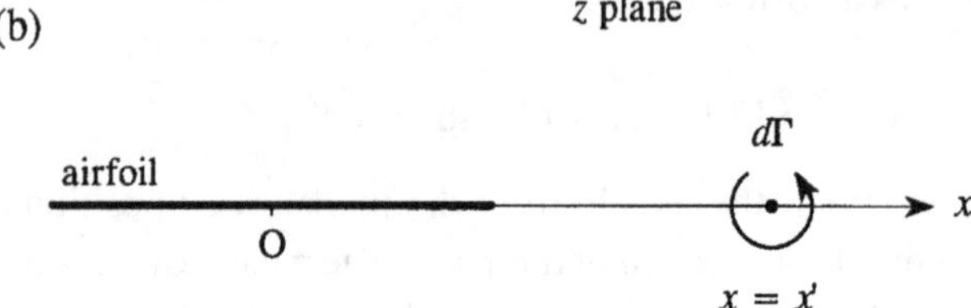

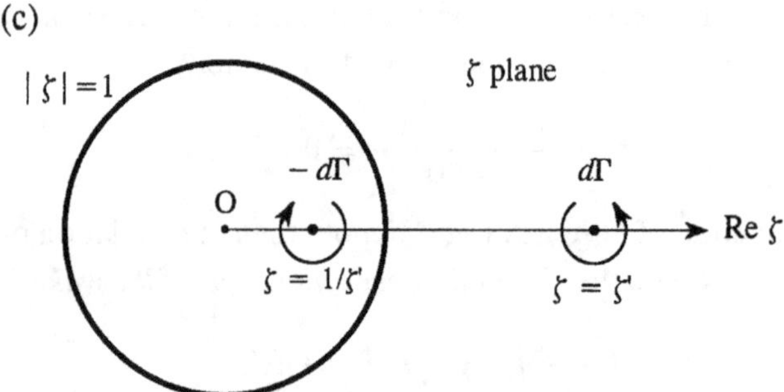

Figure 3.12.1

at O (to maintain vanishing circulation around the circle), because for each element of circulation $d\Gamma$ in the wake, Kelvin's circulation theorem requires there to be an equal and opposite circulation $-d\Gamma$ around the airfoil. Thus,

$$dw_{\mathrm{w}} = \frac{id\Gamma}{2\pi}\left[\ln\left(\zeta - \frac{1}{\zeta'}\right) - \ln(\zeta - \zeta')\right], \tag{3.12.4}$$

where

$$\zeta = \frac{z}{a} + \sqrt{\frac{z^2}{a^2} - 1} \quad (z = x + iy), \quad \text{and} \quad \zeta' = \frac{x'}{a} + \sqrt{\frac{x'^2}{a^2} - 1}. \tag{3.12.5}$$

We obtain the complex potential of the whole wake by summing the contributions from all of the vortex elements $d\Gamma$. The result is an integral involving the circulation per unit length γ_{w}:

$$w_{\mathrm{w}} = \frac{i}{2\pi}\int_a^\infty \gamma_{\mathrm{w}}\left(t - \frac{x'}{U}\right)\left[\ln\left(\zeta - \frac{1}{\zeta'}\right) - \ln(\zeta - \zeta')\right] dx'. \tag{3.12.6}$$

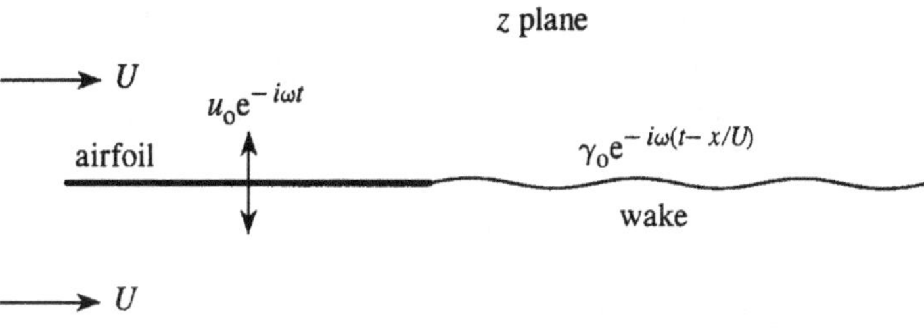

Figure 3.12.2

This and Equations (3.12.5) now supply the following representation of the complex velocity induced by the wake,

$$\frac{dw_{\mathrm{w}}}{dz} \equiv u_{\mathrm{w}} - i v_{\mathrm{w}} = \frac{-i}{2\pi\sqrt{z^2-a^2}} \int_a^\infty \gamma_{\mathrm{w}}\left(t-\frac{x'}{U}\right) \frac{\sqrt{x'^2-a^2}\,dx'}{z-x'}, \qquad (3.12.7)$$

which exhibits the expected singularities at the leading and trailing edges ($z = \pm a$) of the airfoil.

3.12.2 Translational oscillations

Suppose the airfoil is in uniform forward motion at speed U and is also executing small-amplitude vertical translational oscillations at velocity $\mathcal{V}(t)$ (Figure 3.12.2). The velocity potential excluding the contribution from the wake is

$$w_o = Uz + i\mathcal{V}(t)\left(\sqrt{z^2-a^2}-z\right). \qquad (3.12.8)$$

The overall complex potential $w = w_o + w_{\mathrm{w}}$, including the part of (3.12.6) generated by the wake, supplies

$$\frac{dw}{dz} = U + i\mathcal{V}(t)\left(\frac{z}{\sqrt{z^2-a^2}}-1\right) - \frac{i}{2\pi\sqrt{z^2-a^2}} \int_a^\infty \gamma_{\mathrm{w}}\left(t-\frac{x'}{U}\right) \frac{\sqrt{x'^2-a^2}\,dx'}{z-x'}. \qquad (3.12.9)$$

The velocity is finite at the trailing edge, and the flow will therefore leave the edge tangentially, provided the coefficient of $1/\sqrt{z^2-a^2}$ vanishes as $z \to +a$ (Glauert 1929). This condition yields the following integral equation for the circulation γ_{w}:

$$\mathcal{V}(t) = -\frac{1}{2\pi a} \int_a^\infty \gamma_{\mathrm{w}}\left(t-\frac{x'}{U}\right) \sqrt{\frac{x'+a}{x'-a}}\,dx'. \qquad (3.12.10)$$

Let us consider time harmonic oscillations of frequency $\omega > 0$, where

$$\mathcal{V}(t) = \mathrm{Re}\left(u_o e^{-i\omega t}\right), \quad u_o = \text{constant}. \qquad (3.12.11)$$

The wake vorticity must fluctuate at the same frequency, and we can put

$$\gamma_{\mathrm{w}}\left(t-\frac{x}{U}\right) = \mathrm{Re}\left[\gamma_o e^{-i(\omega t-\kappa x)}\right], \quad \gamma_o = \text{constant}, \qquad (3.12.12)$$

where $\kappa = \omega/U > 0$ is called the *hydrodynamic wavenumber*.

Substituting these expressions into integral equation (3.12.10) and making the change of integration variable $\lambda = x'/a$, we then find that the complex amplitude γ_o of the wake vorticity is determined in terms of the amplitude u_o of the airfoil oscillations by

$$\gamma_o = -2\pi u_o \bigg/ \int_1^\infty \sqrt{\frac{\lambda+1}{\lambda-1}}\, e^{i\kappa a\lambda} d\lambda. \qquad (3.12.13)$$

The integral in this formula diverges for real values of the hydrodynamic wavenumber. It must be interpreted in the sense of generalised functions (Lighthill 1958), or equivalently κ must be replaced with $\kappa + i\epsilon$, where ϵ is a small *positive* quantity that is subsequently allowed to vanish, but is sufficient to secure convergence because the modulus of the integrand now behaves like $e^{-\epsilon\lambda}$ as $\lambda \to \infty$. We shall adopt the latter procedure, which can be justified on the grounds that dissipation within a real fluid would ultimately lead to the damping of the vorticity far downstream of the airfoil. We then find (see §3.12.6)

$$\gamma_o = \frac{4iu_o}{[\mathrm{H}_0^{(1)}(\kappa a) + i\mathrm{H}_1^{(1)}(\kappa a)]}, \qquad (3.12.14)$$

where $\mathrm{H}_0^{(1)}$ and $\mathrm{H}_1^{(1)}$ are Hankel functions.

3.12.3 The unsteady lift

The unsteady force $\mathcal{F}$ experienced by the airfoil *per unit span* is calculated using the high-Reynolds-number approximation to formula (4.5.15):

$$\mathcal{F}_i = -\mathrm{M}_{ij}\frac{dU_j}{dt} + \rho_o \int_V \nabla X_i \cdot \boldsymbol{\omega} \wedge \mathbf{v}_{\mathrm{rel}} dx dy. \qquad (3.12.15)$$

The first term on the right-hand side represents the potential flow reaction of fluid in the immediate vicinity of the airfoil. For a thin, two-dimensional airfoil of chord $2a$, $\mathrm{M}_{22} = \rho_o\pi a^2$ is the only the non-zero component of the added-mass tensor. The second term on the right-hand site of (3.12.15) determines the back reaction of the wake on the airfoil.

To evaluate the integral to first order it can be assumed that the airfoil remains in its undisturbed position. Then, for the lift $\mathcal{F}_2$ we have

$$X_2 = \mathrm{Re}\left(-i\sqrt{z^2 - a^2}\right).$$

The vorticity is given in terms of the wake circulation by (3.12.3), and we can take $\mathbf{v}_{\mathrm{rel}} = (U, 0, 0)$. Thus, if $\mathcal{F}_2^{\mathrm{wake}}$ denotes the lift fluctuation per unit span produced by the wake, we can write

$$\mathcal{F}_2 = -\rho_o\pi a^2\frac{d\mathcal{V}}{dt} + \mathcal{F}_2^{\mathrm{wake}}, \qquad (3.12.16)$$

where the first term on the right-hand side is the effect of added mass, and [see (3.12.24)]

$$\mathcal{F}_2^{\mathrm{wake}} = \mathrm{Re}\left[\rho_o U\gamma_o e^{-i\omega t}\int_a^\infty \frac{x' e^{i\kappa x'} dx'}{\sqrt{x'^2 - a^2}}\right] = \mathrm{Re}\left[-\frac{\pi}{2}\rho_o U\gamma_o a\mathrm{H}_1^{(1)}(\kappa a)e^{-i\omega t}\right]$$

$$= -2\pi\,\mathrm{Re}\left[\rho_o u_o U a T(\kappa a)e^{-i\omega t}\right], \qquad (3.12.17)$$

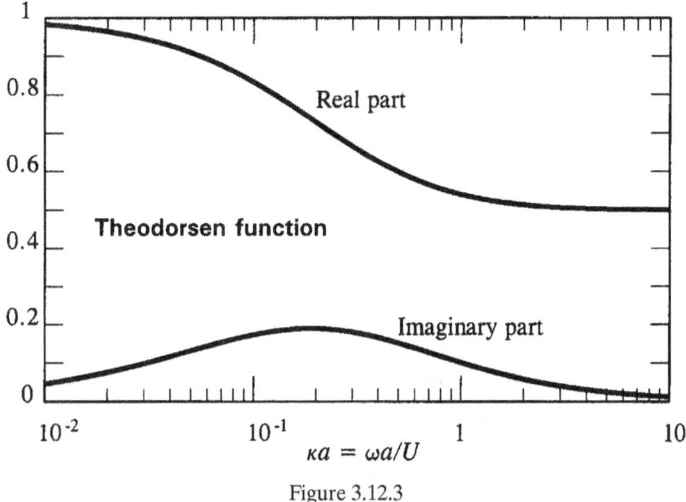

Figure 3.12.3

where $T(\kappa a)$ is the *Theodorsen function*,

$$T(\kappa a) = \frac{i\mathrm{H}_1^{(1)}(\kappa a)}{[\mathrm{H}_0^{(1)}(\kappa a) + i\mathrm{H}_1^{(1)}(\kappa a)]}, \tag{3.12.18}$$

whose real and imaginary parts are plotted in Figure 3.12.3.

The low- and high-frequency asymptotic values of $\mathcal{F}_2^{\text{wake}}$ can be interpreted in terms of solution (3.8.17) of Wagner's problem of the lift generated by an impulsively started airfoil. In the present case the effective (small) angle of attack of the airfoil is

$$\alpha = -\mathrm{Re}\left(u_o e^{-i\omega t}\right)/U.$$

It can be seen from Figure 3.12.3 that $T(\kappa a) \approx 1$ and $\frac{1}{2}$ respectively at low and high nondimensional frequencies $\kappa a \equiv \omega a/U$, and therefore that

$$\mathcal{F}_2^{\text{wake}} \sim \begin{cases} 2\pi\alpha a\rho_o U^2, & \omega \to 0 \\ \pi\alpha a\rho_o U^2, & \omega \to \infty \end{cases}.$$

Recalling that $\Gamma = 2\pi\alpha a U$ in Figure 3.8.4, we see that the low-frequency limit of (3.12.17) corresponds to the quasi-static approximation in which the lift is given at any instant by the steady-state Kutta–Joukowski formula $\rho_o \Gamma U$. At high frequencies, however, the instantaneous unsteady lift is half the steady-state value for the given angle of attack, which is precisely the result obtained from Wagner's solution at the initial instant of the rapid acceleration from rest of the impulsively started airfoil.

3.12.4 Leading-edge suction force

The oscillatory motion of the airfoil produces a reciprocating flow around the leading edge together with a corresponding suction force (§3.3.4). This is the only streamwise force that acts on the airfoil, because it is at zero mean angle of attack, and trailing-edge suction is vitiated by vortex shedding. The oscillating airfoil therefore provides

Thrust producing wake

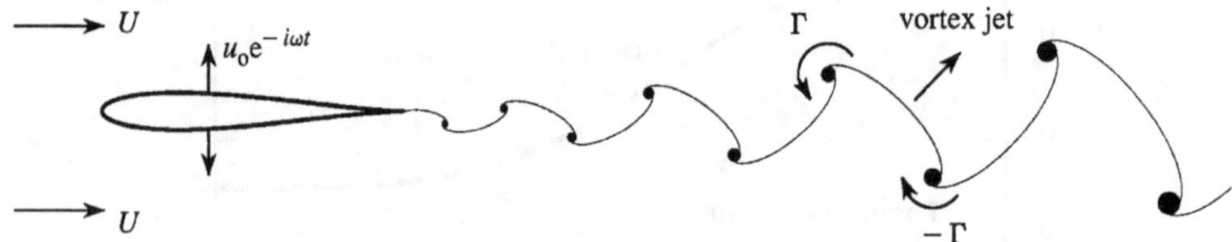

Figure 3.12.4

the simplest illustration of flight (that is, *thrust generation*) by a flapping wing. The force can be determined from general formula (3.12.15) (Question 40, Problems 3) or more simply by calculation of the value of the coefficient C in approximation (3.3.6) of the complex potential near the edge and then by use of formula (3.3.7). We shall discuss the second method.

Expression (3.12.9) for the complex velocity and definitions (3.12.11) and (3.12.12) imply that, near the leading edge $(z = -a)$,

$$\frac{dw}{dz} \sim -\sqrt{2a}\,\mathrm{Re}\left[\left(u_o - \frac{\gamma_o}{2\pi}\int_1^\infty \sqrt{\frac{\lambda-1}{\lambda+1}}\,e^{i\kappa a\lambda}d\lambda\right)e^{-i\omega t}\right]\frac{d}{dz}\sqrt{z+a}.$$

Using (3.12.14) and the formulae (3.12.23), (3.12.24), we then find

$$C = \mathrm{Re}\left[2\sqrt{2a}\,u_o\,T(\kappa a)e^{-i\omega t}\right],$$

where $T(\kappa a)$ is Theodorsen function (3.12.18). It now follows from (3.3.7) that the suction force has both mean and fluctuating parts, the latter varying at frequency 2ω. The mean or time-independent part of the suction force is

$$\mathcal{F}_{\text{suction}} = -\rho_o\pi a|u_o|^2|T(\kappa a)|^2, \tag{3.12.19}$$

where the minus sign shows that the force is in the negative direction of the x axis. According to Figure 3.12.3, $|T(\kappa a)|^2 \sim \frac{1}{4}$ at high frequencies ($\kappa a > 1$). In this limit the suction force $\sim \rho_o\pi a|u_o|^2/4$, the same as that predicted by potential flow formula (3.12.8), which becomes, near the leading edge,

$$w_o \sim -\sqrt{2a}\,V(t)\sqrt{z+a},$$

where $V(t)$ is given by (3.12.11). When $\kappa a \gg 1$ the wake vorticity varies rapidly with position and its aggregate effect at the leading edge of the airfoil is negligible; the main influence of the wake is to remove the singularity at the trailing edge without, however, materially affecting the motion elsewhere on the airfoil.

This theory of thrust generation takes no account of the growth and changes in the wake produced by self-induction of the wake vorticity. A proper analysis of such effects must be based on the nonlinear equations of motion applied to an oscillating airfoil of finite thickness (Figure 3.12.4), and in practice requires the problem to be treated numerically. However, when the wake can be regarded as organised and non-turbulent, the production of thrust demands a growing nonlinear wake that can be considered to consist of a succession of vortex pairs of opposite rotation whose induced velocities tend to augment the streamwise momentum in the wake, i.e. such that the mean momentum of the 'vortex jets' of the vortex pairs (§4.3.1) is in the direction of the mean flow relative to the airfoil, as indicated in the figure.

3.12.5 Energy dissipated by vorticity production

The kinetic energy of vortex flow left in the wake of the airfoil is supplied by the work performed by the unsteady airfoil surface forces.

Using Equation (2.10.1), we find that the kinetic energy T_o of the linear theory motion induced by the wake vorticity per unit span is

$$T_o = \frac{1}{2}\rho_o \int (\nabla \varphi)^2 dx dy \equiv -\frac{1}{2}\rho_o \int \left(\varphi_+ - \varphi_-\right) \frac{\partial \varphi}{\partial y}\, dx, \qquad (3.12.20)$$

where $\varphi_\pm$ denotes the value of the velocity potential on either side $y = \pm 0$ of the vortex sheet wake, across which the normal component of velocity $\partial\varphi/\partial y$ is continuous.

We express the integral in terms of known quantities by noting that the circulation per unit length of the wake of (3.12.12) satisfies

$$\frac{\partial \varphi_-}{\partial x} - \frac{\partial \varphi_+}{\partial x} = \mathrm{Re}\left[\gamma_o e^{-i(\omega t - \kappa x)}\right].$$

Then, because $\varphi(x, y, t)$ satisfies Laplace's equation and $\partial\varphi/\partial y$ is continuous at $y = 0$, we must have, far downstream of the airfoil,

$$\varphi(x, y, t) = \pm \mathrm{Re}\left[\frac{i\gamma_o e^{i(\kappa x - \omega t)\mp \kappa y}}{2\kappa}\right], \quad y \gtrless 0.$$

In this downstream region, we find the kinetic energy per unit length of the wake by performing the final integration in (3.12.20) over a section of the wake of length equal to the hydrodynamic wavelength $2\pi/\kappa$ and dividing by $2\pi/\kappa$. This gives

$$\frac{\rho_o |\gamma_o|^2}{8\kappa}.$$

The rate at which energy is delivered to the wake by the airfoil is equal to this multiplied by the rate of increase of the length of the wake, which is just the airfoil forward velocity U. Hence the

$$\text{power dissipated in the wake} = \frac{\rho_o |\gamma_o|^2 U}{8\kappa}.$$

Now the rate of working of the lift force of (3.12.16) is

$$\langle -\mathcal{V}(t)\mathcal{F}_2\rangle \equiv -\langle \mathrm{Re}\left(u_o e^{-i\omega t}\right) \mathcal{F}_2^{\text{wake}}\rangle, \qquad (3.12.21)$$

where the angle brackets $< >$ denote a time average. Part of this work is dissipated by the vortex motions in the wake and part is compensated for by the negative work done on the fluid by the leading-edge suction force, that is,

$$-\left\langle \mathrm{Re}\left(u_o e^{-i\omega t}\right) \mathcal{F}_2^{\mathrm{wake}}\right\rangle \equiv -U\mathcal{F}_{\mathrm{suction}} + \frac{\rho_0|\gamma_o|^2 U}{8\kappa}. \tag{3.12.22}$$

We can verify the validity of this formula by substituting for $\mathcal{F}_2^{\mathrm{wake}}$ and $\mathcal{F}_{\mathrm{suction}}$ from (3.12.17) and (3.12.19), and then expanding the left-hand side with the help of equations (3.12.25) and (3.12.26) in the next subsection.

A measure of the *efficiency* of the oscillating airfoil in producing forward flight is the following ratio:

$$\text{efficiency} = \frac{\text{power consumed by suction force}}{\text{rate of working of lateral motions}}$$

$$= \frac{-U\mathcal{F}_{\mathrm{suction}}}{-U\mathcal{F}_{\mathrm{suction}} + \frac{\rho_0|\gamma_o|^2 U}{8\kappa}}$$

$$= \frac{\pi\kappa a|\mathrm{H}_1^{(1)}(\kappa a)|^2}{\pi\kappa a|\mathrm{H}_1^{(1)}(\kappa a)|^2 + 2} \rightarrow \begin{cases} 1, & \kappa a \rightarrow 0 \\ \frac{1}{2} & \kappa a \rightarrow \infty \end{cases}.$$

Thus at high frequencies one half of the work performed by the unsteady lift force is used to energise the wake vorticity.

3.12.6 Hankel function formulae

The following formulae involving Hankel functions $\mathrm{H}_0^{(1)}(z)$, $\mathrm{H}_1^{(1)}(z)$ (Abramowitz and Stegun 1970) are set down here for reference:

$$\int_1^\infty \frac{e^{i\lambda z}\,d\lambda}{\sqrt{\lambda^2-1}} = \frac{\pi i}{2}\mathrm{H}_0^{(1)}(z), \quad \mathrm{Im}\,z \geq 0; \tag{3.12.23}$$

$$\int_1^\infty \frac{\lambda e^{i\lambda z}\,d\lambda}{\sqrt{\lambda^2-1}} = -\frac{\pi}{2}\mathrm{H}_1^{(1)}(z), \quad \mathrm{Im}\,z > 0. \tag{3.12.24}$$

$$\mathrm{H}_{\nu+1}^{(1)}(z)\mathrm{H}_\nu^{(2)}(z) - \mathrm{H}_\nu^{(1)}(z)\mathrm{H}_{\nu+1}^{(2)}(z) = -\frac{4i}{\pi z}, \tag{3.12.25}$$

$$\mathrm{H}_\nu^{(1)}(z^*) = [\mathrm{H}_\nu^{(2)}(z)]^*, \quad \mathrm{H}_\nu^{(2)}(z^*) = [\mathrm{H}_\nu^{(1)}(z)]^* \quad (\nu\ \text{real}). \tag{3.12.26}$$

$$\mathrm{H}_\nu^{(1)}(ze^{\pi i}) = -e^{-\nu\pi i}\mathrm{H}_\nu^{(2)}(z), \tag{3.12.27}$$

$$\mathrm{H}_\nu^{(1)}(z) \sim \sqrt{\frac{2}{\pi z}}e^{i\left(z-\frac{\nu\pi}{2}-\frac{\pi}{4}\right)}, \quad \mathrm{H}_\nu^{(2)}(z) \sim \sqrt{\frac{2}{\pi z}}e^{-i\left(z-\frac{\nu\pi}{2}-\frac{\pi}{4}\right)}, \quad |z| \rightarrow \infty. \tag{3.12.28}$$

PROBLEMS 3

1. Verify the two-dimensional formula

$$\lim_{\epsilon \to 0} \frac{\epsilon^2}{\pi (r^2 + \epsilon^2)^2} = \delta(\mathbf{x}), \quad \text{where} \quad r = (x^2 + y^2)^{\frac{1}{2}}, \quad \mathbf{x} = (x, y).$$

2. Show that the suction force exerted on a rigid cylinder of radius a by a line vortex of strength Γ at distance $h > a$ from the centre of the cylinder is $\rho_o \Gamma^2 a^2 / 2\pi h^3$, provided there is no net circulation about the cylinder. (See also the Kirchhoff vector method of Chapter 4).

3. Calculate the added-mass coefficients M_{ij} for an infinite, rigid strip of width $2a$.

4. Calculate the added-mass coefficients M_{ij} for an infinite, rigid cylinder of radius a.

5. Calculate the unsteady lift and drag exerted on a rigid circular cylinder of radius a produced by a parallel line vortex of circulation Γ in the presence of a uniform mean flow normal to the cylinder. Assume the motion is ideal and that the net circulation around the cylinder vanishes.

6. Repeat Question 5 under the assumption that the vortex is convected solely by the mean flow (i.e. when the induced component of the motion of Γ produced by image vortices in the cylinder is neglected).

7. Calculate the path of a line vortex of strength Γ that is parallel to a rigid strip occupying $-a < x_1 < a$, $x_2 = 0$, $-\infty < x_3 < \infty$. Assume the fluid is at rest at infinity and that there is no net circulation around the strip. Determine the unsteady force on the strip.

8. Calculate the path of a line vortex of strength Γ that is parallel to a rigid elliptic cylinder of semi-major and minor axes respectively equal to a and b. Assume the fluid is at rest at infinity and that there is no net circulation around the cylinder.

9. A line vortex of strength Γ is adjacent to a rigid right-angle corner whose edges lie along the positive x_1 and x_2 axes, the vortex being parallel to the edge of the corner. Show that the vortex traverses a path with polar representation $r \sin 2\theta = $ constant.

10. Calculate the trajectories of a vortex pair consisting of two parallel line vortices of strengths $\pm\Gamma$ moving under their mutual induction towards a rigid plane parallel to the line of centres of the vortices.

11. Calculate the trajectory of a line vortex of strength Γ adjacent to the rigid half-plane $x_1 < 0$, $x_2 = 0$ in the presence of a uniform mean flow at speed U in the positive x_1 direction.

12. Show that the transformation $\zeta = \sqrt{z^2/a^2 + 1}$, $a > 0$ maps the upper z plane cut by a thin rigid barrier along the imaginary axis between $z = 0$ and $z = ia$

onto the upper ζ plane. Hence deduce that a line vortex Γ at $z = z_0(t)$ traverses a path determined by the equation

$$\frac{dz_o^*}{dt} = -\frac{i\Gamma}{4\pi}\left[\frac{a^2}{z_o(z_o^2 + a^2)} - \frac{2z_0}{(z_o^2 + a^2 - |z_o^2 + a^2|)}\right],$$

z plane

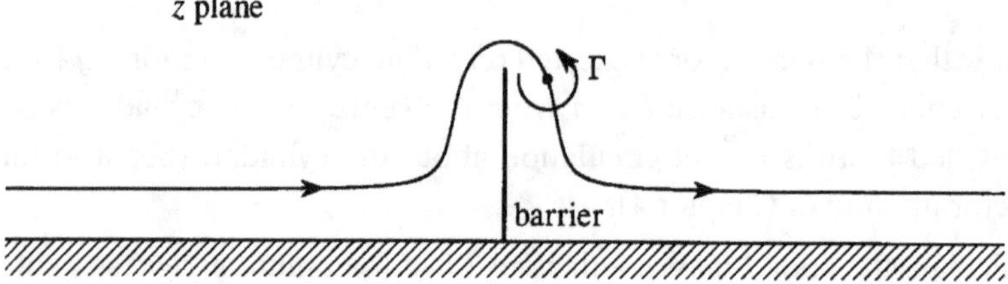

provided the fluid is at rest at infinity.

13. A line source of strength q is located at $z_o = re^{i\theta}$ adjacent to a rigid circular cylinder of radius $a < r$ with centre at the origin. If the fluid is at rest at infinity, derive the complex potential in the form

$$w = \frac{q}{2\pi}\left[\ln(z - z_o) + \ln\left(z - \frac{a^2}{z_o^*}\right) - \ln z\right].$$

Deduce that the force (F_1, F_2) exerted by the cylinder on the fluid is given by

$$F_1 + i F_2 = \frac{-\rho_0 q^2}{2\pi}\frac{a^2 e^{i\theta}}{r(r^2 - a^2)},$$

and that is represents a 'suction' force directed along the radius joining the centre of the cylinder to the source.

14. A rectangle open at infinity in the x direction has solid boundaries along $x = 0$, $y = 0$, $y = h$. Use the method of §3.5, Example 1 to show that the transformation $\zeta = \cosh\frac{\pi z}{h}$, $z = x + iy$, maps the rectangle into the upper half of the ζ plane. Show also that the complex potential of motion created by a source q at $z = 0$ and an equal sink at $z = ih$ is given by

$$w = \frac{2q}{\pi}\ln\tanh\left(\frac{\pi z}{2h}\right).$$

Deduce that half the stream lies between $x = 0$ and the streamline that crosses $y = \frac{1}{2}h$ at $x = \frac{h}{\pi}\ln(1 + 2^{\frac{1}{2}})$.

15. Use the transformation $\zeta = e^{\frac{\pi z}{h}}$, $z = x + iy$, to find the streamlines of the two-dimensional motion produced by a source q midway between the planes $y = 0$, h. If the pressure tends to zero at large distances from the source, show that the planes are forced apart by a force that varies inversely with their distance apart h.

16. A thin layer of water flows two dimensionally and symmetrically around the edge O of a large, thin rigid plate. Far from the edge above and below the plate the water layer has thickness h and the motion is uniform at speed U. The effects of gravity and viscosity are assumed to be negligible.

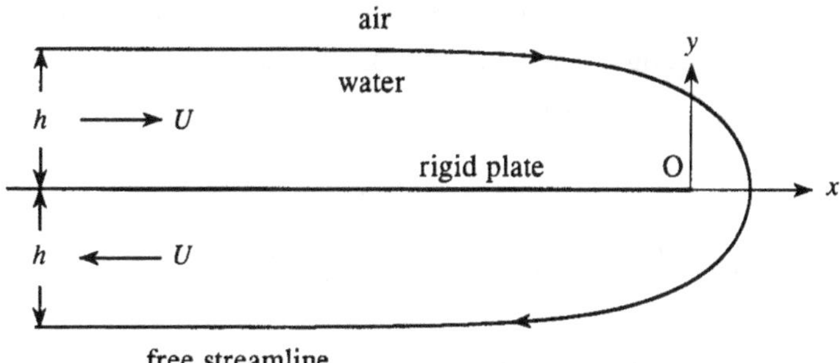

Calculate the steady suction force on the edge O. Hence show that very near the edge the complex potential is given approximately by

$$w = 2iU\sqrt{\frac{2hz}{\pi}} \quad (z = x + iy),$$

where the coordinate origin is taken at O, as shown in the figure.

17. Apply the method of §3.5, Example 1, to derive the transformation $\zeta = i\sqrt{z}$, $z = x + iy$, of the fluid region bounded by the rigid half-plane $x < 0$, $y = 0$ onto the upper half of the ζ plane. Do this by deriving a suitable separable solution of the polar form of Laplace's equation

$$\frac{\partial^2 \eta}{\partial r^2} + \frac{1}{r}\frac{\partial \eta}{\partial r} + \frac{1}{r^2}\frac{\partial^2 \eta}{\partial \theta^2} = 0,$$

where $\zeta = \xi + i\eta$.

18. Show for the irrotational duct flow problem of Figures 3.6.3 and 3.6.4 that the velocity potential in the neighbourhood of the lower lip at $z = -ih$ is given approximately by

$$w = ihU\left[-1 + \sqrt{\frac{2}{\pi h}(z + ih)^{\frac{1}{2}}}\,\right].$$

Hence deduce that the suction force on the lip is $\frac{1}{2}\rho_o hU^2$ per unit span.

19. Show that $w = U\sqrt{z^2 + h^2}$ is the complex potential of flow at speed U parallel to and above a wall at $y = 0$ that supports a thin, vertical barrier between $y = 0$ and $y = h$. Calculate the suction force on the end of the barrier, and verify that it is equal and opposite to the net normal force on $y = 0$.

20. Show that uniform potential flow at speed U over a step of depth h in the bed of a deep river can be represented by

$$w = \frac{Uh\zeta}{\pi}, \quad z = \frac{h}{\pi}\left[\sqrt{\zeta^2 - 1} + \ln\left(\zeta + \sqrt{\zeta^2 - 1}\right)\right].$$

21. If $w = U\sqrt{z^2 + a^2}$, derive the following equation for the streamlines

$$y^2 = \frac{\psi^2[U^2(x^2 + a^2) + \psi^2]}{U^2(U^2x^2 + a^2)}.$$

Show that w represents the flow of a wide stream of velocity U past a thin obstacle of length a projecting normally from a plane wall.

22. Show that the complex potential of the Coanda flow of Figure 3.7.1 is given by

$$\frac{\pi w}{4hU} = \cosh^{-1}\left(e^{-\frac{\pi z}{4h}}\right).$$

23. Derive formula (3.7.13) for the drag coefficient C_D for the separated wake flow of Figure 3.7.7. Let the upper half of the symmetric flow past the plate be mapped onto the upper t plane such that $t = 0$, 1, ∞ correspond respectively to the points O, B and AC in the figure, and show that

$$\Omega \equiv \ln\left(U\frac{dz}{dw}\right) = \frac{i\pi}{2} - i\,\tan^{-1}\left(\sqrt{t-1}\right), \quad w = \frac{U\ell t}{\pi+4}.$$

24. Show for the aperture flow problem of Figure 3.7.3 that the free streamline BC is defined parametrically by

$$\frac{x}{h} = -\frac{(2\cos\theta + \pi)}{\pi+2}, \quad \frac{y}{h} = \frac{2[\sin\theta - \ln(\tan\theta + \sec\theta)]}{\pi+2}, \quad 0 < \theta < \frac{\pi}{2}.$$

25. Water extending to infinity in the upper half of the z plane flows through an opening between the half-lines $x < 0$, $y = 0$ and $x > b$, $y = h$ into the lower z plane in the form of a jet with free streamlines. When gravity is neglected show that the ultimate direction of the jet makes an angle α with the x axis determined by the equation

$$\frac{b}{h} = \frac{1}{2}\tan\alpha + \frac{1}{\pi}\left[\sec\alpha + \ln\left(\tan\frac{\alpha}{2}\right)\right].$$

26. A liquid flows through a two-dimensional aperture between $0 < x < a$ in the lower wall $y = 0$ of a duct bounded on its upper side by $-\infty < x < +\infty$, $y = b > 0$. If the outflow forms a jet of ultimate width c, show that

$$\frac{a}{c} = 1 + \frac{1}{\pi}\left(\frac{2b}{c} + \frac{c}{2b}\right)\ln\frac{2b+c}{2b-c}.$$

Show that the contraction ratio $\sim \frac{4}{7}$ when $a = b$.

27. The two-dimensional jet in the figure emerges from a nozzle of width $2h$. The width of the free jet is increased to $2\sigma h$ ($\sigma > 1$) by the application of suction forces at the lips of the nozzle. Calculate the free streamlines by mapping the lower region AOBCC$'$A$'$ of the symmetric flow onto the upper t plane, such that the points $t = -1$, 1, ∞ correspond to B, CC$'$, O. The flow around the lip O is assumed to have the usual potential flow singularity, and the free streamline separates tangentially from the exterior wall of the nozzle at the point labelled B. Show that $t \equiv t_\sigma = \frac{1}{2}(\sigma + 1/\sigma)$ is the image of AA$'$, and that

$$\Omega \equiv \ln\left(U\frac{dz}{dw}\right) = \ln\left(t - \sqrt{t^2-1}\right), \quad \frac{dw}{dt} = \frac{\sigma hU}{\pi}\left(\frac{1}{t-t_\sigma} - \frac{1}{t-1}\right).$$

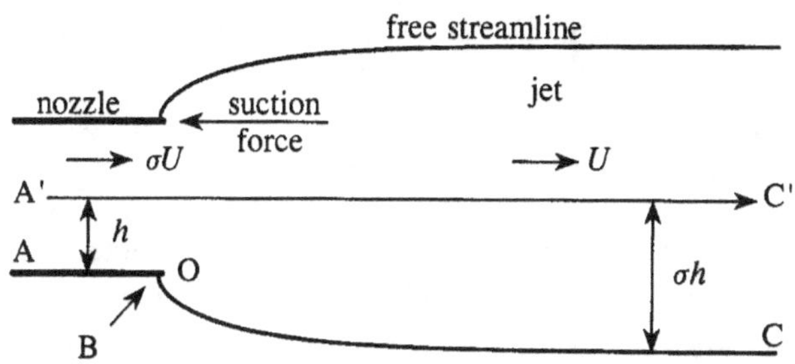

28. The two-dimensional Coanda edge flow around the edge O of the large plate shown in the following figure separates on the underside of the plate at B forming a free jet in a direction χ relative to that of the original flow.

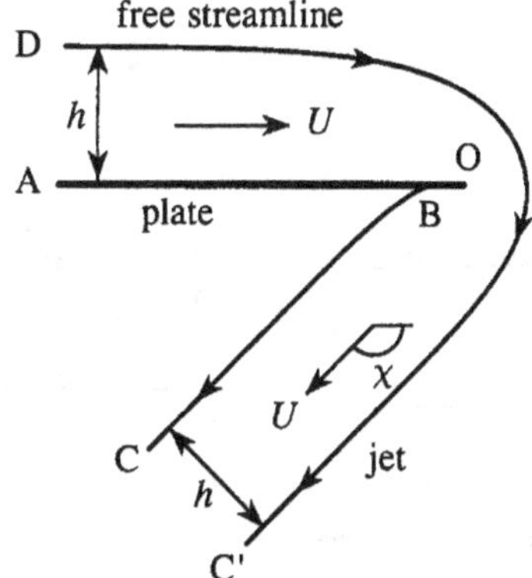

Calculate the free streamlines when all body forces are negligible by mapping the flow region AOBCC′D onto the upper t plane, such that the points $t = -1,\ 1,\ \infty$ correspond to B, DA, O. Show that $t \equiv t_\chi = \cos\chi$ is the image of CC′, and that

$$\Omega \equiv \ln\left(U\frac{dz}{dw}\right) = \ln\left(t - \sqrt{t^2-1}\right), \qquad \frac{dw}{dt} = \frac{hU}{\pi}\left(\frac{1}{t-1} - \frac{1}{t - \cos\chi}\right).$$

Deduce that the separation point B is distance s from O, where

$$\frac{\pi s}{h} = \chi\sin\chi + 2\cos\chi\,\ln\left(2\cos\frac{\chi}{2}\right) - 2\ln 2.$$

29. Use the method of Example 2, §3.7.5, to verify the following special cases for the free-streamline flow of Figure 3.7.12(a):

α°	σ	χ°	t_χ
45	0.4402	37.4	0.9971
135	0.5059	51.9	0.9845

30. Solve the free-streamline problem of Figure 3.7.8. Map the bifurcated flow region ABB′CDEE′F onto the upper half of the t plane such that the points $t = -1,\ 1,\ \infty$

correspond respectively to D, BB′, and C. Show that FA maps into $t = 0$, EE′ into $t = -\sin\chi$, and that

$$\Omega \equiv \ln\left(U\frac{dz}{dw}\right) = -\frac{i\pi}{2} + \ln\left(t + \sqrt{t^2 - 1}\right), \qquad \frac{dw}{dt} = \frac{U}{\pi}\left(\frac{h}{t} - \frac{h_1}{t-1} - \frac{h_2}{t+\sin\chi}\right).$$

31. Investigate by the method of §3.7 the problem depicted in the figure of two-dimensional branching flow of an ideal, incompressible fluid. The channel widths h, h_1, h_2, and the velocity U are given, and the upper branch makes an angle χ with the incoming main flow. There is a stagnation point on the dividing streamline where it meets the corner O; the flow is assumed to remain attached at A, where the velocity becomes infinite.

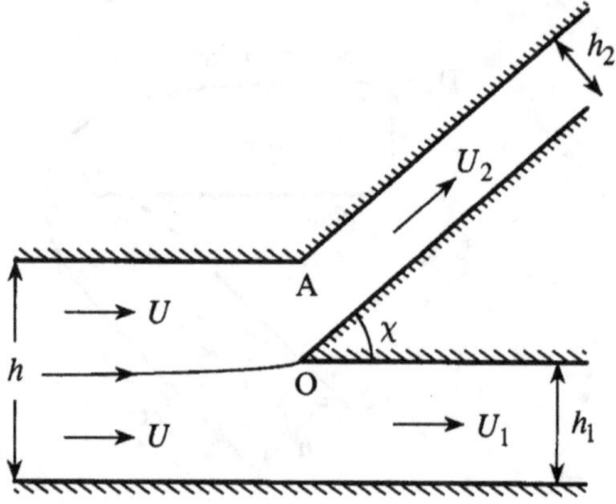

Show that the velocities U_1, U_2 are determined by

$$h_1 U_1 + h_2 U_2 = hU, \qquad h_1 U_1^{1-\frac{\pi}{\chi}} - h_2 U_2^{1-\frac{\pi}{\chi}} = hU^{1-\frac{\pi}{\chi}}.$$

32. Repeat the calculation of Example 3, §3.7, when the point source is replaced with a vortex of strength Γ. Show that the free streamline is given by

$$z = t - (s^* + ih)\left[2\ln\left(\frac{t}{h} + i\right) - i\pi + \frac{it(s^* + ih)}{h(t + ih)}\right],$$

$$-\infty < t < \infty, \qquad \text{where} \quad s = ih\sqrt{1 + \frac{\Gamma}{\pi hU}}.$$

33. Repeat the calculation of Example 4, §3.7, when the vortex is replaced with a point source of strength q.

34. Calculate the net force exerted on the vortex of Example 4, §3.7.

35. The volume flow produced by a point source at $z = ih$, $h > 0$, is assumed to exhaust through an aperture of width $2a$ occupying the section $-a < x < a$ of a thin rigid wall lying along the real axis. In the notation of the following figure, apply Chaplygin's singular point method to show that

$$\frac{dw}{dt} = \frac{-2a\sigma U\ell^2}{\pi t(t^2 + \ell^2)}, \qquad \frac{d\Omega}{dt} = \frac{-2\ell\sqrt{\ell^2 + 1}}{\sqrt{t^2 - 1}(t^2 + \ell^2)} + \frac{K}{\sqrt{t^2 - 1}}, \qquad K = \text{constant},$$

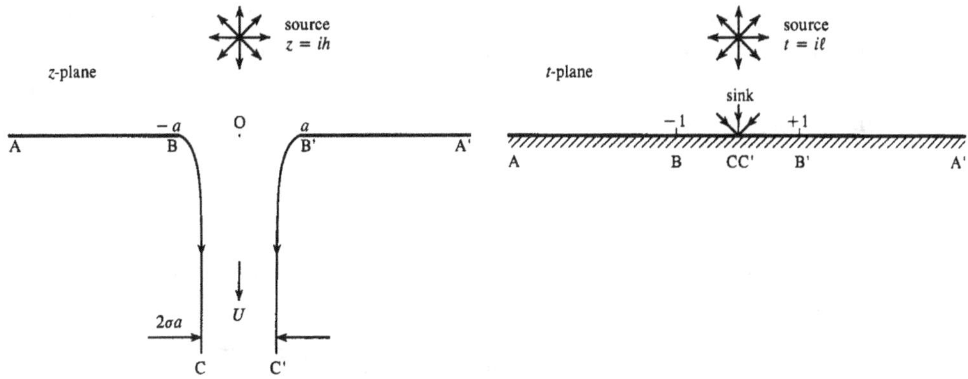

where σ is the contraction ratio of the jet and $t = i\ell$ is the image of the source in the t plane. Deduce that

$$\frac{dz}{dt} = \frac{2a\sigma\ell^2 \left(t + \sqrt{t^2 - 1}\right)^3}{\pi t \left(\ell\sqrt{t^2 - 1} + t\sqrt{\ell^2 + 1}\right)^2},$$

and explain how this equation is used to calculate ℓ and σ.

36. Show that near the leading edge of the flat-plate airfoil in Figure 3.8.4(b) the complex velocity $w \approx U \sin\alpha \sqrt{2a}\sqrt{z + a}$. Hence deduce that the leading-edge suction force is $\rho_o \Gamma U \sin\alpha$ per unit span.

37. Show for the tandem airfoil lifting problem of Figure 3.11.2 (§3.11, Example 4) that the suction forces at the two leading edges are

$$\pi\rho_o U^2 \sin^2\alpha \, \frac{(a_2 - b_1)(b_1 - a_1)}{(b_2 - b_1)} \quad \text{and} \quad \pi\rho_o U^2 \sin^2\alpha \, \frac{(b_2 - a_1)(b_2 - a_2)}{(b_2 - b_1)}.$$

38. Sedov's method applied to evaluate the complex velocity $w'(z)$ of steady irrotational flow through the wall aperture of Figure 3.6.5(a). Use the asymptotic approximation $w'(z) \sim \mp 2hU/\pi z$ on C_∞ respectively in the upper and lower half-planes, take $g(z) = 1/\sqrt{z^2 - h^2}$, and show that

$$w'(z) = \frac{g(z)}{2\pi i} \oint_{C_\infty} \frac{w'(\zeta)d\zeta}{(\zeta - z)g(\zeta)} = \frac{-2hU}{\pi\sqrt{z^2 - h^2}}.$$

39. Consider mean flow at speed U in the x direction in the half-space above the wall aperture of Figure 3.6.5(a). Assume there is no mean flow through the aperture and the fluid is at rest at infinity in the lower region. Write the complex velocity in the form $w'(z) = \frac{1}{2}U + \hat{w}'(z)$, where

$$\hat{w}'(z) \sim \pm\frac{U}{2} + O\left(\frac{1}{z^2}\right) \quad \text{as} \quad |z| \to \infty \text{ respectively in } y \gtrless 0.$$

Use Sedov's method to deduce that

$$w(z) = \frac{U}{2}\left(z + \sqrt{z^2 - h^2}\right).$$

40. The suction force exerted on the oscillating airfoil of Figure 3.12.2 can be calculated from (4.5.15) when $i = 1$, noting that $M_{11} = M_{12} = 0$ and $\nabla X_1 = (1, 0, 0)$. Deduce that only the y component of $\mathbf{v}_{\text{rel}}$ makes a finite contribution and that at position x in the wake it is equal to

$$\frac{-x}{\sqrt{x^2 - a^2}} \operatorname{Re}\left[\left(u_o - \frac{\gamma_o}{2\pi} \int_a^\infty \frac{\sqrt{x'^2 - a^2}\,e^{i\kappa x'}\,dx'}{x - x'}\right) e^{-i\omega t}\right],$$

where the principal value of the integral is to be taken. Hence show that the mean suction force is given by

$$\mathcal{F}_{\text{suction}} = \frac{\rho_o}{2} \operatorname{Re}\left[u_o \gamma_o^* \int_a^\infty \frac{x e^{-i\kappa x}\,dx}{\sqrt{x^2 - a^2}} - \frac{|\gamma_o|^2}{2\pi} \iint_a^\infty \frac{\sqrt{x'^2 - a^2}\,e^{-i\kappa(x - x')}\,dx\,dx'}{(x - x')\sqrt{x^2 - a^2}}\right].$$

The second integral vanishes as $\kappa \to \infty$. Use this fact to evaluate the integral by first differentiating under the integral sign with respect to κ. Hence deduce formula (3.12.19).

41. Verify energy balance equation (3.12.22) for the energy dissipated by vorticity production in the wake of an airfoil executing small-amplitude translational oscillations.

4

Rotational Incompressible Flow

4.1 The vorticity equation

Irrotational flow can be established from a state of rest in an ideal incompressible fluid by the instantaneous transmission throughout the fluid of impulsive pressures from a moving boundary. If the boundary motion is subsequently arrested the motion everywhere ceases immediately. Kelvin's theorem (§2.10), that the kinetic energy of an irrotational flow is always smaller than that of any other flow consistent with the same boundary conditions, is a consequence of the fact that the number of degrees of freedom of irrotational motion is exactly the same as the number of degrees of freedom of the boundary itself. In a real fluid, however, there are typically an unlimited number of degrees of freedom, the flow is rotational, and the motion continues after the boundary stops moving . Kelvin (1867) therefore proposed the following definition of a vortex in a homogeneous incompressible fluid: '... *a portion of fluid having any motion that it could not acquire by fluid pressure transmitted from its boundary*'. Vorticity is actually a derived *kinematic* quantity, but its introduction greatly increases understanding of a complex flow and a knowledge of its distribution frequently permits the description of the fluid motion to be simplified.

When a small fluid particle is imagined to be suddenly solidified without change in its angular momentum, it continues to translate and rotate as a solid body. Its initial angular velocity of rotation is determined by its moment of inertia tensor, which depends on the particle shape. The velocity distribution relative to the centre of mass of the fluid particle just before solidification is given by Equation (1.4.6), and this can be used to calculate its angular momentum. In the special case of a *spherical* fluid particle the irrotational component of (1.4.6) (involving e_{ij}) makes no contribution to the angular momentum. This means that a spherical particle will rotate at angular velocity $\frac{1}{2}\omega$ after solidification – this is not generally the case for an arbitrary fluid element. The vorticity $\omega(\mathbf{x}, t)$ may therefore be interpreted as twice the initial angular velocity of the solid sphere formed when an infinitesimal sphere of fluid at $\mathbf{x}$ is solidified without change of angular momentum. Vorticity is accordingly a measure of the angular momentum within the fluid, which cannot be instantaneously destroyed when the fluid boundaries are brought to rest.

211

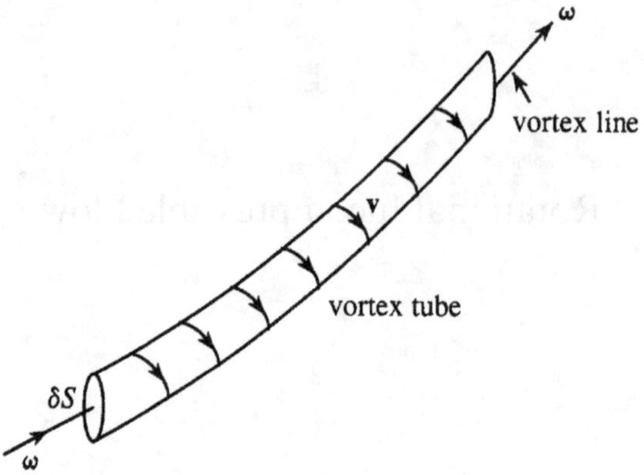

Figure 4.1.1

4.1.1 Vortex lines

A vortex line is tangential to the vorticity vector $\omega(\mathbf{x}, t)$ at all points along its length. The vortex line that passes through the point $\mathbf{x}_0$ at some given instant t can be found by integration of the system of equations

$$\frac{dx_1}{\omega_1(\mathbf{x}, t)} = \frac{dx_2}{\omega_2(\mathbf{x}, t)} = \frac{dx_3}{\omega_3(\mathbf{x}, t)},$$

starting at $\mathbf{x} = \mathbf{x}_0$ and keeping t fixed.

4.1.2 Vortex tubes

Vortex lines that pass through every point of a simple closed curve define the boundary of a *vortex tube*. The vorticity ω may be regarded as constant over the cross section of a narrow tube, and the product $\omega \delta S$ is called the tube strength, where δS is the tube cross-sectional area. The tube strength is constant along the tube because

$$\operatorname{div} \omega = \operatorname{div}(\operatorname{curl} \mathbf{v}) \equiv 0$$

and the divergence theorem therefore implies that $\oint \omega \cdot d\mathbf{S} = 0$ for any closed surface, and in particular for the surface formed by two cross sections of the tube and the tube wall separating them, on the latter of which $\omega \cdot d\mathbf{S} = 0$. The vortex tube of the type formed, for example, by a trailing wing-tip vortex is approximately circular in cross section (Figure 4.1.1), and the flow (streamline pattern) relative to the tube spirals around the tube axis, the fluid particles on the tube surface rotating about the tube axis at angular velocity $\omega/2$, which increases or decreases respectively as the tube cross section δS decreases or increases.

The solenoidal condition $\operatorname{div} \omega = 0$ also implies that vortex tubes and lines cannot begin or end within the fluid. The 'no-slip' condition requires the velocity at a boundary

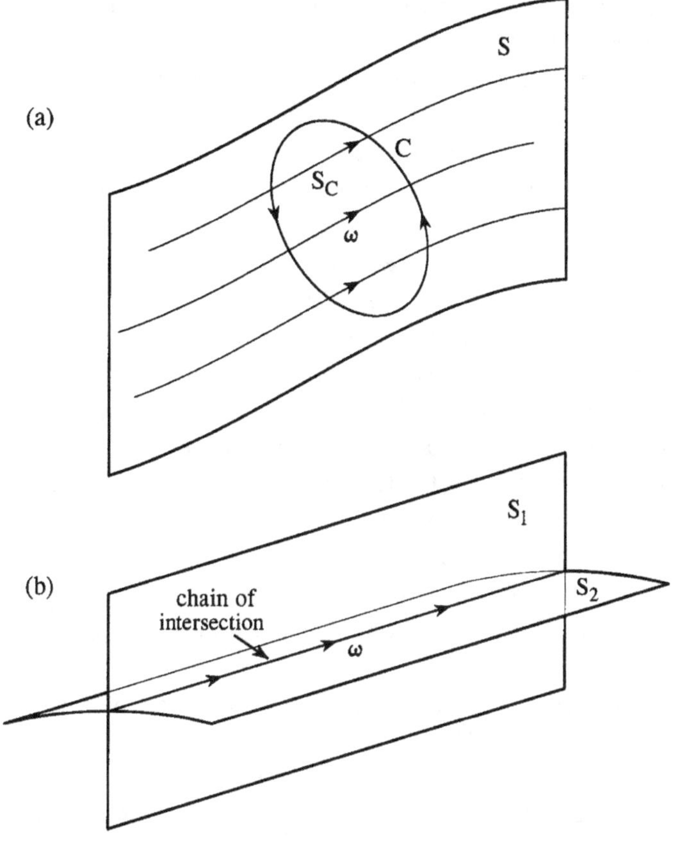

Figure 4.1.2

to be the same as that of the boundary. A vortex line must therefore form a closed loop,
or end on a *rotating* surface S at which

$$\mathbf{n} \cdot \omega = 2\mathbf{n} \cdot \mathbf{\Omega}, \tag{4.1.1}$$

where $\mathbf{\Omega}$ is the angular velocity of S. In an ideal fluid a vortex line can appear to 'end' on
a surface where, in fact, a surface vortex sheet is formed, and at the surface the vortex
line must be supposed to abruptly change direction and enter the vortex sheet.

4.1.3 Movement of vortex lines: Helmholtz's vortex theorem

Let S be a material surface element in an *ideal* fluid that consists at time $t = t_0$ entirely
of vortex lines [Figure 4.1.2(a)]. Let C be a closed contour on S and let S_C denote
the portion of S 'within' C. Because the normal component of vorticity vanishes on S,
Kelvin's circulation theorem (2.2.3) implies that the circulation around C

$$\oint_C \mathbf{v} \cdot d\mathbf{x} = \int_{S_C} \omega \cdot d\mathbf{S} = 0$$

for all time. This is true for any closed contour C on S. Therefore, as S evolves in time, it is always composed of vortex lines.

An instantaneous vortex line (at $t = t_0$, say) can be regarded as the intersection of two such vortex surfaces S_1 and S_2 [Figure 4.1.2(b)]. The curve of intersection consists of a chain of fluid particles that lies on both surfaces. The surfaces move with the fluid and always intersect along the space curve defined by the current position of the chain. At time $t_0 + \delta t$ and at any point on the curve of intersection the vorticity continues to have no component normal to S_1 and S_2, so that the curve of intersection continues to be a vortex line. In other words, the chain of fluid particles that defines a vortex line at any given instant moves in such a way that the same chain of fluid particles continues to define a vortex line at later times; the vortex line is said to *move with the fluid*. The same argument shows that vortex tubes also move with the fluid, and moreover that the tube strength $\omega \delta S$ does not change with time (because a closed material contour on the tube surface that encircles the tube once moves with the tube and, by Kelvin's theorem, has constant circulation precisely equal to $\omega \delta S$). This implies that the magnitude of the vorticity vector tangential to the material chain marking the intersection of S_1 and S_2 increases or decreases in proportion to the 'stretching' or foreshortening of the chain, which causes the cross section δS of an enclosing vortex tube to contract or expand.

These conclusions regarding the motion of line vortices and vortex tubes in an ideal fluid constitute the essence of Helmholtz's vortex theorem: *A vortex tube moves with the fluid and its strength remains constant*. However, this idealised representation of the transport of vorticity by convection by the fluid particles is modified in a real fluid by molecular diffusion, although both mechanisms – convection and diffusion – will be seen to imply that an initially confined region of vorticity can usually be assumed to remain within a bounded region at later times. The details of an evolving field of vorticity are determined by the vorticity equation, considered in §4.1.5.

4.1.4 Crocco's equation

Consider homentropic flow [for which $p = p(\rho)$] and ignore body forces (which is permissible if they are conservative). Momentum equation (1.4.11)

$$\frac{\partial \mathbf{v}}{\partial t} + (\mathbf{v} \cdot \nabla)\mathbf{v} + \nabla \left(\int \frac{dp}{\rho} \right) = -\frac{\eta}{\rho}\operatorname{curl}\omega + \frac{1}{\rho}\left(\eta' + \frac{4}{3}\eta \right)\nabla(\operatorname{div}\mathbf{v})$$

can be rewritten by use of the vector identity

$$(\mathbf{v} \cdot \nabla)\mathbf{v} = \omega \wedge \mathbf{v} + \nabla \left(\frac{1}{2}v^2 \right) \tag{4.1.2}$$

in **Crocco's form**:

$$\frac{\partial \mathbf{v}}{\partial t} + \omega \wedge \mathbf{v} + \nabla B = -\frac{\eta}{\rho}\operatorname{curl}\omega + \frac{1}{\rho}\left(\eta' + \frac{4}{3}\eta \right)\nabla(\operatorname{div}\mathbf{v}), \tag{4.1.3}$$

where

$$B = \int \frac{dp}{\rho} + \frac{1}{2}v^2 \qquad (4.1.4)$$

is the **total enthalpy** in homentropic flow; $\omega \wedge \mathbf{v}$ is called the **Lamb vector**. When the fluid is incompressible,

$$B = \frac{p}{\rho_o} + \frac{1}{2}v^2. \qquad (4.1.5)$$

We obtain a useful simplification of Crocco's equation by discarding the term in div $\mathbf{v}$ in (4.1.3), because it represents a relatively small effect of compressibility on *viscous dissipation*; then

$$\frac{\partial \mathbf{v}}{\partial t} + \omega \wedge \mathbf{v} + \nabla B = -\nu \, \mathrm{curl}\, \omega, \qquad (4.1.6)$$

where $\nu = \eta/\rho$ is the *kinematic viscosity*. Viscosity tends to be important only in thin layers of fluid at solid boundaries of the flow, where $\nu\,\mathrm{curl}\,\omega \equiv (\eta/\rho)\mathrm{curl}\,\omega$ becomes very large and, indeed, usually many orders of magnitude larger than the corresponding terms in (4.1.3) involving div $\mathbf{v}$.

4.1.5 Convection and diffusion of vorticity

We obtain the vorticity equation by taking the curl of Equation (4.1.6). When the kinematic viscosity ν is assumed to be constant we find (using the relation $\mathrm{curl}\,\mathrm{curl}\,\omega \equiv -\nabla^2\omega$)

$$\frac{\partial \omega}{\partial t} + \mathrm{curl}(\omega \wedge \mathbf{v}) = \nu\nabla^2\omega. \qquad (4.1.7)$$

If the fluid is incompressible div ω = div $\mathbf{v}$ = 0 and

$$\mathrm{curl}(\omega \wedge \mathbf{v}) \equiv (\mathbf{v} \cdot \nabla)\omega + \omega\,\mathrm{div}\,\mathbf{v} - (\omega \cdot \nabla)\mathbf{v} - \mathbf{v}\,\mathrm{div}\,\omega = (\mathbf{v} \cdot \nabla)\omega - (\omega \cdot \nabla)\mathbf{v},$$

therefore the vorticity equation can also be written as

$$\frac{D\omega}{Dt} = (\omega \cdot \nabla)\mathbf{v} + \nu\nabla^2\omega. \qquad (4.1.8)$$

The terms on the right-hand side of this equation represent the two principal mechanisms that change the vorticity of a moving fluid particle in incompressible flow: respectively (1) the 'stretching' and rotation of vortex lines by the velocity field (i.e. the effect of convection already discussed for an ideal fluid), and (2) the spreading of vorticity to neighbouring fluid particles by molecular diffusion.

When the motion is inviscid ($\nu = 0$) Equations (4.1.7) and (4.1.8) must be equivalent to a differential form of Kelvin's circulation theorem (see Examples 1 and 2 of this section) and describe the movement of vortex lines with fluid particles. To verify this, consider in Figure 4.1.3 a fluid particle initially at A with velocity $\mathbf{v}$ at time t. Let the vorticity at A be $\omega = \omega\mathbf{n}$, where $\mathbf{n}$ is a unit vector, and consider a neighbouring particle at B a small distance s from A in the direction $\mathbf{n}$, so that $s\mathbf{n}$ is the position of B relative

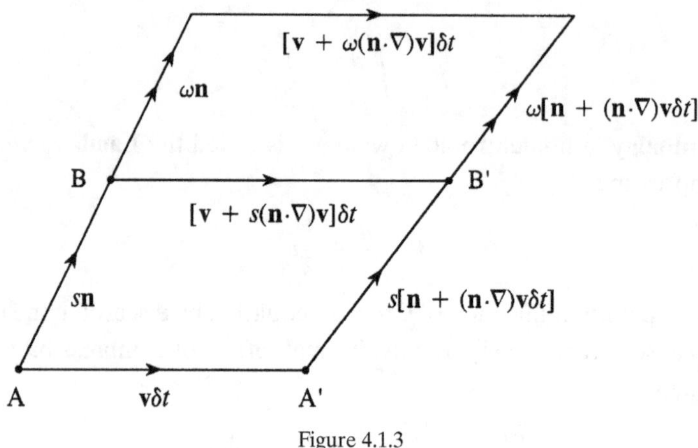

Figure 4.1.3

to A. At time t points A and B lie on the vortex line through A, and the velocity at B is $\mathbf{v} + s(\mathbf{n} \cdot \nabla)\mathbf{v}$.

After a short time δt, A has moved a vector distance $\mathbf{v}\delta t$ to A′ and B has moved to B′ whose position relative to A′ is $s[\mathbf{n} + (\mathbf{n} \cdot \nabla)\mathbf{v}\delta t]$. During this time the term $(\boldsymbol{\omega} \cdot \nabla)\mathbf{v}$ in the vorticity equation causes the vorticity of the fluid particle initially at A to change from $\omega\mathbf{n}$ at A to $\omega[\mathbf{n} + (\mathbf{n} \cdot \nabla)\mathbf{v}\delta t]$ at A′. Thus the vortex line through A′ lies along the relative vector $s[\mathbf{n} + (\mathbf{n} \cdot \nabla)\mathbf{v}\delta t]$ from A′ to B′. Therefore the fluid particles and the vortex line through A and B have deformed and convected in the flow in the same way; in their new positions A and B continue to lie on the same vortex line, and vortex lines therefore 'move with the fluid' when the motion is inviscid; they are rotated and stretched in a manner determined entirely by the relative motions of A and B. The magnitude of ω increases in direct proportion to the stretching of vortex lines. When a vortex tube is stretched, the cross-sectional area δS decreases and therefore ω must increase to preserve the strength of the tube.

The viscous term $\nu\nabla^2\boldsymbol{\omega}$ in (4.1.7) and (4.1.8) is important only in regions of high shear, usually near solid boundaries. Very close to a stationary wall the velocity becomes small and nonlinear terms in the vorticity equation can be neglected, which then reduces to the 'diffusion equation':

$$\frac{\partial \boldsymbol{\omega}}{\partial t} = \nu\nabla^2\boldsymbol{\omega}.$$

Vorticity is generated at solid boundaries, and viscosity is responsible for its diffusion into the body of the fluid, where it can subsequently be convected by the flow.

EXAMPLE 1. EVOLUTION OF A MATERIAL SURFACE ELEMENT Let $d\mathbf{S} = \mathbf{n}\,dS$ be an infinitesimal material surface element in the fluid. Referring to Figure 4.1.4,

$$d\mathbf{S} = \frac{1}{2}\oint_C \mathbf{r}' \wedge d\mathbf{r} = \frac{1}{2}\oint_C (\mathbf{r} - \mathbf{r}_o) \wedge d\mathbf{r} \equiv \frac{1}{2}\oint_C \mathbf{r} \wedge d\mathbf{r},$$

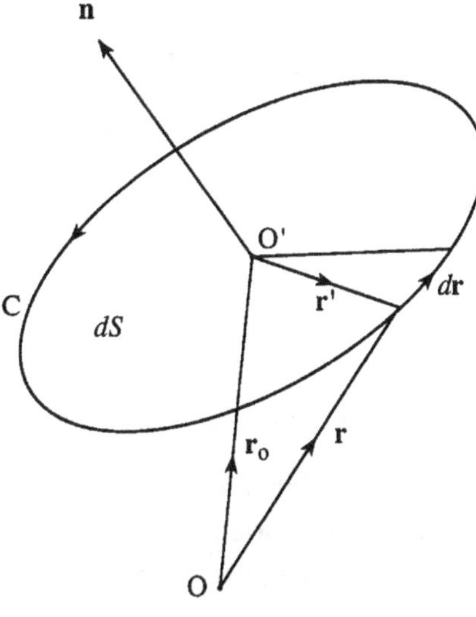

Figure 4.1.4

where $\mathbf{r}'$ is the vector position relative to a moving origin O′ in $d\mathbf{S}$, $\mathbf{r}$ is measured from a fixed origin O, and C is the material boundary of $d\mathbf{S}$. Then

$$\frac{D\mathbf{r}}{Dt} = \mathbf{v} \quad \text{and} \quad \frac{D}{Dt}(d\mathbf{r}) = d\mathbf{v},$$

where $\mathbf{v}$ is the velocity on C at $\mathbf{r}$. Hence

$$\frac{D}{Dt}(d\mathbf{S}) = \frac{1}{2}\oint_C (\mathbf{v} \wedge d\mathbf{r} + \mathbf{r} \wedge d\mathbf{v}) = \frac{1}{2}\oint_C [2\mathbf{v} \wedge d\mathbf{r} + d(\mathbf{r} \wedge \mathbf{v})] \equiv \oint_C \mathbf{v} \wedge d\mathbf{r}.$$

To simplify the last integral, consider the component in the direction of an arbitrary unit vector $\mathbf{i}$ and apply Stokes' theorem:

$$\oint_C \mathbf{i} \cdot \mathbf{v} \wedge d\mathbf{r} = \oint_C \mathbf{i} \wedge \mathbf{v} \cdot d\mathbf{r} = d\mathbf{S} \cdot [\nabla \wedge (\mathbf{i} \wedge \mathbf{v})] \equiv -(d\mathbf{S} \wedge \nabla) \wedge \mathbf{v} \cdot \mathbf{i}.$$

Hence

$$\frac{D}{Dt}(d\mathbf{S}) = -(d\mathbf{S} \wedge \nabla) \wedge \mathbf{v} \equiv d\mathbf{S}\,\mathrm{div}\,\mathbf{v} - dS_j \nabla v_j. \tag{4.1.9}$$

EXAMPLE 2. DIFFERENTIAL FORM OF KELVIN'S CIRCULATION THEOREM By Stokes' theorem the circulation Γ around the boundary C of the material element of Figure 4.1.4 is $d\mathbf{S} \cdot \boldsymbol{\omega}$. Hence, with (4.1.9),

$$\frac{D\Gamma}{Dt} = \boldsymbol{\omega} \cdot \frac{D}{Dt}(d\mathbf{S}) + d\mathbf{S} \cdot \frac{D\boldsymbol{\omega}}{Dt} = d\mathbf{S} \cdot \left[\frac{D\boldsymbol{\omega}}{Dt} + \boldsymbol{\omega}\,\mathrm{div}\,\mathbf{v} - (\boldsymbol{\omega} \cdot \nabla)\mathbf{v}\right].$$

For incompressible, ideal flow the term in the brackets vanishes by the inviscid form of (4.1.8). For compressible flow, it also vanishes by virtue of the compressible, homentropic form of the vorticity equation (see Problems 4, Question 1). In either case we recover Kelvin's circulation theorem $D\Gamma/Dt = 0$.

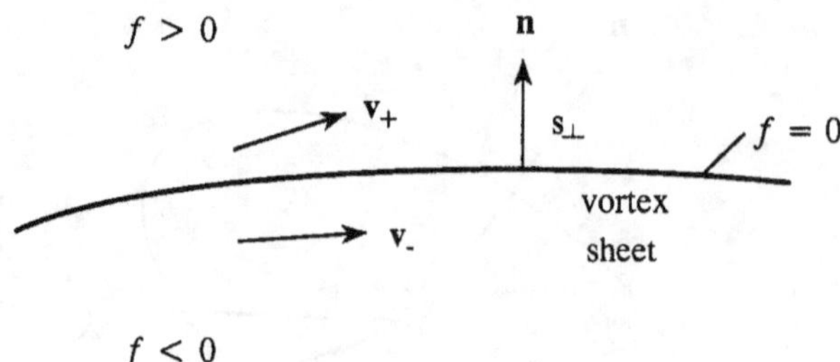

Figure 4.1.5

4.1.6 Vortex sheets

It should be understood that viscosity does not generate the vorticity, but merely serves to diffuse vorticity into the fluid from a solid boundary. In an ideal fluid the slipping of the flow over the surface creates a singular layer of vorticity at the surface called a **vortex sheet** whose strength is determined by the tangential velocity difference between the surface and the ideal exterior flow. This vorticity stays on the surface; it would start to diffuse into the fluid if the fluid were suddenly endowed with viscosity. The *rate* of diffusion would depend on the value of v, but the *amount* of vorticity available for diffusion from the surface is independent of v.

A vortex sheet is a useful model of a thin layer of vorticity when viscous diffusion can be neglected. Imagine a thin shear layer (Figure 4.1.5) across which the velocity changes rapidly from $\mathbf{v}_-$ to $\mathbf{v}_+$. We approximate the layer by a surface $f(\mathbf{x}, t) = 0$ with unit normal $\mathbf{n}$ across which the normal components of velocity are equal ($\mathbf{n} \cdot \mathbf{v}_- = \mathbf{n} \cdot \mathbf{v}_+$), but the tangential components are discontinuous ($\mathbf{n} \wedge \mathbf{v}_- \neq \mathbf{n} \wedge \mathbf{v}_+$). Let $f \gtrless 0$ respectively on the $\pm$ sides of the surface. Near the sheet, on either side, it can be assumed that $\operatorname{curl} \mathbf{v}_\pm = 0$, and we can set

$$\mathbf{v} = \mathrm{H}(f)\mathbf{v}_+ + \mathrm{H}(-f)\mathbf{v}_-,$$
$$\therefore \quad \omega = \nabla\mathrm{H} \wedge (\mathbf{v}_+ - \mathbf{v}_-) = \mathbf{n} \wedge (\mathbf{v}_+ - \mathbf{v}_-)\delta(s_\perp), \tag{4.1.10}$$

where $\nabla\mathrm{H} \equiv \nabla\mathrm{H}(f) = -\nabla\mathrm{H}(-f) = \mathbf{n}\delta(s_\perp)$, and $s_\perp$ is distance measured in the normal direction from the sheet.

In a real fluid the vorticity would diffuse out from the sheet and it could not therefore persist indefinitely. In an ideal fluid the sheet is subject only to convection and stretching by the flow. To calculate the convection velocity $\mathbf{v}_c$ of the vorticity within the vortex sheet we integrate Crocco's equation (4.1.6) (with the viscous term discarded) in the normal direction across the infinitesimal width of the sheet. The velocity is finite but the vorticity is singular, so that only the terms in $\omega \wedge \mathbf{v} \equiv \omega \wedge \mathbf{v}_c$ and ∇B contribute to the integral. Because the pressure must be continuous, we find [from (4.1.10)]

$$\mathbf{n} \cdot [\mathbf{n} \wedge (\mathbf{v}_+ - \mathbf{v}_-)] \wedge \mathbf{v}_c = \frac{1}{2}\left(v_-^2 - v_+^2\right) \equiv \frac{1}{2}(\mathbf{v}_- + \mathbf{v}_+) \cdot (\mathbf{v}_- - \mathbf{v}_+).$$

By using the identity

$$[\mathbf{n} \wedge (\mathbf{v}_+ - \mathbf{v}_-)] \wedge \mathbf{v}_c = (\mathbf{n} \cdot \mathbf{v}_c)(\mathbf{v}_+ - \mathbf{v}_-) - \mathbf{n}[\mathbf{v}_c \cdot (\mathbf{v}_+ - \mathbf{v}_-)]$$

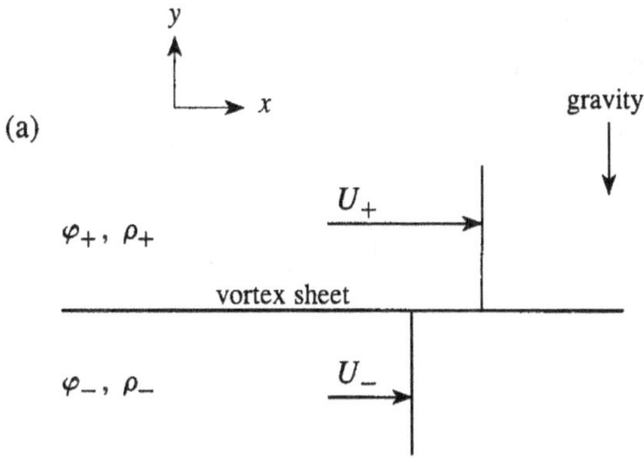

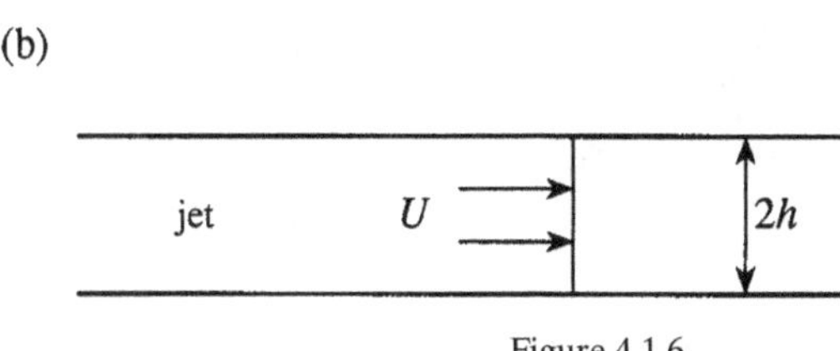

Figure 4.1.6

and recalling that $\mathbf{n} \cdot \mathbf{v}_- = \mathbf{n} \cdot \mathbf{v}_+$ we see that

$$\mathbf{v}_c \cdot (\mathbf{v}_- - \mathbf{v}_+) = \frac{1}{2}(\mathbf{v}_- + \mathbf{v}_+) \cdot (\mathbf{v}_- - \mathbf{v}_+).$$

This determines the component of $\mathbf{v}_c$ parallel to the sheet. Because the normal component of velocity is continuous across the sheet, we therefore conclude that

$$\mathbf{v}_c = \frac{1}{2}\left(\mathbf{v}_+ + \mathbf{v}_-\right), \tag{4.1.11}$$

where $\mathbf{v}_\pm$ are evaluated just above and below the sheet.

A singular vorticity distribution of the type (4.1.10) that defines a vortex sheet is compatible with the equations of ideal flow, but cannot exist in a steady, *stable* motion of the fluid.

EXAMPLE 3. INSTABILITY OF A VORTEX SHEET (KELVIN–HELMHOLTZ INSTABILITY)
Consider the situation illustrated in Figure 4.1.6(a), in which a vortex sheet coincident with the plane $y = 0$ separates upper and lower ideal fluids in steady motions parallel to the x direction at speeds $U_\pm$. Let the interface be disturbed from the equilibrium position by a *small* displacement ζ in the y direction given by

$$\zeta = \zeta_o e^{i(kx - \omega t)},$$

where $\zeta_o = \text{constant}$ and $k > 0$. The motions in the upper and lower regions are irrotational and must decrease in amplitude with distance $|y|$ from the interface. The corresponding *perturbation* velocity potentials satisfy Laplace's equation, and can evidently be taken in the respective forms

$$\varphi_\pm = C_\pm e^{\mp ky + i(kx - \omega t)}, \quad \text{where } C_\pm \text{ are constant.} \tag{4.1.12}$$

For small-amplitude motion all conditions at the interface will be imposed in the *linearised* approximation. Thus, kinematic condition (1.7.1) that the interface moves with the fluid becomes

$$\frac{D}{Dt}\big(\zeta(x,t) - y\big) = 0 \quad \text{at} \quad y \to \pm 0,$$

which yields the two relations

$$\left(\frac{\partial}{\partial t} + U_\pm \frac{\partial}{\partial x}\right)\zeta = \frac{\partial \varphi_\pm}{\partial y} \quad \text{at} \quad y = 0. \tag{4.1.13}$$

Similarly, the pressure perturbations $p_\pm$ in the upper and lower regions are, by Bernoulli's equation,

$$p_\pm = -\rho_\pm \left(\frac{\partial \varphi_\pm}{\partial t} + U_\pm \frac{\partial \varphi_\pm}{\partial x} + gy\right). \tag{4.1.14}$$

Using the definition of ζ and substituting from (4.1.12) into (4.1.13) we find

$$-i(\omega - U_+ k)\zeta_o = -kC_+, \quad -i(\omega - U_- k)\zeta_o = kC_-;$$

also, equating the perturbation pressures at the interface $y = \zeta$, we obtain

$$\rho_+[-i(\omega - U_+ k)C_+ + g\zeta_o] = \rho_-[-i(\omega - U_- k)C_- + g\zeta_o].$$

By eliminating the amplitudes ζ_o, $C_\pm$ from these equations we derive the *dispersion relation* governing the motion of the interface:

$$\rho_+(\omega - U_+ k)^2 + \rho_-(\omega - U_- k)^2 + gk(\rho_+ - \rho_-) = 0. \tag{4.1.15}$$

Consider the particular case $U_+ = U$, $U_- = 0$ and uniform density $\rho_+ = \rho_-$. Then (4.1.15) reduces to the following formula for the complex *phase velocity* ω/k:

$$\frac{\omega}{k} = \frac{U}{2} \pm \frac{iU}{2}. \tag{4.1.16}$$

Thus the disturbance at the interface propagates in the x direction at a phase speed equal to the mean velocity $\frac{1}{2}U$ of the mean velocities on either side of the vortex sheet; but the motion is unstable, increasing in amplitude exponentially with time.

EXAMPLE 4. INSTABILITY OF A TWO-DIMENSIONAL JET The stability of the two-dimensional jet of Figure 4.1.6(b) (where there is no mean flow outside the jet) can also be discussed by the method of Example 3. Consider the case of uniform fluid density (so that gravity can be ignored), and let ζ_+, ζ_- respectively denote the displacements of the upper and lower boundaries of the jet. Then two kinds of motion can be considered in which (1) $\zeta_+ = \zeta_-$ and (2) $\zeta_+ = -\zeta_-$. The motion in case (1) is *asymmetric*, with dispersion relation

$$\omega^2 + (\omega - Uk)^2 \tanh(kh) = 0, \quad 2h = \text{width of the jet.}$$

As $kh \to \infty$ (for very small *wavelength* disturbances of the jet shear layers) we recover the dispersion relation for a vortex sheet. For long-wavelength 'sinuous' motions of the jet ($kh \ll 1$), the complex phase velocity becomes

$$\frac{\omega}{k} \sim Ukh \pm iU\sqrt{kh},$$

corresponding to a slowly propagating, weak instability of the jet.

Similarly, for the symmetric motion of case (2), we find

$$\omega^2 + (\omega - Uk)^2 \coth(kh) = 0,$$

in which case

$$\frac{\omega}{k} \sim U(1 \pm i\sqrt{kh}) \quad \text{when} \quad kh \ll 1,$$

which represents a slowly growing wave that travels at the jet speed U.

EXAMPLE 5. VORTEX SHEET WITH $\rho'/\rho_0 \ll 1$. In the flow in Figure 4.1.6(a) set $U_+ = U$, $U_- = U'$, $\rho_+ = \rho_0$, $\rho_- = \rho'$ and *neglect* gravity. Then

$$\omega = \frac{k(U \pm iU'\sqrt{\rho'/\rho_o})}{(1 \pm i\sqrt{\rho'/\rho_o})}. \tag{4.1.17}$$

Thus vortex sheets of the type occurring in the free-streamline flow of a fluid of density ρ_o (§3.7) are unstable (with growth rate $\propto \sqrt{\rho'/\rho_o}$) unless $\rho' = 0$.

4.2 The Biot–Savart law

In an unbounded fluid the velocity $\mathbf{v}$ can always be expressed in terms of scalar and vector potentials Φ and $\mathbf{A}$ such that

$$\mathbf{v} = \nabla\Phi + \operatorname{curl}\mathbf{A}, \text{ where } \operatorname{div}\mathbf{A} = 0. \tag{4.2.1}$$

We find the equations determining Φ and $\mathbf{A}$ by taking in turn the divergence and curl (using the formula $\operatorname{curl}\operatorname{curl}\mathbf{A} = \operatorname{grad}\operatorname{div}\mathbf{A} - \nabla^2\mathbf{A}$):

$$\nabla^2\Phi = \operatorname{div}\mathbf{v}, \quad \nabla^2\mathbf{A} = -\operatorname{curl}\mathbf{v} \equiv -\boldsymbol{\omega}. \tag{4.2.2}$$

The decomposition therefore associates sources and sinks with the scalar potential Φ and vorticity with the vector potential $\mathbf{A}$. In unbounded, incompressible flow in which $\operatorname{div}\mathbf{v} = q(\mathbf{x}, t)$, and which is at rest at infinity, we can take Φ in the familiar form (2.6.4).

To find $\mathbf{A}$ we first observe that any particular solution $\mathbf{A}'$ can always be augmented by $\nabla\lambda$, where λ is a scalar function of position, because $\operatorname{curl}\mathbf{A} = \operatorname{curl}\mathbf{A}' + \operatorname{curl}(\nabla\lambda) \equiv \operatorname{curl}\mathbf{A}'$. Then λ can be chosen to ensure that $\operatorname{div}\mathbf{A} = \operatorname{div}\mathbf{A}' + \nabla^2\lambda = 0$. However, by

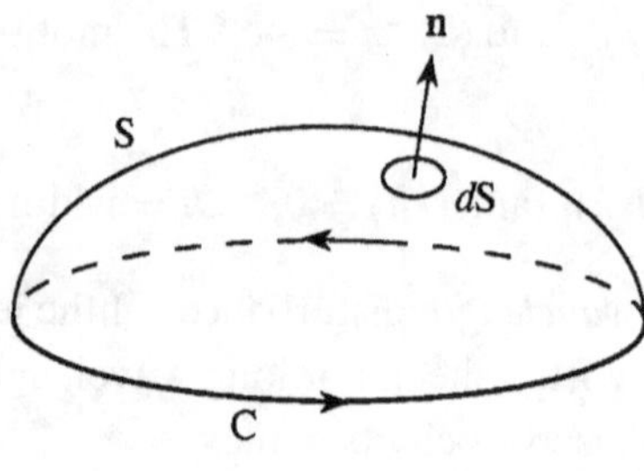

Figure 4.2.1

again using the Green's function (2.6.2) for Laplace's equation, we find from the second of (4.2.2) the particular solution

$$\mathbf{A} = \int \frac{\omega(\mathbf{y}, t) d^3\mathbf{y}}{4\pi |\mathbf{x} - \mathbf{y}|},$$
(4.2.3)

which may be verified to satisfy div $\mathbf{A} = 0$.

Thus, in the important case in which div $\mathbf{v} = 0$ throughout the flow, the velocity is given by the **Biot-Savart** formula

$$\mathbf{v}(\mathbf{x}, t) = \text{curl}\,\mathbf{A} \equiv \text{curl} \int \frac{\omega(\mathbf{y}, t) d^3\mathbf{y}}{4\pi |\mathbf{x} - \mathbf{y}|}.$$
(4.2.4)

Equation (4.2.4) is a purely *kinematic* relation between a vector $\mathbf{v}$ that vanishes at infinity and $\omega = \text{curl}\,\mathbf{v}$. Because div curl $\mathbf{A} \equiv 0$, it is evident that the vector potential constitutes a generalisation of the stream functions of two-dimensional and axisymmetric flows, which are constructed to satisfy identically the continuity equation div $\mathbf{v} = 0$. The geometrical significance of $\mathbf{A}$ is illustrated in Figure 4.2.1. By Stokes' theorem, the volume flux through a two-sided control surface S bounded by the closed contour C is equal to the circulation of the vector potential around C:

$$\int_S \mathbf{v} \cdot d\mathbf{S} = \oint_C \mathbf{A} \cdot d\mathbf{x},$$
(4.2.5)

where the direction of integration on C is in the positive sense with respect to the normal $\mathbf{n}$ on S.

EXAMPLE 1. VELOCITY FIELD OF A VORTEX FILAMENT The velocity field associated with a curvilinear vortex tube of strength γ (= constant) and of infinitesimal cross section is given according to (4.2.4) by

$$\mathbf{v}(\mathbf{x}, t) = \frac{\gamma}{4\pi} \text{curl} \oint \frac{d\mathbf{y}}{|\mathbf{x} - \mathbf{y}|},$$

where the integration is along the axis of the tube. The contribution $\delta\mathbf{v}$ to $\mathbf{v}$ from a vector element $\delta\mathbf{y}$ of the tube is

$$\delta\mathbf{v} = \frac{\gamma\,\delta\mathbf{y} \wedge (\mathbf{x} - \mathbf{y})}{4\pi |\mathbf{x} - \mathbf{y}|^3},$$

which is in a direction normal to the plane of $\delta\mathbf{y}$ and the relative position vector $\mathbf{x} - \mathbf{y}$ (see Figure 4.2.2).

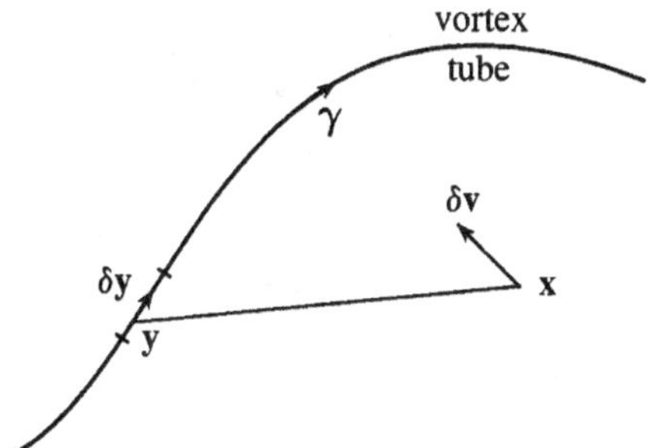

Figure 4.2.2

4.2.1 The far field

Vorticity is transported by convection and diffusion, so that an initially confined region of vorticity will tend to remain within a bounded domain, and it may therefore be assumed that $\omega \to 0$ as $|\mathbf{x}| \to \infty$. The divergence theorem then shows that

$$\int \omega_i(\mathbf{y}, t)d^3\mathbf{y} = -\oint_\Sigma y_i\omega_j(\mathbf{y}, t)n_j dS(\mathbf{y}) \equiv 0,$$

where the closed surface Σ (with inward normal $\mathbf{n}$) is large enough to contain all the vorticity. This conclusion is otherwise obvious when the vortex field is pictured as a distribution of non-intersecting, re-entrant vortex tubes. By use of this result and expansion (2.7.15) when $|\mathbf{x}| \to \infty$, it follows from (4.2.3) that, in the *hydrodynamic far field*:

$$\mathbf{A}(\mathbf{x}, t) \approx \frac{x_j}{4\pi |\mathbf{x}|^3} \int y_j\boldsymbol{\omega}(\mathbf{y}, t)d^3\mathbf{y}, \quad |\mathbf{x}| \to \infty. \tag{4.2.6}$$

Similarly, by the divergence theorem,

$$\int \mathrm{div}\big[y_i y_j\boldsymbol{\omega}(\mathbf{y}, t)\big]d^3\mathbf{y} = 0 \quad \text{and therefore} \quad \int \big[y_i\omega_j(\mathbf{y}, t) + y_j\omega_i(\mathbf{y}, t)\big]d^3\mathbf{y} = 0,$$

which means that (4.2.6) is equivalent to

$$\mathbf{A}(\mathbf{x}, t) \approx \frac{1}{4\pi |\mathbf{x}|^3} \frac{1}{2} \int \big\{(\mathbf{x} \cdot \mathbf{y})\,\boldsymbol{\omega}(\mathbf{y}, t) - [\mathbf{x} \cdot \boldsymbol{\omega}(\mathbf{y}, t)]\,\mathbf{y}\big\} d^3\mathbf{y}$$

$$= \frac{-1}{4\pi |\mathbf{x}|^3} \frac{1}{2} \int \mathbf{x} \wedge [\mathbf{y} \wedge \boldsymbol{\omega}(\mathbf{y}, t)]d^3\mathbf{y}$$

$$\equiv \mathrm{curl}\left[\frac{1}{4\pi |\mathbf{x}|} \frac{1}{2} \int \mathbf{y} \wedge \boldsymbol{\omega}(\mathbf{y}, t)\,d^3\mathbf{y}\right], \quad |\mathbf{x}| \to \infty.$$

Hence

$$\mathbf{A}(\mathbf{x}, t) \approx \mathrm{curl}\left(\frac{\mathbf{I}}{4\pi |\mathbf{x}|}\right), \quad |\mathbf{x}| \to \infty, \tag{4.2.7}$$

where

$$\mathbf{I} = \frac{1}{2} \int \mathbf{y} \wedge \boldsymbol{\omega}(\mathbf{y}, t)d^3\mathbf{y}. \tag{4.2.8}$$

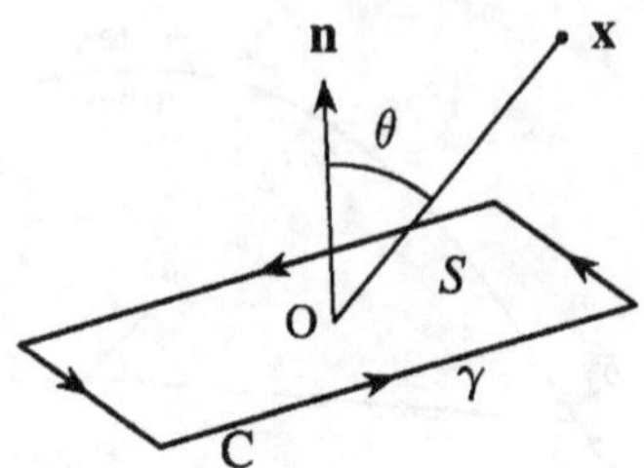

Figure 4.2.3

The vector $\mathbf{I}$ is the *impulse* of the vortex system; it is an *absolute constant* in an unbounded (viscous or inviscid) incompressible flow (Example 5 of this section). We shall subsequently see how it is related to the specific total impulse $\mathbf{I}$ introduced in §2.13. [Batchelor (1967) denotes $\mathbf{I}$ by $\mathbf{P}/\rho_o$; Lighthill (1978, 1986) uses $\mathbf{G}$.]

If we now observe that

$$\nabla^2 \left(\frac{1}{4\pi|\mathbf{x}|} \right) \equiv 0 \quad \text{when} \quad |\mathbf{x}| > 0,$$

it follows from (4.2.4), (4.2.7) and the relation $\operatorname{curl} \operatorname{curl} \mathbf{A} = \operatorname{grad} \operatorname{div} \mathbf{A} - \nabla^2 \mathbf{A}$ that the velocity at large distances from the vorticity distribution can be expressed in either of the equivalent forms

$$\mathbf{v}(\mathbf{x}, t) \approx \operatorname{curl} \operatorname{curl} \left(\frac{\mathbf{I}}{4\pi|\mathbf{x}|} \right) = \operatorname{grad} \operatorname{div} \left(\frac{\mathbf{I}}{4\pi|\mathbf{x}|} \right), \quad |\mathbf{x}| \to \infty. \tag{4.2.9}$$

These formulae supply alternative representations of $\mathbf{v}$ in the hydrodynamic far field (where the motion is entirely irrotational) in terms of either the vector potential $\mathbf{A}$ or the scalar potential φ:

$$\mathbf{v}(\mathbf{x}, t) \approx \begin{cases} \operatorname{curl} \mathbf{A}, & \mathbf{A} = \operatorname{curl}(\mathbf{I}/4\pi|\mathbf{x}|) \\ \nabla \varphi, & \varphi = \operatorname{div}(\mathbf{I}/4\pi|\mathbf{x}|) \end{cases}. \tag{4.2.10}$$

The second of these representations is the field of a dipole source, to which the vorticity is equivalent at large distances. It is also formally identical to far-field formula (2.13.5) for the motion produced in an ideal fluid by rigid-body motion, which should therefore be equivalent to a suitable distribution of vorticity.

EXAMPLE 2. IMPULSE OF A PLANE, RE-ENTRANT VORTEX FILAMENT   A narrow vortex tube C of circulation γ lies in a plane with unit normal $\mathbf{n}$ and encircles $\mathbf{n}$ in the positive direction (Figure 4.2.3). If S is the area of the plane enclosed by the vortex then (§4.1, Example 1)

$$\mathbf{I} = \frac{1}{2} \int \mathbf{y} \wedge \omega d^3\mathbf{y} = \frac{\gamma}{2} \oint_C \mathbf{y} \wedge d\mathbf{y} \equiv \gamma \mathbf{n} S. \tag{4.2.11}$$

The velocity potential at large distances $|\mathbf{x}|$ from the vortex is

$$\varphi(\mathbf{x}) \approx \operatorname{div} \left(\frac{\gamma \mathbf{Sn}}{4\pi|\mathbf{x}|} \right) = -\frac{\gamma S \cos\theta}{4\pi|\mathbf{x}|^2}, \quad |\mathbf{x}| \to \infty. \tag{4.2.12}$$

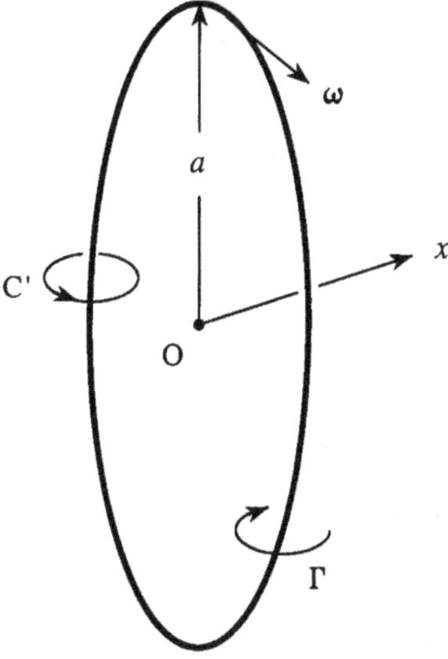

Figure 4.2.4

EXAMPLE 3. VORTEX RING This is a circular vortex tube of radius a and strength Γ (Figure 4.2.4). The vortex strength Γ is equal to the circulation $\oint_{C'} \mathbf{v} \cdot d\mathbf{x}$ around a closed contour C' encircling once the 'core' (tube cross section) of the vortex. Let the core be infinitesimal and lie in the plane $x = 0$ with the axis of symmetry of the vortex along the x axis. This simple model corresponds to the vorticity distribution

$$\boldsymbol{\omega} = \Gamma\delta(x)\delta(\varpi - a)\mathbf{i}_\phi, \quad \varpi = (y^2 + z^2)^{\frac{1}{2}}, \tag{4.2.13}$$

where (x, ϖ, ϕ) are cylindrical polar coordinates, $\mathbf{x} = (x, \varpi\cos\phi, \varpi\sin\phi)$, and $\mathbf{i}_\phi$ is a unit azimuthal vector.

From (4.2.11), the impulse of the vortex ring

$$\mathbf{I} = \frac{1}{2}\int \mathbf{x} \wedge \boldsymbol{\omega}\, d^3\mathbf{x} = \pi a^2\Gamma\, \mathbf{i},$$

where $\mathbf{i}$ is a unit vector in the x direction, normal to the plane of the vortex. The velocity potential of the motion at large distances $|\mathbf{x}|$ from the centre of the vortex is

$$\varphi = \frac{\partial}{\partial x}\left(\frac{a^2\Gamma}{4|\mathbf{x}|}\right) = \frac{-a^2\Gamma\cos\theta}{4|\mathbf{x}|^2}, \quad |\mathbf{x}| \gg a,$$

where θ is the angle between the radius vector $\mathbf{x}$ and the x axis. The streamlines of this far-field motion have the characteristic pattern shown in Figure 2.7.2.

The axisymmetric velocity field induced by the vortex causes the vortex to translate by 'self-induction' in the direction of $\mathbf{I}$. However, the magnitude V_l of this velocity calculated from equation (4.2.4) is infinite, because the integral diverges when the vortex core is infinitesimal and $\mathbf{x}$ lies on the vortex filament. The infinity is not associated with the contribution from a local 'self-potential' discussed in §3.4 (which is easily removed), but with the finite-radius of curvature a of the vortex filament. Real vortices have finite

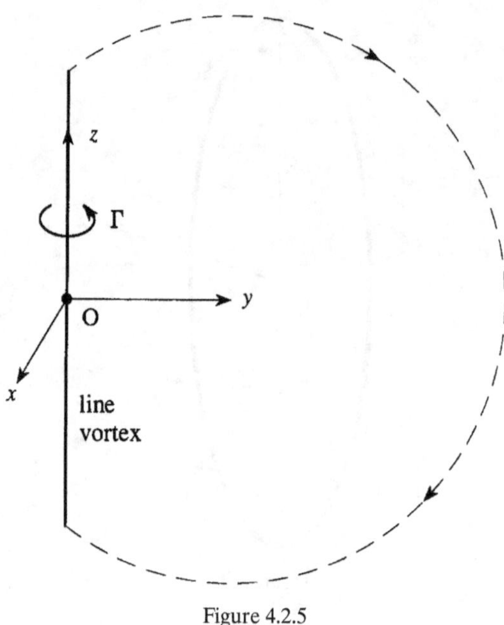

Figure 4.2.5

core diameters, and when this is taken into account V_t is finite. Kelvin showed that when the core is assumed to have circular cross section of radius $\sigma \ll a$,

$$V_t \approx \frac{\Gamma}{4\pi a}\left[\ln\left(\frac{8a}{\sigma}\right) - \frac{1}{4}\right]. \tag{4.2.14}$$

EXAMPLE 4. THE RECTILINEAR VORTEX Let the vortex have circulation (strength) Γ and extend along the z axis (Figure 4.2.5), so that

$$\omega = \Gamma\delta(x)\delta(y)\mathbf{k},$$

where $\mathbf{k}$ is the unit vector in the z direction.

The hypothetical rectilinear vortex is not re-entrant, and integral (4.2.3) defining the vector potential diverges. However, the vortex may be supposed to close at 'infinity', as indicated in the figure, and the precise manner in which this is done has no effect on the velocity in the vicinity of the vortex, as the integral obtained after the curl of the integrand is taken *is* convergent, leading to

$$\mathbf{v} = \frac{\Gamma\mathbf{k}}{4\pi} \wedge \int_{-\infty}^{\infty} \frac{(x, y, 0)d\xi}{(x^2 + y^2 + \xi^2)^{\frac{3}{2}}} = \frac{\Gamma}{2\pi}\left(-\frac{y}{r^2}, \frac{x}{r^2}, 0\right) \equiv \frac{\Gamma}{2\pi}\left(-\frac{\sin\theta}{r}, \frac{\cos\theta}{r}, 0\right),$$

where the last formula is expressed in terms of the polar coordinates $(x, y) = r(\cos\theta, \sin\theta)$ of the two-dimensional theory (§3.4).

The irrotational motion in $r > 0$ is governed by the velocity potential

$$\varphi = \frac{\Gamma\theta}{2\pi}.$$

The flow circulates around the vortex with velocity

$$v_\theta = \frac{1}{r}\frac{\partial\varphi}{\partial\theta} = \frac{\Gamma}{2\pi r},$$

decreasing inversely with distance r from the vortex, the streamlines being circles centred on the vortex axis.

EXAMPLE 5. CONSERVATION OF VORTEX IMPULSE ($I = $ CONSTANT) IN UNBOUNDED FLOW Form the vector product of $\mathbf{x}$ and vorticity equation (4.1.7), and use the relation $\nabla^2\omega = -\mathrm{curl}(\mathrm{curl}\,\omega)$:

$$\frac{\partial}{\partial t}(\mathbf{x}\wedge\omega) + \mathbf{x}\wedge\mathrm{curl}(\omega\wedge\mathbf{v}) = -\nu\mathbf{x}\wedge\mathrm{curl}(\mathrm{curl}\,\omega).$$

Assume that $\omega = 0$ for $|\mathbf{x}| \to \infty$, and use the identity

$$\mathbf{x}\wedge\mathrm{curl}\,\mathcal{A} = 2\mathcal{A} + \nabla(\mathbf{x}\cdot\mathcal{A}) - \frac{\partial}{\partial x_j}(x_j\mathcal{A})$$

satisfied by any vector field $\mathcal{A}$. Then

$$\frac{d\mathbf{I}}{dt} \equiv \frac{\partial}{\partial t}\left(\frac{1}{2}\int \mathbf{x}\wedge\omega\,d^3\mathbf{x}\right) = -\int \omega\wedge\mathbf{v}\,d^3\mathbf{x} - \nu\int \mathrm{curl}\,\omega\,d^3\mathbf{x} \equiv 0,$$

because the volume integrals both vanish [the first by appeal to (4.1.2)].

4.2.2 Kinetic energy

By writing

$$T = \frac{1}{2}\rho_o\int v^2 d^3\mathbf{x} = \frac{1}{2}\rho_o\int \mathbf{v}\cdot\mathrm{curl}\,\mathbf{A}\,d^3\mathbf{x}$$

and integrating by parts, we readily deduce from (4.2.3) that the kinetic energy of an unbounded (three-dimensional) incompressible flow can be cast in the following form:

$$T = \frac{\rho_o}{8\pi}\iint \frac{\omega(\mathbf{x},t)\cdot\omega(\mathbf{y},t)}{|\mathbf{x}-\mathbf{y}|}\,d^3\mathbf{x}d^3\mathbf{y}. \tag{4.2.15}$$

Similarly [from (4.1.2)], integration of the identity

$$\mathbf{x}\cdot\omega\wedge\mathbf{v} = \mathbf{x}\cdot\left[(\mathbf{v}\cdot\nabla)\mathbf{v} - \nabla\left(\frac{1}{2}v^2\right)\right] \equiv \frac{1}{2}v^2 + \frac{\partial}{\partial x_j}\left[x_i\left(v_iv_j - \delta_{ij}\frac{1}{2}v^2\right)\right]$$

yields the following alternative representation:

$$T = \rho_o\int \mathbf{x}\cdot(\omega\wedge\mathbf{v})(\mathbf{x},t)d^3\mathbf{x}. \tag{4.2.16}$$

These results serve to emphasize that in an incompressible, homogeneous, unbounded fluid, motion is possible only if $\omega \neq 0$.

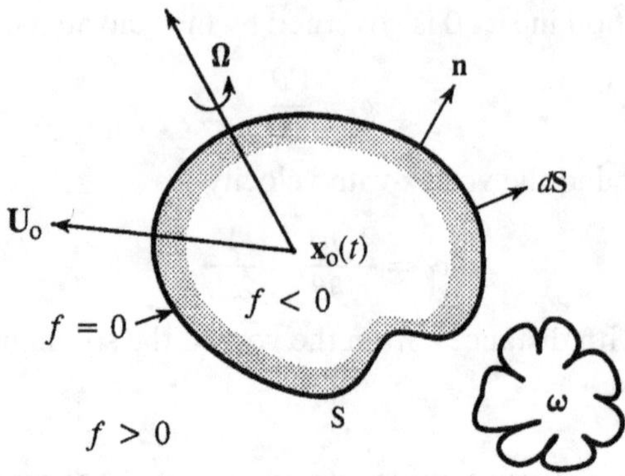

Figure 4.2.6

4.2.3 The Biot–Savart formula in the presence of an internal boundary

The Biot–Savart representation of velocity (4.2.4) is readily extended to account for the presence of internal boundaries. Consider, for example, a rigid body (Figure 4.2.6) moving in a general manner with velocity

$$\mathbf{U} = \mathbf{U}_o + \mathbf{\Omega} \wedge [\mathbf{x} - \mathbf{x}_o(t)], \tag{4.2.17}$$

where $\mathbf{U}_o = d\mathbf{x}_o/dt$ is the velocity of its centre of volume $\mathbf{x}_o(t)$ and $\mathbf{\Omega}(t)$ is its angular velocity. The fluid is incompressible and assumed to be at rest at infinity.

In the usual way let $f(\mathbf{x}, t) = 0$ define the surface S of the body, with $f > 0$ in the fluid. Then $H(f)\mathbf{v} + H(-f)\mathbf{U}$ is the velocity everywhere, in both the fluid and solid, and we can put

$$H(f)\mathbf{v} + H(-f)\mathbf{U} = \nabla\Phi + \operatorname{curl}\mathbf{A}, \quad (\operatorname{div}\mathbf{A} = 0).$$

The body has constant volume ($\operatorname{div}\mathbf{U} = 0$), but $\operatorname{curl}\mathbf{U} = 2\mathbf{\Omega}$; vortex lines continue into the body where $\omega = 2\mathbf{\Omega}$. Continuity of normal velocity on S supplies

$$\nabla^2\Phi = \operatorname{div}[H(f)\mathbf{v} + H(-f)\mathbf{U}] = \nabla H(f) \cdot (\mathbf{v} - \mathbf{U}) \equiv 0,$$

so that $\Phi = 0$. Also, for viscous flow the no-slip condition on S implies that

$$\operatorname{curl}\operatorname{curl}\mathbf{A} = -\nabla^2\mathbf{A} = \operatorname{curl}[H(f)\mathbf{v} + H(-f)\mathbf{U}]$$

$$= \nabla H(f) \wedge (\mathbf{v} - \mathbf{U}) + H(f)\omega + H(-f)2\mathbf{\Omega}$$

$$\equiv H(f)\omega + H(-f)2\mathbf{\Omega}. \tag{4.2.18}$$

Therefore the velocity *everywhere* is given by the following modification of Biot–Savart formula (4.2.4):

$$\mathbf{v}(\mathbf{x}, t) = \operatorname{curl}\int_V \frac{\omega(\mathbf{y}, t)d^3\mathbf{y}}{4\pi|\mathbf{x} - \mathbf{y}|} + \operatorname{curl}\int_\Delta \frac{2\mathbf{\Omega}(t)d^3\mathbf{y}}{4\pi|\mathbf{x} - \mathbf{y}|}, \tag{4.2.19}$$

where the integrations are taken respectively over the volume V of the fluid and the volume Δ of the solid body.

This formula predicts that $\mathbf{v} = \mathbf{U}$ when $\mathbf{x}$ lies within the body. The identity $\int \mathrm{curl}\,[\mathrm{H}(f)\mathbf{v} + \mathrm{H}(-f)\mathbf{U}]\,d^3\mathbf{x} = 0$ implies that $\mathbf{v} \sim O(1/|\mathbf{x}|^3)$ as $|\mathbf{x}| \to \infty$, as for unbounded flow. Similarly, far-field approximations (4.2.10) remain valid provided integral (4.2.8) defining the impulse $\mathbf{I}$ includes the contribution from the region occupied by the body (where $\omega = 2\Omega$).

EXAMPLE 6. CREEPING FLOW (see §4.4.3) A rigid sphere of radius a translates at constant velocity $\mathbf{U}$ without rotation at infinitesimal Reynolds number $\mathrm{Re} = Ua/\nu$. When the origin is at the centre of the sphere, the vorticity distribution is

$$\omega = \mathrm{curl}\left(\frac{3a\mathbf{U}}{2|\mathbf{x}|}\right), \quad |\mathbf{x}| > a. \tag{4.2.20}$$

Verify that the velocity $\mathbf{v}$ predicted by Biot–Savart formula (4.2.19) yields $\mathbf{v} = \mathbf{U}$ for $|\mathbf{x}| < a$.

We need to evaluate the first integral in (4.2.19), which is only conditionally convergent when ω is defined as in (4.2.20). However, the $\mathbf{x}$-dependent part of the integral is well defined, and we can write

$$\mathbf{v}(\mathbf{x}) = \frac{1}{4\pi} \int_V \frac{(\mathbf{y} - \mathbf{x})}{|\mathbf{y} - \mathbf{x}|^3} \wedge \mathrm{curl}\left(\frac{3a\mathbf{U}}{2|\mathbf{y}|}\right) d^3\mathbf{y}.$$

We shall evaluate the integral for $\mathbf{x} = 0$ (at the centre of the sphere); the general case is left as an exercise for the reader. Introduce polar integration coordinates (r, θ, ϕ) with the polar direction parallel to $\mathbf{U}$. Then

$$\mathbf{v}(0) = -\frac{3a}{8\pi} \int_V \frac{\mathbf{y} \wedge (\mathbf{y} \wedge \mathbf{U})}{|\mathbf{y}|^6} d^3\mathbf{y}$$

$$= \frac{3a}{8\pi} \int_V \frac{r^2\mathbf{U} - \mathbf{y}Ur\cos\theta}{r^4} \sin\theta \, dr\,d\theta\,d\phi$$

$$= \frac{3a\mathbf{U}}{4} \int_a^\infty \frac{dr}{r^2} \int_0^\pi (1 - \cos^2\theta)\sin\theta \, d\theta$$

$$= \mathbf{U}.$$

4.2.4 The Biot–Savart formula for irrotational flow

Biot–Savart formula (4.2.19) is applicable also when the motion is inviscid and irrotational. When $\nu \to 0$ vorticity generated at a solid boundary cannot diffuse away from the boundary into the flow. Therefore, in the absence of vorticity in the fluid from other sources, the first integral of (4.2.19) must be evaluated with ω replaced with the 'bound vorticity' in the vortex sheet on the surface of the body, whose strength is determined by the tangential velocity of slip at the surface.

In formulating the generalised Biot–Savart representation of the fluid velocity in the presence of internal boundaries, we are actually free to specify *arbitrarily* conditions within, say, the surface S of Figure 4.2.6. The particular representation (4.2.19) supplies a description of the motion that is correct at all points in space, including the region occupied by the solid. However, by appropriately prescribing *hypothetical* conditions within S, we may possibly derive a more convenient integral representation of the fluid velocity. We have already seen, for example, how Green's formula (§§2.8, 2.9) can be modified to give a representation of irrotational flow in terms of surface distributions of either monopole or dipole sources.

In the case of irrotational motion generated by a moving boundary, the Biot–Savart formula can be manipulated to yield precisely all of the representations obtained previously from Green's formula. To illustrate this, suppose the motion in Figure 4.2.6 is entirely irrotational, with velocity potential φ. Introduce scalar and vector potentials Φ and $\mathbf{A}$ and put

$$H(f)\nabla\varphi = \nabla\Phi + \text{curl}\,\mathbf{A}. \tag{4.2.21}$$

According to this formula Φ and $\mathbf{A}$ determine a velocity field that *vanishes* identically within S and is equal to the flow velocity outside S.

Proceeding in the usual way, we find (because $\nabla^2\varphi = 0$ and $\text{curl}\,\nabla\varphi = 0$)

$$\nabla^2\Phi = \nabla H(f)\cdot\nabla\varphi, \quad \nabla^2\mathbf{A} = -\nabla H(f)\wedge\nabla\varphi \equiv \text{curl}\,[\varphi\nabla H(f)].$$

Hence, setting $H = H(f)$, we find that (4.2.21) becomes

$$H\nabla\varphi = -\nabla\int\frac{\nabla\varphi\cdot\nabla H\,d^3\mathbf{y}}{4\pi|\mathbf{x}-\mathbf{y}|} - \text{curl}\int\frac{\text{curl}(\varphi\nabla H)d^3\mathbf{y}}{4\pi|\mathbf{x}-\mathbf{y}|}$$

$$= -\nabla\int\frac{\nabla\varphi\cdot\nabla H\,d^3\mathbf{y}}{4\pi|\mathbf{x}-\mathbf{y}|} - \text{curl}\,\text{curl}\int\frac{\varphi\nabla H\,d^3\mathbf{y}}{4\pi|\mathbf{x}-\mathbf{y}|}. \tag{4.2.22}$$

Now, on the left-hand side,

$$H\nabla\varphi = \nabla(H\varphi) - \varphi\nabla H,$$

and on the right-hand side,

$$\text{curl}\,\text{curl} = \nabla\text{div} - \nabla^2 \quad\text{and}\quad \nabla^2\left(\frac{1}{4\pi|\mathbf{x}-\mathbf{y}|}\right) = -\delta(\mathbf{x}-\mathbf{y}).$$

Therefore, (4.2.22) becomes

$$\nabla(H\varphi) = -\nabla\left(\int\frac{\nabla\varphi\cdot\nabla H\,d^3\mathbf{y}}{4\pi|\mathbf{x}-\mathbf{y}|} + \text{div}\int\frac{\varphi\nabla H\,d^3\mathbf{y}}{4\pi|\mathbf{x}-\mathbf{y}|}\right),$$

$$\therefore\quad H(f)\varphi(\mathbf{x}) = -\frac{\partial}{\partial x_j}\oint_S\frac{\varphi(\mathbf{y},t)n_j dS}{4\pi|\mathbf{x}-\mathbf{y}|} - \oint_S n_j\frac{\partial\varphi}{\partial y_j}(\mathbf{y},t)\frac{dS}{4\pi|\mathbf{x}-\mathbf{y}|},$$

where an arbitrary constant of integration has been discarded and we have used transformation formula (2.8.2). This is Green's formula (2.8.7) in the absence of a source distribution $q(\mathbf{x},t)$.

In a similar manner, by introducing a velocity potential $\bar{\varphi}$ to represent a possible fluid motion within S, and setting

$$H(f)\nabla\varphi + H(-f)\nabla\bar{\varphi} = \nabla\Phi + \text{curl}\,\mathbf{A},$$

We obtain Green's generalisation (2.9.5) together with the special monopole and dipole cases (2.9.6) and (2.9.7).

EXAMPLE 7. BIOT–SAVART FORMULA APPLIED TO IRROTATIONAL MOTION PRO-DUCED BY A MOVING BODY Let the body translate at velocity $\mathbf{U}$, and take coordinate axes fixed within the body. The velocity potential of the motion is therefore

$$\varphi = U_j\varphi_j^*(\mathbf{x}),$$

where φ_j^* is the velocity potential introduced in §2.13.2 that represents the flow produced by motion of the body at unit speed in the j direction.

To apply generalised Biot–Savart formula (4.2.19) (where, for this problem, $\Omega = \text{curl}\,\mathbf{U} = 0$) we can adopt either of the following approaches:

(1) that ω in (4.2.19) is formally replaced with the vortex sheet strength $\nabla H(f) \wedge (\nabla\varphi - \mathbf{U})$ of the limiting boundary layer on the body;

or

(2) that $\omega \equiv 0$ within the fluid, but because the no-slip condition is not applicable, Equation (4.2.18) for the vector potential must be replaced with

$$-\nabla^2\mathbf{A} = \nabla H(f) \wedge (\nabla\varphi - \mathbf{U}).$$

In either case this is equivalent in (4.2.19) to setting

$$\omega(\mathbf{y}, t) = -U_j\nabla H(f) \wedge (\nabla y_j - \nabla\varphi_j^*) = \text{curl}\,[U_j Y_j \nabla H(f)],$$

where $Y_j \equiv y_j - \varphi_j^*(\mathbf{y})$ is the Kirchhoff vector (2.16.7) for the body. Hence, substituting into the first integral of (4.2.19), we find, for $\mathbf{x}$ within the fluid,

$$\mathbf{v} = \text{curl}\int \frac{\text{curl}\,[U_j Y_j \nabla H(f)]d^3\mathbf{y}}{4\pi|\mathbf{x} - \mathbf{y}|} = \text{curl}\,\text{curl}\int \frac{U_j Y_j \nabla H(f)\,d^3\mathbf{y}}{4\pi|\mathbf{x} - \mathbf{y}|}$$

$$= (\nabla\text{div} - \nabla^2)\int \frac{U_j Y_j \mathbf{n}\,dS}{4\pi|\mathbf{x} - \mathbf{y}|} \equiv \nabla\text{div}\int \frac{U_j Y_j \mathbf{n}\,dS\mathbf{y}}{4\pi|\mathbf{x} - \mathbf{y}|},$$

$$\therefore \quad \varphi(\mathbf{x}) = \text{div}\int \frac{U_j Y_j \mathbf{n}\,dS\mathbf{y}}{4\pi|\mathbf{x} - \mathbf{y}|}, \tag{4.2.23}$$

because $\nabla^2(1/4\pi|\mathbf{x} - \mathbf{y}|) \equiv 0$ for $\mathbf{x} \neq \mathbf{y}$. The reader can verify that this result is equivalent to (2.9.7) when $q = 0$.

For the special case of a sphere of radius a the Kirchhoff vector is (see §2.16.2)

$$Y_j = y_j \left(1 + \frac{a^3}{2|\mathbf{y}|^3}\right) = \frac{3y_j}{2} \quad \text{on } |\mathbf{y}| = a.$$

Equation (4.2.23) now reduces to Green's dipole formula of §2.9, Example 2.

4.3 Examples of axisymmetric vortical flow

The vortex lines of an axisymmetric flow are circular and coaxial with the axis of symmetry, and the velocity is parallel to the meridian planes. If the motion is assumed to be symmetric about the x direction, flow quantities are functions of only the variables x and ϖ of the cylindrical polar coordinates $\mathbf{x} = (x, \varpi \cos \phi, \varpi \sin \phi)$ of §2.12. The vector potential $\mathbf{A}$ defined by (4.2.3) will evidently be perpendicular at any given point to both the x and ϖ directions, so that

$$\mathbf{A} = A(x, \varpi)\mathbf{i}_\phi,$$

where $\mathbf{i}_\phi$ is a unit vector in the azimuthal direction. However, according to (4.2.5) the volume flux in the x direction through a circle of symmetry of radius ϖ is $2\pi \varpi A(x, \varpi)$, so that the Stokes stream function ψ (= volume flux/radian; §2.12) and the vector potential $A(x, \varpi)$ are related by

$$\psi = \varpi A(x, \varpi). \tag{4.3.1}$$

4.3.1 Circular vortex filament

The stream function $\psi(\mathbf{x})$ for the circular vortex filament of Figure 4.2.4 can be evaluated by use of relation (4.3.1) from integral formula (4.2.3) for the vector potential. However, that formula is valid only for the separate orthogonal components of $\mathbf{A} = (A_1, A_2, A_3)$ with respect to rectangular axes (x, y, z). Thus, taking the origin at the centre of the vortex and putting $\mathbf{y} = (0, a \cos \phi', a \sin \phi')$ on the vortex ring core, we find from (4.2.3) at $\mathbf{x} = (x, \varpi \cos \phi, \varpi \sin \phi)$,

$$\mathbf{A} \equiv (0, A_2, A_3) = \frac{\Gamma a}{4\pi} \int_0^{2\pi} \frac{(0, -\sin \phi', \cos \phi')\, d\phi'}{|\mathbf{x} - \mathbf{y}|},$$

where

$$|\mathbf{x} - \mathbf{y}| = \sqrt{z^2 + \varpi^2 + a^2 - 2a\varpi \cos(\phi' - \phi)}.$$

Now in (4.3.1) $A(x, \varpi) = \mathbf{A} \cdot (0, -\sin \phi, \cos \phi)$ and therefore (putting $\xi = \phi' - \phi$)

$$\psi = \varpi A = \frac{\Gamma a \varpi}{4\pi} \int_0^{2\pi} \frac{\cos \xi\, d\xi}{\sqrt{z^2 + \varpi^2 + a^2 - 2a\varpi \cos \xi}}. \tag{4.3.2}$$

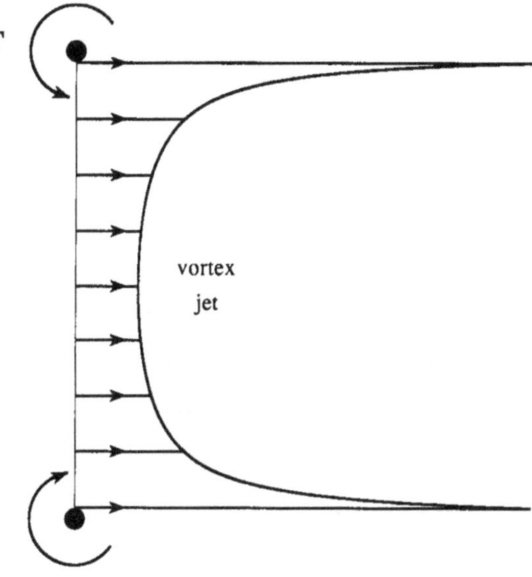

Figure 4.3.1

EXAMPLE 1. VORTEX JET By setting $x = 0$ in (4.3.2) and assuming that ϖ/a is small, we find that in the plane of the vortex, near the axis,

$$\psi = \varpi A = \frac{\Gamma \varpi^2}{4a}\left[1 + O\left(\frac{\varpi^2}{a^2}\right)\right].$$

Therefore near the centre of the vortex the axial flow speed

$$u = \frac{1}{\varpi}\frac{\partial \psi}{\partial \varpi} \sim \frac{\Gamma}{2a} + \cdots +, \qquad (4.3.3)$$

where the omitted terms are smaller by a factor $O(\varpi^2/a^2)$. Thus the flow speed near the centre varies slowly with radial distance ϖ, and the inner flow region may be regarded as a localized 'jet'. Figure 4.3.1 shows the actual axial velocity distribution calculated numerically from (4.3.2). Near the vortex filament the velocity becomes very large (infinite when the core is infinitesimal); but this would remain bounded for a finite core size, and the velocity close to the axis is unchanged for small but finite core diameters.

4.3.2 Rate of production of vorticity at a nozzle

A 'puff' of air blown from a circular nozzle into a nominally quiescent ambient atmosphere will typically evolve into a roughly axisymmetric vortex ring (Figure 4.3.2). Consider a nozzle of radius a, and suppose the flow starts impulsively at time $t = 0$. We neglect viscosity except insofar as it is responsible for the release of vorticity from the circular rim of the duct, initially in the form of a circular cylindrical vortex sheet (cf. §2.12.5). Let the jet velocity be uniform across the duct with speed $V(t)$ and take the x axis along the duct axis, with the origin O in the exit plane. The circulation of

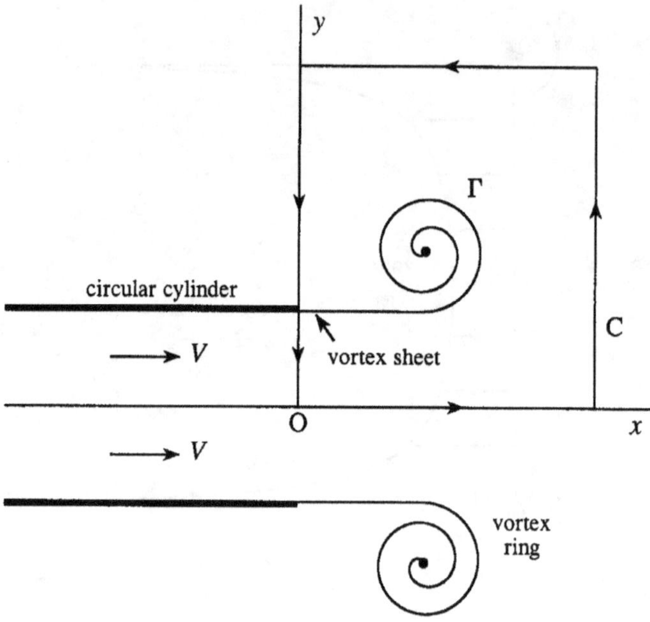

Figure 4.3.2

the vortex sheet per unit length of the jet at the exit plane is $V(t)$ and the vorticity is convected at the local mean stream velocity $\frac{1}{2}V$, equal to the mean of the values within and outside the jet. The rate of increase $d\Gamma/dt$ of the circulation Γ of the axisymmetric flow outside the nozzle is therefore $\frac{1}{2}V^2$.

A formal derivation of this result is obtained by integration of the inviscid form of vorticity equation (4.1.7) over the area of a radial rectangular section of the exterior flow with boundary C, as indicated in Figure 4.3.2. When the rectangle is large enough it will be pierced by all of the circular vortex lines of the exterior flow. Without loss we can take the rectangle to lie in the xy plane with unit normal $\mathbf{k}$. Then $\Gamma = \int_A \boldsymbol{\omega} \cdot \mathbf{k}\,dxdy$, where the integration is taken over the area $\mathcal{A}$ of the rectangle. Thus, using the inviscid form of (4.1.7), we obtain

$$\frac{d\Gamma}{dt} = -\int_A \mathbf{k} \cdot \mathrm{curl}\,(\boldsymbol{\omega} \wedge \mathbf{v})\,dxdy = -\oint_C \boldsymbol{\omega} \wedge \mathbf{v} \cdot d\mathbf{x}.$$

The only contribution to the final integral is from the intersection of C with the vortex sheet where it emerges from the nozzle. At that point

$$\boldsymbol{\omega} = V(t)\mathbf{k}\delta(y - a), \quad \mathbf{v} = \frac{1}{2}V(t)\mathbf{i}, \quad d\mathbf{x} = -\mathbf{j}dy,$$

where $\mathbf{i}$, $\mathbf{j}$, $\mathbf{k}$ are the usual unit vectors in the x, y, and z directions.

Hence at time t, the vortex ring circulation is given approximately by

$$\Gamma = \frac{1}{2}\int_0^t V^2(t)\,dt. \tag{4.3.4}$$

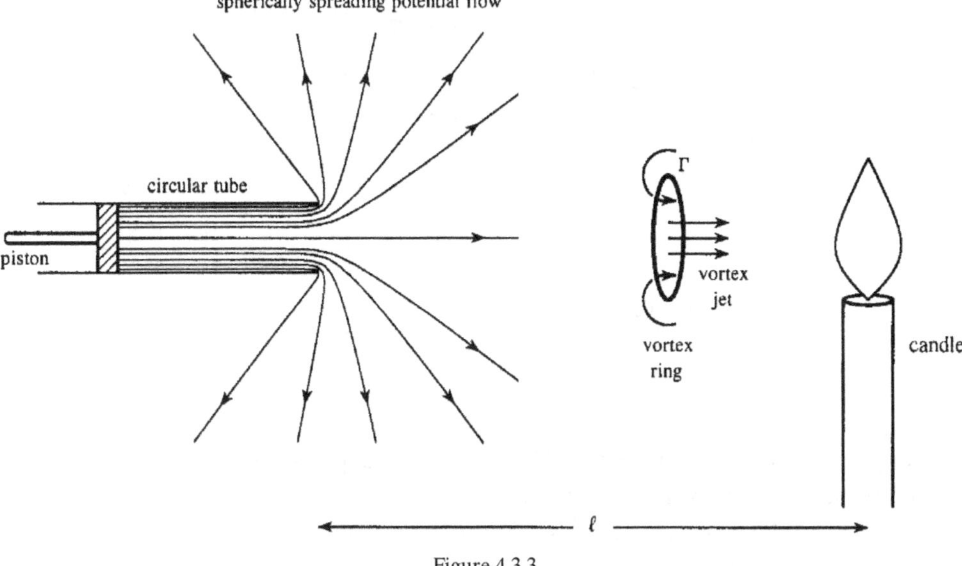

Figure 4.3.3

4.3.3 Blowing out a candle

A candle is 'blown out' by the action of a vortex jet (Lighthill 1963). In Figure 4.3.3 a puff of air is ejected from the tube and directed at a candle flame. Let the tube have radius a, and suppose air is forced out at constant speed V by rapid movement of the piston over a distance L. In an ideal fluid the motion outside at distance r is irrotational and resembles a radially symmetric source flow of strength $q = \pi a^2 V$ and velocity potential $\varphi \sim -Va^2/4r$ that persists only while the piston is moving, causing the air to blow against the flame at distance $r = \ell$ at speed

$$V_\varphi \sim \frac{V}{4}\frac{a^2}{\ell^2}.$$

The fluid leaving the tube is initially confined to a cylindrical slug of length L and radius a, whose displacement from the tube generates the potential flow φ. In reality, of course, vorticity is generated at the tube exit, where it emerges in the form of a circular cylindrical vortex sheet of circulation V per unit length. The sheet may be pictured as a succession of vortex rings of radius a and infinitesimal core radii, and when the piston comes to rest after a time L/V their total circulation $\Gamma = \frac{1}{2}LV$ [by (4.3.4)].

The cylindrical vortex 'rolls up' to form a vortex ring of circulation Γ that proceeds towards the flame at speed given by Kelvin's formula (4.2.14):

$$V_t \sim \frac{VL}{8\pi a_o}\left[\ln\left(\frac{8a_o}{\sigma}\right) - \frac{1}{4}\right],$$

where, according to experiment, $a_o \sim 1.2a$ is the effective radius of the vortex ring and $\sigma \sim 0.2a$ is the radius of its core (assumed to be of circular cross section). The ring arrives at the flame after a time $t_\ell \sim \ell/V_t$, where the flame is extinguished provided the

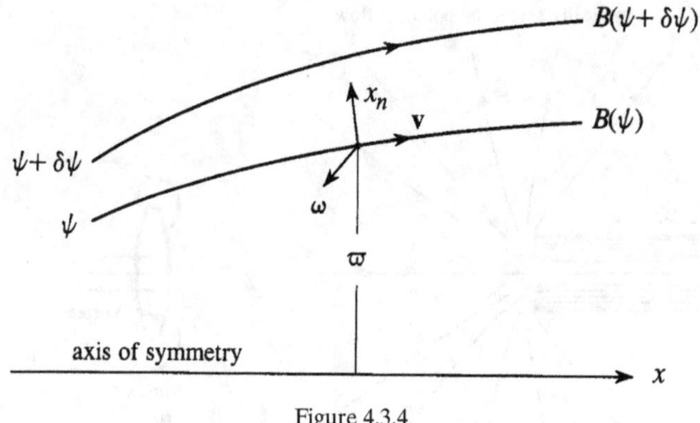

Figure 4.3.4

'vortex jet' [of speed $V_j \sim \Gamma/2a_o \approx VL/4a_o$, see (4.3.3)] is large enough to blow away the hot combusting gases from newly vaporized wax.

Thus the candle is blown out only because of the existence of the vortex. In its absence the flame would barely flicker under the influence of blowing by the potential flow velocity V_φ. The reader may verify, for example, that if $V = 10\,\text{m/s}$, $L = 1\,\text{cm}$, $a = 0.5\,\text{cm}$, and $\ell = 0.3\,\text{m}$, then $V_\varphi \approx 0.0007\,\text{m/s}$, which is negligible compared with the vortex jet speed $V_j \approx 4\,\text{m/s}$. Furthermore, $V_t \sim 2.4\,\text{m/s}$, so that the vortex takes about 0.13 s to reach the flame.

4.3.4 Axisymmetric steady flow of an ideal fluid

In steady, ideal flow, Crocco's equation (4.1.6) reduces to

$$\boldsymbol{\omega} \wedge \mathbf{v} + \nabla B = 0.$$

Hence $\mathbf{v} \cdot \nabla B = \boldsymbol{\omega} \cdot \nabla B = 0$, which means that B is constant along the fixed streamlines and vortex lines of the motion, and therefore a surface on which

$$B \equiv \frac{p}{\rho_o} + \frac{1}{2}v^2 = \text{constant} \tag{4.3.5}$$

is spanned by intersecting families of streamlines and vortex lines.

In axisymmetric flow (Figure 4.3.4) $\boldsymbol{\omega} = \omega_\phi \mathbf{i}_\phi$, and these surfaces coincide with the stream surfaces on which the Stokes stream function $\psi = \text{constant}$, so that $B = B(\psi)$. The streamlines and vortex lines cut at right angles and the *dynamic* condition for steady flow becomes

$$\omega_\phi v + \frac{\partial B}{\partial x_n} = 0,$$

$$\text{i.e., } \omega_\phi v + B'(\psi)\frac{\partial \psi}{\partial x_n} = 0, \tag{4.3.6}$$

where x_n is normal to the stream surface in the direction of $\boldsymbol{\omega} \wedge \mathbf{v}$ and $B'(\psi) = \partial B/\partial \psi$.

The continuity equation requires

$$v = \frac{1}{\varpi} \frac{\partial \psi}{\partial x_n},$$

where the differentiation is 'at right angles to the left' of the direction of motion, in accordance with the rule of §2.12. Therefore, condition (4.3.6) can be written as

$$\frac{\omega_\phi}{\varpi} = -B'(\psi) = \text{constant on a streamline.} \tag{4.3.7}$$

By combining this result with Equation (2.12.4) we conclude that $\psi(x, \varpi)$ satisfies

$$\frac{\partial^2 \psi}{\partial x^2} + \frac{\partial^2 \psi}{\partial \varpi^2} - \frac{1}{\varpi} \frac{\partial \psi}{\partial \varpi} = \varpi^2 B'(\psi), \tag{4.3.8}$$

and we can investigate different flows by assigning different functional forms to $B'(\psi)$.

Equation (4.3.7) can also be derived directly from vorticity equation (4.1.8) for ideal, axisymmetric flow. First, we have

$$(\boldsymbol{\omega} \cdot \nabla)\mathbf{v} = \frac{\omega_\phi}{\varpi} \frac{\partial}{\partial \phi} \left(v_x \mathbf{i} + v_\varpi \mathbf{i}_\varpi \right) \equiv \frac{\omega_\phi v_\varpi}{\varpi} \frac{\partial}{\partial \phi} \mathbf{i}_\varpi,$$

where $\mathbf{i}_\varpi$ is the unit vector in the direction of increasing ϖ. Now

$$\frac{\partial}{\partial \phi} \mathbf{i}_\varpi = \mathbf{i}_\phi.$$

Hence, for steady, axisymmetric motion the vorticity equation becomes

$$\frac{D}{Dt} \omega_\phi = \frac{\omega_\phi v_\varpi}{\varpi}, \quad \text{i.e.,} \quad \frac{D}{Dt} \left(\frac{\omega_\phi}{\varpi} \right) = 0, \tag{4.3.9}$$

that is, for a dynamically possible steady flow ω_ϕ/ϖ must assume a constant value on the stream surface $\psi = \text{constant}$ on which it lies.

This is precisely the condition that the strength $\omega_\phi \delta S$ of a vortex tube of cross-sectional area δS should remain constant as the tube is convected and stretched by the flow. In axisymmetric flow the tube forms a circular vortex ring of circumferential length $2\pi \varpi$ and for incompressible flow its volume $2\pi \varpi \delta S$ is constant; it follows that the ratio ω_ϕ/ϖ of the tube strength and volume must also be invariant.

4.3.5 Hill's spherical vortex

Hill's spherical vortex corresponds to the simplest case in which $B'(\psi) = \text{constant}$ within a sphere of radius a, say, with centre at the origin. If we put $B'(\psi) = -5A$ for some suitable constant A, then $\omega_\phi = 5A\varpi$, and we can easily verify that (4.3.8) is satisfied by

$$\psi = \frac{A}{2} \varpi^2 (a^2 - r^2), \quad \text{for} \quad r < a, \quad \text{where} \quad r^2 = x^2 + \varpi^2. \tag{4.3.10}$$

By writing $\varpi = r \sin \theta$, where θ is the polar angle measured from the x direction, we can also verify that the normal component of velocity on the sphere

$$\frac{1}{\varpi} \frac{\partial \psi}{r \partial \theta} \equiv 0 \quad \text{at} \quad r = a,$$

i.e., that the sphere is a stationary stream surface of the motion.

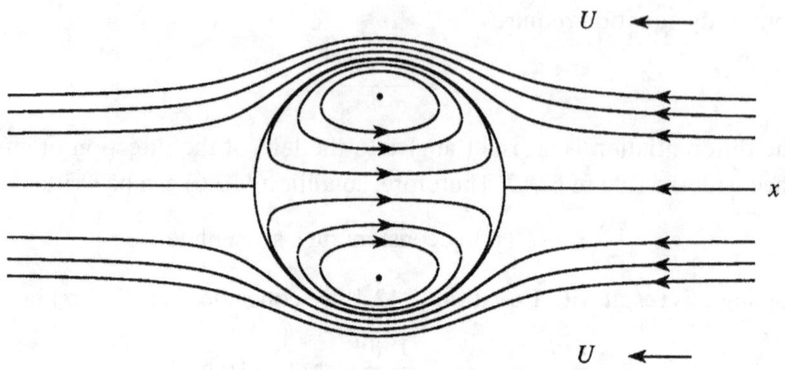

Figure 4.3.5

Let us compare (4.3.10) with the stream function for steady irrotational flow of the ambient fluid past this stationary sphere at speed U. The stream function for a sphere with centre at the origin moving in the positive x direction at speed U is given by (2.12.7) in a reference frame fixed relative to the fluid at infinity. We obtain the stream function with respect to a frame moving with the sphere is by imposing a flow at speed U in the $-x$ direction, i.e., by adding $-\frac{1}{2}U\varpi^2$ to stream function (2.12.7) [cf. Figure 2.12.2(a)], so that

$$\psi = -\frac{U}{2}\varpi^2\left(1 - \frac{a^3}{r^3}\right), \quad r > a. \tag{4.3.11}$$

There will be a smooth transition between the interior and exterior flows described by (4.3.10) and (4.3.11) provided the tangential velocity is continuous, i.e., provided $\partial\psi/\partial r$ is continuous at $r = a$. This condition requires $A = \frac{3}{2}U/a^2$, so that, within the sphere

$$\psi = \frac{3U\varpi^2}{4a^2}\left[a^2 - (x^2 + \varpi^2)\right]. \tag{4.3.12}$$

The system of closed stream surfaces $\psi = $ constant inside the sphere is illustrated in Figure 4.3.5. The vorticity is

$$\omega_\phi = \frac{15U\varpi}{2} \quad \text{in} \quad \sqrt{x^2 + \varpi^2} < a,$$

and vanishes elsewhere in the fluid. The pressure within the vortex can be calculated from (4.3.5), wherein $B = -5A\psi + $ constant for $r < a$, i.e., putting $A = \frac{3}{2}U/a^2$, from

$$\frac{p}{\rho_0} + \frac{1}{2}v^2 = -\frac{15U\psi}{2a^2} + \frac{1}{2}U^2,$$

this being continuous with the pressure in the ambient flow at $r = a$.

By imposing a uniform velocity U in the x direction (i.e. by adding $\frac{1}{2}U\varpi^2$ to both stream functions), we obtain the case of a spherical vortex advancing at speed U into a stationary medium. The figure illustrates the entire streamline pattern in a frame of reference fixed relative to the vortex.

4.4 Some viscous flows

The study of the behaviour of real fluids near solid boundaries leads to the concept of the 'boundary layer' and the vortex wake. However, most problems involving viscous flow are not amenable to exact analytical treatments of the kind already given. Approximate analytical discussions are sometimes possible in two limiting extremes in which either (1) viscous forces in the momentum equation greatly exceed the effects of fluid inertia, and (2) viscous effects can be assumed to be small except in the immediate neighbourhoods of 'thin' boundary layers on solid boundaries. Roughly speaking, these extremes correspond to very small and large values of the Reynolds number $\mathrm{Re} = U\ell/v$, where U and ℓ are the characteristic velocity and length scales of the problem. There is therefore an important intermediate range of Reynolds numbers within which analytical methods are effectively useless, and progress in analysis is possible only by means of lengthy numerical treatment of the equations. Even at high Reynolds numbers, analytical methods are limited to flows that essentially do not involve 'separation' (i.e. the movement of vorticity away from a boundary).

4.4.1 Diffusion of vorticity from an impulsively started plane wall

A rigid plane boundary at $y = 0$ of a viscous fluid in $y > 0$ starts moving at constant speed U in the x direction at time $t = 0$. The fluid is set into motion entirely through the action of vorticity diffusing from the wall, from the initial vortex sheet of circulation U per unit length. All motion is parallel to the wall, $\mathbf{v} = (u, 0, 0)$, and can depend on only the time t and distance y from the wall; in particular the pressure is uniform everywhere. The only non-trivial component of the momentum equation is

$$\frac{\partial u}{\partial t} = v\frac{\partial^2 u}{\partial y^2} \quad \text{where} \quad u = \begin{cases} 0 & \text{for } t = 0, \ y > 0 \\ U & \text{for } t \geq 0, \ y = 0 \end{cases}. \tag{4.4.1}$$

Because there are no externally imposed length or time scales in this problem, the solution must be a function of the *similarity* variable $y/\sqrt{vt}$ alone, and it is easily deduced that

$$u(y, t) = U\left[1 - \mathrm{erf}\left(\frac{y}{\sqrt{4vt}}\right)\right], \quad y > 0, \ t > 0, \tag{4.4.2}$$

where $\mathrm{erf}(x) = \frac{2}{\sqrt{\pi}}\int_0^x \mathrm{e}^{-\xi^2}\,d\xi$ is the *error function*.

The impulsive start of the motion generates a vortex sheet at the wall. In an ideal fluid the vorticity would remain at the wall and there would be no motion in the fluid [as $v \to 0$ the error function in (4.4.2) $\to 1$ for $y, \ t > 0$]. The diffusion is illustrated in Figure 4.4.1 for three different characteristic times $U^2 t/v$. The points labelled $y = (4vt)^{\frac{1}{2}}$ are representative of the effective *penetration depth* of the influence of wall motion on the fluid at time t, where the fluid has acquired about 15% of the wall speed.

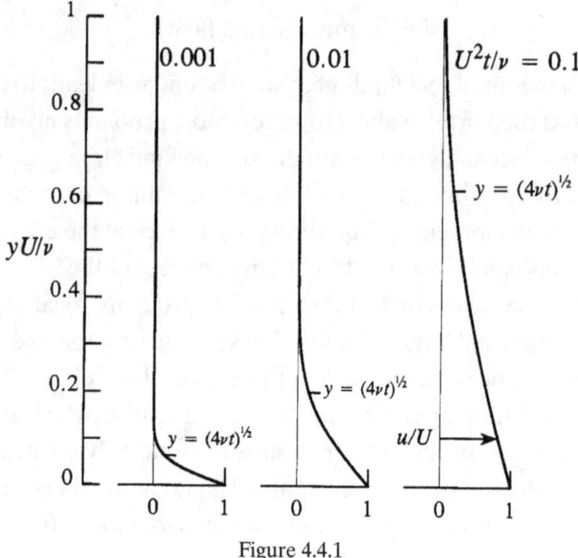

Figure 4.4.1

The frictional drag exerted on unit area of the wall is equal to the value of $\eta\,\partial u/\partial y$ at $y = 0$:

$$\text{wall drag per unit area} = \frac{\rho_0 U^2}{\sqrt{\pi U^2 t/v}}.$$

It is infinitely large at the instant at which the motion starts, but rapidly relaxes as the fluid near the wall is accelerated into motion.

EXAMPLE 1. When the wall $y = 0$ executes periodic oscillations at speed $u = U\cos(\omega t)$ in the x direction, the steady-state solution of Equation (4.4.1) that vanishes at large distances from the wall becomes

$$u = U\cos\left(\omega t - y\sqrt{\frac{\omega}{2v}}\right)e^{-y\sqrt{\frac{\omega}{2v}}}.$$

The penetration depth of the motion into the fluid $\sim \sqrt{2v/\omega}$.

4.4.2 Diffusion of vorticity from a line vortex

Suppose at time $t = 0$ a vortex filament of circulation Γ lies along the z axis in an unbounded incompressible fluid. The streamlines are circles with centres on the z axis, and all physical quantities are evidently functions of t and the radial distance $r = \sqrt{x^2 + y^2}$ alone.

The vorticity $\omega = (0, 0, \omega_3)$ and remains in the z direction. The vorticity equation is

$$\frac{\partial\omega_3}{\partial t} = v\left(\frac{\partial^2}{\partial x^2} + \frac{\partial^2}{\partial y^2}\right)\omega_3 \quad \text{where} \quad \omega_3 = \Gamma\delta(x)\delta(y) \text{ at } t = 0,$$

with solution

$$\omega_3 = \frac{\Gamma}{4\pi \nu t} e^{-r^2/4\nu t}, \quad t > 0. \tag{4.4.3}$$

Vorticity diffuses outwards from the centre $r = 0$, such that at time t the total circulation around a circle S of radius r is

$$\int_S \omega_3 dx dy = \Gamma \left(1 - e^{-r^2/4\nu t} \right)$$

which ultimately vanishes as $t \to \infty$. The circulation is also equal to $2\pi r v_\phi$, where v_ϕ is the circulatory flow velocity at distance r from the axis, so that

$$v_\phi = \frac{\Gamma}{2\pi r} \left(1 - e^{-r^2/4\nu t} \right).$$

Thus in the region where $e^{-r^2/4\nu t}$ is non-zero the motion is rotational and the flow speed is smaller than in the initial potential flow. We can say that the motion at distance r from the axis remains essentially the same as in irrotational flow until $t \sim r^2/4\nu$, after which the vortex core may be said to have diffused out a distance r. The solid-body rotational flow velocity $v_\phi \sim \Gamma r/8\pi \nu t$ in the core gradually decreases to zero at later times.

We can remove the apparent artificiality of this problem by considering first the steady flow produced by a circular cylinder of radius a with centre along the z axis rotating at constant angular velocity Ω, say. Again, v_ϕ is the only non-zero component of velocity, and the corresponding component of the momentum equation, namely

$$\frac{\partial v_\phi}{\partial t} + v_r \frac{\partial v_\phi}{\partial r} + v_z \frac{\partial v_\phi}{\partial z} + \frac{v_r v_\phi}{r} = -\frac{1}{\rho_o} \frac{\partial p}{r \partial \phi} + \nu \left[\frac{\partial^2 v_\phi}{\partial r^2} + \frac{\partial}{\partial r} \left(\frac{v_\phi}{r} + \frac{\partial^2 v_\phi}{\partial z^2} \right) \right],$$

reduces to

$$\frac{\partial^2 v_\phi}{\partial r^2} + \frac{\partial}{\partial r} \left(\frac{v_\phi}{r} \right) = 0, \quad r > a. \tag{4.4.4}$$

Hence,

$$v_\phi = \frac{A}{r} + Br$$

where A and B are constants. The no-slip condition at $r = a$ and the requirement that v_ϕ should vanish at large distances yield $A = \Omega a^2$. The fluid motion is now identical with that produced by a line vortex at the origin of strength $\Gamma = 2\Omega\pi a^2$. The steady flow is maintained by the energy supplied by the rotating cylinder and dissipated by viscous action in the flow, which, according to (1.5.12), occurs at a rate

$$2\eta \int e_{ij} e_{ij} \, dx dy \equiv 2\eta \int \left(\frac{\partial^2 \varphi}{\partial x_i \partial x_j} \right)^2 dx dy$$

$$= 4\pi a^2 \eta \Omega^2 \quad \text{per unit length of the cylinder,} \tag{4.4.5}$$

where the integration is over the fluid and the velocity potential is given by

$$\varphi = \text{Re} \left(-\frac{i\Gamma}{2\pi} \ln z \right) \equiv \text{Re} \left(-i\Omega a^2 \ln z \right), \quad z = x + iy.$$

If the cylinder is imagined to be 'dissolved' at time $t = 0$ the subsequent decay of the potential flow vortex motion is governed by the solution to the initial-value problem previously discussed.

4.4.3 Creeping flow

At very small Reynolds numbers (corresponding to very small velocities and body dimensions or to very large viscosities), frictional forces are so large that fluid inertia can be ignored. This is called the creeping flow approximation. We shall subsequently examine the classic problem of steady creeping flow past a sphere. In steady flows the inertia terms in the momentum equation are second order in the velocity, frictional terms are first order, and their ratio $\sim$ Re, so that the creeping flow approximation is strictly one in which Re is small. For steady flow (in the absence of body forces) we therefore have the **creeping flow approximation**

$$\nabla p = \eta \nabla^2 \mathbf{v} \equiv -\eta \operatorname{curl} \boldsymbol{\omega},$$
$$\operatorname{div} \mathbf{v} = 0. \tag{4.4.6}$$

The first of these equations demonstrates how a field of vorticity generates a stress distribution that in a viscous fluid must be balanced by the creation of a pressure gradient. For example, if a small vortex ring is imagined to be embedded in the flow, the tendency to form a vortex jet sets up a viscous drag that produces a pressure force across the circular plane of the ring equal to the area of the ring times $\eta \operatorname{curl} \boldsymbol{\omega}$. Solid surfaces in the flow will experience a tangential viscous drag, but in addition the pressure distribution created by the viscous stress field exerts a normal component of force on the surface, whose magnitude is typically of the same order as that produced by the surface shear stress.

4.4.4 Motion of a sphere at very small Reynolds number

Consider a rigid sphere of radius a in steady motion at speed U and very small Reynolds number $\mathrm{Re} = Ua/\nu$ in the direction of the positive x axis. Take the coordinate origin at the centre of the sphere and observe that the vortex lines are circles with centres on the x axis in planes perpendicular to the x direction, so that

$$\mathbf{x} \cdot \boldsymbol{\omega} = 0 \quad \text{and} \quad \mathbf{U} \cdot \boldsymbol{\omega} = 0,$$

which imply that $\boldsymbol{\omega} \propto \mathbf{x} \wedge \mathbf{U}$, and moreover $\boldsymbol{\omega}$ must decay in magnitude at large distances from the sphere. Now by taking the curl of the first of Equations (4.4.6) (and using the identity $\operatorname{curl} \operatorname{curl} = \nabla \operatorname{div} - \nabla^2$), we see that

$$\nabla^2 \boldsymbol{\omega} = 0,$$

and the simplest solution of this equation with the required properties can be written as

$$\boldsymbol{\omega} = A \frac{\mathbf{U} \wedge \mathbf{x}}{|\mathbf{x}|^3} \equiv A \operatorname{curl}\left(\frac{\mathbf{U}}{|\mathbf{x}|}\right), \tag{4.4.7}$$

where A is a constant.

Similarly, the excess pressure p is expected to be positive ahead of the sphere and negative to the rear, and must also decay at large distances from the sphere. Now by taking the divergence of the first of Equations (4.4.6), we find it follows that

$$\nabla^2 p = 0,$$

and the simplest solution of this equation with the required properties is

$$p = B \frac{\mathbf{x} \cdot \mathbf{U}}{|\mathbf{x}|^3} \equiv -B \operatorname{div}\left(\frac{\mathbf{U}}{|\mathbf{x}|}\right),$$

where B is a constant. By substituting from this formula and from (4.4.7) into the first of (4.4.6), we see that $B = \eta A$, i.e., that

$$p = -\eta A \operatorname{div}\left(\frac{\mathbf{U}}{|\mathbf{x}|}\right). \tag{4.4.8}$$

To determine A we first write

$$\omega = \frac{AU \sin \theta}{r^2} \mathbf{i}_\phi, \quad r = |\mathbf{x}|,$$

where θ is the usual polar angle between $\mathbf{x}$ and the positive x axis. Then Equation (2.12.4), which expresses the Stokes stream function in terms of the vorticity ω_ϕ, becomes

$$\left(\frac{\partial^2}{\partial x^2} + \frac{\partial^2}{\partial \varpi^2} - \frac{1}{\varpi}\frac{\partial}{\partial \varpi}\right)\psi = -\frac{\varpi\, AU \sin \theta}{r^2}.$$

However,

$$x = r \cos \theta \quad \text{and} \quad \varpi = r \sin \theta,$$

and therefore the equation for ψ becomes

$$\left(\frac{\partial^2}{\partial r^2} + \frac{1}{r^2}\frac{\partial^2}{\partial \theta^2} - \frac{\cot \theta}{r^2}\frac{\partial}{\partial \theta}\right)\psi = -\frac{AU \sin^2 \theta}{r}. \tag{4.4.9}$$

To solve this equation, set $\psi = f(r) \sin^2 \theta$. Then

$$\frac{d^2 f}{dr^2} - \frac{2f}{r^2} = -\frac{AU}{r}.$$

This has the particular integral $\frac{1}{2} AUr$, and solutions of the homogeneous equation (which represent *irrotational* components of the flow) are r^2 and $1/r$. Hence the general solution is

$$f = Cr^2 + \frac{D}{r} + \frac{AUr}{2},$$

where $C,\ D$ are constants, and therefore

$$\psi = \left(Cr^2 + \frac{D}{r} + \frac{AUr}{2}\right) \sin^2 \theta. \tag{4.4.10}$$

The term $Cr^2 \sin^2 \theta$ is the stream function of a uniform flow and must be absent if there is no mean flow at infinity. The values of D and A are now determined from the no-slip condition on the sphere, on which the velocities in the radial and θ directions are respectively

$$v_r = U \cos\theta, \quad v_\theta = -U \sin\theta.$$

Thus, by using

$$v_r = \frac{1}{r^2 \sin\theta} \frac{\partial \psi}{\partial \theta} = \frac{2\cos\theta}{r^2} \left(\frac{D}{r} + \frac{AUr}{2} \right),$$

$$v_\theta = -\frac{1}{r \sin\theta} \frac{\partial \psi}{\partial r} = -\frac{\sin\theta}{r} \left(-\frac{D}{r^2} + \frac{AU}{2} \right)$$

at $r = a$, we find $A = 3a/2$, $D = -a^3 U/4$. Hence (4.4.7), (4.4.8), and (4.4.10) become

$$\omega_\phi = \frac{3aU \sin\theta}{2r^2}, \quad p = \frac{3\eta aU \cos\theta}{2r^2},$$

$$\psi = \frac{3aUr \sin^2\theta}{4} \left(1 - \frac{a^2}{3r^2} \right). \tag{4.4.11}$$

These results may now be used to evaluate the drag force on the sphere. If we write the momentum equation for incompressible flow in the form

$$\rho_o \frac{D\mathbf{v}}{Dt} = -\nabla p - \eta \operatorname{curl}\omega + \mathbf{F},$$

the force on the sphere is equal and opposite to the net pressure and viscous forces applied to the fluid: $\int (\nabla p + \eta \operatorname{curl}\omega)\, d^3\mathbf{x}$, where the integration is over the fluid. Thus, by the divergence theorem,

$$\text{drag force} = -\oint_S (p\mathbf{n} + \eta\mathbf{n} \wedge \omega)\, dS, \tag{4.4.12}$$

where S is the surface of the sphere and $\mathbf{n}$ is the unit normal directed into the fluid. The pressure force is in the normal direction on S and the viscous drag is locally tangential to the surface, and it is clear by symmetry that their net line of action is in the $-x$ direction; they are easily evaluated by use of (4.4.11):

$$\text{pressure drag} = 2\pi\eta Ua, \quad \text{viscous drag} = 4\pi\eta Ua. \tag{4.4.13}$$

The net drag force on the sphere is therefore $6\pi\eta Ua$. This result was first obtained by Stokes (1851) and is usually referred to as the Stokes drag. It is applicable for only very small values of the Reynolds number, $\mathrm{Re} = Ua/\nu < 1$.

The streamlines of the creeping flow, on each of which $\psi = \text{constant}$, can be plotted by use of the formula on the second line of (4.4.11). They are illustrated in Figure 4.4.2 for a series of equidistant values of ψ and represent the pattern relative to a reference frame fixed with respect to the fluid at large distances from the sphere. The corresponding pattern of streamlines in the opposite extreme of $\mathrm{Re} \gg 1$, where frictional forces are negligible compared with the inertia of the fluid, is shown in Figure 2.12.3.

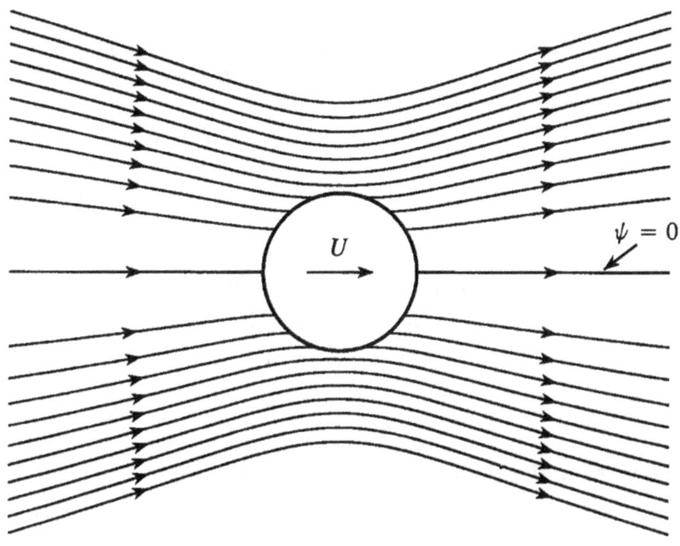

Figure 4.4.2

4.4.5 The Oseen approximation

The application of the creeping flow approximation to steady flow at speed U past an infinite cylinder of radius a, in a direction at right angles to the cylinder generators, leads to a solution in which the perturbation velocity grows like $U \ln(r/a)$ at large distances r from the cylinder axis. This happens because frictional forces produced by an infinitely long cylinder are capable of producing a finite influence on the flow at infinite distances from the cylinder axis. Similarly, although the Stokes solution for the sphere is well behaved at infinity, the low-Reynolds-number approximation on which it is based is not valid at large distances.

At large distances from the sphere, the velocity $v \sim Ua/r$ and viscous terms in the momentum equation are of the order of $\eta Ua/r^3$, whereas inertial terms (of the form $\rho_o v \partial v/\partial x$) are of the order of $\rho_o U^2 a/r^2$. Thus the ratio of the inertia to the viscous terms $\sim Ur/v$. This Reynolds number based on r evidently becomes arbitrarily large at large distances from the sphere, even when $\mathrm{Re} = Ua/v \ll 1$, so that at sufficient distance from the sphere the assumption that inertial forces are negligible ceases to be true. The first successful attempt to correct this deficiency and to extend the range of validity of the low-Reynolds-number approximation, was by Oseen (1910). Adopting a reference frame in which the sphere is at rest, Oseen suggested that a suitable correction to the creeping flow equations can be obtained by retaining in the momentum equation the *linear* inertial term $\rho_o U \partial \mathbf{v}/\partial x$. The corrected equation is obviously applicable at large distances from the sphere, but the low-Reynolds-number limit also means that it is applicable near the sphere, where the additional inertia term is actually negligible compared with the viscous force. This procedure leads to the following corrected form for the Stokes drag on a sphere:

$$\text{Oseen drag} = 6\pi\eta Ua \left(1 + \frac{3\,\mathrm{Re}}{8}\right), \quad \mathrm{Re} = \frac{Ua}{v}.$$

Table 4.4.1

	ρ/ρ_o	v, cm^2/s	Re $= Ua/v$	$a_{\max}$, cm
Sand in water	~ 2	0.01	$2.2 \times 10^6 a^3$	~ 0.008
Water droplet in air	~ 780	0.15	$7.5 \times 10^6 a^3$	~ 0.005

It is perhaps indicative of the difficulties involved in deriving useful analytic representations of viscous flows that, whereas the Stokes drag formula tends to underestimate the drag when Re $\sim O(1)$, this first-order correction supplies an equally erroneous *overestimate* of the drag!

EXAMPLE 2. The terminal velocity U of a sphere of mass density ρ and radius a falling vertically under gravity in a fluid of density ρ_o and shear viscosity η is given by

$$U = \frac{2(\rho - \rho_o)ga^2}{9\eta}$$

when the Stokes approximation $6\pi \eta U a$ is used to estimate the drag. Relevant numerical estimates for a sand particle in water and a water droplet in air are given in Table 4.4.1.

The final column headed $a_{\max}$ is the maximum radius of the particle for which Re ≤ 1. Experiment shows that Re ~ 1 is the *practical* upper limit of validity of the Stokes drag approximation, although it supplies an excellent first approximation for Re $\sim$ 5–10.

EXAMPLE 3. MOTION PRODUCED BY A ROTATING SPHERE Consider the steady rotary motion produced by rotation at angular velocity $\boldsymbol{\Omega} = \Omega\mathbf{k}$ about the z direction of a sphere of radius a with centre at the origin. In the creeping flow approximation we can set $\mathbf{v} = f(r)\mathbf{k} \wedge \mathbf{x}$, where $r = |\mathbf{x}|$, so that

$$\nabla p = \eta\nabla^2[f(r)\mathbf{k} \wedge \mathbf{x}]$$

$$= \left[-\eta y\left(\frac{d^2 f}{dr^2} + \frac{4}{r}\frac{df}{dr}\right),\ \eta x\left(\frac{d^2 f}{dr^2} + \frac{4}{r}\frac{df}{dr}\right),\ 0\right].$$

This is satisfied by $p = $ constant and

$$\frac{d^2 f}{dr^2} + \frac{4}{r}\frac{df}{dr} = 0,$$

so that

$$f = \frac{A}{r^3} + B, \quad A,\ B \text{ constant.}$$

If the motion vanishes at infinity $B = 0$, and the no-slip condition gives $A = \Omega a^3$,

$$\therefore \qquad f = \Omega\frac{a^3}{r^3}, \quad \mathbf{v} = \frac{a^3\boldsymbol{\Omega} \wedge \mathbf{x}}{r^3}. \tag{4.4.14}$$

4.4.6 Laminar flow in a tube (Hagen–Poiseuille flow)

A flow is said to be *laminar* when only one velocity component, say u in the x direction, is non-zero and depends only on coordinates transverse to the flow. In such flows, fluid particles having the same value of u lie on a single cylindrical surface parallel to the x axis, and the motion consists of these cylinders sliding over one another without change, each with its own velocity. Such flows automatically satisfy the continuity equation.

One of the simplest illustrations is that of a steady, laminar flow within a smooth, circular cylindrical pipe. Let the pipe be horizontal and have radius a, and take the x axis along the axis of the pipe in the direction of flow. The three components of the momentum equation in terms of cylindrical polar coordinates (x, ϖ, ϕ) reduce to the following simple forms for laminar flow:

$$\frac{\partial p}{\partial x} = \eta \left(\frac{\partial^2}{\partial \varpi^2} + \frac{1}{\varpi}\frac{\partial}{\partial \varpi} \right) u, \quad \frac{\partial p}{\partial \varpi} = 0, \quad \frac{1}{\varpi}\frac{\partial p}{\partial \phi} = 0. \qquad (4.4.15)$$

The pressure p is therefore uniform over a cross section of the pipe and, moreover, because u does not depend on x, the pressure gradient

$$\frac{dp}{dx} = \text{constant} = -\frac{(p_1 - p_2)}{\ell},$$

where p_1 and p_2 are the pressures at the ends of a section of the pipe of length ℓ (Figure 4.4.3). The first of (4.4.15) therefore becomes

$$\frac{d}{d\varpi}\left(\varpi \frac{du}{d\varpi} \right) = -\frac{\varpi(p_1 - p_2)}{\eta\ell},$$

with the general solution

$$u = -\frac{\varpi^2(p_1 - p_2)}{4\eta\ell} + A\ln\varpi + B, \quad (A, B \text{ constant}).$$

The velocity must remain finite on the axis where $\varpi = 0$, so that $A = 0$; B is then determined by the no-slip condition that $u = 0$ at $\varpi = a$. Hence

$$u = \frac{(p_1 - p_2)}{4\eta\ell}\left(a^2 - \varpi^2\right), \quad \varpi < a. \qquad (4.4.16)$$

The velocity has the parabolic profile shown in Figure 4.4.3; the mean flow speed

$$\bar{u} = \frac{1}{\pi a^2}\int_0^a u(\varpi)\, 2\pi\varpi\, d\varpi = \frac{a^2(p_1 - p_2)}{8\eta\ell},$$

which is half the peak velocity, which occurs on the axis.

The volume flux Q, say, along the pipe is given by the Hagen–Poiseuille formula:

$$Q = \pi a^2 \bar{u} \equiv \frac{\pi a^4(p_1 - p_2)}{8\eta\ell}. \qquad (4.4.17)$$

This formula provides a simple means of determining η from measurements of Q for known values of the pressure drop $p_1 - p_2$ between the ends of a pipe. When verifying (4.4.17) in practice we must introduce a correction to allow for the acceleration of fluid

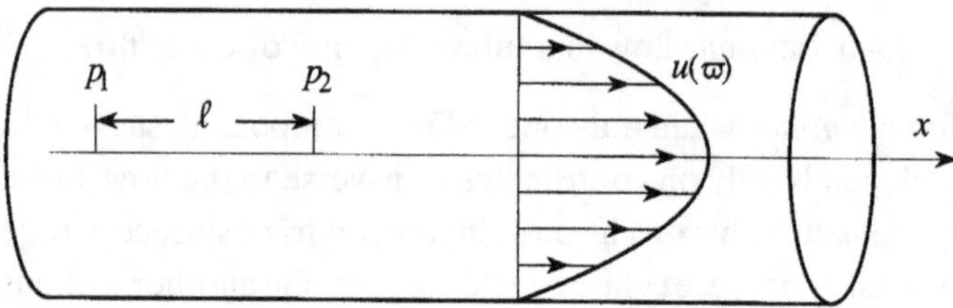

Figure 4.4.3

entering the pipe, say from a large reservoir; but when this and other possible errors are properly accounted for, the law embodied in (4.4.17) is found to be valid for most 'normal' fluids.

However, when the mean velocity $\bar{u}$ is large enough, or more exactly, when the Reynolds number $\bar{u}a/\nu$ exceeds a certain value, the postulated steady flow in the pipe becomes unstable. According to Hagen–Poiseuille formula (4.4.17), the pressure drop along the pipe should vary linearly with the mean velocity. In practice, however, the pressure drop varies more nearly in proportion to $\bar{u}^2$. Experiments performed in glass tubes with coloured dye to visualise the streamlines indicate that, except at very small Reynolds numbers, the actual flow is far from being laminar, fluid particles are seen to move along chaotic or 'turbulent' paths superimposed on the local mean flow. When the flow becomes turbulent a coloured thread of fluid persists for only a short distance into the tube before rapidly being dispersed over the cross section. This causes lateral 'mixing' of fluid particles and their momenta, the latter being averaged out over the cross section of the tube. In consequence the parabolic velocity profile of Figure 4.4.3 is replaced in this turbulent state with a much flattened *time-averaged* velocity distribution of the type shown in Figure 4.4.4. Inspection of this figure makes it clear that the velocity gradient at the wall is very much larger than for laminar flow with the same volume flux, so that a correspondingly larger pressure gradient is required for maintaining the flow.

Reynolds (1883) found that the transition from laminar to turbulent flow always occurs at approximately the same value of the Reynolds number $\mathrm{Re} = \bar{u}d/\nu \sim 2000$ ($d = 2a$), although transition could be delayed to much higher values of Re by use of a pipe with a specially rounded inlet and by ensuring that the inflow reservoir was sufficiently 'calm'.

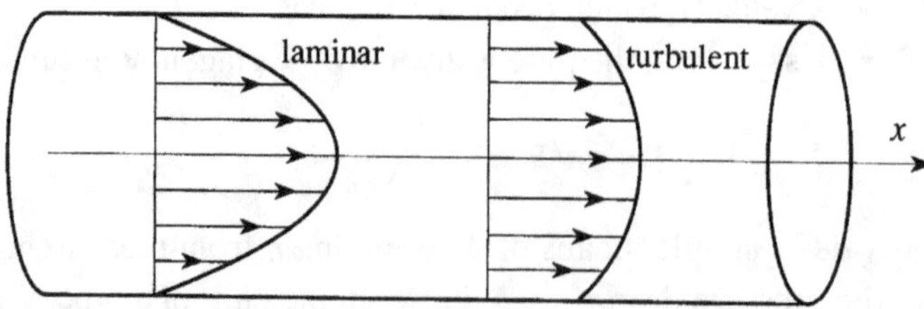

Figure 4.4.4

4.4.7 Boundary layer on a flat plate; Kármán momentum integral method

In high-Reynolds-number flow past a solid the effects of viscosity can frequently be ignored except in regions very close to the surface where the influence of the no-slip condition becomes paramount. When the flow is past a solid obstacle there is a thin layer of fluid adjacent to the solid surface in which viscosity is important and from which vorticity can emerge and be 'shed' into a wake. The limiting form of such flows as $\mathrm{Re} \to \infty$ does not, however, necessarily correspond to the flow that would be found according to potential flow theory, because nominally small viscous terms in the equations of motion always become significant very close to the wall for arbitrarily large values of Re.

Consider two-dimensional flow in the x direction over a rigid, flat plate of infinitesimal thickness placed parallel and edgewise to the mean stream. A boundary layer begins to form at the leading edge and grows in thickness with distance downstream. If the mean stream velocity outside the boundary layer is U, experiment indicates that provided $\mathrm{Re}_x = Ux/v$ does not exceed about 10^5 the boundary-layer flow remains steady and locally laminar, inasmuch as the streamlines are effectively parallel to the plate. A fluid particle in the boundary layer is carried from the leading edge at $x = 0$, say, to a position x downstream in a time $\sim O(x/U)$. One might then expect from the solution in §4.4.1 of the problem of an impulsively started wall that the boundary layer at x will have acquired a 'thickness' $\delta(x) \sim \sqrt{vx/U}$.

Taking the origin at the leading edge, with y in the normal direction from the plate, we find that the relevant steady flow equations in two dimensions are

$$\left.\begin{aligned} u\frac{\partial u}{\partial x} + v\frac{\partial u}{\partial y} &= -\frac{1}{\rho_o}\frac{\partial p}{\partial x} + v\left(\frac{\partial^2 u}{\partial x^2} + \frac{\partial^2 u}{\partial y^2}\right) \\ u\frac{\partial v}{\partial x} + v\frac{\partial v}{\partial y} &= -\frac{1}{\rho_o}\frac{\partial p}{\partial y} + v\left(\frac{\partial^2 v}{\partial x^2} + \frac{\partial^2 v}{\partial y^2}\right) \end{aligned}\right\} \mathbf{v} = (u, v, 0), \qquad (4.4.18)$$

and

$$\frac{\partial u}{\partial x} + \frac{\partial v}{\partial y} = 0. \qquad (4.4.19)$$

The continuity equation implies that, if $\partial u/\partial x$ is $O(U/\ell)$, where $\ell \gg \delta$, then so is $\partial v/\partial y$, and therefore, because $v = 0$ at $y = 0$, we must have $v \sim O(\delta/\ell)u$ in the neighbourhood of the boundary layer. Similarly, in the first of (4.4.18) $\partial^2 u/\partial x^2 \sim (\delta/\ell)^2 \partial^2 u/\partial y^2$, and may be neglected. The frictional term in this equation is therefore of the order of Uv/δ^2 and can balance the $O(U^2/\ell)$ inertial terms on the left-hand side provided $\delta \sim \sqrt{v\ell/U} \sim \sqrt{vx/U}$, as previously foreshadowed.

In the second of Equations (4.4.18) the smallness of v implies that p is effectively independent of y, and therefore also of x, because p is constant in the mean stream. The governing equations accordingly reduce to (4.4.19) and

$$u\frac{\partial u}{\partial x} + v\frac{\partial u}{\partial y} = v\frac{\partial^2 u}{\partial y^2}. \qquad (4.4.20)$$

Kármán investigated the consequences of these equations indirectly by integrating momentum equation (4.4.20) with respect to y over the interval $0 < y < \delta(x)$ within which viscous effects are important, obtaining in the first instance, using the continuity equation (4.4.19), and multiplying by ρ_o,

$$\rho_o \int_0^\delta \frac{\partial u^2}{\partial x}\, dy + \rho_o\, [vu]_0^\delta = \left[\eta \frac{\partial u}{\partial y} \right]_0^\delta. \tag{4.4.21}$$

Next, because $u(x, \delta) = U$,

$$\int_0^\delta \frac{\partial u^2}{\partial x}\, dy = \frac{\partial}{\partial x} \int_0^\delta u^2\, dy - u^2(x, \delta)\frac{\partial \delta}{\partial x} = \frac{\partial}{\partial x} \int_0^\delta u^2\, dy - U^2\frac{\partial \delta}{\partial x},$$

and

$$[vu]_0^\delta = U \int_0^\delta \frac{\partial v}{\partial y}\, dy = -U \int_0^\delta \frac{\partial u}{\partial x}\, dy$$

$$= -U\frac{\partial}{\partial x} \int_0^\delta u\, dy + U^2\frac{\partial \delta}{\partial x}.$$

Therefore (4.4.21) becomes

$$\eta \left(\frac{\partial u}{\partial y} \right)_{y=0} = \frac{\partial}{\partial x} \int_0^\infty \rho_o u(U - u)\, dy, \tag{4.4.22}$$

where the upper limit of integration has been extended to $y = \infty$ because $u \to U$ for $y > \delta$.

To interpret this equation, consider the steady form of Reynolds momentum equation (1.4.14) applied to unit span of the flow above the plate bounded by a surface S consisting of the semi-infinite planes AA′ upstream of the edge O of the plate and BB′ at x (Figure 4.4.5), and the section AB of the plane $y = +0$ that passes just above the plate.

For steady flow, subject to no body forces and vanishing excess pressure, the x momentum satisfies

$$\oint_S \left(-\rho v_1 v_j dS_j + \sigma_{1j} dS_j \right) = 0,$$

where the surface element $d\mathbf{S}$ is assumed to be directed *out* of S. Only the planes AA′, BB′ contribute to the integral $\oint_S -\rho v_1 v_j dS_j$. The mass flux at BB′ (i.e., at x) through the interval dy at distance y above the plate is $\rho_o u\, dy$ per unit span. This fluid mass had previously a velocity U upstream of the leading edge O. Its velocity is now $u \equiv u(x, y)$. Hence

$$\oint_S -\rho v_1 v_j dS_j = \int_0^\infty \rho_o u U\, dy - \int_0^\infty \rho_o u^2\, dy.$$

The viscous force $\oint_S \sigma_{1j} dS_j = -\int_0^x \eta(\partial u/\partial y)_{y=0}\, dx$ is supplied by friction at the upper surface of the plate.

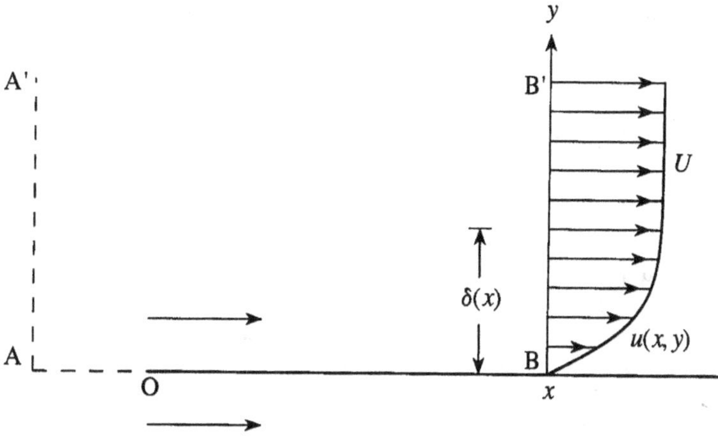

Figure 4.4.5

Hence between O and x, and per unit span,

$$\int_0^\infty \rho_o u(U - u)\, dy = \text{net surface resistance}$$

$$= \int_0^x \eta \left(\frac{\partial u}{\partial y}\right)_{y=0} dx, \qquad (4.4.23)$$

and differentiation of this formula yields (4.4.22).

When the velocity profile $u(x, y)$ is known this formula can be used to evaluate the frictional drag on the plate. Suppose, for example, that

$$u = Uf\left(\frac{y}{\delta}\right), \qquad (4.4.24)$$

where $f(y/\delta)$ is some suitable approximation to the velocity profile across the boundary layer [so that $f(y/\delta) = 1$ for $y > \delta$]. Then

$$\int_0^\infty \rho_o u(U - u)\, dy = \alpha \rho_o U^2 \delta,$$

where

$$\alpha = \int_0^\infty f(\xi)[1 - f(\xi)]\, d\xi = \text{numerical constant.}$$

Similarly,

$$\eta \left(\frac{\partial u}{\partial y}\right)_{y=0} = \beta \frac{\eta U}{\delta}, \quad \text{where} \quad \beta = f'(0).$$

Thus, substituting these approximations into (4.4.22), we obtain the following differential equation governing the growth of the boundary-layer thickness $\delta = \delta(x)$:

$$\beta \frac{\eta U}{\delta} = \alpha \rho_o U^2 \frac{d\delta}{dx}, \qquad (4.4.25)$$

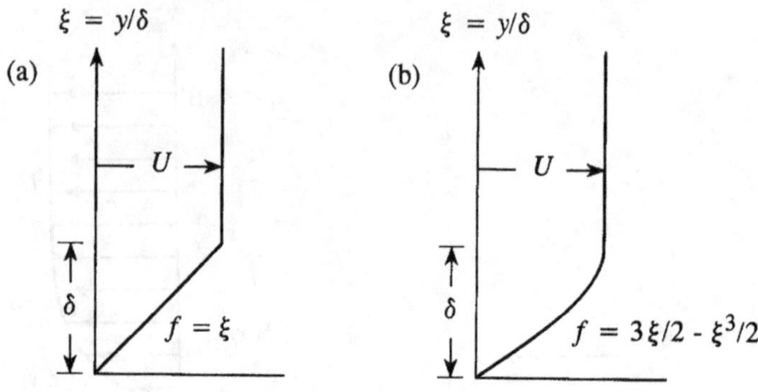

Figure 4.4.6

with solution

$$\delta = \sqrt{\frac{2\beta}{\alpha}\frac{\nu x}{U}} \equiv x\sqrt{\frac{2\beta}{\alpha}\frac{1}{\mathrm{Re}_x}}, \quad \mathrm{Re}_x = \frac{Ux}{\nu}. \tag{4.4.26}$$

Therefore, the surface stress on the plate $\eta(\partial u/\partial y)_{y=0} = \rho_o U^2\sqrt{\alpha\beta/2\mathrm{Re}_x}$, and the total drag on *both sides* of the section OB of the plate is

$$\text{drag per unit span} = 2\int_0^x \eta\left(\frac{\partial u}{\partial y}\right)_{y=0} dx = 2\rho_o U^2 x\sqrt{\frac{2\alpha\beta}{\mathrm{Re}_x}}. \tag{4.4.27}$$

The reduction in flow speed within the boundary layer causes the main stream to be thrust aside from the plate, by a distance normally denoted by δ_* called the *displacement thickness*. We calculate it by noting that

$$\int_0^h u(x, y)\,dy = U(h - \delta_*),$$

where $h > \delta$. Therefore

$$\delta_* = \frac{1}{U}\int_0^h [U - u(x, y)]\,dy \equiv \frac{1}{U}\int_0^\infty [U - u(x, y)]\,dy$$

$$= \delta \int_0^\infty [1 - f(\xi)]\,d\xi. \tag{4.4.28}$$

EXAMPLE 4. LINEAR BOUNDARY-LAYER VELOCITY PROFILE Consider [Figure 4.4.6(a)] in (4.4.24):

$$f(\xi) = \begin{cases} \xi & \text{for } \xi < 1 \\ 1 & \text{for } \xi > 1 \end{cases}.$$

Then $\alpha = \frac{1}{6}$ and $\beta = 1$, and

$$\delta = 3.464x/\sqrt{\mathrm{Re}_x},$$

$$\text{drag} = 1.155\rho_o U^2 x/\sqrt{\mathrm{Re}_x}, \tag{4.4.29}$$

$$\delta_* = 1.732x/\sqrt{\mathrm{Re}_x}.$$

EXAMPLE 5. CUBIC BOUNDARY-LAYER VELOCITY PROFILE Consider [Figure 4.4.6(b)] in (4.4.24):

$$f(\xi) = \begin{cases} \frac{3}{2}\xi - \frac{1}{2}\xi^3 & \text{for } \xi < 1 \\ 1 & \text{for } \xi > 1 \end{cases}.$$

Then

$$\delta = 4.64x/\sqrt{\text{Re}_x},$$

$$\text{drag} = 1.29\rho_o U^2 x/\sqrt{\text{Re}_x}, \tag{4.4.30}$$

$$\delta_* = 1.740x/\sqrt{\text{Re}_x}.$$

The results in Examples 4 and 5 are consistent with the numerical solution of boundary-layer equations (4.4.19), (4.4.20) that is due to Blasius (1907), who found

$$\delta_{99\%} = 4.92x/\sqrt{\text{Re}_x},$$

$$\text{drag} = 1.328\rho_o U^2 x/\sqrt{\text{Re}_x}, \tag{4.4.31}$$

$$\delta_* = 1.73x/\sqrt{\text{Re}_x}.$$

In this case, however, the boundary-layer velocity profile varies smoothly, and the boundary-layer thickness $\delta = \delta_{99\%}$ is defined as the distance from the plate at which $u = 0.99U$.

4.5 Force on a rigid body

The component F_i of the force $\mathbf{F}$ on a body with surface S and normal $\mathbf{n}$ (directed into the fluid) is given in general form by (see §1.4)

$$F_i = \oint_S \left(-p\delta_{ij} + \sigma_{ij} \right) n_j dS. \tag{4.5.1}$$

For incompressible flow

$$\mathbf{F} = -\oint_S (p\mathbf{n} + \eta\mathbf{n} \wedge \boldsymbol{\omega}) \, dS, \tag{4.5.2}$$

and $\mathbf{F}$ may be considered to consist of

> (1) the pressure force associated with the added mass (§2.14),
>
> (2) the component of the pressure force induced by the vorticity,   (4.5.3)
>
> (3) the surface friction (second term in the integrand).

The force exerted on a body by a neighbouring distribution of vorticity is very closely related to the net potential flow disturbance in the far field of the body. In irrotational flow the far field is determined by the total impulse of the body (§2.13.1); the far field for an isolated distribution of vorticity is determined by the vortex impulse (§4.2.1). The two definitions of impulse are effectively identical, because it is always possible to express the irrotational motion produced by a moving body in terms of a distribution of surface vorticity (§4.2, Example 7). Such a representation of surface force in terms

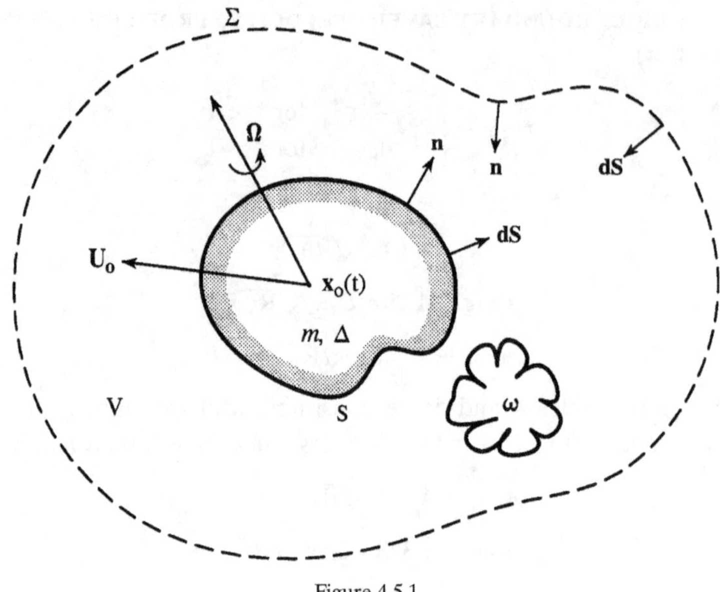

Figure 4.5.1

of the vorticity is sometimes useful because it avoids the need to evaluate the surface pressure, which may be difficult or inconvenient to calculate.

4.5.1 Surface force in terms of the impulse

The force $\mathbf{F}$ on a rigid body in incompressible fluid at rest at infinity is given by

$$\mathbf{F} = m_o \frac{d\mathbf{U}_o}{dt} - \rho_o \frac{d\mathbf{I}}{dt}$$

$$\equiv m_o \frac{d\mathbf{U}_o}{dt} - \frac{\rho_o}{2} \frac{d}{dt} \int \mathbf{x} \wedge \boldsymbol{\omega}(\mathbf{x}, t) d^3\mathbf{x}, \tag{4.5.4}$$

where $\mathbf{U}_o$ is the velocity of the centre of volume of the body (Figure 4.5.1), $\mathbf{I}$ is the impulse defined by the integral in (4.2.8), including any contributions from bound vorticity within and on the surface S of the body, and m_o is the mass of fluid displaced by the body.

To prove (4.5.4), suppose the body has mass m and let

$$\mathcal{F} = \text{external force applied to the body to maintain its motion.}$$

The fluid volume V between a large closed surface Σ and the surface S of the body is assumed to contain all of the vorticity; let V_+ denote the interior of Σ including the region of volume Δ occupied by the body. The body is in arbitrary translational and rotational motion with velocity $\mathbf{U}$ given by (4.2.17) with angular velocity $\boldsymbol{\Omega}$. Because the vorticity vanishes on Σ, the global momentum equations are

$$m \frac{d\bar{\mathbf{U}}}{dt} + \rho_o \frac{d}{dt} \int_V \mathbf{v}(\mathbf{x}, t) d^3\mathbf{x} = \mathcal{F} + \oint_\Sigma p(\mathbf{x}, t) d\mathbf{S},$$

$$m \frac{d\bar{\mathbf{U}}}{dt} = \mathcal{F} + \mathbf{F},$$

where $\bar{\mathbf{U}}$ is the velocity of the centre of mass of the body. Subtracting these equations and extending the volume integral to include the volume Δ occupied by the rigid body (where $\int_\Delta \mathbf{v}d^3\mathbf{x} = \Delta\mathbf{U}_o$), we have

$$\mathbf{F} - m_o\frac{d\mathbf{U}_o}{dt} = -\rho_o\frac{d}{dt}\int_{V_+} \mathbf{v}(\mathbf{x}, t)d^3\mathbf{x} + \oint_\Sigma p(\mathbf{x}, t)d\mathbf{S}. \qquad (4.5.5)$$

Put $\mathbf{v}(\mathbf{x}, t) = \operatorname{curl}\mathbf{A}$, where $\mathbf{A}(\mathbf{x}, t)$ is defined by the integral (4.2.4) over the region V_+, with $\omega = 2\Omega$ within S. When Σ recedes to infinity, $\mathbf{A} = \operatorname{curl}\left[\mathbf{I}(t)/4\pi|\mathbf{x}|\right] + \cdots +$, where the terms omitted decrease faster than $1/|\mathbf{x}|^2$ as $|\mathbf{x}| \to \infty$, and therefore

$$\int_{V_+} \mathbf{v}(\mathbf{x}, t)d^3\mathbf{x} \equiv \int_{V_+} \operatorname{curl}\mathbf{A}d^3\mathbf{x} = -\int_\Sigma \mathbf{n}\wedge\mathbf{A}d\mathbf{S} \to -\int_\Sigma \mathbf{n}\wedge\operatorname{curl}\left[\frac{\mathbf{I}(t)}{4\pi|\mathbf{x}|}\right]d\mathbf{S}.$$

Similarly, the excess pressure $p = -\rho_o\partial\varphi/\partial t + \cdots +$ on Σ, where $\varphi = \operatorname{div}\left[\mathbf{I}(t)/4\pi|\mathbf{x}|\right]$ [see (4.2.10)]. Hence,

$$\oint_\Sigma p(\mathbf{x}, t)d\mathbf{S} \to -\rho_o\frac{d}{dt}\oint_\Sigma \varphi\mathbf{n}d\mathbf{S} = -\rho_o\frac{d}{dt}\oint_\Sigma \mathbf{n}\operatorname{div}\left[\frac{\mathbf{I}(t)}{4\pi|\mathbf{x}|}\right]d\mathbf{S}.$$

The right-hand side of (4.5.5) can therefore be written as

$$\rho_o\frac{d}{dt}\oint_\Sigma \left\{\mathbf{n}\wedge\operatorname{curl}\left[\frac{\mathbf{I}(t)}{4\pi|\mathbf{x}|}\right] - \mathbf{n}\operatorname{div}\left[\frac{\mathbf{I}(t)}{4\pi|\mathbf{x}|}\right]\right\}d\mathbf{S}.$$

By the divergence theorem, the integral in this expression over the large but arbitrary surface Σ can be replaced with an integration over the surface of a large sphere $|\mathbf{x}| = R$, because

$$(\operatorname{curl}\operatorname{curl} - \nabla\operatorname{div})\left[\mathbf{I}(t)/4\pi|\mathbf{x}|\right] = -\nabla^2\left[\mathbf{I}(t)/4\pi|\mathbf{x}|\right] \equiv 0 \text{ for } |\mathbf{x}| > 0.$$

On the sphere $\mathbf{n} = -\mathbf{x}/|\mathbf{x}|$, and the integrand equals $-\mathbf{I}(t)/4\pi R^2$; the integral is therefore just equal to $-\mathbf{I}(t)$. Definition (4.2.8) of $\mathbf{I}(t)$ now shows that (4.5.5) is equivalent to the second line of (4.5.4).

EXAMPLE 1. FORCE IN IRROTATIONAL FLOW In irrotational flow the force on a body in fluid at rest at infinity is given by (§2.14)

$$F_i = \rho_o\frac{d}{dt}\oint \varphi n_i\, d\mathbf{S} = -\frac{d}{dt}\left(U_{oj}\mathrm{M}_{ij} + \Omega_j\mathrm{C}_{ij}\right),$$

where M_{ij}, C_{ij} are the inertia coefficients of (2.14.3) and (2.13.12). To derive this formula from (4.5.4) we must first evaluate the singular distribution (a vortex sheet) of *bound vorticity* on the surface S of the body (Figure 4.5.2).

To do this we need an expression for the velocity *everywhere*, including the region occupied by the body: In the fluid $\mathbf{v} = \nabla\varphi$, and within S $\mathbf{v} = \mathbf{U} \equiv \mathbf{U}_o + \Omega\wedge(\mathbf{x} - \mathbf{x}_o)$. Therefore, if S is the surface $f(\mathbf{x}, t) = 0$, with $f > 0$ in the fluid and $f < 0$ within S, the required formula for the velocity is

$$\mathbf{v} = \mathrm{H}(f)\nabla\varphi + \mathrm{H}(-f)\mathbf{U}.$$

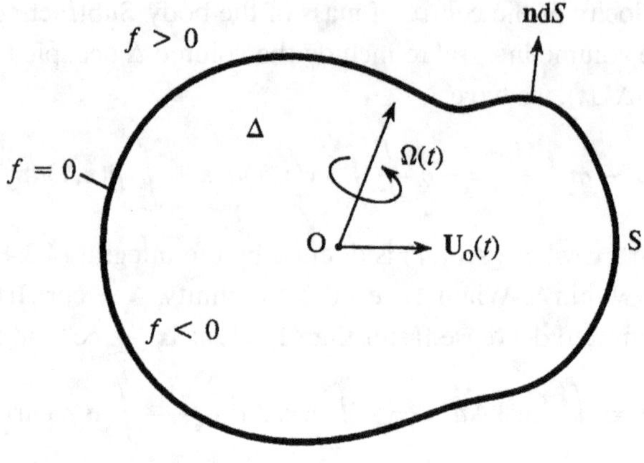

Figure 4.5.2

By noting that $\mathbf{U}_o = U_{oj}\nabla x_j$, we can write the vorticity as

$$\omega = \operatorname{curl}\left[(U_{oj}x_j - \varphi)\nabla \mathrm{H}(f)\right] + \operatorname{curl}\left[\Omega \wedge (\mathbf{x} - \mathbf{x}_o)\mathrm{H}(-f)\right].$$

Therefore (4.5.4) gives

$$\mathbf{F} = m_o\frac{d\mathbf{U}}{dt} - \frac{\rho_o}{2}\frac{d}{dt}\int \left\{\mathbf{x} \wedge \operatorname{curl}\left[(U_{oj}x_j - \varphi)\nabla \mathrm{H}(f)\right]\right.$$
$$\left. + \mathbf{x} \wedge \operatorname{curl}\left[\Omega \wedge (\mathbf{x} - \mathbf{x}_o)\mathrm{H}(-f)\right]\right\} d^3\mathbf{x}.$$

The integral is evaluated by use of the identities

$$\mathbf{x} \wedge \operatorname{curl}\mathcal{A} = 2\mathcal{A} + \nabla(\mathbf{x} \cdot \mathcal{A}) - \frac{\partial}{\partial x_j}(x_j\mathcal{A}), \quad \int (\cdot)\nabla \mathrm{H}(\pm f)d^3\mathbf{x} = \pm \oint_S (\cdot)d\mathbf{S}. \quad (4.5.6)$$

Because $\nabla \mathrm{H}(\pm f)$ vanishes at infinity, only the terms in the integrand corresponding to $2\mathcal{A}$ can make non-trivial contributions. The second such term gives zero, and we are left with

$$F_i = m_o\frac{dU_{oi}}{dt} - \rho_o\frac{d}{dt}\int (U_{oj}x_j - \varphi)\frac{\partial \mathrm{H}}{\partial x_i}(f)\,d^3\mathbf{x}$$

$$= m_o\frac{dU_{oi}}{dt} - \rho_o\frac{d}{dt}\int (U_{oj}x_j - \varphi)n_i\,d\mathrm{S}.$$

However, $\quad \rho_o\oint_S x_jn_i\,d\mathrm{S} = \rho_o\Delta\delta_{ij} \equiv m_o\delta_{ij}, \quad \therefore \quad \rho_o\frac{d}{dt}\oint_S U_{oj}x_jn_i\,d\mathrm{S} = m_o\frac{dU_{oi}}{dt},$

$$\therefore \quad \text{in irrotational flow} \quad \mathbf{F} = \rho_o\frac{d}{dt}\int \varphi\mathbf{n}\,d\mathrm{S}.$$

4.5.2 The Kirchhoff vector force formula

The impulse force formula (4.5.4) is sometimes inconvenient because it requires a knowledge of the vorticity distribution on the surface S of the body. Thus, to calculate the force at high Reynolds number produced by a prescribed distribution of vorticity close to the body, it would first be necessary to evaluate the *induced* vorticity in the boundary layer on S. In the limit of infinite Reynolds number, when it might reasonably be assumed that the surface boundary-layer thickness is negligible, it would still be necessary to include in the integral the contribution from the bound vorticity on S. This difficulty can sometimes be avoided (for the general case illustrated in Figure 4.5.1) by use of the following property of the Kirchhoff vector X_i: $\mathbf{n} \cdot \nabla X_i = 0$ on S, $\nabla^2 X_i = 0$ in V, to express the component F_i of the force in the form (Howe 1995)

$$F_i = \frac{d}{dt} \oint_S \rho_o \varphi_i^* \mathbf{v} \cdot d\mathbf{S} + \oint_S \rho_o \frac{DX_i}{Dt} \mathbf{v} \cdot d\mathbf{S} + \rho_o \int_V \nabla X_i \cdot \boldsymbol{\omega} \wedge \mathbf{v} \, d^3\mathbf{x} - \eta \oint_S \boldsymbol{\omega} \wedge \nabla X_i \cdot d\mathbf{S}.$$

$$(4.5.7)$$

PROOF OF (4.5.7) Rewrite (4.5.5) in the form

$$F_i = -\rho_o \frac{d}{dt} \int_V v_i(\mathbf{x}, t) d^3\mathbf{x} + \oint_\Sigma p(\mathbf{x}, t) n_i \, dS. \qquad (4.5.8)$$

We transform the right-hand side into (4.5.7) by considering the divergence of Crocco's equation (4.1.6) (wherein $B = p/\rho_o + \frac{1}{2}v^2$) and thence the pair of equations

$$\nabla^2 \left(p + \frac{1}{2}\rho_o v^2 \right) = -\mathrm{div}\,(\rho_o \boldsymbol{\omega} \wedge \mathbf{v}), \qquad (4.5.9)$$

$$\nabla^2 X_i = 0. \qquad (4.5.10)$$

Multiply these equations respectively by X_i and $(p + \frac{1}{2}\rho_o v^2)$ and apply Green's theorem to obtain

$$\oint_{S+\Sigma} \left\{ X_i \left[\nabla \left(p + \frac{1}{2}\rho_o v^2 \right) + \rho_o \boldsymbol{\omega} \wedge \mathbf{v} \right] - \left(p + \frac{1}{2}\rho_o v^2 \right) \nabla X_i \right\} \cdot d\mathbf{S}$$

$$= -\rho_o \int_V \nabla X_i \cdot \boldsymbol{\omega} \wedge \mathbf{v} \, d^3\mathbf{x}. \qquad (4.5.11)$$

This is simplified as follows.

By use of Crocco's equation (4.1.6), the first term in the brackets on the left-hand side yields

$$\oint_{S+\Sigma} X_i \left[\nabla \left(p + \frac{1}{2}\rho_o v^2 \right) + \rho_o \boldsymbol{\omega} \wedge \mathbf{v} \right] \cdot d\mathbf{S} = - \oint_{S+\Sigma} X_i \left(\rho_o \frac{\partial \mathbf{v}}{\partial t} + \eta\,\mathrm{curl}\,\boldsymbol{\omega} \right) \cdot d\mathbf{S}$$

$$= - \oint_{S+\Sigma} X_i \rho_o \frac{\partial \mathbf{v}}{\partial t} \cdot d\mathbf{S} - \eta \oint_S \boldsymbol{\omega} \wedge \nabla X_i \cdot d\mathbf{S},$$

$$(4.5.12)$$

because $X_i\,\mathrm{curl}\,\boldsymbol{\omega} = \mathrm{curl}\,(X_i\boldsymbol{\omega}) + \boldsymbol{\omega} \wedge \nabla X_i$, and $\boldsymbol{\omega} = 0$ on Σ. Furthermore,

$$\oint_{S+\Sigma} X_i \rho_o \frac{\partial \mathbf{v}}{\partial t} \cdot d\mathbf{S} = \oint_{S+\Sigma} \frac{\partial}{\partial t} \left(\rho_o X_i \mathbf{v} \right) \cdot d\mathbf{S} - \oint_S \rho_o \frac{\partial X_i}{\partial t} \mathbf{v} \cdot d\mathbf{S},$$

and

$$\oint_{S+\Sigma} \frac{\partial}{\partial t}\left(\rho_o X_i \mathbf{v}\right) \cdot d\mathbf{S} = -\frac{d}{dt}\int_V \operatorname{div}\left(\rho_o X_i \mathbf{v}\right) d^3\mathbf{x} - \oint_{S+\Sigma} \operatorname{div}\left(\rho_o X_i \mathbf{v}\right)\mathbf{v} \cdot d\mathbf{S}$$

$$= -\frac{d}{dt}\int_V \rho_o v_i \, d^3\mathbf{x} - \frac{d}{dt}\oint_S \rho_o \varphi_i^* \mathbf{v} \cdot d\mathbf{S} - \oint_S \rho_o (\mathbf{v} \cdot \nabla X_i)\mathbf{v} \cdot d\mathbf{S}.$$

Therefore, (4.5.12) becomes

$$\oint_{S+\Sigma} X_i \left[\nabla\left(p + \frac{1}{2}\rho_o v^2 \right) + \rho_o \boldsymbol{\omega} \wedge \mathbf{v} \right] \cdot d\mathbf{S} = \frac{d}{dt}\int_V \rho_o v_i \, d^3\mathbf{x} + \frac{d}{dt}\oint_S \rho_o \varphi_i^* \mathbf{v} \cdot d\mathbf{S}$$

$$+ \oint_S \rho_o \frac{DX_i}{Dt} \mathbf{v} \cdot d\mathbf{S} - \eta \oint_S \boldsymbol{\omega} \wedge \nabla X_i \cdot d\mathbf{S}.$$

$$(4.5.13)$$

The contribution from S of the second term in the brackets of (4.5.11) vanishes identically because $\nabla X_i \cdot \mathbf{n} = 0$. On Σ this term supplies $-\oint_\Sigma pn_i dS$, and we finally obtain Equation (4.5.7) by inserting this and (4.5.13) into (4.5.11) and using (4.5.8).

Q. E. D.

4.5.3 The Kirchhoff vector force formula for irrotational flow

When $\omega = 0$ only the first integral on the right-hand side of (4.5.7) is non-zero. The vanishing of the second integral

$$\oint_S \rho_o \frac{DX_i}{Dt} \mathbf{v} \cdot d\mathbf{S} \equiv \rho_o \oint_S \left(\frac{\partial X_i}{\partial t} + \nabla\varphi \cdot \nabla X_i \right) \mathbf{U} \cdot d\mathbf{S}$$

is discussed in §4.5.10.

 Hence

$$F_i = \frac{d}{dt}\oint_S \rho_o \varphi_i^* \mathbf{v} \cdot d\mathbf{S} \equiv \frac{d}{dt}\oint_S \rho_o \varphi n_i dS = -\frac{d}{dt}\left(U_{oj} M_{ij} + \Omega_j C_{ij} \right),$$

as in §2.14.

4.5.4 Arbitrary motion in a viscous fluid

The no-slip condition on S and the relation (4.5.35) permits (4.5.7) to be written as

$$F_i = \frac{d}{dt}\oint_S \rho_o \varphi_i^* \mathbf{v} \cdot d\mathbf{S} + \oint_S \rho_o (\boldsymbol{\Omega} \wedge \mathbf{X})_i \mathbf{U} \cdot d\mathbf{S} + \rho_o \int_V \nabla X_i \cdot \boldsymbol{\omega} \wedge \mathbf{v} \, d^3\mathbf{x}$$

$$- \eta \oint_S \boldsymbol{\omega} \wedge \nabla X_i \cdot d\mathbf{S}. \tag{4.5.14}$$

The first integral here can be expressed in terms of the added-mass coefficients as for irrotational flow. The second integral involves the vorticity $2\boldsymbol{\Omega}$ *within* the solid body, the normal component of which is continuous with the vorticity in the fluid across the interface S between the solid and fluid media.

4.5.5 Body moving without rotation

For a body in translational motion $\mathbf{U} = \mathbf{U}_o$ and $\mathbf{\Omega} = 0$. Then the no-slip condition and the identity

$$\nabla X_i \cdot \boldsymbol{\omega} \wedge \mathbf{U}_o = \operatorname{div}\left[\mathbf{U}_o(\mathbf{v} \cdot \nabla X_i) - \mathbf{v}(\mathbf{U}_o \cdot \nabla X_i) - (\mathbf{v} \cdot \mathbf{U}_o)\nabla X_i\right]$$

imply that $\int_V \nabla X_i \cdot \boldsymbol{\omega} \wedge \mathbf{U}_o \, d^3\mathbf{x} = 0$, so that (4.5.14) can be written as

$$F_i = -\mathsf{M}_{ij}\frac{dU_{oj}}{dt} + \rho_o \int_V \nabla X_i \cdot \boldsymbol{\omega} \wedge \mathbf{v}_{\mathrm{rel}}\, d^3\mathbf{x} + \eta \oint_S \nabla X_i \wedge \boldsymbol{\omega} \cdot d\mathbf{S}, \quad \mathbf{v}_{\mathrm{rel}} = \mathbf{v} - \mathbf{U}_o,$$
$$(4.5.15)$$

where $\mathbf{v}_{\mathrm{rel}}$ is the fluid velocity relative to the translational velocity of S, and we have used the definition $\mathsf{M}_{ij} = \mathsf{M}_{ji} = -\rho_o \oint_S n_j \varphi_i^* dS$ of the added-mass coefficient.

The first term on the right-hand side of (4.5.15) represents the *inviscid* component of the surface pressure on the body associated with the added mass. The volume integral furnishes the component produced by 'free' vorticity *within* the fluid; the final surface integral represents the net contribution from frictional effects on S. Now $\mathbf{v}_{\mathrm{rel}} = 0$ on S, and therefore the contribution to the volume integral from vorticity close to and on S is negligible; indeed, even in the inviscid limit, there is no contribution to the integral from the surface vortex sheet forming the bound vorticity, because ∇X_i and the relative Lamb vector $\boldsymbol{\omega} \wedge \mathbf{v}_{\mathrm{rel}}$ are orthogonal on S.

The contribution from surface friction has two components. Indeed,

$$\eta \nabla X_i \wedge \boldsymbol{\omega} \cdot d\mathbf{S} = \eta(\boldsymbol{\omega} \wedge d\mathbf{S})_i - \eta \nabla \varphi_i^* \cdot \boldsymbol{\omega} \wedge d\mathbf{S}.$$

The first term on the right-hand side corresponds to the usual surface friction, tangential to S [see (4.5.2)], whereas $-\eta \oint_S \nabla \varphi_i^* \cdot \boldsymbol{\omega} \wedge d\mathbf{S}$ may be attributed to the normal pressure forces on S produced by the frictional stress. The first term on the right-hand side of (4.5.15) is the reaction to the force necessary to accelerate the added mass of the body. The viscous 'skin friction' in the i direction is $\eta \oint_S (\boldsymbol{\omega} \wedge d\mathbf{S})_i \equiv \eta \oint_S \nabla x_i \cdot \boldsymbol{\omega} \wedge d\mathbf{S}$. Thus (because $X_i = x_i - \varphi_i^*$) the net contribution of the normal pressure forces on S associated with vorticity is represented in (4.5.15) by the terms

$$\rho_o \int_V \nabla X_i \cdot \boldsymbol{\omega} \wedge \mathbf{v}_{\mathrm{rel}} d^3\mathbf{x} - \eta \oint_S \nabla \varphi_i^* \cdot \boldsymbol{\omega} \wedge d\mathbf{S}.$$

The viscous component is comparable in magnitude with the skin friction and is produced by the pressure field established by the surface shear stress.

EXAMPLE 2. STOKES DRAG The necessity for such a term is vividly illustrated by the Stokes drag on a sphere. In the notation of §4.4.4, let the sphere have radius a and translate at constant velocity $\mathbf{U} = (U, 0, 0)$, $U > 0$, along the x_1 axis. At very small Reynolds numbers $\mathrm{Re} = aU/\nu \ll 1$, 'inertia' terms are unimportant and the equations of motion reduce to the creeping flow approximation (4.4.6), according to which

$$\boldsymbol{\omega} = \operatorname{curl}\left(\frac{3a\mathbf{U}}{2|\mathbf{x}|}\right),$$

where the origin is taken at the centre of the moving sphere.

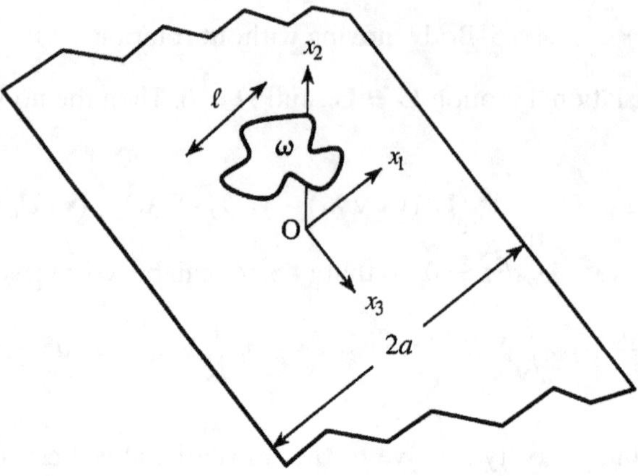

Figure 4.5.3

The net force F_1 on the sphere is parallel to the x_1 axis and given by the final integral on the right-hand side of (4.5.15); $-F_1$ is equal to $D_s + D_p$, where D_s and D_p are the respective magnitudes of the Stokes drag components produced by the skin friction and the viscous surface pressure. For the sphere $\varphi_1^* = -a^3 x_1 / 2|\mathbf{x}|^3$, and direct calculation shows that $F_1 = -6\pi\eta U a$ and that

$$D_s = -\eta \oint_S (\omega \wedge d\mathbf{S})_1 = 4\pi\eta U a, \quad D_p = \eta \oint_S \nabla\varphi_1^* \cdot \omega \wedge d\mathbf{S} = 2\pi\eta U a.$$

The pressure drag is therefore equal to half the skin friction drag (see §4.4.4).

To confirm this interpretation of D_p as the drag attributable to the normal pressure force on the sphere, recall that $n_1 = \mathbf{n} \cdot \nabla\varphi_1^*$ on S (§2.13.2); then, by application of the divergence theorem,

$$D_p = \oint_S p n_1 dS \equiv \oint_S p\nabla\varphi_1^* \cdot d\mathbf{S} = -\int_V \nabla p \cdot \nabla\varphi_1^* d^3\mathbf{x}$$

$$= \eta \int_V \operatorname{curl}\omega \cdot \nabla\varphi_1^* d^3\mathbf{x} = \eta \oint_S \nabla\varphi_1^* \cdot \omega \wedge d\mathbf{S},$$

where (4.4.6) and the identity $\operatorname{div}(\mathcal{A} \wedge \mathcal{B}) = \operatorname{curl}\mathcal{A} \cdot \mathcal{B} - \mathcal{A} \cdot \operatorname{curl}\mathcal{B}$ have been used on the second line.

EXAMPLE 3. VORTEX FORCE ON A LARGE, RIGID PLATE Let the plate occupy the strip $|x_1| < a$, $-\infty < x_3 < \infty$ of the plane $x_2 = 0$. A region of vorticity of length scale $\ell \ll a$ and is located above the plate ($x_2 > 0$) near the coordinate origin O (Figure 4.5.3).

The unsteady normal force (in the $-x_2$ direction) is equal to $-F_2$, where

$$F_2 = \rho_o \int_V \nabla X_2 \cdot \omega \wedge \mathbf{v} d^3\mathbf{x} + \eta \oint_S \nabla X_2 \cdot \omega \wedge d\mathbf{S},$$

where $X_2 = \operatorname{Re}\left[-i(z^2 - a^2)^{1/2}\right]$, where $z = x_1 + ix_2$ (2.17.7). Now $\nabla X_2 \approx \mathbf{Z}/a$, where $\mathbf{Z} = (-x_1, x_2, 0)$, when $\ell \ll a$, so that

$$F_2 \approx \frac{\rho_o}{a} \int \mathbf{Z} \cdot \omega \wedge \mathbf{v} d^3\mathbf{x} + \frac{\eta}{a} \oint_S \mathbf{Z} \cdot \omega \wedge d\mathbf{S}, \quad \ell \ll a.$$

The vorticity in these integrals must include that generated by the interaction of the vorticity with the plate. This additional contribution must be finite, and therefore $F_2 \sim (\ell/a)\rho_o v^2 \ell^2 f(t)$ when $\ell \ll a$, where $f(t)$ is a dimensionless function of the time that does not depend on the width a of the plate. The force vanishes for an infinitely large plate.

EXAMPLE 4. The force formula (4.5.7) is valid also when the body is deformable.

4.5.6 Surface force in two dimensions

The force formulae (4.5.4) and (4.5.7) are readily extended to two-dimensional flows, involving a cylindrical body, provided the circulation around a contour enclosing the body and all of the vorticity vanishes. This ensures that $\mathbf{v}$ decreases like $1/|\mathbf{x}|^2$ at large distances. Therefore non-zero circulation about a closed contour that just encloses the body must be countered by an equal and opposite circulation within the flow, i.e. the circulation about the cylinder must be a consequence of vortex shedding at some earlier time. In these circumstances integrations parallel to the body should be omitted, and the formulae then give the force per unit span. The factor of $\frac{1}{2}$ in the integral of (4.5.4) should be omitted, so that in two dimensions the impulse $\mathbf{I}$ is defined by

$$\mathbf{I} = \int \mathbf{x} \wedge \omega \, d^2\mathbf{x}.$$

Formula (4.5.7) is unchanged.

4.5.7 Bluff body drag at high Reynolds number

Surface forces at high Reynolds numbers are dominated by surface pressures produced by free-stream vorticity, and it is usually permissible to ignore the explicit contribution of frictional effects when estimating the drag experienced by a body placed in a high-speed flow (Figure 4.5.4). Let the incident flow be nominally steady at speed U in the x_1 direction. Vorticity is shed continually from the body and swept downstream to form a wake where the flow is randomly irregular but statistically steady. The kinetic energy of the unsteady vortex motions is extracted from the mean flow, which is therefore retarded in the vicinity of the wake, the loss of mean stream momentum being balanced by a drag force exerted on the body.

The motion in the wake becomes approximately steady at large distances from the disturbed zone just to the rear of the body, and the streamlines become approximately parallel to the mean stream outside the wake. In this region the vorticity is governed by the *linearised* vorticity equation (4.1.7). Because the flow is roughly parallel to the undisturbed flow convection predominates in the mean stream direction, but the gradual lateral spreading of wake vorticity is dominated by diffusion.

At large downstream distances the width of the wake is sufficiently large that the momentum deficit is spread over a large cross-sectional area, so that the mean streamwise

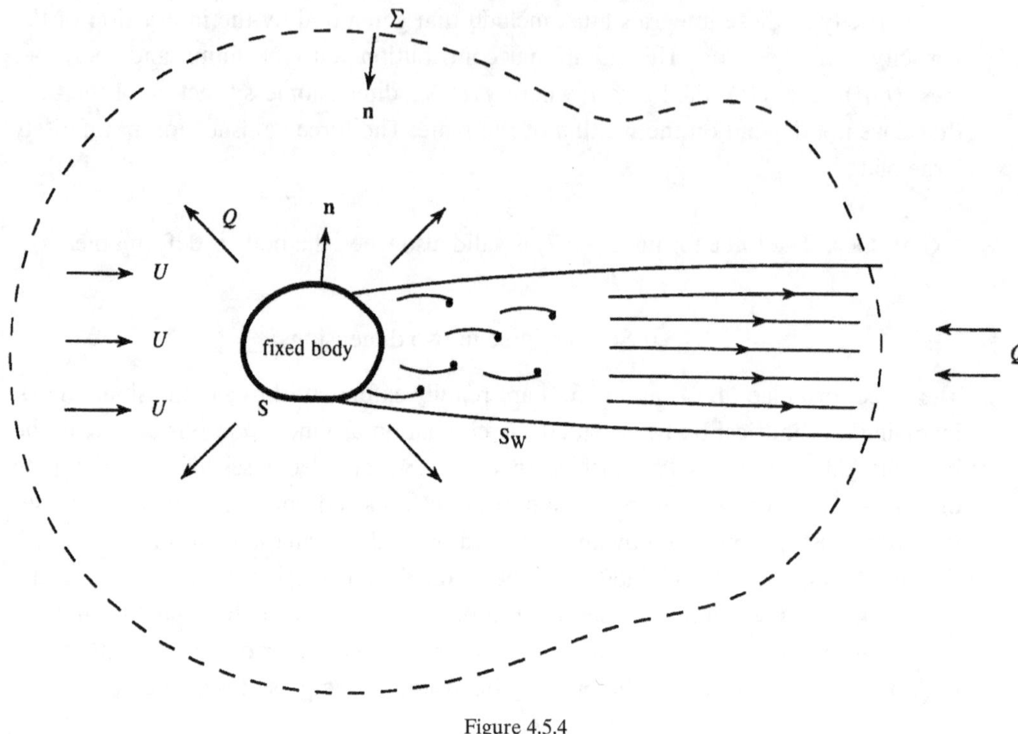

Figure 4.5.4

velocity v_1 in the wake differs by a small amount from the undisturbed mean stream velocity U. The pressure in the far wake is therefore the same as in the free stream, and the steady component of the streamwise velocity satisfies the following linearised version of momentum equation (1.4.10),

$$U\frac{\partial v_1}{\partial x_1} = \nu\left(\frac{\partial^2}{\partial x_2^2} + \frac{\partial^2}{\partial x_3^2}\right)v_1, \qquad (4.5.16)$$

subject to the condition that $v_1 \to U$ as $(x_2^2 + x_3^2)^{\frac{1}{2}} \to \infty$ outside the wake (the 'axis of symmetry' of the wake is assumed to coincide with the x_1 axis). This equation represents a local balance between streamwise convection on the left-hand side and lateral spreading by molecular diffusion on the right-hand side. When $x_1 \to \infty$, the solution can be written as (Batchelor 1967)

$$v_1 \sim U\left[1 - \frac{Q}{4\pi\nu x_1}\exp\left(\frac{-Ur^2}{4\nu x_1}\right)\right], \qquad r = \sqrt{x_2^2 + x_3^2}.$$

Q is a constant that satisfies

$$Q = \iint_{-\infty}^{\infty}(U - v_1)dx_2dx_3,$$

and $\rho_o Q$ is therefore just equal to the momentum deficit in the wake. 90% of this deficit is confined to a cross-sectional area of the wake of radius R, say, determined by

$$0.9 = \frac{U}{4\pi \nu x_1} \int_0^R 2\pi r e^{-Ur^2/4\nu x_1} dr = 1 - e^{-UR^2/4\nu x_1},$$

so that $R \propto \sqrt{x_1}$, the wake profile being parabolic at large distances.

The drag D is easily calculated in terms of Q by use of the integral form of Reynolds momentum equation (1.4.14), which at high Reynolds numbers implies that

$$D = \oint_\Sigma (pn_1 + \rho_o v_1 \mathbf{v} \cdot \mathbf{n}) \, dS, \tag{4.5.17}$$

where $\mathbf{n}$ is the unit *inward* normal on the distant control surface Σ of Figure 4.5.4.

By continuity $\oint_\Sigma \rho_o \mathbf{v} \cdot \mathbf{n} \, dS = 0$. Therefore

$$\oint_\Sigma \rho_o v_1 \mathbf{v} \cdot \mathbf{n} \, dS = \oint_\Sigma \rho_o (v_1 - U) \mathbf{v} \cdot \mathbf{n} \, dS.$$

The integrand of the term on the right-hand side vanishes except where Σ cuts the wake, where to a first approximation we can take $\mathbf{v} \cdot \mathbf{n} \, dS = -U dx_2 dx_3$. Thus, with the pressure taken to be constant on Σ, (4.5.17) becomes

$$D \approx \oint_{\text{wake}} \rho_o U(U - v_1) \, dx_2 dx_3 = \rho_o U Q$$

$$= \text{flux of momentum deficit in the wake.} \tag{4.5.18}$$

The volume flux deficit Q in the wake can be regarded as produced by a volume flow towards the body within the wake region S_W of Figure 4.5.4. This volume flux is compensated for by a source-like, radial outflow from the vicinity of the body of strength Q, so that at large distances from S that are small compared with the length of the wake S_W, the mean perturbation velocity produced by the body in the irrotational zone is radially directed and decreases slowly, with monopole dependence $\sim 1/|\mathbf{x}|^2$.

Conclusion (4.5.18) that $D = \rho_o U Q$ depends on the high-Reynolds-number consideration that the motion in the near wake is dominated by the convection of vorticity, and that viscous diffusion can be ignored. In this case the drag can evidently be expressed in the form

$$D = C_D \frac{1}{2} \rho_o U^2 A, \tag{4.5.19}$$

where A is the projected cross-sectional area of S in the streamwise direction and C_D is a dimensionless *drag coefficient*. This formula provides an empirical representation of the drag for all flow speeds. The drag coefficient C_D depends on body shape and Reynolds number $\mathrm{Re} \sim U\sqrt{A/\pi}/\nu$. The Stokes' drag formula $D = 6\pi \eta U R$ shows that $C_D = 12/\mathrm{Re}$ is large for a sphere at very low velocities and that it initially decreases with increasing Re. This decrease is continued at higher Reynolds numbers in a manner that depends critically on body shape (Figure 4.5.5). At low speeds the flow over S is 'laminar', but becomes unstable as U increases, ultimately causing the

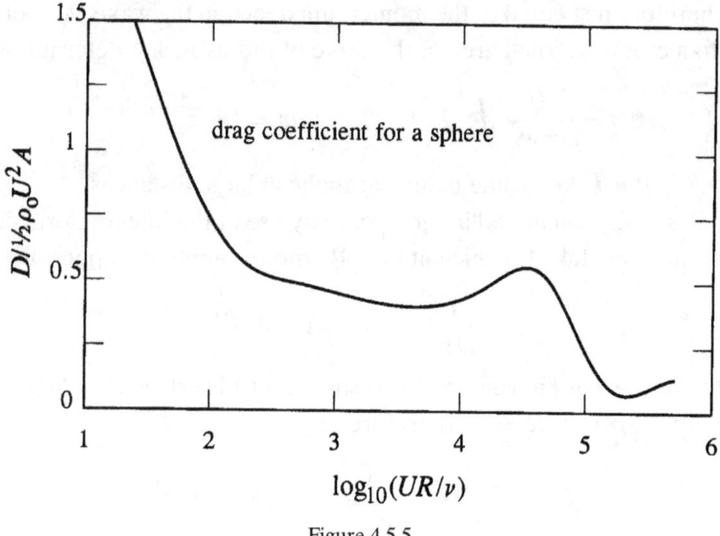

Figure 4.5.5

boundary layer to separate from S along a curvilinear arc, resulting in the discharge of a 'street' of discrete vortices that promotes the 'forward' flow in the wake towards the body characteristic of a 'drag wake' (Figure 4.5.6). At higher speeds the vortices evolve into turbulent eddies accompanied by a large increase in drag (even though C_D is decreasing) because a greatly increased quantity of mean stream kinetic energy is transferred to the unsteady and disorderly motions in the wake. Further increases in the Reynolds number result in the surface boundary layer's becoming turbulent, and a corresponding increase in the diffusion of mean flow momentum towards S. This brings higher-speed flow nearer to the surface, causing a shift in the line of separation downstream to the rear of S, and eventually results in the formation of a wake of greatly reduced cross section. At a critical Reynolds number (of the order of 10^5 for a sphere) the wake suddenly contracts, and there is a precipitous reduction in C_D (to about 0.2 for the sphere) that causes a local net decrease in the drag experienced by the body.

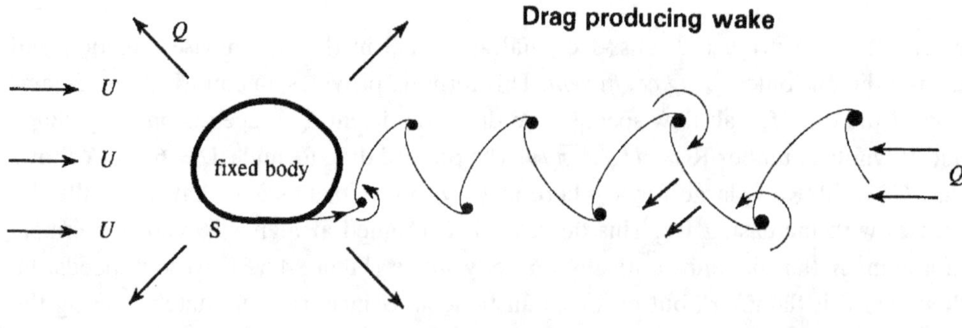

Figure 4.5.6

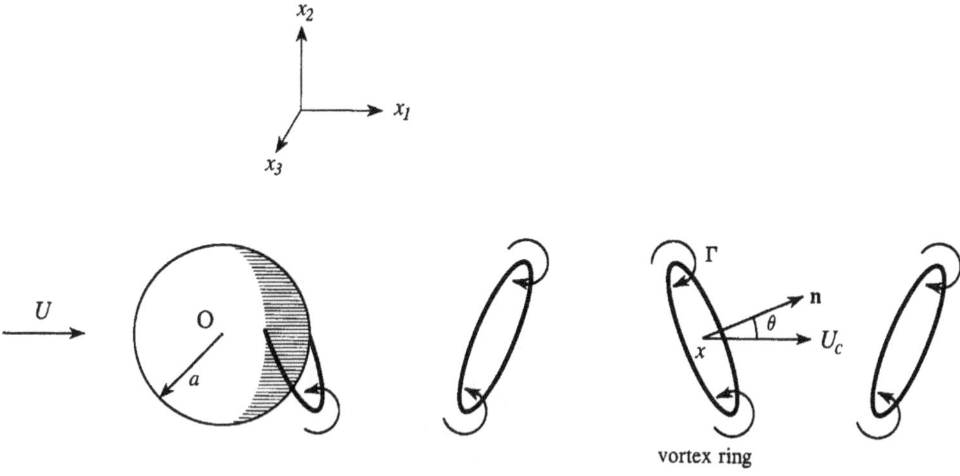

Figure 4.5.7

4.5.8 Modelling vortex shedding from a sphere

A useful model of high-Reynolds-number vortex shedding from a sphere (Howe *et al.* 2001) is illustrated in Figure 4.5.7. The large-scale coherent behaviour of the wake is represented by a sequence of vortex rings that are imagined to be shed from the sphere quasi-periodically at frequency $\sim f_o$ and *Strouhal number* $f_o \ell / U \sim 0.2$, where $\ell = 2a$ and a is the radius of the sphere. The vortex rings convect in the mean stream direction with their centres on the x_1 axis (the origin being at the centre of the sphere) at a constant velocity $U_c \approx 0.7U$, in rough accord with observation.

Each ring has circulation Γ and radius R and is orientated with the normal $\mathbf{n}$ to the plane of the ring at an angle θ to the positive x_1 direction. The azimuthal angle between the x_2 direction and the plane defined by the x_1 axis and the normal is denoted by ϕ. The values of both of these angles vary randomly from ring to ring. Changes in the circular shape of a ring during shedding from the sphere are ignored, so that shedding consists simply of the translation of each ring through the surface of the sphere and occurs during a time $\delta t = 2R \sin \theta / U_c$. If we consider a ring that begins to form at $t = 0$, its formation is complete when $t = \delta t$, and the position $x_1 = x(t)$ of the centre of the ring on the x_1 axis is

$$x(t) = U_c t + \sqrt{a^2 - R^2 \cos^2 \theta} - R \sin \theta, \quad t > 0. \tag{4.5.20}$$

In the figure about one half of a nascent vortex ring has been formed; the semicircular arc of the ring already in the flow has circulation in the anticlockwise sense. There must therefore be a clockwise circulation around the sphere that produces an overall 'lift' or side force on the sphere. This force is in the x_2 direction when the latter is taken to lie in the plane of the x_1 axis and the vortex normal $\mathbf{n}$ (so that $\phi = 0$ for this vortex). The net circulation around the sphere decreases as more of the vortex is shed and vanishes when the vortex separates and moves away into the wake. In this latter phase there is

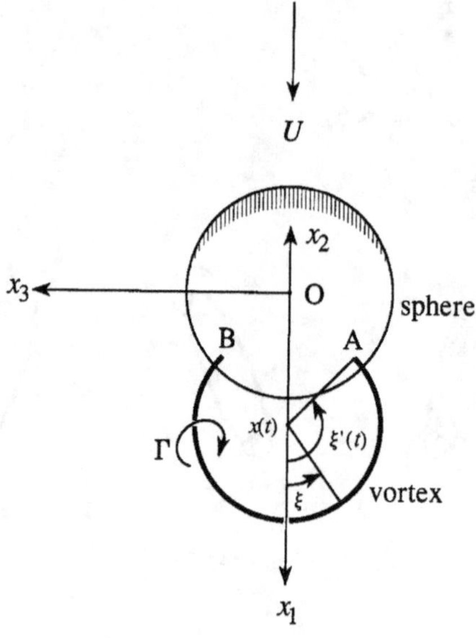

Figure 4.5.8

no mean circulation around the sphere, and any surface forces will then tend to be small until a new vortex begins to form.

For this vortex the high-Reynolds-number force on the sphere is

$$F_i(t) = 0, \quad t < 0$$
$$= \rho_0 \int \nabla X_i \cdot \boldsymbol{\omega} \wedge \mathbf{v}\, d^3\mathbf{x}, \quad t > 0,$$

where $\boldsymbol{\omega}(\mathbf{x}, t)$ is the vorticity of the ring within the flow at time t, and

$$X_i = x_i \left(1 + \frac{a^3}{2|\mathbf{x}|^3}\right), \quad \mathbf{v} = (U_c, 0, 0).$$

The integral is easily evaluated when the core of the vortex ring is assumed to have infinitesimal cross section.

To fix ideas, consider a ring that begins to form at $t = 0$ orientated with its normal $\mathbf{n}$ in the $x_1 x_2$ plane ($\phi = 0$). Let $s = R\xi$ denote the curvilinear distance along the axis of the core in the right-handed sense with respect to $\mathbf{n}$, where the angle ξ is measured from the 'lower' point of intersection of the ring and the $x_1 x_2$ plane (see Figure 4.5.8). Then

$$\boldsymbol{\omega} \wedge \mathbf{v}\, d^3\mathbf{x} = \Gamma U_c (0, \cos\xi, \cos\theta \sin\xi)\, R\, d\xi,$$

and

$$F_i(t) = \rho_0 U_c \Gamma R \int_{-\xi'(t)}^{\xi'(t)} \left(\cos\xi \frac{\partial X_i}{\partial x_2} + \cos\theta \sin\xi \frac{\partial X_i}{\partial x_3}\right) d\xi, \quad t > 0. \quad (4.5.21)$$

In this formula $\xi'(t) = \pi$ for $t > \delta t$ after the vortex is released by the sphere. At earlier times $\xi'(t)$ is the angle illustrated in Figure 4.5.8, determined by the points of intersection (A and B in the figure) of the vortex with the surface of the sphere:

$$\xi'(t) = \frac{\pi}{2} + \sin^{-1}\left[\frac{x^2(t) + R^2 - a^2}{2Rx(t)\sin\theta}\right], \qquad 0 < t < \delta t, \qquad (4.5.22)$$

where $x(t)$ is the x_1 coordinate of the centre of the vortex ring given by (4.5.20). The derivatives $\partial X_i/\partial x_2$, $\partial X_i/\partial x_3$ are evaluated on the vortex ring at the integration point given parametrically by

$$\mathbf{x} = \left[x(t) + R\sin\theta\cos\xi, \, -R\cos\theta\cos\xi, \, -R\sin\xi\right]. \qquad (4.5.23)$$

When $\phi = 0$ the force $\mathbf{F}$ that is due to the vortex ring can be resolved into a 'lift' $\mathcal{L}(t)$ in the x_2 direction and a drag $\mathcal{D}(t)$ in the x_1 direction, although the mean lift produced by the wake vanishes. Using the preceding relations, we find

$$\mathcal{L} = \rho_0 U_c \Gamma R \mathcal{F}(t), \qquad \mathcal{D} = \rho_0 U_c \Gamma R \mathcal{G}(t),$$

where

$$\mathcal{F}(t) = 0, \quad t < 0$$

$$= \int_{-\xi'(t)}^{\xi'(t)} \cos\xi \left\{ 1 + \frac{a^3\left[x^2(t) + 2Rx(t)\cos\xi\sin\theta + R^2(1 - 3\cos^2\theta)\right]}{2\left[x^2(t) + 2Rx(t)\cos\xi\sin\theta + R^2\right]^{\frac{5}{2}}} \right\} d\xi, \quad t > 0.$$

and

$$\mathcal{G}(t) = 0, \quad t < 0$$

$$= \frac{3Ra^3\cos\theta}{2} \int_{-\xi'(t)}^{\xi'(t)} \frac{\left[x(t) + R\cos\xi\sin\theta\right] d\xi}{\left[x^2(t) + 2Rx(t)\cos\xi\sin\theta + R^2\right]^{\frac{5}{2}}}, \quad t > 0.$$

The integrals must be evaluated numerically. Following the release of the vortex from the sphere [after which $\xi'(t) = \pi$] the lift and drag decay very rapidly and satisfy

$$\frac{\mathcal{L}}{\rho_0 U_c \Gamma R} \sim \frac{\pi a^3}{U_c^3 t^3}, \qquad \frac{\mathcal{D}}{\rho_0 U_c \Gamma R} \sim \frac{3\pi Ra^3\cos\theta}{U_c^4 t^4} \qquad \text{when } U_c t \gg a.$$

These results are illustrated in Figure 4.5.9 for $R = 0.7a$ and $\theta = \frac{\pi}{8}$. The lift grows rapidly during the initial stages of shedding, attaining a maximum when roughly half the ring vortex has been formed; the subsequent release of vorticity of opposite sign reduces the net circulation around the sphere and causes the lift to decrease. After release the lift force slowly decays and is negligible when the ring has convected about a sphere diameter into the wake. On the other hand, the drag increases monotonically until the vortex is released, following which it decreases slowly and becomes negligible when the ring is about three diameters downstream. These conclusions indicate that the fluctuations in the lift or side force on the sphere are comparable with the fluctuations in the drag, although the mean lift in any direction must vanish. Similarly, the results

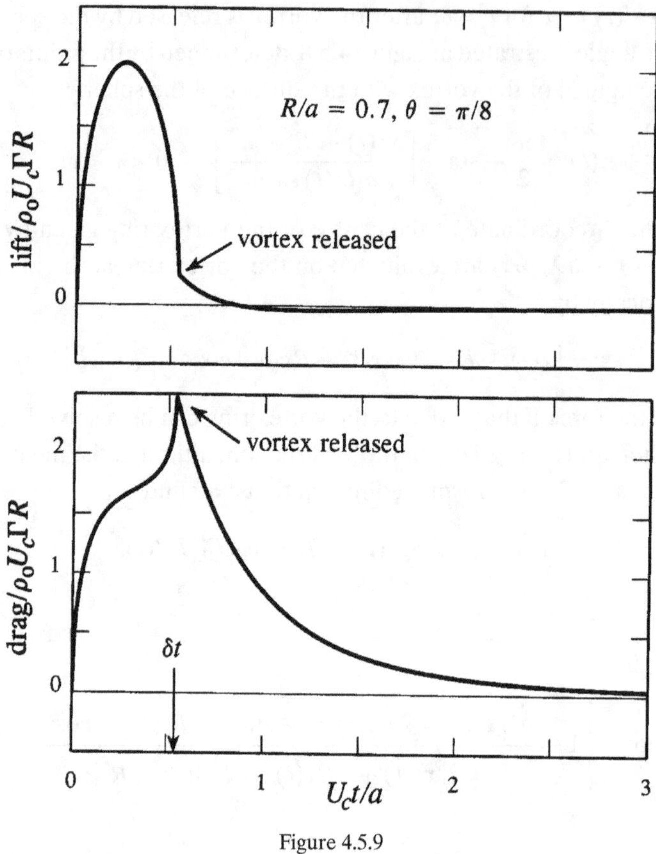

Figure 4.5.9

imply that the behaviour of a vortex in the distant wake (where it is not permissible
to ignore the influences of turbulence convection of vorticity and the misalignment of
the vortex centres produced by self-induction) makes a very limited contribution to the
unsteady surface force.

EXAMPLE 5. VORTEX DRAG IN TWO DIMENSIONS Consider the quasi-periodic drag
on a fixed, cylindrical body in a steady cross-flow with a two-dimensional wake (Figure 4.5.10, where the motion is parallel to the x_1x_2 plane). The uniform mean flow is
in the x_1 direction at speed U. Introduce a cylindrical control surface Σ that is large
enough to contain all of the shed vorticity when the motion is assumed to start from a
state of rest at some time in the distant past. The interior of Σ is partitioned into regions
V_1, V_2 by a plane S_o perpendicular to the mean flow. The figure shows the body in
cross section together with a vortex wake modelled as a Kármán vortex street (§3.4.3),
the vortices in the upper/lower rows being of strength $\pm\Gamma$ ($\Gamma > 0$). At a large distance
downstream within V_1 the vortex spacing is uniform, equal to a in each row, and the
rows are separated by a constant transverse distant b, each vortex being opposite the
midpoint between neighboring vortices in the opposite row. The plane S_o is assumed to

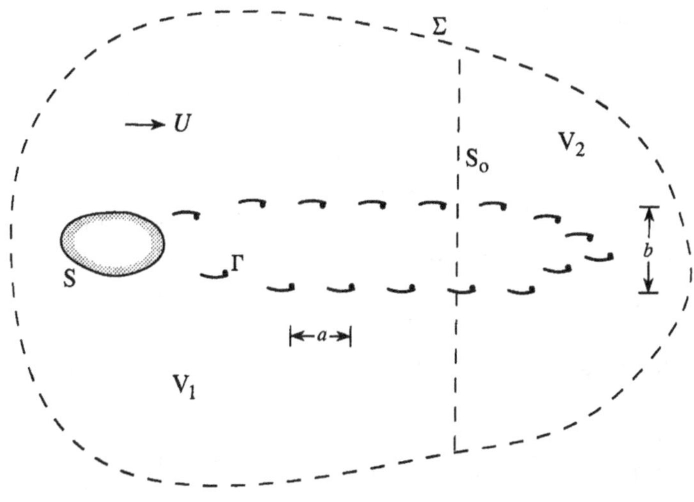

Figure 4.5.10

intersect the vortex street in this uniform region and convects in the x_1 direction at the mean translational velocity V of the vortices

$$ V = U - \frac{\Gamma}{2a} \tanh\left(\frac{\pi b}{a}\right). $$

The motion in V_1 upstream of S_o may be assumed to have period $\tau = a/V$. Further downstream (in V_2) the vortex motion is assumed to dissipate by viscous action or, alternatively, to be terminated in an irregular domain of 'starting vortices'.

The mean drag per unit span is $D = \frac{1}{\tau}\int_0^\tau F_1 dt$, where

$$ F_1 = \rho_o \int_{V_1+V_2} \nabla X_1 \cdot \omega \wedge \mathbf{v} d^2\mathbf{x} $$

$$ = \rho_o \sum_{V_1} \Gamma_n X_1 \cdot \mathbf{i}_3 \wedge \mathbf{v} + \rho_o \int_{V_2} \nabla X_1 \cdot \omega \wedge \mathbf{v} d^2\mathbf{x}, $$

the summation being over the vortices in V_1 of strengths $\Gamma_n = \pm\Gamma$ respectively in the upper/lower row. The time average of the sum can be evaluated in terms of the stream function ψ conjugate to X_1, for which the Cauchy–Riemann equations can be combined in the form $\nabla X_1 \wedge \mathbf{i}_3 = -\nabla\psi$. Then the time average of the summation yields

$$ -\frac{\rho_o}{\tau}\int_0^\tau \sum_{V_1}\Gamma_n\nabla\psi\cdot\mathbf{v}\, dt = -\frac{\rho_o}{\tau}\sum_{V_1}\Gamma_n\big[\psi_n(\tau) - \psi_n(0)\big], \qquad (4.5.24) $$

where $\psi_n(t)$ is the stream function evaluated at the nth vortex at time t. During the time interval $(0,\tau)$ two new vortices enter V_1, but as $x_1 \to \infty$, $\psi \to x_2 +$ constant, and when S_o is sufficiently far downstream the final term in (4.5.24) becomes constant and equal to $\rho_o\Gamma Vb/a$.

To evaluate the integral over V_2, where $X_1 \sim x_1$, we have

$$\rho_o \int_{V_2} \nabla X_1 \cdot \boldsymbol{\omega} \wedge \mathbf{v} d^2\mathbf{x} \sim \rho_o \int_{V_2} (\boldsymbol{\omega} \wedge \mathbf{v})_1 \, d^2\mathbf{x} = \frac{\rho_o}{2} \int_{S_o} \left(v_2^2 - v_1^2\right) dx_2,$$

where (v_1, v_2) is the velocity relative to the translating vortex street. The value of this integral does not depend on t because S_o is fixed with respect to the advancing street. The integration contour may be assumed to lie within irrotational fluid and may be shown to equal $b(\Gamma/a)^2[a/\pi b - \tanh(\pi b/a)]$ by use of the known form of the velocity potential (§3.4.3).

Combining these results for V_1 and V_2, we find the drag per unit span

$$D = \frac{1}{\tau} \int_0^\tau F_1 \, dt = \frac{\rho_o b \Gamma U}{a} + \frac{\rho_o \Gamma^2}{2\pi a} \left[1 - \frac{2\pi b}{a} \tanh\left(\frac{\pi b}{a}\right)\right]. \qquad (4.5.25)$$

4.5.9 Force and impulse in fluid of non-uniform density

In an unbounded fluid of non-uniform mean density ρ, the far-field behaviour continues to be given by asymptotic formula (4.2.9), but the total impulse $\mathbf{I}$ is not now an invariant of the motion. To see what happens in this case, suppose that $\rho \to \rho_o = \text{constant}$, at large distances from the vortex flow. Write the momentum equation in the form

$$\frac{\partial \mathbf{v}}{\partial t} + \nabla\left(\frac{1}{2}v^2 + \frac{p}{\rho_o}\right) = -\boldsymbol{\omega} \wedge \mathbf{v} + \left(\frac{1}{\rho_o} - \frac{1}{\rho}\right)\nabla p - \frac{\eta}{\rho}\,\text{curl}\,\boldsymbol{\omega}.$$

Take the curl and form the vector product with $\mathbf{x}$; integrate over the volume of the fluid, and use the first of Equations (4.5.6) to obtain

$$\frac{\partial}{\partial t}\left(\frac{1}{2}\int \mathbf{x} \wedge \boldsymbol{\omega} \, d^3\mathbf{x}\right) = \int \left(\frac{1}{\rho_o} - \frac{1}{\rho}\right)\nabla p \, d^3\mathbf{x} - \eta \int \frac{\text{curl}\,\boldsymbol{\omega}}{\rho} \, d^3\mathbf{x}$$

$$= \int \left(\frac{1}{\rho_o} - \frac{1}{\rho}\right)\left(\nabla p + \eta\,\text{curl}\,\boldsymbol{\omega}\right) d^3\mathbf{x}, \qquad (4.5.26)$$

where the variation of η has been neglected (although this does not affect the final conclusion), and the second line follows from the observation that $\boldsymbol{\omega} = 0$ in the far field.

The final integral in (4.5.26) is generally non-zero when the density is non-uniform in the region where $\nabla p + \eta\,\text{curl}\,\boldsymbol{\omega} = -\rho D\mathbf{v}/Dt \neq 0$. The integrand represents a 'dipole' source whose strength is determined by the acceleration of the density inhomogeneities relative to the ambient fluid when subject to the same pressure gradient and viscous stress.

To see this more clearly we can consider the limit in which the density inhomogeneities are in the form of immiscible lumps of fluid of uniform density ρ_1 immersed in homogeneous fluid of density ρ_o (Figure 4.5.11). The integral in the final member of

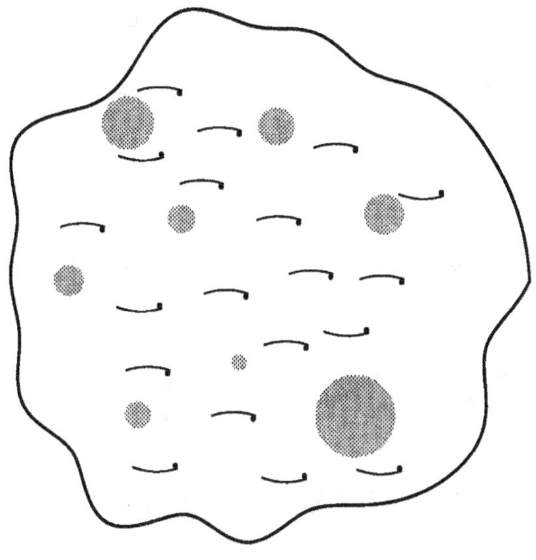

Figure 4.5.11

(4.5.26) is then confined to the volumes occupied by the inhomogeneities. For one such inhomogeneity of volume V_n and surface S_n the integral becomes

$$\left(\frac{1}{\rho_o} - \frac{1}{\rho_1}\right)\int_{V_n}\left(\nabla p + \eta\,\mathrm{curl}\,\omega\right)d^3\mathbf{x} \equiv \left(\frac{1}{\rho_o} - \frac{1}{\rho_1}\right)\int_{V_n} -\rho_1\frac{D\mathbf{v}}{Dt}\,d^3\mathbf{x}$$

$$= \frac{m_n}{\rho_o}\frac{d\mathbf{U}_n}{dt} + \frac{\mathbf{F}_n}{\rho_o},$$

where $\mathbf{U}_n$ is the velocity of the centre of volume of V_n, $\mathbf{F}_n$ is the force exerted on the ambient fluid by V_n, and m_n is the mass of ambient fluid displaced by V_n. We accordingly recover the following generalised form of Equation (4.5.4):

$$\rho_o\frac{d\mathbf{I}}{dt} = \rho_o\frac{\partial}{\partial t}\left(\frac{1}{2}\int \mathbf{x}\wedge\omega\,d^3\mathbf{x}\right) = \sum_n\left(m_n\frac{d\mathbf{U}_n}{dt} + \mathbf{F}_n\right).$$

4.5.10 Integral identities

Various integral identities used previously in relation to the configuration of Figure 4.5.1 are given here for reference.

DEFINITIONS:

$$\mathbf{U} = \mathbf{U}_o + \mathbf{\Omega}\wedge(\mathbf{x} - \mathbf{x}_o), \tag{4.5.27}$$

$$\frac{D}{Dt} = \frac{\partial}{\partial t} + \mathbf{v}\cdot\nabla, \tag{4.5.28}$$

$$\frac{D^S}{Dt} = \frac{\partial}{\partial t} + \mathbf{U}\cdot\nabla. \tag{4.5.29}$$

Differential relations (dS on S; φ = potential of ideal flow produced by motion of S):

$$\frac{D^S}{Dt}(\mathbf{U} \cdot d\mathbf{S}) \equiv \frac{D^S}{Dt}(\nabla\varphi \cdot d\mathbf{S}) = \frac{\partial \mathbf{U}}{\partial t} \cdot d\mathbf{S}, \qquad (4.5.30)$$

$$\frac{D^S}{Dt}(d\mathbf{S}) = \mathbf{\Omega} \wedge d\mathbf{S}. \qquad (4.5.31)$$

Integral relations [$\Phi \equiv \Phi(\mathbf{x}, t)$, arbitrary function; φ = potential of ideal flow produced by motion of S; X_i = Kirchhoff vector]:

$$\oint_S \Phi \frac{\partial \mathbf{v}}{\partial t} \cdot d\mathbf{S} = \frac{\partial}{\partial t} \oint_S \Phi\mathbf{U} \cdot d\mathbf{S} - \oint_S \left[\frac{\partial \Phi}{\partial t} + \mathrm{div}(\mathbf{v}\Phi)\right] \mathbf{U} \cdot d\mathbf{S}, \qquad (4.5.32)$$

$$\oint_S \nabla X_i \cdot (\mathbf{U} - \nabla\varphi)\mathbf{U} \cdot d\mathbf{S} = -\oint_S X_i \frac{\partial}{\partial t}(\mathbf{U} - \nabla\varphi) \cdot d\mathbf{S}, \qquad (4.5.33)$$

$$\oint_S \left(\frac{\partial X_i}{\partial t} + \nabla\varphi \cdot \nabla X_i\right)\mathbf{U} \cdot d\mathbf{S} = 0, \qquad (4.5.34)$$

$$\oint_S \left(\frac{\partial X_i}{\partial t} + \mathbf{U} \cdot \nabla X_i\right)\mathbf{U} \cdot d\mathbf{S} = \oint_S (\mathbf{\Omega} \wedge \mathbf{X})_i \mathbf{U} \cdot d\mathbf{S}. \qquad (4.5.35)$$

EXAMPLE 6. Proof of (4.5.34):

$$\oint_S \left(\frac{\partial X_i}{\partial t} + \nabla\varphi \cdot \nabla X_i\right)\mathbf{U} \cdot d\mathbf{S}$$

$$= \oint_S \frac{D^S X_i}{Dt}\mathbf{U} \cdot d\mathbf{S} - \oint_S \nabla X_i \cdot (\mathbf{U} - \nabla\varphi)\mathbf{U} \cdot d\mathbf{S}$$

$$= \frac{\partial}{\partial t} \oint_S X_i \nabla\varphi \cdot d\mathbf{S} - \oint_S X_i \frac{D^S}{Dt}(\mathbf{U} \cdot d\mathbf{S}) + \oint_S X_i \frac{\partial}{\partial t}(\mathbf{U} - \nabla\varphi) \cdot d\mathbf{S}$$

$$= \frac{\partial}{\partial t} \oint_S X_i \nabla\varphi \cdot d\mathbf{S} - \oint_S X_i \frac{\partial \mathbf{U}}{\partial t} \cdot d\mathbf{S} + \oint_S X_i \frac{\partial}{\partial t}(\mathbf{U} - \nabla\varphi) \cdot d\mathbf{S}$$

$$= \frac{\partial}{\partial t} \oint_S X_i \nabla\varphi \cdot d\mathbf{S} - \oint_S X_i \nabla\frac{\partial\varphi}{\partial t} \cdot d\mathbf{S}$$

$$\equiv \frac{\partial}{\partial t} \oint_S (X_i \nabla\varphi - \varphi\nabla X_i) \cdot d\mathbf{S} - \oint_S \left(X_i \nabla\frac{\partial\varphi}{\partial t} - \frac{\partial\varphi}{\partial t}\nabla X_i\right) \cdot d\mathbf{S}.$$

The divergence theorem permits both of these integrals over S to be replaced with integrals over a *fixed* surface Σ at 'infinity', on which $\varphi \sim 1/|\mathbf{x}|^2$, $X_i \sim x_i$, i.e., by

$$\frac{\partial}{\partial t} \oint_\Sigma (x_i \nabla\varphi - \varphi\nabla x_i) \cdot d\mathbf{S} - \oint_\Sigma \left(x_i \nabla\frac{\partial\varphi}{\partial t} - \frac{\partial\varphi}{\partial t}\nabla x_i\right) \cdot d\mathbf{S} \equiv 0.$$

4.6 Surface moment

The moment $\mathbf{M}$ exerted on a rigid body in arbitrary motion in incompressible flow (Figure 4.5.1) is given in terms of the vorticity (including bound vorticity) by the following analogue of force formula (4.5.4):

$$M_i + \frac{d}{dt}(\mathcal{I}_{ij}^o \Omega_j) = \frac{\rho_o}{3}\frac{d}{dt}\int \left\{ (\mathbf{x} - \mathbf{x}_o) \wedge \left[(\mathbf{x} - \mathbf{x}_o) \wedge \boldsymbol{\omega}(\mathbf{x}, t) \right] \right\}_i d^3\mathbf{x}. \tag{4.6.1}$$

The body has angular velocity $\boldsymbol{\Omega}$, and the moment $\mathbf{M}$ is about its centre of volume $\mathbf{x}_o$; $\mathcal{I}_{ij}^o$ is the moment of inertia about this point of the fluid displaced by the body (that is, of the body when its density is regarded as uniform and equal to the density ρ_o of the fluid), and $\boldsymbol{\omega}$ is the generalized vorticity, including bound vorticity within and on the surface S of the body.

The formula is proved by the method used in §4.5.1 for (4.5.4). Consider first the equation for the angular momentum

$$M_i + \frac{d}{dt}(\mathcal{I}_{ij}^o \Omega_j) = \rho_o \frac{d}{dt}\int_{V_+} \left[(\mathbf{x} - \mathbf{x}_o) \wedge \mathbf{v}(\mathbf{x}, t) \right]_i d^3\mathbf{x} - \oint_\Sigma p(\mathbf{x}, t)[(\mathbf{x} - \mathbf{x}_o) \wedge \mathbf{n}]_i dS. \tag{4.6.2}$$

The right-hand side is expressed in terms of the vorticity by means of the substitution

$$\rho_o(\mathbf{x} \wedge \mathbf{v})_i = \frac{\rho_o}{3}\left[\mathbf{x} \wedge (\mathbf{x} \wedge \boldsymbol{\omega}) \right]_i + \frac{\rho_o}{3}\epsilon_{ijk}\left[\frac{\partial}{\partial x_k}(x_j x_l v_l) - \frac{\partial}{\partial x_l}(x_j x_l v_k) \right], \tag{4.6.3}$$

in the first integral on the right-hand side, where the coordinate origin is temporarily placed at $\mathbf{x}_o$. This yields (4.6.1) because the subsequent application of the divergence theorem to the first integral and use of asymptotic formula (4.2.9) reveal that the term involving ϵ_{ijk} just cancels the final surface integral in (4.6.2) of the pressure over Σ.

It follows that in an unbounded, incompressible fluid the second moment

$$\int_{-\infty}^{\infty} \mathbf{x} \wedge (\mathbf{x} \wedge \boldsymbol{\omega})\, d^3\mathbf{x} = \text{constant}. \tag{4.6.4}$$

4.6.1 Moment for a non-rotating body

The following alternative formula for the moment is applicable to the special case of a rigid body in translational motion at velocity $\mathbf{U}_o(t)$ (Howe 1995):

$$M_i = -C_{ji}\frac{\partial U_{oj}}{\partial t} + \rho_o(\mathbf{I} \wedge \mathbf{U}_o)_i + \rho_o \int_V [\mathbf{i} \wedge (\mathbf{x} - \mathbf{x}_o) - \nabla \chi_i^*] \cdot \boldsymbol{\omega} \wedge \mathbf{v}_{\text{rel}}\, d^3\mathbf{x}$$

$$+ \eta \oint_S \boldsymbol{\omega} \wedge [\mathbf{i} \wedge (\mathbf{x} - \mathbf{x}_o) - \nabla \chi_i^*] \cdot d\mathbf{S}, \tag{4.6.5}$$

where C_{ij} is defined as in (2.13.12) and (2.14.3), $\mathbf{i}$ is a unit vector in the i direction, $\mathbf{I}$ is the total impulse of the motion determined by (4.5.4), and χ_i^* is the velocity potential satisfying condition (2.13.9) on S. This result should be compared with formula (2.14.6) for irrotational motion.

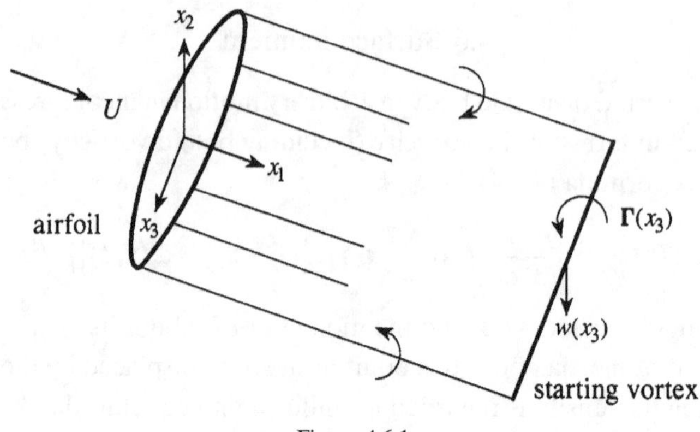

Figure 4.6.1

4.6.2 Airfoil lift, drag, and moments

Spanwise discontinuities in velocity occur when the airstreams above and below an airfoil of finite span and angle of attack meet at the trailing edge. On the lower or 'pressure' side the mean streamlines diverge because of the rise in pressure (causing flow tubes to expand) and converge on the upper or 'suction' side of the airfoil. Therefore, at the trailing edge (except at centre span) the flow velocity has a component towards the side edge on the pressure side and a component towards the centre span on the suction side. When the flow is steady the pressure is continuous through the wake, so that the flow speed must be the same just above and below the airfoil at the trailing edge. Thus a discontinuity occurs in the direction at right angles to the main stream, and vortex lines nominally parallel to the mean flow direction are shed from the edge, forming a system of 'horseshoe' vortex loops in the fluid that are closed (in accordance with the kinematic theorem that a vortex tube must be re-entrant) by continuation along the span of the airfoil. The surface of discontinuity is unstable, and observation reveals that it rolls up, starting from the side edges, and gives rise to to 'trailing vortices' far to the rear that extend back along the path of the airfoil through the air.

Consider an airfoil (Figure 4.6.1) of large span ℓ relative to its maximum chord (i.e. of large 'aspect ratio') that translates at constant speed U in the negative x_1 direction. Neglect viscous stresses, and let the coordinate axes be configured as depicted in the figure. According to (4.5.15) the airfoil lift F_2 is given for inviscid motion by

$$F_2 = \rho_o \int_V \nabla X_2 \cdot \omega \wedge \mathbf{v}_{\mathrm{rel}} \, d^3\mathbf{x}.$$

In a first approximation the wake can be modelled as a vortex sheet formed by the system of horseshoe vortices shed from the trailing edge. At time t the sheet is terminated by the starting vortex

$$\Gamma(x_3)\delta[x_1 - s_1(t)]\delta[x_2 - s_2(t)]\mathbf{i}_3 \quad \text{at} \quad x_1 = s_1(t),\ x_2 = s_2(t),\ \ -\ell/2 < x_3 < \ell/2,$$

where $\Gamma(x_3)$ is the circulation about the airfoil at the spanwise distance x_3 from the airfoil centroid.

In a reference frame fixed relative to the airfoil the streamwise vorticity is aligned with the local mean flow and therefore satisfies to first order $\boldsymbol{\omega} \wedge \mathbf{v}_{\text{rel}} = 0$. Thus, to first order,

$$\boldsymbol{\omega} \wedge \mathbf{v}_{\text{rel}} \approx \boldsymbol{\omega} \wedge U\mathbf{i}_1 = U\Gamma(x_3)\delta[x_1 - s_1(t)]\delta[x_2 - s_2(t)]\mathbf{i}_2, \quad -\ell/2 < x_3 < \ell/2.$$

The lift force therefore becomes

$$\text{lift} = \rho_o U \int_{-\frac{\ell}{2}}^{\frac{\ell}{2}} \Gamma(x_3)\left(1 - \frac{\partial \varphi_2^*}{\partial x_2}\right) dx_3 \tag{4.6.6}$$

The derivative $\partial \varphi_2^*/\partial x_2$ is evaluated at the starting vortex, its value being dependent on the planform and thickness distribution of the airfoil. This formula is applicable for arbitrary locations of the starting vortex and accordingly gives an approximate representation of initial growth of the lift when the airfoil starts from rest (see §3.8.4); the usual Kutta–Joukowski lift formula (lift $\sim \rho_o U\Gamma \ell$) is recovered at large time, when the starting vortex is far downstream and $\partial \varphi_2^*/\partial x_2 \sim 0$.

The induced drag is equal to F_1 and is a *second-order* quantity, because the relative convection velocity of the vorticity can be written as

$$v_{\text{rel}} = U\nabla X_1 + \mathbf{v}'$$

where $\mathbf{v}'$ is the induced velocity of the wake and bound vorticity, so that

$$F_1 = \rho_o \int_V \nabla X_1 \cdot \boldsymbol{\omega} \wedge \mathbf{v}' \, d^3\mathbf{x}.$$

Also, $X_1 = x_1 - \varphi_1^*$, where $\nabla \varphi_1^*$ is typically a small quantity whose magnitude is proportional to airfoil thickness and angle of attack. It follows that, if $v_2' = -w(x_3)$ at the starting vortex, where w is the *downwash* velocity, then the

$$\text{drag} = \rho_o \int_{-\frac{\ell}{2}}^{\frac{\ell}{2}} w(x_3)\Gamma(x_3) \, dx_3, \tag{4.6.7}$$

which is valid (to second order) for any location of the starting vortex.

The *rolling moment* is the x_1 component of $\mathbf{M}$ and is determined by Equation (4.6.5) (in which $x_o = 0$). Because $\mathbf{i}_1$ is parallel to the mean velocity, only the first integral on the right-hand side makes a finite contribution at high Reynolds numbers, and this comes from the starting vortex:

$$\text{rolling moment} = \rho_o U \int_{-\frac{\ell}{2}}^{\frac{\ell}{2}} \Gamma(x_3)\left(x_3 - \frac{\partial \chi_1^*}{\partial x_3}\right) dx_3. \tag{4.6.8}$$

The derivative $\partial \chi_1^*/\partial x_3$ can be neglected when the starting vortex is several chord lengths downstream of the airfoil.

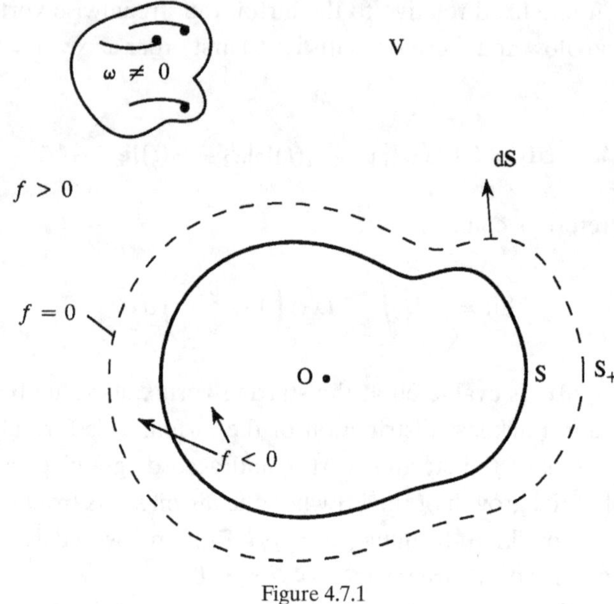

Figure 4.7.1

The main contribution to the *yawing* moment M_2 is again supplied by the starting vortex, there being no contributions from the impulse terms in (4.6.5). Correct to second order, we find

$$\text{yawing moment} = \rho_o \int_{-\frac{\ell}{2}}^{\frac{\ell}{2}} x_3 w(x_3) \Gamma(x_3) \, dx_3. \tag{4.6.9}$$

4.7 Vortex–surface interactions

Let us consider the motion produced by vorticity in the neighbourhood of a body with surface S (Figure 4.7.1). The vorticity may be said to interact with the surface, and we shall examine the problem of calculating directly the influence of this interaction in the far field. The problem is therefore analogous to that discussed in connection with Biot–Savart formula (§4.2), but the viewpoint is sufficiently different to be of interest in its own right. The body may be at rest in a uniform stream containing convected vorticity or moving in some arbitrary manner, and in addition vorticity is formed in the boundary layers on S and shed into a vortical wake.

4.7.1 Pressure expressed in terms of the total enthalpy

The total enthalpy

$$B = \frac{p}{\rho_o} + \frac{1}{2} v^2 \tag{4.7.1}$$

is a convenient independent variable to define the far field of a vortex flow, where the flow becomes irrotational and momentum equation (4.1.6) reduces to

$$\frac{\partial \mathbf{v}}{\partial t} = -\nabla B.$$

In other words,

$$B = -\frac{\partial \varphi}{\partial t} \quad \text{in regions where } \omega = 0, \tag{4.7.2}$$

where $\varphi(\mathbf{x}, t)$ is the velocity potential that determines the whole motion in the irrotational regions of the fluid. B is therefore *constant* in steady irrotational flow.

If the mean flow is at rest in the far field, the unsteady component of the pressure is given there by

$$p = \rho_o B \equiv -\rho_o \frac{\partial \varphi}{\partial t}. \tag{4.7.3}$$

To calculate the pressure in terms of B elsewhere in the flow, differentiate (4.7.1) with respect to time and use Crocco's equation (4.1.6):

$$\frac{1}{\rho_o} \frac{\partial p}{\partial t} = \frac{\partial B}{\partial t} - \mathbf{v} \cdot \frac{\partial \mathbf{v}}{\partial t}$$

$$= \frac{\partial B}{\partial t} - \mathbf{v} \cdot \left(-\nabla B - \omega \wedge \mathbf{v} - \nu \operatorname{curl} \omega \right)$$

$$= \frac{DB}{Dt} + \nu \mathbf{v} \cdot \operatorname{curl} \omega.$$

The viscous term can be ignored in high-Reynolds-number flows, where p and B are therefore related by

$$\frac{1}{\rho_o} \frac{\partial p}{\partial t} = \frac{DB}{Dt}. \tag{4.7.4}$$

4.7.2 Equation for B

Take the divergence of Crocco's equation (4.1.6) to obtain (for incompressible flow)

$$\nabla^2 B = -\operatorname{div}(\omega \wedge \mathbf{v}). \tag{4.7.5}$$

The 'source' term on the right-hand side vanishes where the flow is irrotational. If $\omega = 0$ everywhere, if there are no moving boundaries, and if the fluid is at rest or in uniform motion at speed U at infinity with uniform pressure p_∞, we recover Bernoulli's equation for steady flow:

$$B \equiv \frac{p}{\rho_o} + \frac{1}{2} v^2 = \frac{p_\infty}{\rho_o} + \frac{1}{2} U^2.$$

It is convenient to set $B = 0$ at infinity by introducing the modified definition $B' = B - B_\infty$, where $B_\infty = p_\infty/\rho_o + \frac{1}{2}U^2$. This has no effect on Equation (4.7.5), and we therefore discard the prime on B' by adopting the convention that B vanishes at infinity.

4.7.3 Solution of the B equation

Introduce a *stationary*, closed control surface S_+ on which $f(\mathbf{x}) = 0$, such that $f(\mathbf{x}) \gtrless 0$ according to whether $\mathbf{x}$ lies outside or within S_+. The rigid body lies within S_+, and S_+ is ultimately allowed to shrink down to coincide with the instantaneous body surface S. Multiply Equation (4.7.5) by $H \equiv H(f)$ and form the inhomogeneous equation for the new variable HB.

Using the transformations

$$H\nabla^2 B \equiv H\,\mathrm{div}\,(\nabla B) = \mathrm{div}\,(H\nabla B) - \nabla H \cdot \nabla B$$
$$= \nabla^2(HB) - \mathrm{div}\,(B\nabla H) - \nabla H \cdot \nabla B,$$
$$\text{and} \quad H\,\mathrm{div}\,(\omega \wedge \mathbf{v}) = \mathrm{div}\,(H\omega \wedge \mathbf{v}) - \nabla H \cdot \omega \wedge \mathbf{v},$$

we find

$$\nabla^2(HB) = \mathrm{div}\,(B\nabla H) + \nabla H \cdot (\nabla B + \omega \wedge \mathbf{v}) - \mathrm{div}\,(H\omega \wedge \mathbf{v}). \tag{4.7.6}$$

However, by Crocco's equation (4.1.6)

$$\nabla H \cdot (\nabla B + \omega \wedge \mathbf{v}) = -\nabla H \cdot \left(\frac{\partial \mathbf{v}}{\partial t} + \nu\,\mathrm{curl}\,\omega \right) \equiv -\nabla H \cdot \frac{\partial \mathbf{v}}{\partial t} + \nu\,\mathrm{div}\,(\nabla H \wedge \omega).$$

Hence (4.7.6) becomes

$$\nabla^2(HB) = \mathrm{div}\,(B\nabla H) - \nabla H \cdot \frac{\partial \mathbf{v}}{\partial t} - \mathrm{div}\,(H\omega \wedge \mathbf{v}) + \nu\,\mathrm{div}\,(\nabla H \wedge \omega). \tag{4.7.7}$$

This equation is formally valid everywhere, including the region within S_+ where $HB \equiv 0$. The source terms involving ∇H are concentrated on the control surface. When $\mathbf{x}$ lies in the exterior region these surface terms take account of the presence of the solid body inside S_+; if the body is absent (so that S_+ is filled with fluid) the surface sources are a representation 'to the outside world' ($f > 0$) of hydrodynamic events occurring within S_+.

The 'solution' of (4.7.7) within the fluid is obtained by use of any Green's function $G(\mathbf{x}, \mathbf{y})$ that satisfies

$$\nabla^2 G = \delta(\mathbf{x} - \mathbf{y}), \quad \text{and} \quad G \to 0 \text{ as } |\mathbf{x}| \to \infty,$$

for $\mathbf{x}$ and $\mathbf{y}$ anywhere within the fluid. In Figure 4.7.1 the fluid occupies the region V *outside* the surface S of the solid body; the control surface S_+ [$f(\mathbf{x}) = 0$] therefore lies *within* V.

Thus, for points $\mathbf{x}$ within the fluid

$$HB(\mathbf{x}) = \int_V G(\mathbf{x}, \mathbf{y}) \left[\mathrm{div}\,(B\nabla H) - \nabla H \cdot \frac{\partial \mathbf{v}}{\partial t} - \mathrm{div}\,(H\omega \wedge \mathbf{v}) + \nu\,\mathrm{div}\,(\nabla H \wedge \omega) \right] d^3\mathbf{y},$$

where the 'sources' in the integrand depend on $\mathbf{y}$ and the time t. Those involving ∇H vanish except on the control surface S_+. The divergence terms are removed by application of the divergence theorem, and the formula

$$\int_V (\cdot)\nabla H d^3\mathbf{y} = \oint_{S_+} (\cdot)d\mathbf{S}$$

(§2.8.1), leading to

$$B(\mathbf{x}, t) = -\oint_{S_+}\left[B(\mathbf{y})\frac{\partial G}{\partial \mathbf{y}}(\mathbf{x}, \mathbf{y}) + G(\mathbf{x}, \mathbf{y})\frac{\partial \mathbf{v}}{\partial t}(\mathbf{y})\right]\cdot d\mathbf{S}(\mathbf{y})$$
$$+ \int_V H[f(\mathbf{y})](\omega\wedge\mathbf{v})(\mathbf{y})\cdot\frac{\partial G}{\partial \mathbf{y}}(\mathbf{x}, \mathbf{y})d^3\mathbf{y}$$
$$- \nu\oint_{S_+}\omega(\mathbf{y})\wedge\frac{\partial G}{\partial \mathbf{y}}(\mathbf{x}, \mathbf{y})\cdot d\mathbf{S}(\mathbf{y}). \tag{4.7.8}$$

We now choose G to have a vanishing normal derivative on the moving surface S of the body. When this is done the control surface S_+ is allowed to shrink down onto S, and the general solution reduces to

$$B(\mathbf{x}) = \int_V (\omega\wedge\mathbf{v})(\mathbf{y})\cdot\frac{\partial G}{\partial \mathbf{y}}(\mathbf{x}, \mathbf{y})d^3\mathbf{y} - \nu\oint_S\omega(\mathbf{y})\wedge\frac{\partial G}{\partial \mathbf{y}}(\mathbf{x}, \mathbf{y})\cdot d\mathbf{S}(\mathbf{y})$$
$$- \frac{\partial}{\partial t}\oint_S G(\mathbf{x}, \mathbf{y})\mathbf{v}(\mathbf{y})\cdot d\mathbf{S}(\mathbf{y}) + \oint_S\frac{DG}{Dt}(\mathbf{x}, \mathbf{y})\mathbf{v}(\mathbf{y})\cdot d\mathbf{S}(\mathbf{y}),$$
$$\text{where}\quad \frac{\partial G}{\partial y_n}(\mathbf{x}, \mathbf{y}) = 0\ \text{ on } S, \tag{4.7.9}$$

and relation (4.5.32) has been used for $\operatorname{div}\mathbf{v} = 0$. In the far field ($|\mathbf{x}| \to \infty$) we can replace $B(\mathbf{x})$ with $p(\mathbf{x})/\rho_o$.

4.7.4 The far field

Far-field Green's function (2.18.3) can be used to evaluate (4.7.9) at large distances from the body. When $|\mathbf{x}| \to \infty$ and the origin is taken at the centre of volume of S, we proceed in the way described in §2.18.1, by expanding the Green's function to first order in $\mathbf{Y}(\mathbf{y})$:

$$G(\mathbf{x}, \mathbf{y}) = \frac{-1}{4\pi|\mathbf{X} - \mathbf{Y}|} \approx \frac{-1}{4\pi|\mathbf{x}|} - \frac{x_j Y_j}{4\pi|\mathbf{x}|^3}, \quad |\mathbf{x}| \to \infty.$$

The first term in this approximation makes no contribution to the integrals in (4.7.9), except possibly to the first integral of the second line if the volume of the body is pulsating. When this happens the resulting monopole pressure field is usually large compared with all other sources. We therefore ignore this possibility and consider only surface vibrations for which the volume of S is constant. In this case therefore the first term in the Green's function expansion can be discarded.

Substituting into (4.7.9), we obtain the following general formula in the far field:

$$B \approx \frac{p}{\rho_o} \approx \frac{-x_j}{4\pi |\mathbf{x}|^3} \left[\int_V (\omega \wedge \mathbf{v}) \cdot \nabla Y_j d^3\mathbf{y} - \nu \oint_S \omega \wedge \nabla Y_j \cdot d\mathbf{S} \right.$$

$$\left. - \frac{\partial}{\partial t} \oint_S (y_j - \varphi_j^*)\mathbf{v} \cdot d\mathbf{S} + \oint_S (\omega \wedge \mathbf{Y})_j \mathbf{v} \cdot d\mathbf{S} \right]$$

$$= \frac{-x_j}{4\pi |\mathbf{x}|^3} \left[-\frac{m_o}{\rho_o} \frac{dU_{oj}}{dt} + \int_V (\omega \wedge \mathbf{v}) \cdot \nabla Y_j d^3\mathbf{y} - \nu \oint_S \Omega \wedge \nabla Y_j \cdot d\mathbf{S} \right.$$

$$\left. + \frac{\partial}{\partial t} \oint_S \varphi_j^* \mathbf{v} \cdot d\mathbf{S} + \oint_S (\Omega \wedge \mathbf{Y})_j \mathbf{v} \cdot d\mathbf{S} \right]$$

$$\equiv \frac{-x_j}{4\pi |\mathbf{x}|^3 \rho_o} \left[F_j - m_o \frac{dU_{oj}}{dt} \right], \quad |\mathbf{x}| \to \infty, \tag{4.7.10}$$

where $\mathbf{F}$ is the unsteady force on the body, $\mathbf{U}_o$ is the velocity of its centre of volume, Ω is its angular velocity of rotation, and m_o is the mass of fluid displaced by the body. For a non-rotating body we can replace $\mathbf{v}$ by $\mathbf{v}_{rel}$ in the volume integral.

Equation (4.5.4) shows that this result can be expressed in the following alternative form

$$B = -\frac{\partial \varphi}{\partial t} \approx \frac{1}{\rho_o} \text{div} \left[\frac{1}{4\pi |\mathbf{x}|} \left(\mathbf{F} - m_o \frac{d\mathbf{U}_o}{dt} \right) \right] = -\frac{\partial}{\partial t} \text{div} \left(\frac{\mathbf{I}}{4\pi |\mathbf{x}|} \right),$$

which is equivalent to asymptotic formula (4.2.10).

The following high-Reynolds-number form of (4.7.10) may be noted for a stationary body or for one in uniform translational motion,

$$p(\mathbf{x}, t) \approx \frac{-\rho_o x_j}{4\pi |\mathbf{x}|^3} \int \omega \wedge \mathbf{v}_{rel} \cdot \nabla Y_j d^3\mathbf{y}, \quad |\mathbf{x}| \to \infty. \tag{4.7.11}$$

The velocity $\mathbf{v}$ has been replaced with $\mathbf{v}_{rel}$ (see §4.5.5) to emphasize that there is no contribution to the integral from bound vorticity on S.

EXAMPLE 1. PRESSURE FLUCTUATIONS NEAR A WALL APERTURE Calculate the pressure fluctuations at large distance $|\mathbf{x}| \gg a$ from a circular aperture of radius a in a thin rigid wall coinciding with the plane $x_1 = 0$ (Figure 4.7.2) produced by high-Reynolds-number vorticity near the aperture.

Take the coordinate origin at the centre of the aperture. At large distances from the aperture the pressure $p \sim \rho_o B$, where B is determined by the following reduced form of (4.7.9):

$$B(\mathbf{x}, t) = \int_V (\omega \wedge \mathbf{v})(\mathbf{y}, t) \cdot \frac{\partial G}{\partial \mathbf{y}}(\mathbf{x}, \mathbf{y}) d^3\mathbf{y}.$$

Suppose $x_1 > 0$ and also, for simplicity, that the vorticity lies in $y_1 > 0$, as in the figure. Using far-field Green's function (2.19.7), we find

$$p(\mathbf{x}, t) \approx \frac{\rho_o}{2\pi |\mathbf{x}|} \int (\omega \wedge \mathbf{v})(\mathbf{y}, t) \cdot \frac{\partial \Phi}{\partial \mathbf{y}}(\mathbf{y}) d^3\mathbf{y}, \quad |\mathbf{x}| \gg a.$$

Figure 4.7.2

This is the field of a *monopole source* whose strength is the unsteady induced volume flux through the aperture.

PROBLEMS 4

1. Show that in inviscid, homentropic flow (where div $\mathbf{v} \neq 0$) vorticity equation (4.1.8) takes the form

$$\frac{D}{Dt}\left(\frac{\omega}{\rho}\right) = \left(\frac{\omega}{\rho}\cdot\nabla\right)\mathbf{v}.$$

2. Prove that for a material surface element $d\mathbf{S}$ (Figure 4.1.4) and an arbitrary vector field $\mathcal{A}(\mathbf{x}, t)$,

$$\frac{D}{Dt}(\mathcal{A}\cdot d\mathbf{S}) = d\mathbf{S}\cdot\left[\frac{\partial\mathcal{A}}{\partial t} + \mathbf{curl}(\mathcal{A}\wedge\mathbf{v}) + \mathbf{v}\,\mathrm{div}\mathcal{A}\right].$$

3. Show that, in an incompressible fluid with constant properties and subject to conservative body forces, the rate of change of circulation Γ around a material fluid contour C is

$$\frac{D\Gamma}{Dt} = -\nu\oint_C \mathbf{curl}\,\omega\cdot d\mathbf{x}.$$

4. Show for the vortex sheet problem of Figure 4.1.6(a) that the motion is unstable provided that

$$(U_+ - U_-)^2 > \frac{g}{k}\frac{\rho_-^2 - \rho_+^2}{\rho_-\rho_+}.$$

5. A rigid sphere of radius a translates at constant velocity $\mathbf{U}$ without rotation at infinitesimal Reynolds number $\mathrm{Re} = Ua/\nu$ (§§4.2, 4.4). When the origin is at the centre of the sphere, the vorticity distribution is

$$\boldsymbol{\omega} = \mathbf{curl}\left(\frac{3a\mathbf{U}}{2|\mathbf{x}|}\right), \quad |\mathbf{x}| > a.$$

Verify that the velocity $\mathbf{v}$ predicted by Biot–Savart formula (4.2.19) yields $\mathbf{v} = \mathbf{U}$ for $|\mathbf{x}| < a$.

6. Show that Green's dipole formula (2.9.7) and Biot–Savart representation (4.2.23) are equivalent when $q \equiv 0$.

7. Show that in steady motion in two dimensions of an ideal, incompressible fluid, the stream function $\psi(x, y)$ satisfies

$$\nabla^2\psi = B'(\psi),$$

where $B = p/\rho_o + \frac{1}{2}v^2$ is constant on stream surfaces $\psi = \mathrm{constant}$.

8. Consider the motion produced when Hill's spherical vortex of §4.3.5 advances into stationary fluid. Show that the vortex specific impulse $\mathbf{I} = 2\pi a^3 U\mathbf{i}$, that the total kinetic energy is $T = \frac{10}{7}\pi\rho_o a^3 U^2$ and that net the circulation of the vortex is $5Ua$.

9. A rigid plane at $y = 0$ oscillates parallel to itself and the x axis at speed $U\cos(\omega t)$. If viscous fluid above the wall (in $y > 0$) is at rest at large distances from the wall, show that the flow is in the x direction at velocity

$$u = U\cos\left(\omega t - y\sqrt{\frac{\omega}{2\nu}}\right)e^{-y\sqrt{\frac{\omega}{2\nu}}}.$$

Show also that the unsteady drag on unit area of the wall is $\rho_o U\sqrt{\omega\nu}\cos(\omega t + \frac{\pi}{4})$.

10. Steady flow is maintained in an infinite expanse of viscous incompressible fluid by rotation of a rigid cylinder of radius a and infinite length immersed in the fluid. If the cylinder rotates at constant angular velocity Ω show that the energy supplied by the cylinder to maintain the motion is $4\pi a^2\eta\Omega^2$ per unit length of the cylinder.

11. Use the creeping flow approximation to show that the terminal velocity U of a sphere of mass density ρ and radius a falling vertically under gravity in a fluid of density ρ_o and shear viscosity η is given by

$$U = \frac{2(\rho - \rho_o)ga^2}{9\eta}.$$

12. For the rotating sphere of §4.4, Example 3, evaluate the surface shear stress and show that the moment on the sphere about the axis of rotation is given by

$$2\pi a^4 \eta \int_0^\pi \sin^3\theta \left(\frac{df}{dr}\right)_{r=a} d\theta = -8\pi\eta\Omega a^3.$$

13. Show that formula (4.4.16) for the velocity distribution for steady viscous flow in a pipe becomes

$$u = \left(\frac{p_1 - p_2}{\ell} + \rho_0 g \cos\vartheta\right)\frac{(a^2 - \varpi^2)}{4\eta}, \qquad \varpi < a,$$

for flow under gravity down a pipe inclined at angle ϑ to the vertical.

14. Show that the velocity distribution for steady viscous flow in a horizontal annular pipe with inner and outer radii b, a is given by the following modification of (4.4.16):

$$u = \frac{p_1 - p_2}{4\eta\ell}\left[a^2 - \varpi^2 + \frac{b^2 - a^2}{\ln(b/a)}\ln\frac{\varpi}{a}\right], \qquad b < \varpi < a.$$

15. Use the von Kármán momentum integral theory of §4.4.7 to show that when, in (4.4.24),

$$f(\xi) = \begin{cases} \sin(\pi\xi/2), & \text{for } \xi < 1 \\ 1, & \text{for } \xi > 1 \end{cases} \quad \text{then} \quad \begin{cases} \delta = 4.80x/\sqrt{\mathrm{Re}_x}, \\ \text{drag} = 1.312\rho_0 U^2 x/\sqrt{\mathrm{Re}_x}, \\ \delta_* = 1.742x/\sqrt{\mathrm{Re}_x}. \end{cases}$$

16. A rigid sphere of radius a translates at constant velocity $\mathbf{U} = (U, 0, 0)$, $U > 0$, along the x axis in fluid at rest at infinity. In the creeping flow approximation the vorticity $\boldsymbol{\omega} = \omega_\phi \mathbf{i}_\phi$ satisfies $\nabla^2(\omega_\phi \mathbf{i}_\phi) = 0$. Deduce that in terms of spherical polar coordinates, the Stokes stream function $\psi(r, \theta)$ is determined by

$$\left(\frac{\partial^2}{\partial r^2} + \frac{1}{r^2}\frac{\partial^2}{\partial\theta^2} - \frac{\cot\theta}{r^2}\frac{\partial}{\partial\theta}\right)^2 \psi = 0.$$

By setting $\psi = f(r)\sin^2\theta$, show that

$$\left(\frac{d^2}{dr^2} - \frac{2}{r^2}\right)^2 f = 0,$$

with the general solution

$$f = \frac{A}{r} + Br + Cr^2 + Dr^4, \quad \text{for constant } A, \ B, \ C, \ D.$$

Apply the no-slip condition on the sphere and the no-flow condition at infinity to show that $A = -\frac{1}{4}Ua^3$, $B = \frac{3}{4}Ua$, $C = D = 0$, and hence that $\psi = \frac{3}{4}aUr\sin^2\theta(1 - a^2/3r^2)$, $r > a$, as in (4.4.11).

17. Show that the force on the rigid body S of Figure 4.5.1 in translational motion at velocity $\mathbf{U}_o$ in *inviscid, rotational* flow can be expressed in the form

$$F_i = \frac{d}{dt}\oint_S \rho_0\varphi_i^* \mathbf{U}_o \cdot d\mathbf{S} + \oint_S \rho_0(\mathbf{v} - \mathbf{U}_o)\cdot\nabla X_i\mathbf{U}_o\cdot d\mathbf{S} + \rho_0\int_V \nabla X_i\cdot\boldsymbol{\omega}\wedge\mathbf{v}\,d^3\mathbf{x}.$$

Hence deduce the following inviscid form of (4.5.15):

$$F_i = -\mathbf{M}_{ij}\frac{dU_{oj}}{dt} + \rho_o \int_V \nabla X_i \cdot \boldsymbol{\omega} \wedge \mathbf{v}_{\mathrm{rel}}\, d^3\mathbf{x}, \quad \mathbf{v}_{\mathrm{rel}} = \mathbf{v} - \mathbf{U}_o.$$

18. Use impulse formula (4.5.4) to derive the asymptotic limit of formula (4.6.6) (as $t \to \infty$) for the lift experienced by an airfoil in incompressible flow. Establish by the same means drag formula (4.6.7).

19. Use moment formula (4.6.1) to derive the asymptotic limit of the rolling moment formula (4.6.8) and formula (4.6.9) for the yawing moment.

20. Consider a spherical bubble of radius a in steady translational motion at speed U (in the x_1 direction). If the bubble encloses fluid of viscosity $\bar{\eta}$, the creeping flow approximation for the surface vorticity is given by (4.4.7) with $A = \frac{1}{2}(2\eta + 3\bar{\eta})/(\eta + \bar{\eta})$. Deduce that the drag at very small Reynolds number is given by

$$D = 4\pi C\eta Ua = 2\pi\frac{(2\eta + 3\bar{\eta})Ua}{(\eta + \bar{\eta})}.$$

21. For a gas bubble of radius a moving slowly at speed U in the x_1 direction we can take

$$\boldsymbol{\omega} = -\frac{2}{a}\frac{U\mathbf{x} \wedge \nabla X_1}{|\mathbf{x}|}.$$

Show that the drag $D = 12\pi\eta Ua$ (Levich 1962).

22. A vortex ring of radius R and circulation Γ and core radius $\sigma \ll R$ translates in the x direction with its axis of symmetry coinciding with the x axis. The ring passes over a fixed rigid sphere of radius $a \ll R$ whose centre is at the origin. Calculate the unsteady force exerted on the sphere when its influence on the trajectory of the vortex can be ignored.

23. Repeat the calculation of Question 22 for a vortex ring translating at uniform velocity in the x direction in the presence of a rigid blade occupying $-a < x < a$, $y = 0$, $-\infty < z < \infty$. Assume the axis of symmetry of the vortex to be the line $y = h > 0$, $z = 0$ and examine the cases $h > R$ and $h < R$.

24. A circular vortex ring of radius $R < a$, of small core and of circulation Γ translates axisymmetrically by self-induction through the aperture in the plane wall of Figure 4.7.2. Show that the volume flux $q(t)$ through the aperture (in the $+x_1$ direction) is given by

$$q(t) = 2\pi R\Gamma \int_{-\infty}^{t} U_\Gamma(t)\mathrm{sgn}(y_1)\frac{\partial\Phi}{\partial r}(y_1, R)\, dt$$

where $U_\Gamma(t)$ is the velocity of translation of the vortex at time t, $y_1 = \int_{-\infty}^{t} U_\Gamma(t)\, dt$, and Φ is defined as in §2.19.5.

25. A line vortex of strength Γ is in two-dimensional motion in an ideal incompressible fluid of density ρ_o adjacent to the semi-infinite, rigid half-plane $x < 0$, $y = 0$ (see Figure 3.5.5). The position of the vortex at time t is given in terms of the polar coordinates (r, θ) shown in the figure by

$$r = \ell\sqrt{1 + \left(\frac{Ut}{\ell}\right)^2}, \quad \theta = 2\tan^{-1}\left(-\frac{Ut}{\ell}\right), \quad \text{where} \quad U = \frac{\Gamma}{8\pi\ell}$$

and ℓ is the distance of closest approach of the vortex to the edge.

Calculate the suction force exerted on the edge of the half-plane

(i) by finding the complex potential near the edge and using suction force formula (3.3.7),

(ii) by means of Kirchhoff vector force formula (4.5.15)

26. Deduce from (4.5.15) the following asymptotic limits for the drag coefficient C_D for a bluff body in incompressible flow:

$$C_D \sim \begin{cases} O(1), & \text{Re} \equiv U\ell/\nu \to \infty \\ O(1/\text{Re}), & \text{Re} \to 0. \end{cases}$$

5

Surface Gravity Waves

5.1 Introduction

Steady free-streamline flows of water when gravitational forces can be neglected have been discussed in §3.7. Most *unsteady* free-streamline problems are intractable except by numerical means and generally become more so when gravitational forces are important. However, flows involving gravity where the unsteady motion is a 'small' perturbation of a relatively simple mean state occur frequently in the form of surface waves. In the absence of motion the free surface of a liquid in equilibrium under gravity is often 'horizontal'. A disturbance applied locally that distorts the surface brings into play gravitational restoring forces that cause the disturbance to spread out over the surface in the form of 'waves'. The waves carry energy away from the source region, propagating parallel to the mean free surface. The agitation produced by a passing wave and the energy flux is generally in the form of a transient disturbance of the fluid particles (around approximately closed particle paths), which are not in themselves transported to any great extent by the wave, and the influence of the wave on fluid at depths exceeding a characteristic wavelength tends to be negligible. In this section these general properties of surface gravity waves are discussed and illustrated by simple examples.

5.1.1 Conditions at the free surface

Consider the simplest case of water whose free surface in equilibrium can be regarded as horizontal and in the plane $z = 0$ of the coordinate axes (x, y, z), where z increases vertically upwards (Figure 5.1.1). Let the motion be inviscid and be started from rest by the action of surface forces or normal surface displacements. The initial motion is therefore irrotational with velocity potential φ, and if viscous dissipation is ignored in the first instance it will continue to be irrotational.

Bernoulli's equation (2.3.3) can be taken in the form

$$\frac{\partial \varphi}{\partial t} + \frac{p}{\rho_o} + \frac{1}{2}v^2 + gz = 0, \tag{5.1.1}$$

286

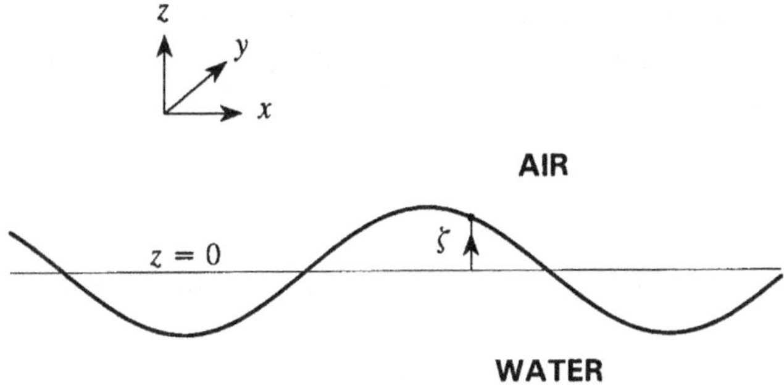

Figure 5.1.1

where $\Phi \equiv -gz$ is the gravitational potential (the acceleration due to gravity g being regarded as constant throughout the depth of the fluid). The inertia of the air above the free surface is neglected, so that (neglecting also small effects of surface tension), the pressure p should normally vanish at the instantaneous position of the free surface

$$z = \zeta \equiv \zeta(x, y, t).$$

However, (irrotational) waves generated by surface forcing can be attributed to an externally applied pressure distribution $p_o(x, y, \zeta, t)$ acting on the instantaneous position $z = \zeta(x, y, t)$ of the free surface. The *dynamical* condition $p = p_o$ at the surface therefore supplies the relation

$$\frac{\partial \varphi}{\partial t} + \frac{1}{2}(\nabla \varphi)^2 + g\zeta = -\frac{p_o}{\rho_o}, \quad \text{at } z = \zeta(x, y, t). \tag{5.1.2}$$

An additional *kinematic* relation between ζ and φ is obtained from the observation that the equation $f \equiv z - \zeta(x, y, t) = 0$ must satisfy $Df/Dt = 0$ (the condition that the free surface moves with the fluid particles; §1.7), which yields

$$\frac{\partial \zeta}{\partial t} + \nabla \varphi \cdot \nabla \zeta = \frac{\partial \varphi}{\partial z}, \quad \text{at } z = \zeta(x, y, t). \tag{5.1.3}$$

5.1.2 Wave motion within the fluid

The velocity potential is a solution of Laplace's equation $\nabla^2 \varphi = 0$ in $z < \zeta$. The solution must satisfy (5.1.2), (5.1.3) at the free surface and also the condition that the normal derivative $\partial \varphi / \partial x_n = 0$ at the fixed boundaries. Now solutions of Laplace's equation produced by localised excitation decay rapidly with distance from the source region and are not normally associated with waves. This means that motions within the body of the water produced by a surface disturbance must actually decrease very rapidly with depth. However, this is not incompatible with surface wave motions, because Laplace's equation for a disturbance on deep water that decreases exponentially fast, such that $\varphi \propto e^{kz}$, $k > 0$ as $z \to -\infty$, becomes

$$\frac{\partial^2 \varphi}{\partial x^2} + \frac{\partial^2 \varphi}{\partial y^2} + k^2 \varphi = 0.$$

This is the two-dimensional *Helmholtz* equation that has 'plane-wave' solutions proportional to $e^{i(k_1 x + k_2 y)}$, where $k_1^2 + k_2^2 = k^2$, describing oscillatory disturbances in planes parallel to the free surface (the xy plane).

5.1.3 Linearised approximation

When the surface waves represent a small perturbation of the water about its undisturbed state, the quantities φ, ζ, and p_o in (5.1.2), (5.1.3) are of the first order of smallness. The *linearised* counterparts of these conditions are therefore

$$\frac{\partial \varphi}{\partial t} + g\zeta = -\frac{p_o}{\rho_o}, \quad \frac{\partial \zeta}{\partial t} = \frac{\partial \varphi}{\partial z}, \quad \text{at } z = 0. \tag{5.1.4}$$

By combining these equations, we find the linearised surface condition in terms of φ alone to be

$$\partial^2 \varphi / \partial t^2 + g\partial \varphi / \partial z = (-1/\rho_o)\partial p_o / \partial t, \quad \text{at} \quad z = 0.$$

On deep water (with no bottom boundary) the linearised *boundary-value problem* governing the wave motion accordingly becomes

$$\nabla^2 \varphi = 0, \qquad z < 0;$$
$$\frac{\partial^2 \varphi}{\partial t^2} + g\frac{\partial \varphi}{\partial z} = -\frac{1}{\rho_o}\frac{\partial p_o}{\partial t}, \quad z = 0. \tag{5.1.5}$$

5.1.4 Time harmonic, plane waves on deep water

Consider 'unforced' plane-wave motion (for which $p_o \equiv 0$) in the x direction where

$$\varphi = \hat{\varphi}(z)e^{i(kx - \omega t)}, \quad z < 0,$$

and where the 'wavenumber' k and radian frequency ω are real. Laplace's equation reduces to $d^2\hat{\varphi}/dz^2 - k^2\hat{\varphi} = 0$. The solution that remains bounded as $z \to -\infty$ is $\hat{\varphi} = Ae^{|k|z}$ where A is a constant, and therefore $\varphi = Ae^{i(kx-\omega t)+|k|z}$ $(z < 0)$. This will also satisfy the surface boundary condition [the second of (5.1.5) with $p_o = 0$] provided k and ω are related by the *deep-water dispersion relation*

$$\omega^2 = g|k| \quad \text{or} \quad \omega = \pm\sqrt{g|k|}. \tag{5.1.6}$$

The simple solution $\varphi = Ae^{i(kx-\omega t)+|k|z}$ represents a disturbance of wavelength $\lambda = 2\pi/|k|$ parallel to the undisturbed surface of the water, with maximum amplitude at the surface $z = 0$. The first of equations (5.1.4) shows that the surface elevation

$$\zeta = \zeta_o e^{i(kx-\omega t)}, \quad \text{where} \quad \zeta_o = \frac{i\omega A}{g}.$$

The potential can therefore be expressed in the form

$$\varphi = -\frac{ig\zeta_o}{\omega}e^{i(kx-\omega t)+|k|z}, \quad z < 0.$$

It is frequently more convenient to have a solution in real form. If, for example, we take the wave amplitude ζ_o to be a real quantity, and put

$$\zeta = -\frac{1}{g}\frac{\partial\varphi}{\partial t} \equiv \zeta_o\sin(kx - \omega t),$$

$$\text{then}\quad \varphi = -\frac{g\zeta_o}{\omega}\cos(kx - \omega t)e^{|k|z}, \quad z < 0.$$

(5.1.7)

The 'phase' $kx - \omega t$ varies with position and time but takes constant values on, for example, the 'wave crests' where $kx - \omega t = \frac{\pi}{2} + 2n\pi$, $n =$ integer; constant values of the phase propagate in the x direction at the *phase speed* ω/k. At a depth of one wavelength $z = -\lambda = -2\pi/|k|$ the amplitude of the motion $\sim e^{-2\pi} = 0.002$ relative to its value at the surface. This means that the effective 'penetration' of the wave motion into the water does exceed about one wavelength. The following table indicates how

$$\text{phase speed} = \sqrt{\frac{g\lambda}{2\pi}} \approx 1.25\sqrt{\lambda}\ (\lambda,\ \text{m}), \quad \text{wave period} = \frac{2\pi}{\omega} \equiv \frac{\text{wavelength}}{\text{phase speed}},$$

vary with the wavelength on deep water:

λ, m	phase speed, m/s	period, s
1	1.25	0.80
10	3.95	2.53
50	8.84	5.66
100	12.5	8.00

EXAMPLE 1. PARTICLE PATHS Equations (5.1.7) with $k, \omega > 0$ define wave motion in the positive x direction. Let $\hat{\mathbf{x}} = (\hat{x}, \hat{z})$ denote the components of the small displacement of a fluid particle from its equilibrium position at (x, z). Then $d\hat{\mathbf{x}}/dt = \nabla\varphi$, i.e.,

$$\frac{d\hat{x}}{dt} = \frac{gk\zeta_o}{\omega}\sin(kx - \omega t)e^{kz}, \quad \frac{d\hat{z}}{dt} = -\frac{gk\zeta_o}{\omega}\cos(kx - \omega t)e^{kz},$$

$$\therefore\quad \hat{x} = \frac{gk\zeta_o}{\omega^2}\cos(kx - \omega t)e^{kz} \equiv \zeta_o e^{kz}\cos(kx - \omega t)$$

$$\hat{z} = \frac{gk\zeta_o}{\omega^2}\sin(kx - \omega t)e^{kz} \equiv \zeta_o e^{kz}\sin(kx - \omega t)$$

$$\left.\right\}\quad \text{because}\quad \omega^2 = gk.$$

A fluid particle at depth $|z|$ therefore moves around a circular path of radius $\zeta_o e^{kz}$, which becomes negligible at depths of a wavelength or more. There is no mean translational motion of the particle – a consequence of linear theory. At the free surface $\hat{z} = \zeta_o\sin(kx - \omega t)$ is just the elevation of the free surface, where

$$\frac{d\hat{x}}{dt} = \frac{gk\hat{z}}{\omega} \equiv \frac{gk}{\omega}\zeta(kx - \omega t)$$

The fluid particle motion at the surface is therefore in the direction of propagation of the wave for particles at a wave crest ($\zeta > 0$) and in the opposite direction at a trough. Figure 5.1.2 illustrates various phases of particle trajectories at different points over a complete wavelength.

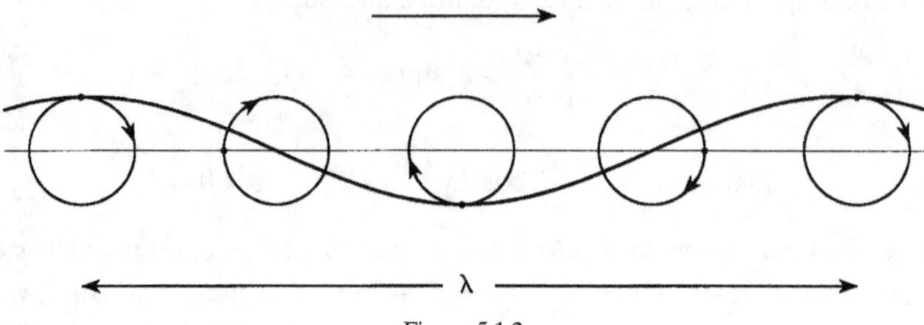

Figure 5.1.2

5.1.5 Water of finite depth

Unforced small-amplitude waves $\propto e^{-i\omega t}$ propagating in water of uniform depth h (Figure 5.1.3) are governed by

$$\nabla^2 \varphi = 0, \qquad -h < z < 0,$$

$$g\frac{\partial \varphi}{\partial z} = \omega^2 \varphi, \qquad z = 0,$$

$$\frac{\partial \varphi}{\partial z} = 0, \qquad z = -h, \tag{5.1.8}$$

$$\zeta = \frac{i\omega\varphi}{g}, \qquad z = 0.$$

The condition at $z = -h$ is satisfied by

$$\varphi = \Psi(x, y)\cosh[k(z+h)]e^{-i\omega t},$$

where the first and second of Equations (5.1.8) require that

$$\left(\frac{\partial^2}{\partial x^2} + \frac{\partial^2}{\partial y^2} + k^2\right)\Psi = 0 \tag{5.1.9}$$

and that ω and k satisfy the dispersion relation

$$\omega^2 = gk\tanh(kh). \tag{5.1.10}$$

When $\omega > 0$ is prescribed the dispersion relation defines two equal and opposite values for the wavenumber k. Then for waves propagating in the x direction we can take solutions in real form of the type

$$\varphi = -\frac{g\zeta_o}{\omega}\frac{\cosh[k(z+h)]}{\cosh(kh)}\cos(kx - \omega t), \qquad -h < z < 0,$$

$$\zeta = \zeta_o \sin(kx - \omega t). \tag{5.1.11}$$

This reverts to the deep-water limit as $h \to \infty$ when, also, dispersion formula (5.1.10) reduces to $\omega^2 = g|k|$. By putting $c = \omega/k$ we can rewrite this formula in the form

$$\omega\sqrt{\frac{h}{g}} = \left[\frac{\omega c/g}{2}\ln\left(\frac{1 + \omega c/g}{1 - \omega c/g}\right)\right]^{\frac{1}{2}}. \tag{5.1.12}$$

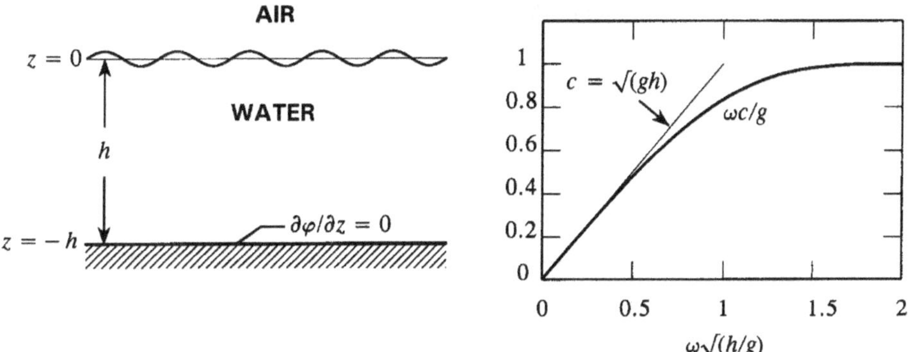

Figure 5.1.3

The non-dimensional phase speed $\omega c/g$ is plotted against $\sqrt{\text{depth}}$ in Figure 5.1.3, showing how c decreases rapidly with h towards the constant value $\sqrt{gh}$ (when $\omega^2 \approx ghk^2$). This *shallow*-water limit is attained when the wavelength $\lambda \gg h$ ($kh < 1$) and when, indeed, the wave period $2\pi/\omega$ exceeds about twice the period of a simple pendulum of length h (viz., $2\pi\sqrt{h/g}$). This decrease in phase speed with depth is responsible for the lining-up of wavefronts with a shore line of long waves approaching the shore from a deep ocean. The wave frequency does not change for such waves, but the phase velocity decreases with depth, causing the more rapidly moving wavefronts further from the shore to 'swing round' towards the shore (see Figures 5.10.2 and 5.10.3 in §5.10).

It is clear from the shallow-water limit $kh \to 0$ of (5.1.10) that long waves propagating in arbitrary directions in the xy plane are governed by the classical wave equation

$$\frac{1}{c_o^2}\frac{\partial^2 \varphi}{\partial t^2} - \left(\frac{\partial^2}{\partial x^2} + \frac{\partial^2}{\partial y^2}\right)\varphi = 0, \quad c_o = \sqrt{gh}. \tag{5.1.13}$$

EXAMPLE 2. PARTICLE PATHS FOR SHALLOW-WATER WAVES Finite depth solution (5.1.11) and the method of Example 1 yield the fluid particle path equations:

$$\hat{x} = \frac{\zeta_o \cosh[k(z+h)]}{\sinh(kh)}\cos(kx - \omega t), \quad \hat{z} = \frac{\zeta_o \sinh[k(z+h)]}{\sinh(kh)}\sin(kx - \omega t).$$

The fluid particle trajectory is therefore an ellipse $\hat{x}^2/a^2 + \hat{z}^2/b^2 = 1$ with semi-major and semi-minor axes $a = \zeta_o \cosh[k(z+h)]/\sinh(kh)$, $b = \zeta_o \sinh[k(z+h)]/\sinh(kh)$.

As $kh \to 0$ the trajectory becomes infinitely elongated in the x direction, with $a \to \infty$, $b \sim \zeta_o(1 + z/h)$, and the fluid particle motion becomes essentially a reciprocating motion parallel to the bed of the fluid (see Figure 5.1.4).

5.2 Surface wave energy

The energy equation for surface waves over an incompressible fluid like water subject to viscous dissipation is conveniently derived from Equation (1.5.12) in which $\Phi = -gz$ and $\rho = \rho_o$. When quantities of third order and higher are discarded, we have

$$\frac{\partial}{\partial t}\left(\frac{1}{2}\rho_o v^2\right) + \frac{\partial}{\partial x_j}\left[(p + \rho_o gz)v_j - v_i\sigma_{ij}\right] = -2\eta e_{ij}^2. \tag{5.2.1}$$

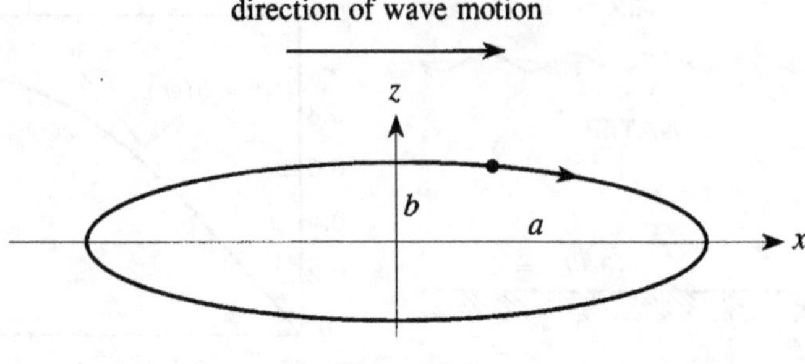

Figure 5.1.4

Let the waves propagate in the x direction and integrate the equation over the interior of a thin, vertical slice of fluid, aligned with the wave crests, of thickness δx and of unit lateral span (Figure 5.2.1). For viscous flow the velocity $\mathbf{v} = (v_x, 0, v_z)$ must vanish at the bottom $z = -h$, and the stress-free condition at the free surface must be applied in the form $(p\delta_{ij} - \sigma_{ij})n_j \equiv pn_i - \sigma_{ij}n_j = 0$ at $z = \zeta$, where $v_z \approx \partial\zeta/\partial t$. Retaining only second-order terms and considering the limit as $\delta x \to 0$, we find

$$\frac{\partial}{\partial t}\left(\int_{-h}^{0}\frac{1}{2}\rho_o v^2 dz + \frac{1}{2}\rho_o g\zeta^2\right) + \frac{\partial}{\partial x}\int_{-h}^{0}\left(\hat{p}v_x - v_i\sigma_{ix}\right)dz = -2\eta\int_{-h}^{0}e_{ij}^2\,dz, \qquad (5.2.2)$$

where

$$\int_{-h}^{0}\frac{1}{2}\rho_o v^2 dz + \frac{1}{2}\rho_o g\zeta^2 = (\text{kinetic} + \text{potential})\text{ energy per unit surface area,}$$

$$\int_{-h}^{0}\left(\hat{p}v_x - v_i\sigma_{ix}\right)dz = \text{energy flux in the }x\text{ direction,}$$

$$2\eta\int_{-h}^{0}e_{ij}^2\,dz = \text{rate of viscous dissipation per unit surface area,}$$

and

$$\hat{p} = p + \rho_o gz \qquad\qquad (5.2.3)$$

is the pressure *reduced* by the mean hydrostatic pressure $-\rho_o gz$.

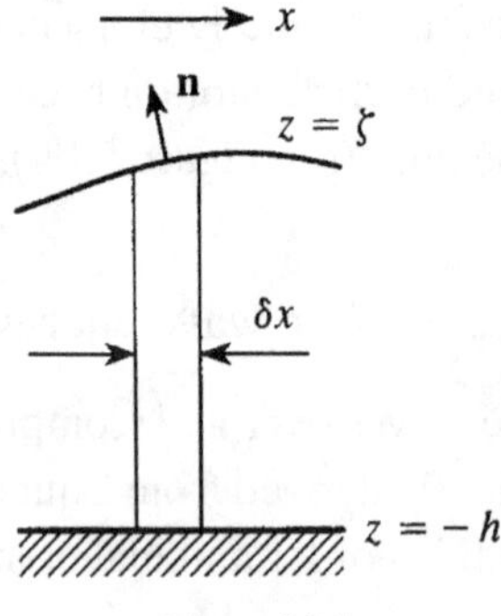

Figure 5.2.1

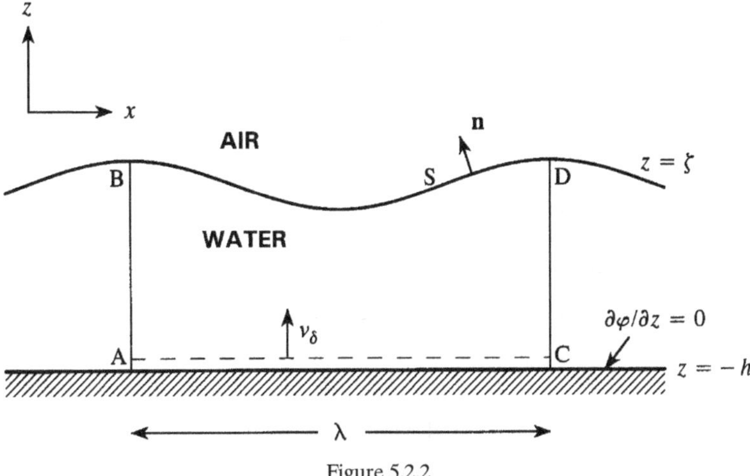

Figure 5.2.2

5.2.1 Wave-energy density

The wave energy E per unit surface area for the wave (5.1.11) on water of finite depth consists of kinetic energy T and potential energy Q, whose values we first calculate for the fluid region ABCD of Figure 5.2.2 beneath *one complete wave* of wavelength λ and of unit span. First

$$T = \frac{1}{2}\rho_o \int_{ABCD} (\nabla\varphi)^2 d^3\mathbf{x} = \frac{1}{2}\rho_o \oint_S \varphi\, \mathbf{n} \cdot \nabla\varphi \, dS$$

$$\approx \frac{1}{2}\rho_o \int_0^\lambda \left(\varphi\frac{\partial\zeta}{\partial t} \right)_{z=0} dx = \frac{\lambda}{4}\rho_o g\zeta_o^2,$$

where in passing from the first to the second line, the periodicity of the wave ensures that there are no contributions to the surface integral from the vertical faces AB and CD. The result is actually independent of the depth h, because the final integral involves terms evaluated only at the free surface $z = 0$ [where, for example, expressions (5.1.11) for finite h are identical with (5.1.7) for infinite depth].

Similarly,

$$Q = \frac{1}{2}\rho_o g \int_0^\lambda \zeta^2 \, dx = \frac{\lambda}{4}\rho_o g\zeta_o^2.$$

This coincides, of course, with the value obtained from the usual definition of potential energy in terms of the mean change in the height of the centre of gravity of the fluid, as

$$\int_0^\lambda \left(\int_{-h}^\zeta \rho_o gz\,dz - \int_{-h}^0 \rho_o gz\,dz \right) dx = \frac{1}{2}\rho_o g \int_0^\lambda \zeta^2 \, dx \equiv Q.$$

Hence, the wave-energy density (*per unit surface area*) $E = (T + Q)/\lambda$ is given by

$$E = \frac{1}{2}\rho_o g\zeta_o^2, \tag{5.2.4}$$

a result that does not depend on the depth of the water.

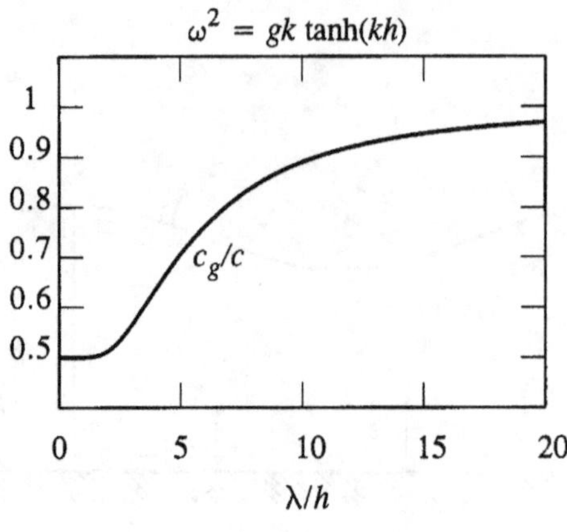

Figure 5.2.3

5.2.2 Wave-energy flux

We can evaluate similarly the energy flux $I = \int_{-h}^{0}(\hat{p}v_x - v_i\sigma_{ix})\,dz$ in the x direction for the wave (5.1.11). Because the motion is inviscid only the pressure term need be considered, and we calculate the perturbation pressure using the linearised form of Bernoulli's equation (5.1.1):

$$\hat{p} \equiv p + \rho_0 gz = -\rho_0 \frac{\partial\varphi}{\partial t}. \tag{5.2.5}$$

Then,

$$I = -\rho_o \int_{-h}^{0} \frac{\partial\varphi}{\partial t}\frac{\partial\varphi}{\partial x}\,dz = \left(\frac{1}{2}\rho_o g\zeta_o^2\right)\frac{2gk}{\omega}\sin^2(kx - \omega t)\int_{-h}^{0}\frac{\cosh^2[k(z + h)]}{\cosh^2(kh)}\,dz$$

$$= \left(1 + \frac{2kh}{\sinh(2kh)}\right)\frac{g\tanh(kh)}{\omega}E\sin^2(kx - \omega t).$$

However, only the mean value of the energy flux is relevant, because local reciprocating fluctuations average to zero over a wave period. Therefore $\sin^2(kx - \omega t)$ may be replaced with its mean value $\frac{1}{2}$, and dispersion formula (5.1.10) can be used to make the substitution $\tanh(kh) = \omega^2/gk$. Hence,

$$I = \frac{1}{2}\left[1 + \frac{2kh}{\sinh(2kh)}\right]\left(\frac{\omega}{k}\right)E.$$

This result shows that the *wave-energy propagation velocity* $c_g = I/E$ generally differs from the phase velocity $c = \omega/k$:

$$c_g = \frac{1}{2}\left[1 + \frac{2kh}{\sinh(2kh)}\right]c.$$

The ratio c_g/c is plotted as a function of λ/h ($\lambda = 2\pi/k$) in Figure 5.2.3. The wave crests and wave energy propagate at the same speed $\sqrt{gh}$ in shallow water, when $\lambda \gg h$. As the wavelength decreases the deep ocean limit $c_g = \frac{1}{2}c = \frac{1}{2}\sqrt{g/k} = \frac{1}{2}\sqrt{g\lambda/2\pi}$ is attained when the wavelength becomes smaller than the depth h. Thus wave energy and the wave crests propagate at the same speed ($\sqrt{gh}$) in very shallow water, whereas energy is transported by short waves at precisely *half* the speed of advance of the wave crests.

5.2.3 Group velocity

A simple calculation shows that the energy propagation velocity c_g is also given by the formula

$$c_g = \frac{\partial \omega}{\partial k}(k), \tag{5.2.6}$$

where ω is defined as a function of the wavenumber k by the dispersion relation $\omega^2 = gk\tanh(kh)$. Thus

$$\frac{\partial \omega}{\partial k} = \frac{g\tanh(kh)}{2\omega}\left[1 + \frac{2kh}{\sinh(2kh)}\right] = \frac{1}{2}\left[1 + \frac{2kh}{\sinh(2kh)}\right]c,$$

the same as before. The velocity defined by (5.2.6) is called the *group velocity*, and the assertion that wave energy propagates at the group velocity turns out to be true quite generally for waves described by a linear system of equations. It is an entirely *kinematic* result that actually derives from the proposition that any general propagating disturbance can be represented as a superposition of elementary time harmonic waves of the type (5.1.7), say, each of which propagates independently. To reconstitute the wave at a later time it is necessary to account for the *dispersion* in space of the components of different wavenumbers, which propagate differently because of the variation of phase speed with k. At any instant the shape of the wave profile is determined by interference between the sinusoidal components, and points of constructive interference, where the overall wave amplitude is large, are associated with interfering waves of the same phase. This interference causes dispersion of an initially confined wave region whenever the phase speed c depends on k. For shallow-water waves $c = \sqrt{gh}$, and any general surface wave of arbitrary profile propagates without change of form. On deep water, however, $c = \sqrt{g/k}$ and an initially coherent wave profile breaks up into an extended surface disturbance.

For an infinitely extended sinusoidal wave such as (5.1.7) defined by a single wavenumber, the energy density is the same throughout the wave and there is no way to *observe* the propagation of wave energy. However, the plane wave can be regarded as a local approximation to a more general wave profile produced by a distant source (Stokes 1876), whose amplitude is determined by interference between components whose wavenumbers k_1 and k_2, say, differ by a small amount. A constructive interference peak can then be identified as a 'marker' whose motion determines the energy propagation velocity. Thus, consider

$$\zeta = \frac{\zeta_o}{2}\left[\sin(k_1 x - \omega_1 t) + \sin(k_2 x - \omega_2 t)\right]$$

$$= \zeta_o \cos\left[\frac{(k_1 - k_2)}{2}x - \frac{(\omega_1 - \omega_2)}{2}t\right]\sin\left[\frac{(k_1 + k_2)}{2}x - \frac{(\omega_1 + \omega_2)}{2}t\right], \tag{5.2.7}$$

where $\omega_1 = \omega(k_1)$, $\omega_2 = \omega(k_2)$.

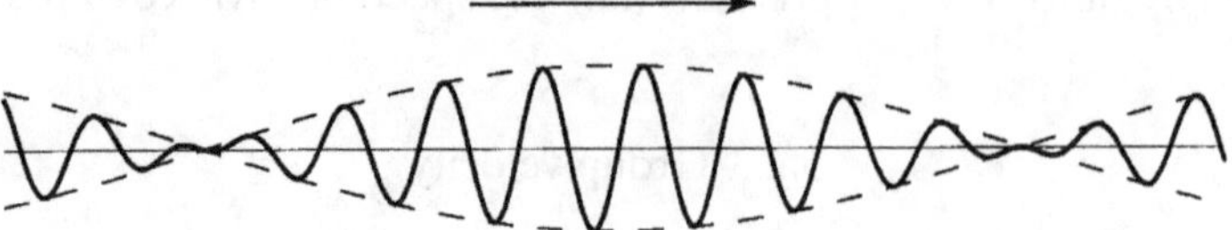

Figure 5.2.4

If k_1 is very nearly equal to k_2 the resultant wave profile corresponds to the desired surface wave $\zeta = \zeta_o \sin(kx - \omega t)$, where $k = k_1 \approx k_2$ and $\omega = \omega(k)$, subject to amplitude modulation by the factor $\cos[\frac{1}{2}(\delta k\, x - \delta\omega\, t)]$ (where $\delta k = k_1 - k_2$, $\delta\omega = \omega_1 - \omega_2$), as indicated in Figure 5.2.4. We can make the modulation rate arbitrarily small by taking δk to be sufficiently small, so that the wave differs negligibly from plane wave (5.1.7). The amplitude is a maximum when $\delta k\, x - \delta\omega\, t = $ a multiple of 2π, each maximum being found at position x and time t that increase in a fixed ratio (equal to c_g) in order that $\delta k\, x - \delta\omega\, t$ remains constant. Destructive interference occurs at intermediate locations between the maxima, so that the whole surface motion is partitioned into a succession of energetic wave groups separated by quiescent intervals at which the surface displacement and wave energy are negligible. Each group advances in the x direction at speed $\delta\omega/\delta k \sim \partial\omega/\partial k$, and the wave energy associated with each group evidently advances at the same speed.

For waves on deep water $c_g = \frac{1}{2}c$. Therefore wave crests travel forward at twice the speed of the wave group. The crests in a particular group advance through the group at speed $\frac{1}{2}c$ relative to an observer moving with the group, appearing to be formed at the rear of the group and 'disappearing' at the front.

The relation between the phase and group velocities can be represented graphically as in Figure 5.2.5. Let point A lie on the dispersion curve $\omega = \omega(k)$. Phase velocity c is

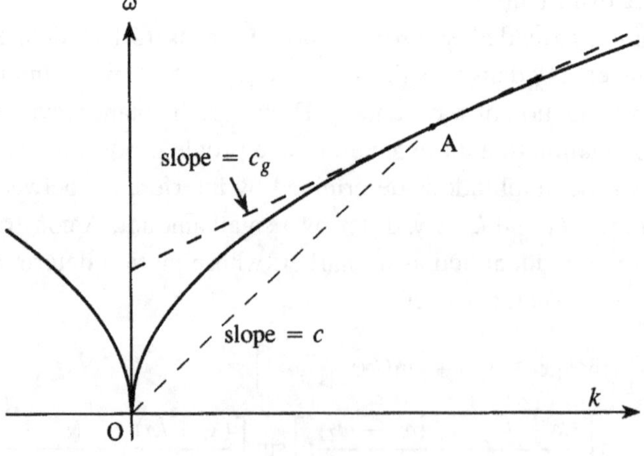

Figure 5.2.5

the slope ω/k of the chord OA; the group velocity c_g is the slope of the tangent to the dispersion curve at A.

5.3 Viscous damping of surface waves

Let us now consider the dissipation term $-2\eta \int_{-h}^{0} e_{ij}^2 \, dz$ on the right-hand side of energy equation (5.2.2). The dissipation *per unit surface area* is just equal to

$$\frac{2\eta}{\lambda} \int_{\text{ABCD}} e_{ij}^2 \, dx dz, \tag{5.3.1}$$

where the integration is over the two-dimensional region beneath one wave, as in Figure 5.2.2. The motion has thus far been assumed to be entirely irrotational, and in this case the rate of strain e_{ij} decreases rapidly with distance from the free surface, so that the main contribution to the dissipation integral is the *interior damping* that occurs within a layer of fluid of width $\sim \lambda$ near the surface. However, viscous effects cannot strictly be neglected, and for a wave of frequency ω the motion is actually rotational within a boundary layer of width $\delta \sim \sqrt{\nu/\omega}$ on the bottom boundary at $z = -h$ (§4.4, Example 1), which is typically very small compared with both the wavelength λ and the depth h. The velocity gradient and e_{ij} can become very large within the boundary layer provided that $\lambda \geq h$ (when the wave experiences significant bottom drag), and it is then necessary to consider separately the contribution of the boundary layer, to overall damping.

There is a similar narrow boundary layer at the free surface, where irrotational and rotational components of the flow are coupled by the linearised stress-free boundary conditions

$$p - 2\eta \frac{\partial v_z}{\partial z} = 0; \quad \frac{\partial v_z}{\partial x} + \frac{\partial v_x}{\partial z} = 0, \quad \text{at} \quad z = 0.$$

However, there is no rapid variation in velocity across this layer, and the contribution of the thin surface boundary layer to dissipation integral (5.3.1) is negligible compared with that from the irrotational interior motions.

5.3.1 The interior damping

From what has been said previously, we determine the interior damping by approximating the rate of strain e_{ij} in (5.3.1) by its irrotational value $\partial^2\varphi/\partial x_i \partial x_j$, where φ is given by (5.1.11). The integral can be evaluated over the whole of the interior of ABCD, as small errors incurred by including the regions occupied by the boundary layers are negligible:

$$\left.\begin{array}{r}\text{interior dissipation rate}\\\text{per unit surface area}\end{array}\right\} = \frac{2\eta}{\lambda} \int_{\text{ABCD}} \left(\frac{\partial^2\varphi}{\partial x_i \partial x_j}\right)^2 dx dz$$

$$= \frac{\eta}{\lambda} \int_0^{\lambda} \left[\frac{\partial}{\partial z}(\nabla\varphi)^2\right]_{z=0} dx = 2\eta g k^2 \zeta_o^2 = 4\nu k^2 E. \tag{5.3.2}$$

5.3.2 Boundary-layer damping

To calculate the damping produced by the bottom boundary layer we make use of a formula given in Chapter 6 (§6.10.2) for the normal component of velocity v_δ at the outer edge $z = -h + \delta$ of the boundary layer. The boundary layer is driven by the tangential pressure gradient of the irrotational wave (5.1.11). Just above the bottom wall,

$$\hat{p} = \frac{\rho_0 g \zeta_0}{\cosh(kh)} \sin(kx - \omega t) \quad \text{at} \quad z = -h + \delta \approx -h \qquad (5.3.3)$$

and the theory of §6.10.2 [Equation (6.10.5)] shows that

$$v_\delta = \frac{g k^2 \zeta_0}{\omega \cosh(kh)} \sqrt{\frac{\nu}{\omega}} \cos\left(kx - \omega t + \frac{\pi}{4}\right). \qquad (5.3.4)$$

The integration procedure leading from general energy equation (5.2.1) to wave-energy equation (5.2.2) is now repeated, but with the region of integration within the fluid of Figure 5.2.1 restricted to the irrotational domain *above* the outer edge $z = -h + \delta$ of the bottom boundary layer. Then the integral of the divergence term in (5.2.1) supplies an additional contribution $-\hat{p} v_\delta \delta x$ from the short segment of length δx at the foot $z = -h + \delta$ of the the slice, such that (5.2.2) assumes the revised form

$$\frac{\partial}{\partial t}\left(\int_{-h}^{0} \frac{1}{2}\rho_0 v^2 dz + \frac{1}{2}\rho_0 g \zeta^2\right) + \frac{\partial}{\partial x} \int_{-h}^{0} \left(\hat{p} v_x - v_i \sigma_{ix}\right) dz = -2\eta \int_{-h}^{0} e_{ij}^2 \, dz + \left(\hat{p} v_\delta\right)_{z=-h+\delta}.$$

$$(5.3.5)$$

In a first approximation the integrals on the left-hand side take the values calculated previously in §§5.2.1, 5.2.2. The new term $\hat{p} v_\delta$ is just the rate at which the boundary layer is doing work on the wave per unit surface area and must be negative on average to account for the boundary-layer losses. Its mean value (averaged over a wavelength) is evaluated by use of (5.3.3) and (5.3.4), giving

$$\left.\begin{array}{c}\text{boundary layer dissipation}\\ \text{rate per unit surface area}\end{array}\right\} = -\frac{1}{\lambda} \int_0^{\lambda} \left(\hat{p} v_\delta\right)_{z=-h+\delta} dx$$

$$= \frac{\rho_0 g^2 k^2 \zeta_0^2}{2\omega \cosh^2(kh)} \sqrt{\frac{\nu}{2\omega}} = \frac{\omega}{h}\left[\frac{2kh}{\sinh(2kh)}\right]\sqrt{\frac{\nu}{2\omega}} E. \quad (5.3.6)$$

We may now form the complete wave-energy equation with both dissipation mechanisms:

$$\frac{\partial E}{\partial t} + \frac{\partial}{\partial x}(c_g E) = -4\nu k^2 E - \frac{\omega}{h}\sqrt{\frac{\nu}{2\omega}}\left[\frac{2kh}{\sinh(2kh)}\right] E. \qquad (5.3.7)$$

This equation governs the temporal and spatial variations in the wave-energy density E produced by damping. Undamped energy propagates in the x direction at the group velocity c_g. The space and time averages used to evaluate the various components of the equation are valid only when E changes 'slowly' when viewed on the time scale of the wave period $2\pi/\omega$ and on distances $\sim O(\lambda)$. In other words, the damping must affect the wave only over many wave periods and over distances large compared with the wavelength. The interior damping [first term on the right-hand side of (5.3.7)] is dominant at short wavelengths, for which $kh \gg 1$ (however, *very* short surface waves, which would normally be called 'ripples', are in practice strongly influenced also by *surface tension* forces that have not been considered). Boundary-layer damping is important when $\lambda > h$, as otherwise the wave motion near the bottom 'driving' the boundary layer is exponentially small.

5.3.3 Comparison of boundary-layer and internal damping for long waves

When $kh \to 0$ the relation $\omega \sim k\sqrt{gh}$ implies that

$$\frac{\text{internal dissipation}}{\text{boundary layer dissipation}} \sim \frac{(\omega/h)\sqrt{\nu/2\omega}}{4\nu\omega^2/gh} = 16\pi^{\frac{3}{2}}\left(\frac{\nu}{g^2\tau^3}\right)^{\frac{1}{2}}$$

where $\tau = 2\pi/\omega$ is the wave period. Values of the dissipation ratio are given in the following table for different values of τ, taking $\nu = 1.2 \times 10^{-6}\,\mathrm{m}^2/\mathrm{s}$ for water, and $g = 10\,\mathrm{m/s}^2$:

Wave period τ, s	0.1	1	10
Internal/boundary-layer dissipation	0.3	0.01	0.0003

5.4 Shallow-water waves

The distinguishing feature of long waves on shallow water is that the fluid motions are predominantly parallel with the solid bottom. Fluid particles rise and fall during the passage of a wave through a distance comparable with the surface wave amplitude ζ_o (Figure 5.1.4), but at the same time they perform very much larger horizontal excursions over distances $\sim \zeta_o\lambda/h$ ($\lambda \gg h$). This observation can be made the basis of the derivation of a simplified equation of motion for long waves on shallow water, based on the hypothesis that the rate of change of the vertical component of fluid momentum (normal to the bottom) can be neglected.

Indeed, let us integrate the z component of the linearised momentum equation from a general depth z to the surface at $z = \zeta$,

$$\int_z^\zeta \rho_o \frac{\partial v_z}{\partial t}\,dz = (p + \rho_o gz) - p_o - \rho_o g\zeta, \tag{5.4.1}$$

where p is the pressure in the fluid at z and $p_o \equiv p_o(x, y, t)$ is the applied surface pressure (if any). The maximum vertical acceleration of a fluid particle $\sim \omega^2\zeta$, so that the maximum value of the left-hand side $\sim \omega^2\zeta h$. This is small compared with the term in ζ on the right-hand side, provided that

$$\frac{\omega^2 h}{g} \ll 1.$$

For shallow-water waves we anticipate that the wavelength $\lambda \sim (2\pi/\omega)\sqrt{gh}$, so that this condition is equivalent to

$$\frac{h^2}{\lambda^2} \ll 1. \tag{5.4.2}$$

When this condition is satisfied Equation (5.4.1) shows that the perturbation pressure $\hat{p} \equiv -\rho_o \partial\varphi/\partial t$ is *independent of z* and given by

$$\hat{p} \equiv p + \rho_o g z = p_o + \rho_o g \zeta. \tag{5.4.3}$$

Thus the linearised momentum equation governing the horizontal motion becomes

$$\frac{\partial}{\partial t}(v_x, v_y) = -g\nabla\zeta - \frac{1}{\rho_o}\nabla p_o, \tag{5.4.4}$$

which is again independent of z, so that the horizontal flow is locally uniform.

5.4.1 Waves on water of variable depth

Let us consider a generalised geometry where the depth $h = h(x, y)$ is permitted to vary with position, but sufficiently slowly that the preceding approximation continues to be valid. The condition to be satisfied on the rigid surface $z = -h(x, y)$ is therefore $\mathbf{v} \cdot \nabla[z + h(x, y)] = 0$, i.e.,

$$v_z = -\mathbf{v} \cdot \nabla h \quad \text{at} \quad z = -h, \tag{5.4.5}$$

so that the approximation leading to (5.4.3) remains valid provided that $|\nabla h| < h/\lambda$ [in which case the maximum value of v_z is still $O(\partial\zeta/\partial t)$, and the change in h is small over a distance of one wavelength].

Next, integrate the continuity equation $\mathrm{div}\,\mathbf{v} = 0$ over the depth $-h(x, y) < z < 0$ then, because

$$\int_{-h}^{0} \frac{\partial v_z}{\partial z} dz = \frac{\partial\zeta}{\partial t} - (v_z)_{z=-h} \equiv \frac{\partial\zeta}{\partial t} + \mathbf{v} \cdot \nabla h$$

and, because v_x, v_y do not depend on z,

$$\int_{-h}^{0} \left(\frac{\partial v_x}{\partial x} + \frac{\partial v_y}{\partial y}\right) dz = \frac{\partial}{\partial x}(hv_x) + \frac{\partial}{\partial y}(hv_y) - \mathbf{v} \cdot \nabla h$$

we obtain the kinematic constraint

$$\frac{\partial}{\partial x}(hv_x) + \frac{\partial}{\partial y}(hv_y) + \frac{\partial\zeta}{\partial t} = 0. \tag{5.4.6}$$

The elimination of v_x and v_y between this and Equation (5.4.4) supplies the shallow-water wave equation for fluid of variable depth:

$$\frac{\partial^2\zeta}{\partial t^2} - \frac{\partial}{\partial x_j}\left(gh\frac{\partial\zeta}{\partial x_j}\right) = \frac{\partial}{\partial x_j}\left(\frac{h}{\rho_o}\frac{\partial p_o}{\partial x_j}\right), \tag{5.4.7}$$

where the repeated suffix j implies summation over the horizontal coordinates x, y.

5.4.2 Shallow-water Green's function

Green's function for the shallow-water wave equation for uniform depth h is the *outgoing* wave solution $G(\mathbf{x}, \mathbf{y}, t - \tau)$ of

$$\left(\frac{1}{c^2}\frac{\partial^2}{\partial t^2} - \nabla^2\right) G = \delta(\mathbf{x} - \mathbf{y})\delta(t - \tau), \quad c = \sqrt{gh}, \tag{5.4.8}$$

where $\nabla^2 = \partial^2/\partial x^2 + \partial^2/\partial y^2$ and $\mathbf{x}$, $\mathbf{y}$ are two-dimensional coordinates in the xy plane.

Let us determine G by the method of Fourier transforms. The space–time Fourier transform $f(\mathbf{k}, \omega)$ of a function $f(\mathbf{x}, t)$ satisfies the reciprocal relations

$$f(\mathbf{k}, \omega) = \frac{1}{(2\pi)^3} \iint_{-\infty}^{\infty} f(\mathbf{x}, t) e^{-i(\mathbf{k}\cdot\mathbf{x} - \omega t)} d^2x\, dt,$$

$$f(\mathbf{x}, t) = \iint_{-\infty}^{\infty} f(\mathbf{k}, \omega) e^{i(\mathbf{k}\cdot\mathbf{x} - \omega t)} d^2k\, d\omega. \tag{5.4.9}$$

The function $f(\mathbf{x}, t)$ and its transform $f(\mathbf{k}, \omega)$ are distinguished by their arguments, $\mathbf{x}$ and t for the space–time function, and the *wavenumber* $\mathbf{k}$ and frequency ω for the Fourier transform. This convention conveniently avoids an awkward proliferation of symbolic notations.

Take the Fourier transform of Equation (5.4.8) and solve the resulting algebraic equation to obtain

$$G(\mathbf{k}, \omega) = \frac{e^{-i(\mathbf{k}\cdot\mathbf{y} - \omega\tau)}}{(2\pi)^3(k^2 - \omega^2/c^2)}.$$

The inverse transform gives

$$G(\mathbf{x}, \mathbf{y}, t - \tau) = \frac{1}{(2\pi)^3} \iint_{-\infty}^{\infty} \frac{e^{-i\{\mathbf{k}\cdot(\mathbf{x}-\mathbf{y}) - \omega(t-\tau)\}}}{[k^2 - (\omega + i0)^2/c^2]} d^2k\, d\omega. \tag{5.4.10}$$

The notation $\omega + i0$ is introduced to indicate that the path of integration in the complex ω plane should pass *above* the poles at $\omega = \pm ck$. In that case the integration contour can be shifted to $\omega \sim +i\infty$ without crossing any singularities when $t - \tau < 0$, where the integrand becomes exponentially small and therefore $G(\mathbf{x}, \mathbf{y}, t - \tau) = 0$. Thus the fluid is undisturbed for $t < \tau$ and any disturbances at later times must satisfy the outgoing wave condition (the *causality* condition).

When $t > \tau$ the contour in the ω plane is shifted to $\omega \sim -i\infty$, thereby capturing residue contributions from the two poles. By transforming to polar coordinates in the wavenumber plane, so that $d^2k = k\, dk\, d\theta$, and performing the integration with respect to k, we find

$$G(\mathbf{x}, \mathbf{y}, t - \tau) = -\frac{c\,H(t - \tau)}{(2\pi)^2|\mathbf{x} - \mathbf{y}|} \oint_0^{2\pi} \frac{d\theta}{[\cos\theta - c(t - \tau)/|\mathbf{x} - \mathbf{y}|]}$$

$$= \frac{H[t - \tau - |\mathbf{x} - \mathbf{y}|/c]}{2\pi\sqrt{(t - \tau)^2 - |\mathbf{x} - \mathbf{y}|^2/c^2}}, \tag{5.4.11}$$

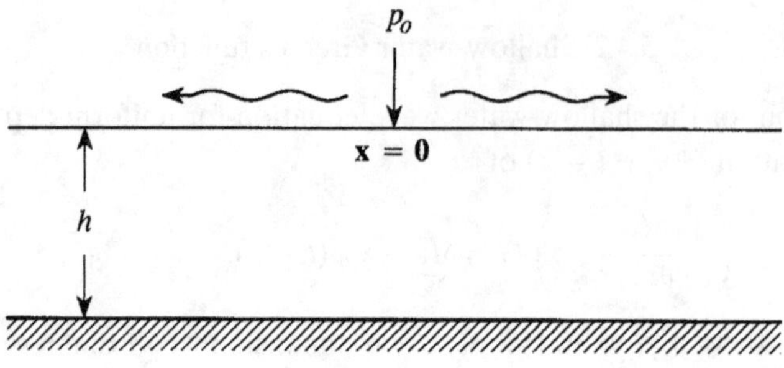

Figure 5.4.1

because the principal value integral in the first line vanishes identically when

$$c(t - \tau)/|\mathbf{x} - \mathbf{y}| < 1.$$

Green's function (5.4.11) is a circular-symmetric wave spreading out from the source at $\mathbf{y}$ at the shallow-water wave phase speed $c = \sqrt{gh}$. The wavefront at $|\mathbf{x} - \mathbf{y}| = c(t - \tau)$ exhibits an integrable, inverse-square-root singularity whose strength decreases like $1/\sqrt{|\mathbf{x} - \mathbf{y}|}$, and the motion decays slowly behind the wavefront like the inverse square root of the distance from the front.

5.4.3 Waves generated by a localised pressure rise

Green's function is now used to solve the elementary problem of Figure 5.4.1 where the surface waves are generated by a surface pressure rise $p_o(x, y, t)$ concentrated in the neighbourhood of the origin and represented by

$$p_o(x, y, t) = P_o \mathcal{A} \mathcal{H}(t) \delta(x) \delta(y), \quad \text{where} \quad \mathcal{H}(t) = \frac{1}{2} + \frac{1}{\pi} \tan^{-1}\left(\frac{t}{\tau_o}\right).$$

In this formula P_o is a constant with the dimensions of pressure, $\mathcal{A}$ is an area over which the pressure P_o may be regarded as applied, and $\mathcal{H}(t)$ is a smooth unit 'step function' that effectively vanishes for $t < -\tau_o$ and rises to a constant value of 1 when $t > \tau_o$.

For uniform depth it is convenient to recast Equation (5.4.7) in the form

$$\left(\frac{1}{c^2}\frac{\partial^2}{\partial t^2} - \nabla^2\right)\left(\zeta + \frac{p_o}{\rho_o g}\right) = \frac{1}{c^2 \rho_o g}\frac{\partial^2 p_o}{\partial t^2}, \quad c = \sqrt{gh}. \tag{5.4.12}$$

Because ζ and p_o both vanish at large distances from the pressure source the surface waves are still determined by the outgoing wave solution of (5.4.12) that, by use of the Green's function (5.4.11), is given by

$$\zeta + \frac{p_o}{\rho_o g} = \frac{1}{c^2 \rho_o g} \iint_{-\infty}^{\infty} \frac{\partial^2 p_o}{\partial \tau^2}(\mathbf{y}, \tau) G(\mathbf{x}, \mathbf{y}, t - \tau)\, d^2y d\tau.$$

Integrating by parts once with respect to τ and noting that there is no contribution from $\tau = \pm\infty$, we find

$$\zeta(\mathbf{x}, t) = -\frac{p_o}{\rho_o g} + \frac{P_o \mathcal{A}\tau_o}{2\pi^2 \rho_o g c^2} \frac{\partial}{\partial t} \int_{-\infty}^{[t]} \frac{d\tau}{(\tau^2 + \tau_o^2)\sqrt{(t - \tau)^2 - |\mathbf{x}|^2/c^2}}, \qquad (5.4.13)$$

where $\quad [t] = t - \dfrac{|\mathbf{x}|}{c}\quad$ is the *retarded time.*

The first term on the right-hand side of (5.4.13) is the local surface depression beneath the pressure source and is not part of the radiation field. The latter is the symmetric disturbance represented by the integral, which ranges over all $\tau < [t] = t - |\mathbf{x}|/c$ and therefore includes all radiating effects of the source that occurred at time t diminished by the minimum time of travel $|\mathbf{x}|/c$ of waves from the source to an observer at $\mathbf{x}$.

However, the maximum contribution at time t to the surface elevations at distance $|\mathbf{x}|$ occur near $t - \tau \sim |\mathbf{x}|/c$, and when $|\mathbf{x}|$ is large, we can evaluate the integral in simple analytic form by confining attention to this neighbourhood of retarded times τ by making the approximation

$$\frac{1}{\sqrt{(t - \tau)^2 - |\mathbf{x}|^2/c^2}} \approx \sqrt{\frac{c}{2|\mathbf{x}|}} \frac{1}{\sqrt{t - |\mathbf{x}|/c - \tau}}. \qquad (5.4.14)$$

Then

$$\zeta \sim \frac{1}{4\pi} \frac{P_o}{\rho_o g} \frac{\mathcal{A}}{(c\tau_o)^2} \left(\frac{c\tau_o}{|\mathbf{x}|}\right)^{\frac{1}{2}} \Psi'\left(\frac{[t]}{\tau_o}\right), \quad |\mathbf{x}| \to \infty,$$

where the prime denotes differentiation with respect to the argument, and

$$\Psi(x) = \left[\frac{\sqrt{1 + x^2} + x}{(1 + x^2)}\right]^{\frac{1}{2}}.$$

The surface wave at distance $|\mathbf{x}|$ therefore has the profile illustrated in Figure 5.4.2. The surface elevation grows to a large positive maximum over a time interval $\sim 5\tau_o$. The elevation rapidly falls to the rear of this maximum and forms an extensive *negative* tail, wherein the maximum depression is no more than about a third of the positive peak, yet the integrated surface displacement $\int \zeta([t])\,d[t]$ across the wave profile vanishes identically.

EXAMPLE 1. SCATTERING BY A CHANGE IN DEPTH A plane shallow-water wave $\zeta = \zeta_o e^{i(k_1 x - \omega t)}$ is normally incident from $x < 0$ upon a step change in depth from h_1 in $x < 0$ to h_2 in $x > 0$ (Figure 5.4.3), where $k_1 = \omega/\sqrt{gh_1}$. Transmitted and reflected waves are produced at the step, and we put

$$\zeta = \begin{cases} \zeta_o\left(e^{ik_1 x} + Re^{-ik_1 x}\right)e^{-i\omega t}, & x < 0, \quad k_1 = \omega/\sqrt{gh_1} \\[2mm] \zeta_o T e^{i(k_2 x - \omega t)}, & x > 0, \quad k_2 = \omega/\sqrt{gh_2} \end{cases},$$

where R and T are suitable reflection and transmission coefficients.

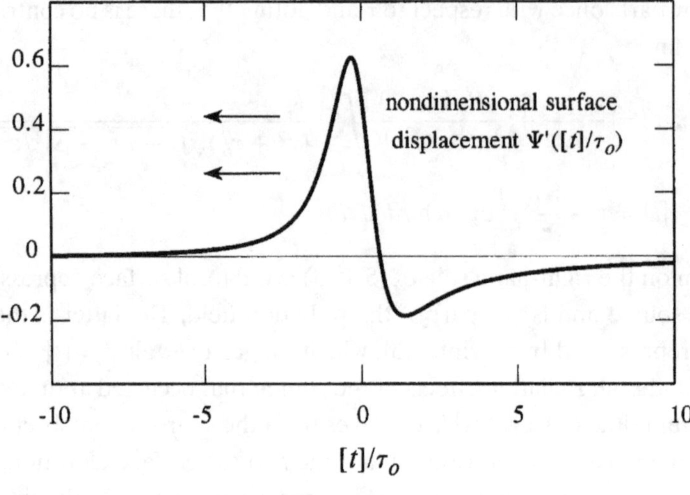

Figure 5.4.2

The shallow-water approximation fails across the step within an interval of length ℓ, say, that is small compared with the wavelength. In that case the inertia of this region and the net volume flow into it are both negligible, and this permits conditions at $x = \pm\ell/2$ to be matched. Thus, in the integral of x-momentum equation (5.4.4) (with $p_o = 0$) over $-\ell/2 < x < \ell/2$, the left-hand side $\sim \omega\ell v_x$, where v_x varies between values $\sim \omega\zeta/kh$ at the ends of the interval. Therefore $\omega\ell v_x \sim O(k\ell g\zeta) \ll g\zeta$, and to a first approximation it may be assumed that there is no jump in the excess pressure $\rho_o g\zeta$ across the step. Similarly, integration of continuity equation (5.4.6) across the transition region and application of the relation $v_x \sim gk\zeta/\omega$ at either end implies that the jump in the volume flux hv_x across the step can be neglected.

These conditions of continuity of pressure and volume flux can be applied at $x = 0$ because $k_1\ell \sim k_2\ell \ll 1$ and yield the equations

$$1 + R = T, \quad k_1 h_1(1 - R) = k_2 h_2 T,$$

so that

$$R = \frac{1 - \sqrt{h_2/h_1}}{1 + \sqrt{h_2/h_1}}, \quad T = \frac{2}{1 + \sqrt{h_2/h_1}}.$$

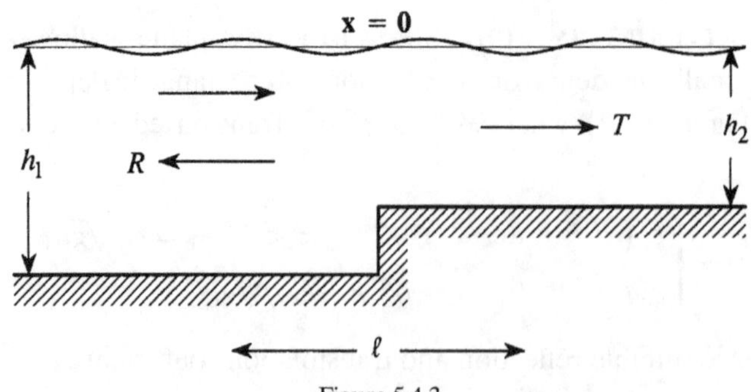

Figure 5.4.3

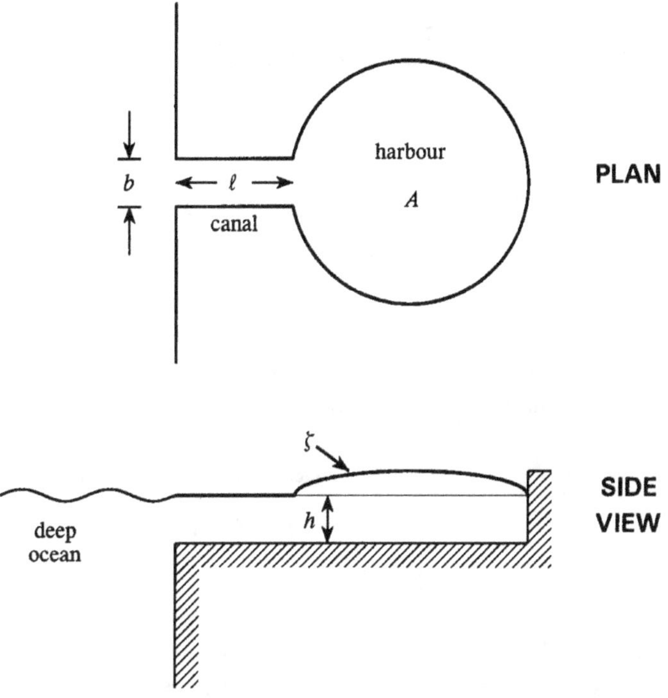

Figure 5.4.4

Therefore, when the incident wave approaches a step up to shallower water ($h_2 < h_1$), as when approaching a shore, the transmitted wave amplitude is increased. These results also show that wave energy is conserved at the step, because the net flux of energy in the x direction (propagated at the local phase speed $\sqrt{gh}$) is preserved across the step, because

$$\sqrt{gh_1}(1 - |R|^2) = \sqrt{gh_2}|T|^2.$$

EXAMPLE 2. RESONANT OSCILLATIONS IN A HARBOUR (Miles & Lee 1975) Large-amplitude oscillations in the water level in the idealized harbour of Figure 5.4.4 are produced by low-frequency surface waves incident from the open ocean. The harbour communicates with the ocean by means of a short canal of length ℓ and width b, and the undisturbed depth of the water in the canal and harbour is h.

At very low frequencies the water level rises by an approximately uniform amount $\zeta = \zeta(t)$ across the surface area A of the water in the harbour, so that the volume flux through the canal into the harbour is $A d\zeta/dt$. The pressure rise at the inner end of the canal is therefore $\rho_0 g\zeta$, and the mean *inward* velocity in the canal is $(A/hb)d\zeta/dt$. By contrast the rise in water level just outside the ocean end of the canal is negligible. Thus, when frictional resistance along the sides and bottom of the canal is ignored, the equation of motion of the mass $\rho_0 hb\ell$ of water oscillating back and forth in the canal is

$$(\rho_0 hb\ell)\frac{d}{dt}\left(\frac{A}{hb}\frac{d\zeta}{dt}\right) = -(hb)\rho_0 g\zeta,$$

$$\therefore \quad \frac{d^2\zeta}{dt^2} + \Omega^2\zeta = 0,$$

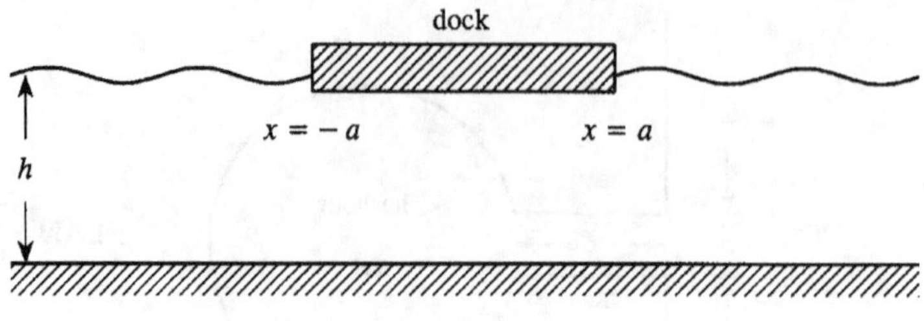

Figure 5.4.5

where $\Omega = \sqrt{ghb/\ell A}$ is the frequency of the lowest-order resonance mode of the harbour. The frequency formula can also be written as

$$\Omega = c\sqrt{\frac{S}{\ell V}}, \quad \text{where} \quad \begin{cases} S = \text{cross-sectional area of the canal} \\ V = \text{volume of water in the harbour} \\ c = \sqrt{gh} \end{cases},$$

which is the analogue of the formula for *acoustic* resonant oscillations in a *Helmholtz resonator* whose interior volume is V and where S, ℓ are respectively the cross-sectional area and length of the resonator 'neck' (see §6.6.4).

EXAMPLE 3. SCATTERING BY A DOCK (John 1949) A two-dimensional rigid dock of width $2a$ is firmly anchored in water of uniform depth h with its flat underside just below the mean water level (Figure 5.4.5). The sides of the dock are at $x = \pm a$, and a *shallow-water* plane wave

$$\zeta = \zeta_o e^{i(kx-\omega t)}, \quad k = \omega/\sqrt{gh}$$

is incident from $x < -a$.

Introduce reflected and transmitted waves respectively in $x < -a$ and $x > a$ as in Example 1, and write for the velocity potential

$$\varphi = \begin{cases} -\dfrac{ig\zeta_o}{\omega}\left(e^{ikx} + Re^{-ikx}\right)e^{-i\omega t}, & x < -a \\[4mm] -\dfrac{ig\zeta_o}{\omega}Te^{i(kx-\omega t)}, & x > a \end{cases},$$

where R and T are reflection and transmission coefficients.

There is no vertical motion beneath the dock, where continuity implies that $v_x = \partial\varphi/\partial x = \text{constant}\times e^{-i\omega t}$, i.e.,

$$\varphi = -\frac{ig\zeta_o}{\omega}(\alpha x + \beta)e^{-i\omega t}, \quad |x| < a, \quad (\alpha, \beta = \text{constant}).$$

The volume flux $hv_x = h\partial\varphi/\partial x$ and excess pressure $\hat{p} = -\rho_o\partial\varphi/\partial t$ are continuous at the ends $x = \pm a$ of the dock, and $R,\ T,\ \alpha,\ \beta$ accordingly satisfy

$$e^{-ika} + Re^{ika} = -\alpha a + \beta,$$

$$ik\left(e^{-ika} - Re^{ika}\right) = \alpha,$$

$$Te^{ika} = \alpha a + \beta,$$

$$ikTe^{ika} = \alpha.$$

(5.4.15)

Hence

$$R = \frac{-ika e^{-2ika}}{1 - ika}, \quad T = \frac{e^{-2ika}}{1 - ika},$$

$$\alpha = \frac{ike^{-ika}}{1 - ika} \qquad \beta = e^{-ika}.$$

These results show that $|R|^2 + |T|^2 = 1$, and therefore that wave energy is conserved during the interaction with the dock. The relative magnitudes of the transmitted and reflected wave energies are illustrated in the table for different values of the ratio $2a/\lambda$ ($\lambda = 2\pi/k$):

Dock width/λ	$|T|^2$	$|R|^2$
0.5	0.288	0.712
1	0.092	0.908
2	0.025	0.975

The pressure distribution on the underside of the dock is

$$p = -\rho_o\frac{\partial\varphi}{\partial t} \equiv \rho_o g\zeta_o(\alpha x + \beta), \quad |x| < a.$$

Therefore the

$$\text{vertical force on unit span of dock} = \int_{-a}^{a} p\,dx = 2a\rho_o g\zeta_o e^{-i(ka+\omega t)}.$$

The peak force is just equal to the weight of water occupying a volume equal to the wave amplitude $\zeta_o \times$ (surface area of dock).

5.4.4 Waves approaching a sloping beach

Figure 5.4.6 shows a simple example involving waves on water of continuously variable depth $h(x)$. A beach starting at the origin $x = 0$, $z = 0$ slopes down uniformly into the sea, attaining a uniform depth h_o at $x = \ell$. A plane shallow-water wave $\zeta = \zeta_o e^{-i[k(x-\ell)-\omega t]}$ (of wavelength $\lambda = 2\pi/k \gg h_o$, where $k = \omega/\sqrt{gh_o}$) approaches the beach from the open sea. A wave of amplitude $R\zeta_o$ is reflected back into $x > \ell$, such that

$$\zeta = \zeta_o\left[e^{-ik(x-\ell)} + Re^{ik(x-\ell)}\right]e^{-i\omega t}, \quad x > \ell. \tag{5.4.16}$$

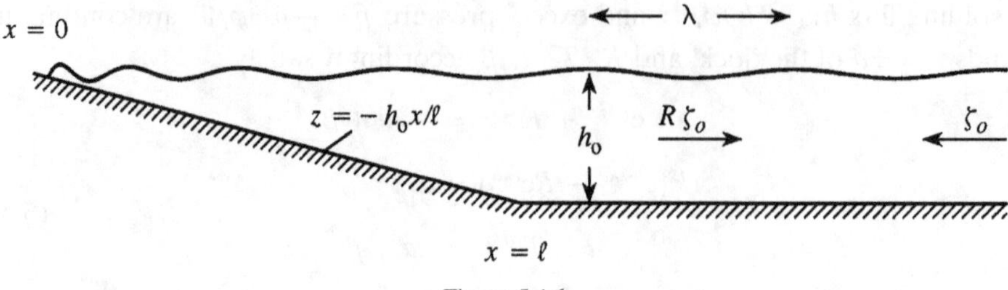

Figure 5.4.6

The behaviour is more complicated in the sloping-beach domain, where the depth $h(x) = h_o x/\ell$ and the waves are governed by the following reduced form of Equation (5.4.7):

$$\frac{\partial}{\partial x}\left(x\frac{\partial \zeta}{\partial x}\right) + \left(\frac{\omega^2 \ell}{gh_o}\right)\zeta = 0, \quad \text{for } 0 < x < \ell.$$

By setting $x = 2\hat{x}^2$ [so that $\partial/\partial x = (1/4\hat{x})\partial/\partial\hat{x}$], we transform this equation into *Bessel's equation* of the order of 0:

$$\frac{\partial^2 \zeta}{\partial \hat{x}^2} + \frac{1}{\hat{x}}\frac{\partial \zeta}{\partial \hat{x}} + \left(\frac{8\omega^2 \ell}{gh_o}\right)\zeta = 0,$$

which has the general solution

$$\zeta = \left[A\mathrm{J}_0\left(\hat{x}\sqrt{\frac{8\omega^2 \ell}{gh_o}}\right) + B\mathrm{Y}_0\left(\hat{x}\sqrt{\frac{8\omega^2 \ell}{gh_o}}\right)\right]\mathrm{e}^{-i\omega t},$$

where J_0, Y_0 are respectively Bessel functions of the first and second kind and A, B are arbitrary constants.

We must take $B = 0$, because $\mathrm{Y}_0(\hat{x}\sqrt{8\omega^2\ell/gh_o})$ is unbounded as $\hat{x} \to 0$. Therefore, putting $\sqrt{\omega^2/gh_o} = k$, we have

$$\zeta = A\mathrm{J}_0\left(2k\sqrt{x\ell}\right)\mathrm{e}^{-i\omega t}, \quad 0 < x < \ell. \tag{5.4.17}$$

We obtain the values of A and R by imposing the conditions of continuity of pressure and volume flux at $x = \ell$, i.e., by requiring ζ and $\partial\zeta/\partial x$ to be continuums. Using (5.4.16), (5.4.17) and recalling that $d\mathrm{J}_0(x)/dx = -\mathrm{J}_1(x)$, we find

$$1 + R = A\mathrm{J}_0(2k\ell), \quad \text{and} \quad -1 + R = i\,A\mathrm{J}_1(2k\ell),$$

$$\therefore \quad A = \frac{2}{\mathrm{J}_0(2k\ell) - i\mathrm{J}_1(2k\ell)}, \qquad R = \frac{\mathrm{J}_0(2k\ell) + i\mathrm{J}_1(2k\ell)}{\mathrm{J}_0(2k\ell) - i\mathrm{J}_1(2k\ell)}. \tag{5.4.18}$$

This simple model of reflection implies that wave energy is conserved, because $|R|^2 = 1$. The coefficient A is finite for all real values of the wavenumber k, so that the wave amplitude on the beach remains finite, although it increases significantly as $x \to 0$.

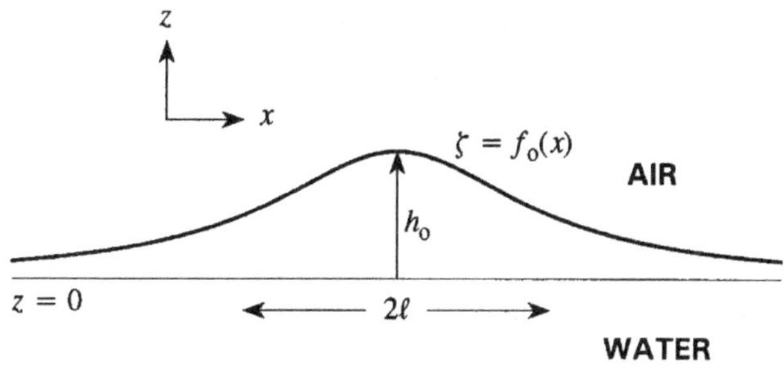

Figure 5.5.1

As an illustration, take

$$\ell = 500\,\text{m}, \quad h_o = 10\,\text{m}, \quad g \approx 10\,\text{m/s}^2, \quad \sqrt{gh_o} = 10\,\text{m/s}, \quad kh_o = 0.3.$$

Then

$$\lambda \approx 200\,\text{m}, \quad \text{wave period} \approx 20\,\text{s}, \quad \frac{\text{shoreline amplitude}}{\text{open-sea amplitude}} \approx \frac{A}{2} \sim \sqrt{\pi k\ell} \approx 7.$$

The wave profile for this case is plotted in Figure 5.4.6 (defined by Re ζ at a fixed value of t, the vertical scale for the beach profile being greatly exaggerated). As the shoreline is approached, there is a dramatic reduction in wavelength accompanying the increase in wave amplitude. In practice, of course, the linearised approximation would fail long before the waves reach the shoreline and there would be significant dissipation through the wave 'breaking'.

5.5 Method of stationary phase

The mathematical treatment of 'dispersive' waves depends on specialised methods involving Fourier transforms and is closely related to the notion of *group velocity*. It provides a simple description of how, for example, an initially coherent free surface wave profile evolves with position and time because of the different phase velocities ($c = \omega/k$) of the different wavenumber components in the Fourier integral representation of the initial profile. To facilitate the solution of such problems by Fourier transforms it is convenient to adapt the formal method of §2.8 by introducing the Heaviside step function H(t) in order to transform the initial-value problem into one defined over all times $-\infty < t < \infty$. The procedure will first be described in detail.

5.5.1 Formulation of initial-value dispersive-wave problems

Suppose an elevated section of the free surface of water of infinite depth is released from rest at $t = 0$ (Figure 5.5.1). To fix ideas let

$$\left. \begin{array}{l} \zeta = f_o(x) \equiv h_o/[1 + (x/\ell)^2] \\[2mm] \partial\zeta/\partial t = 0 \end{array} \right\} \quad \text{for} \quad -\infty < x < \infty, \ t = 0. \qquad (5.5.1)$$

The subsequent motion is described by a velocity potential φ that satisfies Laplace's equation $\nabla^2\varphi \equiv (\partial^2/\partial x^2 + \partial^2/\partial z^2)\varphi = 0$, subject to the following form of surface condition (5.1.5):

$$\frac{\partial^2\varphi}{\partial t^2} + g\frac{\partial\varphi}{\partial z} = 0, \quad \text{at } z = 0. \tag{5.5.2}$$

Multiplying by $H(t)$ we find, using the free surface relation $\partial\varphi/\partial t = -g\zeta$,

$$\frac{\partial^2(\mathrm{H}\varphi)}{\partial t^2} + g\frac{\partial(\mathrm{H}\varphi)}{\partial z} = \delta(t)\frac{\partial\varphi}{\partial t} + \frac{\partial}{\partial t}\Big(\delta(t)\varphi\Big)$$

$$\equiv -g\delta(t)f_o(x) + \frac{\partial}{\partial t}\Big(\delta(t)\varphi_o\Big), \quad \text{at } z = 0, \tag{5.5.3}$$

where φ_o is the velocity potential at $t = 0$ and $\mathrm{H} \equiv \mathrm{H}(t)$. The quantities φ_o and $f_o(x)$ respectively define the dynamic and kinematic states of the water at $t = 0$. The water is initially at rest and therefore $\varphi_o = \text{constant}$.

If Laplace's equation is similarly transformed, we now see that the original initial-value problem can be replaced with

$$\left.\begin{array}{l} \nabla^2(\mathrm{H}\varphi) = 0, \quad z < 0 \\[2mm] \dfrac{\partial^2(\mathrm{H}\varphi)}{\partial t^2} + g\dfrac{\partial(\mathrm{H}\varphi)}{\partial z} = -g\delta(t)f_o(x) + \dfrac{\partial}{\partial t}\big[\delta(t)\varphi_o\big], \quad z = 0 \end{array}\right\} -\infty < t < \infty. \tag{5.5.4}$$

where $\mathrm{H}\varphi = \varphi$ when $t > 0$.

We solve this by writing the solution of Laplace's equation as the Fourier integral

$$\mathrm{H}\varphi = \int_{-\infty}^{\infty} \mathcal{A}(k,\omega)e^{i(kx-\omega t)+z|k|}\, dk d\omega,$$

where $\mathcal{A}(k,\omega)$ is the Fourier transform of $\mathrm{H}\varphi$ at $z = 0$ [defined as in (5.4.9)]. Substitute into the second of (5.5.4) and take the Fourier transform to obtain

$$\mathcal{A}(k,\omega)(\omega^2 - g|k|) = \frac{i\varphi_o}{2\pi}\omega\delta(k) + \frac{g}{2\pi}f_o(k),$$

where

$$f_o(k) = \frac{1}{2\pi}\int_{-\infty}^{\infty} f_o(x)e^{-ikx}dx = \frac{h_o\ell}{2}e^{-|k|\ell}. \tag{5.5.5}$$

Hence, because $f_o(k)$ is an even function,

$$\mathrm{H}\varphi = \frac{1}{2\pi}\int_{-\infty}^{\infty} \frac{[i\varphi_o\omega\delta(k) + gf_o(k)]}{(\omega+i0)^2 - \Omega^2(k)}e^{i(kx-\omega t)+z|k|}\, dk d\omega$$

$$= \mathrm{H}\varphi_o + \mathrm{H}g\int_{-\infty}^{\infty} \frac{f_o(k)}{\Omega(k)}\sin[kx - \Omega(k)t]\, e^{z|k|}\, dk,$$

where $\Omega(k) = \sqrt{g|k|}$, and in the first line the notation $\omega + i0$ indicates that the integration path in the ω plane passes *above* the poles at $\omega = \pm\Omega$ (ensuring that the integral vanishes for $t < 0$; see §5.4.2). Therefore, using $\zeta = -\partial\varphi/g\partial t$ at $z = 0$, we obtain

$$\left.\begin{aligned}
\varphi &= \varphi_0 + g \int_{-\infty}^{\infty} \frac{f_o(k)}{\Omega(k)} \sin\left[kx - \Omega(k)t\right] e^{z|k|} \, dk \\
\zeta &= \int_{-\infty}^{\infty} f_o(k) \cos\left[kx - \Omega(k)t\right] \, dk
\end{aligned}\right\} \quad \text{for } t > 0. \qquad (5.5.6)$$

The constant initial value φ_o of the velocity potential is evidently of no physical significance; it could have been set equal to zero at the outset.

5.5.2 Evaluation of Fourier integrals by the method of stationary phase

It is only in very rare circumstances that Fourier integrals of the type (5.5.6) can be evaluated in closed form. However, the main features of the solution can be obtained at large times by Kelvin's method of stationary phase (Havelock 1914; Kelvin 1887), which asserts that in a dispersive medium the component waves within any interval dk of the integrand are mutually destructive, except when they have the same phase and can reinforce one another. This reinforcement occurs when the argument of the circular functions in (5.5.6) is stationary.

We shall describe the details of the method for the surface displacement

$$\zeta = \int_{-\infty}^{\infty} f_o(k) \cos\left[kx - \Omega(k)t\right] \, dk.$$

The surface profile is a superposition of an infinite number of plane harmonic waves of varying wavelengths and amplitudes. At $t = 0$ the waves reinforce one another near the origin to form the initial surface elevation of (5.5.1); phase differences at large values of x cause destructive interference and give zero surface displacement. The integral determines the surface profile at a later time t after each of the harmonic components has translated a distance $t\Omega(k)/k$ at its phase speed $\Omega(k)/k$. This destroys the initial phase relation between the waves, and at most places the displacement is then small because of wave interference. However, there obviously exists a range of positions and times at which the surface displacement is large, and these occur where a large number of the component waves have approximately the same phase and are mutually reinforcing.

When t is large the surface disturbance has spread to large distances x, so that x as well as t is usually large at places where the motion is significant. Destructive interference is caused by the cosine factor in the integrand, which therefore oscillates increasingly rapidly with k compared with the variations of $f_o(k)$. Constructive reinforcement will occur, however, for wavenumbers clustered around a central wavenumber $k_o = k_o(x, t)$ at which the phase $kx - \Omega(k)t$ is stationary. The slow variation of the cosine factor near this point is illustrated in Figure 5.5.2; the contributions to the integral from the waves in the neighbourhood of k_o have the same sign and therefore combine to produce a finite net displacement of the surface.

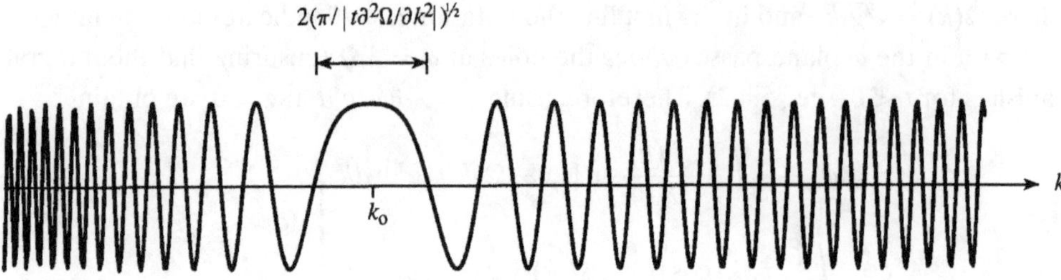

Figure 5.5.2

For given values of x and t the wavenumber $k = k_o$ occurs where the group velocity satisfies

$$\frac{\partial\Omega}{\partial k}(k) = \frac{x}{t}. \tag{5.5.7}$$

To determine the dominating contribution from the waves centred on k_o the phase is expanded to second order in $\xi = k - k_o$:

$$kx - \Omega(k)t \approx [k_ox - \Omega(k_o)t] + \xi\left[x - \frac{\partial\Omega}{\partial k}t\right] - \frac{\xi^2}{2}t\frac{\partial^2\Omega}{\partial k^2}$$

$$\equiv [k_ox - \Omega(k_o)t] - \frac{\xi^2}{2}t\frac{\partial^2\Omega}{\partial k^2},$$

where the derivatives are evaluated at $k = k_o$. The function $f_o(k)$ in the integrand can be replaced with $f_o(k_o)$, provided the variation of $f_o(k)$ is negligible within the interval of width

$$\sim 2\left(\frac{\pi}{t|\partial^2\Omega/\partial k^2|}\right)^{\frac{1}{2}}$$

where the phase varies slowly. The integral accordingly becomes

$$\zeta \approx f_o(k_o)\mathrm{Re}\left\{e^{i[k_ox - \Omega(k_o)t]}\int_{-\infty}^{\infty}e^{-i\frac{t}{2}\frac{\partial^2\Omega}{\partial k^2}\xi^2}\,d\xi\right\},$$

where the limits of integration have been extended to $\xi = \pm\infty$ because of cancellation of contributions from outside the region of slowly varying phase. The remaining integral is calculated from the formula

$$\int_{-\infty}^{\infty}e^{i\alpha\xi^2}\,d\xi = \sqrt{\frac{\pi}{|\alpha|}}\,e^{\frac{i\pi}{4}\mathrm{sgn}(\alpha)}, \quad (\alpha \text{ real}). \tag{5.5.8}$$

Hence, the stationary phase approximation to the surface displacement is

$$\zeta \approx f_o(k)\sqrt{\frac{2\pi}{t|\partial^2\Omega/\partial k^2|}}\cos\left[kx - \Omega(k)t - \frac{\pi}{4}\mathrm{sgn}\left(\frac{\partial^2\Omega}{\partial k^2}\right)\right]. \tag{5.5.9}$$

The suffix 'o' on the wavenumber k has been discarded in this final formula, it being understood that the value determined by the group velocity condition (5.5.7) is to be used. In general (in applications to waves in more general dispersive media), for any given value of the *group velocity* x/t there can be several real wavenumbers satisfying

Equation (5.5.7). We then obtain the net field by combining the separate contributions of the type (5.5.9) from each of the stationary points.

Group velocity equation (5.5.7) determines the wavenumber in (5.5.9) as a function of x/t. Thus, at any particular position x and time t, stationary phase approximation (5.5.9) constitutes a local representation of the surface motion as a harmonic wave train of wavelength $2\pi/k$, frequency $\Omega(k)$ and *slowly varying*

$$\text{amplitude} = \frac{\mathcal{F}(x/t)}{\sqrt{t}}, \quad \text{where} \quad \mathcal{F}(x/t) = f_o(k)\sqrt{\frac{2\pi}{|\partial^2\Omega/\partial k^2|}}. \tag{5.5.10}$$

For an observer starting at the origin at $t = 0$ and travelling at the group velocity $x/t = \partial\Omega/\partial k$, the amplitude factor $\mathcal{F}(x/t)$ remains constant and the surface disturbance has the form of a simple wave of length $2\pi/k$, with 'wave crests' [maxima of the cosine in (5.5.9)] moving forward at the phase velocity Ω/k. This is precisely the behaviour discussed in §5.2.3 for the succession of wave groups of Figure 5.2.4 formed by combination (5.2.7) of *two* harmonic surface waves with neighbouring wavenumbers. Any prominent feature of the function $f_o(k)/\sqrt{|\partial^2\Omega/\partial k^2|}$ as a function of k will determine a corresponding group velocity $x/t = \partial\Omega/\partial k$ and a corresponding prominence in the amplitude function $\mathcal{F}(x/t)$. This prominence will propagate over the surface as an identifiable wave group (or 'wave packet'). In particular, the dominant surface motion at later times is determined by the maximum absolute value of $\mathcal{F}(x/t)$ and in cases in which $\mathcal{F}(x/t)$ has multiple well-defined maxima, the stationary phase formula represents the surface motion as a succession of wave groups with amplitudes determined by the maxima.

Equation (5.5.9) also shows that the overall amplitude of each wave group decreases like $1/\sqrt{t}$ at large time. This is because, as time progresses, the energy within the group is spread over an increasingly larger stretch of the free surface or equivalently, as indicated in Figure 5.5.2, over a decreasing range ($\sim 2\sqrt{\pi/t|\partial^2\Omega/\partial k^2|}$) of wavenumbers. Indeed, if at some initial time the wave group is spread over an interval δk of wavenumbers and over a distance ℓ_o on the free surface, then the length of the free surface occupied by the same waves at a later large time t is equal to $\ell_o + t(\text{change in group velocity}$ across the group$) = \ell_o + \delta k\,t\,\partial^2\Omega/\partial k^2 \approx \delta k\,t\,\partial^2\Omega/\partial k^2$. Because the wave energy ($\propto |\zeta|^2$) is conserved, the wave amplitudes must therefore decrease like $1/\sqrt{t\partial^2\Omega/\partial k^2}$.

5.5.3 Numerical results for the surface displacement

Let us now evaluate explicitly stationary phase formula (5.5.9) for the initial-value problem of Figure 5.5.1. The wavefield is obviously symmetric in x, which may therefore be taken to be positive. The value of k at the stationary point is determined by group velocity formula (5.5.7), with $\Omega = \sqrt{g|k|}$ and $\partial\Omega/\partial k = \frac{1}{2}\text{sgn}(k)\sqrt{g/|k|}$. For x, $t > 0$ we have

$$k = \frac{gt^2}{4x^2}, \quad \Omega = \frac{gt}{2x}, \quad \frac{\partial\Omega}{\partial k} = \frac{x}{t}, \quad \frac{\partial^2\Omega}{\partial k^2} = -\frac{\sqrt{g}}{4}\left(\frac{4x^2}{gt^2}\right)^{\frac{3}{2}}. \tag{5.5.11}$$

Hence, (5.5.9) becomes

$$\frac{\zeta}{h_o} = \frac{\sqrt{\pi}}{2} \left(\frac{\ell}{gt^2}\right)^{\frac{1}{4}} \left(\frac{gt^2\ell}{x^2}\right)^{\frac{3}{4}} e^{-\frac{gt^2\ell}{4x^2}} \cos\left(\frac{gt^2}{4x} - \frac{\pi}{4}\right). \qquad (5.5.12)$$

The amplitude function of (5.5.10) is

$$\mathcal{F}(x/t) = \text{constant} \times \left(\frac{gt^2\ell}{x^2}\right)^{\frac{3}{4}} e^{-\frac{gt^2\ell}{4x^2}} \equiv \text{constant} \times \left(k^{\frac{3}{4}} e^{-k\ell}\right),$$

which has a single maximum at

$$\frac{x}{\ell} = \sqrt{\frac{gt^2/\ell}{3}} \approx 0.577\sqrt{gt^2/\ell}, \quad \text{or} \quad k\ell = \frac{3}{4},$$

which travels out at the group velocity $x/t = 0.577\sqrt{g\ell}$. $\mathcal{F}(x/t)$ is ultimately exponentially small at smaller values of x/ℓ, and becomes less than 10^{-3} of its maximum value when $x/\ell = 0.16\sqrt{gt^2/\ell}$.

According to (5.5.11) the waves observed by an observer fixed at x progressively increase in wavenumber with increasing time, long waves with the larger group velocities are the first to arrive. For fixed t long waves are found at large x and the wavelength progressively decreases as x moves towards the origin of the disturbance; the surface motion is maximal in the vicinity of $x/\ell = 0.577\sqrt{gt^2/\ell}$, and the surface becomes quiescent when x/ℓ is less than about $0.16\sqrt{gt^2/\ell}$, which effectively marks the inner boundary of the wave.

These conclusions are confirmed in Figure 5.5.3, which shows the surface profile at increasing times from $\sqrt{gt^2/\ell} = 0$ to 100. The solid curves are plotted from an 'exact' numerical evaluation of the surface displacement integral; the dots are the predictions of stationary phase formula (5.5.12), which is seen to supply a faithful representation of the surface displacement down to small times when the main group has travelled a relatively short distance of $x/\ell \sim 10$. The inverted arrows mark the centre of the main group at

$$x/\ell = 0.577\sqrt{gt^2/\ell},$$

where the amplitude function is a maximum. The wavelength within the main group is constant and equal to $8\pi\ell/3 \sim 8\ell$, although it is clear that the number of waves involved is relatively small at these small times and that the wavelength varies rapidly with position along the wave profile. Note also the increase in length of the main group with time as it comes to be dominated evermore closely by a diminishing band of wavenumbers centred on the maximum of the amplitude function; it is this dispersion of wave energy over the free surface that is responsible for the decay in amplitude in proportion to $1/\sqrt{t}$.

The preceding calculations are repeated in Figure 5.5.4 to reveal the nature of the free surface motion at smaller times. The stationary phase approximation clearly fails at small times near the origin, before the component waves are sufficiently dispersed. However, the main features of the motion are well represented, and the behaviour of the stationary phase approximation at the origin becomes acceptable for $\sqrt{gt^2/\ell}$ greater than about 5.

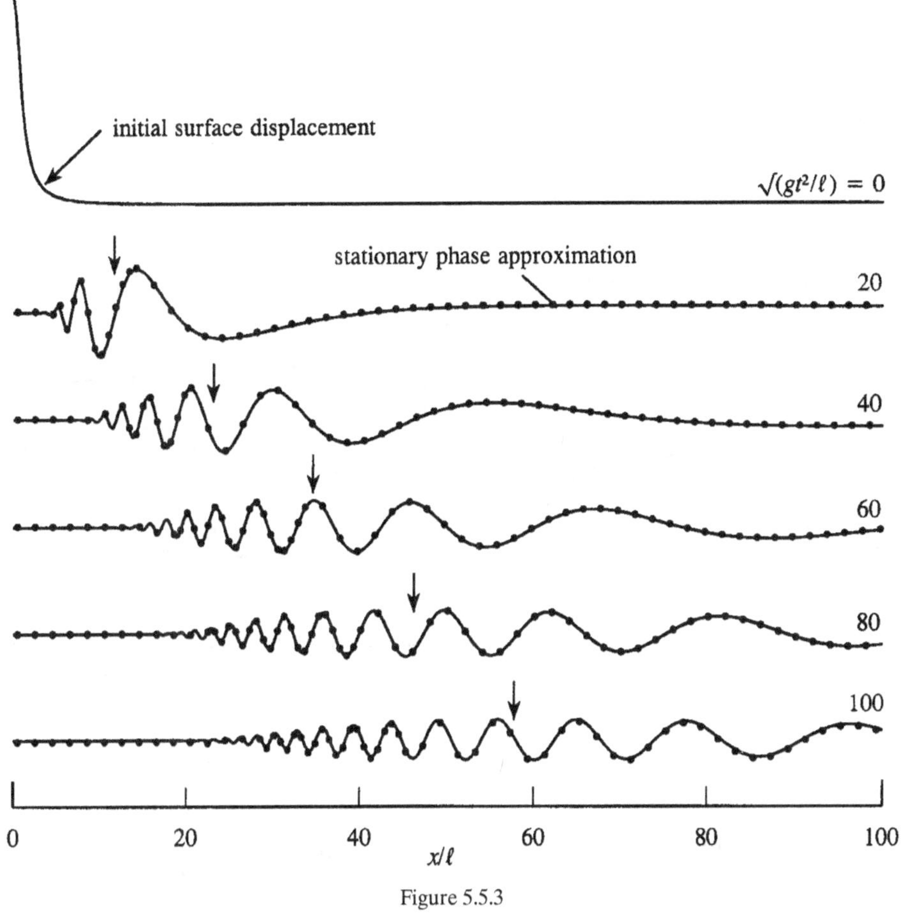

Figure 5.5.3

5.5.4 Conservation of energy

The perturbation energy at $t = 0$ for the initial-value problem of (5.5.1) is entirely
potential and is given, per unit span, by

$$Q_o = \frac{\rho_o g}{2} \int_{-\infty}^{\infty} f_o^2(x)\, dx = \frac{\rho_o g}{2} \iint_{-\infty}^{\infty} f_o(k) f_o(k') \cos(kx) \cos(k'x)\, dkdk'dx$$

$$= \frac{\rho_o g}{4} \iint_{-\infty}^{\infty} f_o(k) f_o(k') \Big\{ \cos[(k - k')x] + \cos[(k + k')x] \Big\}\, dkdk'dx$$

$$= \frac{\pi \rho_o g}{2} \int_{-\infty}^{\infty} f_o(k) f_o(k') \Big[\delta(k - k') + \delta(k + k') \Big]\, dkdk',$$

$$\text{i.e.,} \quad Q_o = \pi \rho_o g \int_{-\infty}^{\infty} |f_o(k)|^2\, dk. \tag{5.5.13}$$

Although this has been proved for only a symmetric disturbance [for which $f_o(k)$ is real
and satisfies $f_o(k) = f_o(-k)$], the final formula (which is just Parseval's theorem) is true
for any general initial wave profile with a complex-valued Fourier transform $f_o(k)$.

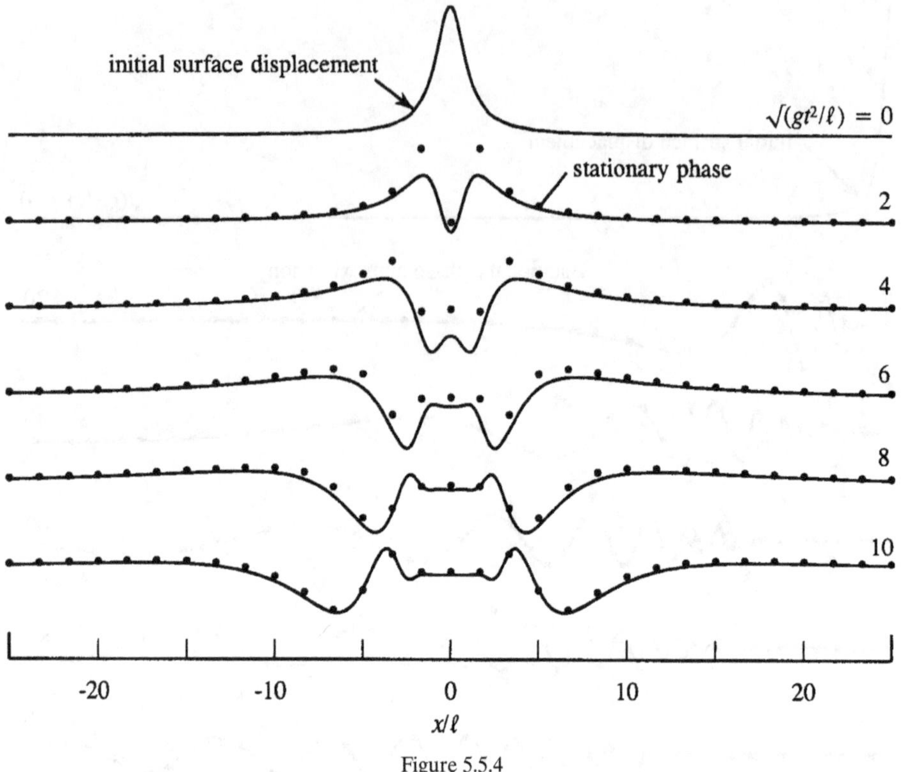

Figure 5.5.4

The potential energy Q at any time $t > 0$ can be calculated similarly:

$$Q = \frac{\rho_o g}{2} \iint_{-\infty}^{\infty} f_o(k) f_o(k') \cos[kx - \Omega(k)t] \cos[k'x - \Omega(k')t] \, dk \, dk' \, dx$$

$$= \frac{\pi \rho_o g}{2} \operatorname{Re} \int_{-\infty}^{\infty} f_o(k) f_o(k') \left[\delta(k - k') + \delta(k + k')e^{-2i\Omega(k)t} \right] dk \, dk',$$

$$\therefore \quad Q = \frac{\pi \rho_o g}{2} \int_{-\infty}^{\infty} |f_o(k)|^2 \left\{ 1 + \cos[2\Omega(k)t] \right\} dk. \tag{5.5.14}$$

The kinetic energy $T = \frac{1}{2}\rho_o \int_{-\infty}^{\infty} (\varphi \partial\varphi/\partial z)_{z=0} \, dx$ for $t > 0$, which gives [by use of (5.5.6)]

$$T = \frac{\rho_o g^2}{2} \iint_{-\infty}^{\infty} \frac{|k|}{\Omega(k)\Omega(k')} f_o(k) f_o(k') \sin[kx - \Omega(k)t] \sin[k'x - \Omega(k')t] \, dk \, dk' \, dx$$

$$= \frac{\pi \rho_o g^2}{2} \operatorname{Re} \int_{-\infty}^{\infty} \frac{|k|}{\Omega(k)\Omega(k')} f_o(k) f_o(k') \left[\delta(k - k') - \delta(k + k')e^{-2i\Omega(k)t} \right] dk \, dk',$$

$$\therefore \quad T = \frac{\pi \rho_o g}{2} \int_{-\infty}^{\infty} |f_o(k)|^2 \left\{ 1 - \cos[2\Omega(k)t] \right\} dk. \tag{5.5.15}$$

Equations (5.5.13)–(5.5.15) confirm the exact equality of the initial potential energy and the subsequent wave energy:

$$Q + T = Q_o.$$

5.5.5 Rayleigh's proof that energy propagates at the group velocity

Rayleigh's proof depends on noticing that, for two waves whose wavenumbers and frequencies differ by $\delta\omega$, δk, the corresponding rates at which energy is dissipated by the waves in the presence of a small amount of damping must differ by terms of second order.

In the absence of viscous forces the plane surface wave

$$\zeta = \mathrm{Re}\left[\zeta_o e^{i\{kx - \Omega(k)t\}}\right]$$

is an exact solution of the linearised equations in which $\Omega(k)$ is real for real values of the wavenumber k. The wave-energy density (per unit surface area) $E = \frac{1}{2}\rho_o g |\zeta_o|^2$.

When there is a small amount of damping because the viscosity $\eta \neq 0$, the equations of motion are modified and the dispersion relation becomes $\omega = \Omega(k, \eta)$, say. For real values of k we then have to first order in η

$$\omega = \Omega(k, 0) + \eta \left[\frac{\partial\Omega(k, \eta)}{\partial\eta}\right]_{\eta=0} = \Omega(k) - i\epsilon, \tag{5.5.16}$$

where $\epsilon \ll \Omega(k)$ is real and positive, so that $E = \frac{1}{2}\rho_o g |\zeta_o|^2 e^{-2\epsilon t}$ and the

$$\text{energy dissipation rate per unit area} \equiv -\frac{\partial E}{\partial t} = 2\epsilon E. \tag{5.5.17}$$

When $\eta > 0$ there is also a small $O(\epsilon^2)$ change in the real part of the frequency. This can be ignored and makes no difference to the argument.

Now consider the wave energy within a short interval $\delta x > 0$ of the x axis. In a *steady state* the wave energy $E = E(k, x)$, and it propagates in the x direction at some velocity $U(k)$. Within the interval the

$$\text{rate of energy dissipation} = -U(k)\delta x \frac{\partial E}{\partial x} = 2\epsilon\delta x E. \tag{5.5.18}$$

However, in order that the wave energy should not depend on time, the frequency in (5.5.16) must be *real*. This requires that k in (5.5.16) be replaced with $k + i\mu$, where the value of $\mu \ll k$ is chosen so that

$$\omega = \Omega(k + i\mu) - i\epsilon \approx \Omega(k) + i\left(\mu\frac{\partial\Omega}{\partial k} - \epsilon\right).$$

Therefore wave energy will flow into δx to maintain the energy constant, provided that

$$\mu\frac{\partial\Omega}{\partial k} = \epsilon, \tag{5.5.19}$$

in which case $E = \frac{1}{2}\rho_o g |\zeta_o|^2 e^{-2\mu x}$ and $\partial E/\partial x = -2\mu E$.

Hence (5.5.18) and (5.5.19) imply that

$$U(k) = \frac{\epsilon}{\mu} = \frac{\partial \Omega}{\partial k}(k).$$

The proof is clearly valid for all dispersive systems governed by linear equations with little or no damping.

5.5.6 Surface wave-energy equation

The idea that energy propagates at the group velocity supplies a simple interpretation of predictions made by the method of stationary phase.

Let $E(x, t)$ denote the wave energy per unit surface area generated by a disturbance at the origin at $t \sim 0$. We shall consider the case of waves on deep water, but this is no restriction of the applicability of the method. At large distances the motion at (x, t) consists of a wave train having group velocity x/t, whose wavelength (or wavenumber k) is determined as a function of x/t by the equation

$$\frac{\partial \Omega}{\partial k} \equiv \frac{1}{2}\mathrm{sgn}(k)\sqrt{\frac{g}{|k|}} = \frac{x}{t}, \quad \text{i.e.,} \quad k = \frac{gt^2}{4x|x|}.$$

The energy equation satisfied by E, which describes energy conservation and propagation at velocity x/t, is

$$\frac{\partial E}{\partial t} + \frac{\partial}{\partial x}\left(\frac{x}{t} E\right) = 0. \tag{5.5.20}$$

This can be corrected to take account of viscous attenuation by the addition of small negative terms on the right-hand side [as in Equation (5.3.7)]. By expanding the derivative, we can also write

$$\frac{\partial E}{\partial t} + \frac{x}{t}\frac{\partial E}{\partial x} = -\frac{E}{t}.$$

The term on the right-hand side describes the decay in wave-energy density caused by the steady growth in the region occupied by wave groups because of the variation in group velocity across a group. This simple first-order partial differential equation is solved by integration along characteristics:

$$\frac{dx}{dt} = \frac{x}{t}, \quad \frac{dE}{dt} = -\frac{E}{t},$$

$$\therefore \quad \frac{x}{t} = \alpha, \quad Et = \beta,$$

where α, β are constants. We obtain the general solution by putting $\beta = \Psi(\alpha)$ for any arbitrary function Ψ:

$$\therefore \quad E = \frac{1}{t}\Psi\left(\frac{x}{t}\right).$$

To find $\Psi(x/t)$ we equate the total energy to the initial wave energy of the source. For the problem of (5.5.1) (Figure 5.5.1) this is the potential energy of the initial displacement of the free surface given by (5.5.13). Thus

$$\int_{-\infty}^{\infty} \frac{1}{t} \Psi\left(\frac{x}{t}\right) dx = \pi \rho_o g \int_{-\infty}^{\infty} |f_o(k)|^2 dk.$$

However, on the left-hand side we can set

$$\frac{x}{t} = \frac{\partial \Omega}{\partial k} \quad \text{and} \quad \frac{dx}{t} = \left|\frac{\partial^2 \Omega}{\partial k^2}\right| dk$$

and then equate the integrands to find $\Psi(x/t) = \pi \rho_o g |f_o(k)|^2 / |\partial^2 \Omega/\partial k^2|$, so that

$$E(x,t) = \frac{\pi \rho_o g |f_o(k)|^2}{t|\partial^2 \Omega/\partial k^2|}, \quad \text{where} \quad k = \frac{gt^2}{4x|x|}. \tag{5.5.21}$$

If locally we put $\zeta = \operatorname{Re}\left\{\zeta_o e^{i[k-\Omega(k)t]}\right\}$, then $E = \frac{1}{2}\rho_o g |\zeta_o|^2$, and (5.5.21) yields

$$|\zeta_o| = |f_o(k)| \sqrt{\frac{2\pi}{t|\partial^2 \Omega/\partial k^2|}}, \quad k = \frac{gt^2}{4x|x|}, \tag{5.5.22}$$

which is precisely the modulus of the wave amplitude in stationary phase formula (5.5.9).

5.5.7 Waves generated by a submarine explosion

A two-dimensional underwater explosion occurring at $x = 0$, $z = -\ell$ in deep water can be modelled by a line source of strength $q(t)$, in terms of which the velocity potential is determined by

$$\nabla^2 \varphi = q(t)\delta(x)\delta(z+\ell).$$

The explosion produces an effective volume expansion and subsequent contraction (see Figure 5.5.5). We ignore oscillations of the explosive 'bubble' during the contraction phase and examine only the simplest case in which the net outflow and inflow of fluid volume Δ per unit span of the source occur instantaneously at times $t = 0$ and $t = \tau$, so that

$$q(t) = \Delta \frac{\partial}{\partial t}\Big[\mathrm{H}(t) - \mathrm{H}(t-\tau)\Big].$$

Let $\varphi = \varphi_o + \varphi_s$, where φ_o is the particular integral of Laplace's equation describing the motion when the presence of the free surface is ignored, and φ_s is the correction needed to satisfy the surface boundary conditions. By taking Fourier transforms with respect to x, z and t [defined as in (5.4.9)] we find

$$\varphi_o = -\frac{1}{(2\pi)^2} \int_{-\infty}^{\infty} \frac{q(\omega)}{k_x^2 + k_z^2} e^{i[k_x x + k_z(z+\ell) - \omega t]} dk_x dk_z d\omega$$

$$= -\frac{1}{4\pi} \int_{-\infty}^{\infty} \frac{q(\omega)}{|k|} e^{i(kx-\omega t)-|k|(z+\ell)} dk d\omega, \quad \text{for } z > -\ell,$$

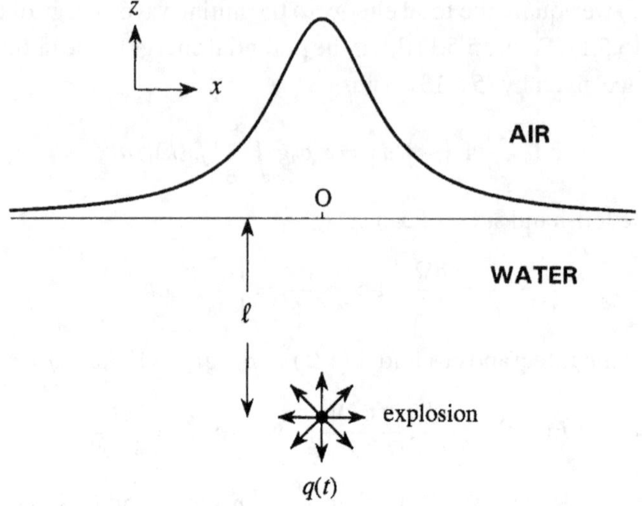

Figure 5.5.5

where in the second line we have dropped the subscript and written $k = k_x$, and

$$q(\omega) = \frac{\Delta}{2\pi}\left(1 - e^{i\omega\tau}\right).$$

If we put $\varphi_s = \int_{-\infty}^{\infty} A(k, \omega)e^{i(kx-\omega t)+z|k|}\, dk d\omega$ $(z < 0)$, then surface condition (5.5.2) satisfied by $\varphi = \varphi_o + \varphi_s$ supplies

$$A(k, \omega) = \frac{q(\omega)(\omega^2 + \Omega^2)e^{-|k|\ell}}{4\pi |k|(\omega^2 - \Omega^2)}, \quad \Omega = \sqrt{g|k|}.$$

The displacement ζ of the surface can now be found from the relation $\zeta = -\partial\varphi/g\partial t$ at $z = 0$:

$$\zeta = \frac{i}{2\pi}\int_{-\infty}^{\infty} \frac{\omega q(\omega)}{(\omega + i0)^2 - \Omega^2}\, e^{-|k|\ell+i(kx-\omega t)}\, dk d\omega.$$

This vanishes for $t < 0$. After the start of the explosion we can evaluate the ω integral by residues, using the definition of $q(\omega)$, and express the result in the form

$$\zeta = \frac{\Delta}{2\pi}H(t)\int_{-\infty}^{\infty} e^{-|k|\ell}\cos(kx - \Omega t)\, dk - \frac{\Delta}{2\pi}H(t - \tau)\int_{-\infty}^{\infty} e^{-|k|\ell}\cos[kx - \Omega(t - \tau)]\, dk.$$

$$(5.5.23)$$

Apart from the numerical factors, these integrals are identical with the second of (5.5.6) [where $f_o(k) = (h\ell/2)e^{-|k|\ell}$], and the corresponding stationary phase approximations are similar to expression (5.5.12). We can write the result in the compact form:

$$\frac{\ell\zeta}{\Delta} \approx \Psi(x, t) - \Psi(x, t - \tau),$$

$$(5.5.24)$$

$$\Psi(x, t) = \frac{1}{2\sqrt{\pi}}\left(\frac{\ell}{gt^2}\right)^{\frac{1}{4}}\left(\frac{gt^2\ell}{x^2}\right)^{\frac{3}{4}} e^{-\frac{gt^2\ell}{4x^2}}\cos\left(\frac{gt^2}{4x} - \frac{\pi}{4}\right).$$

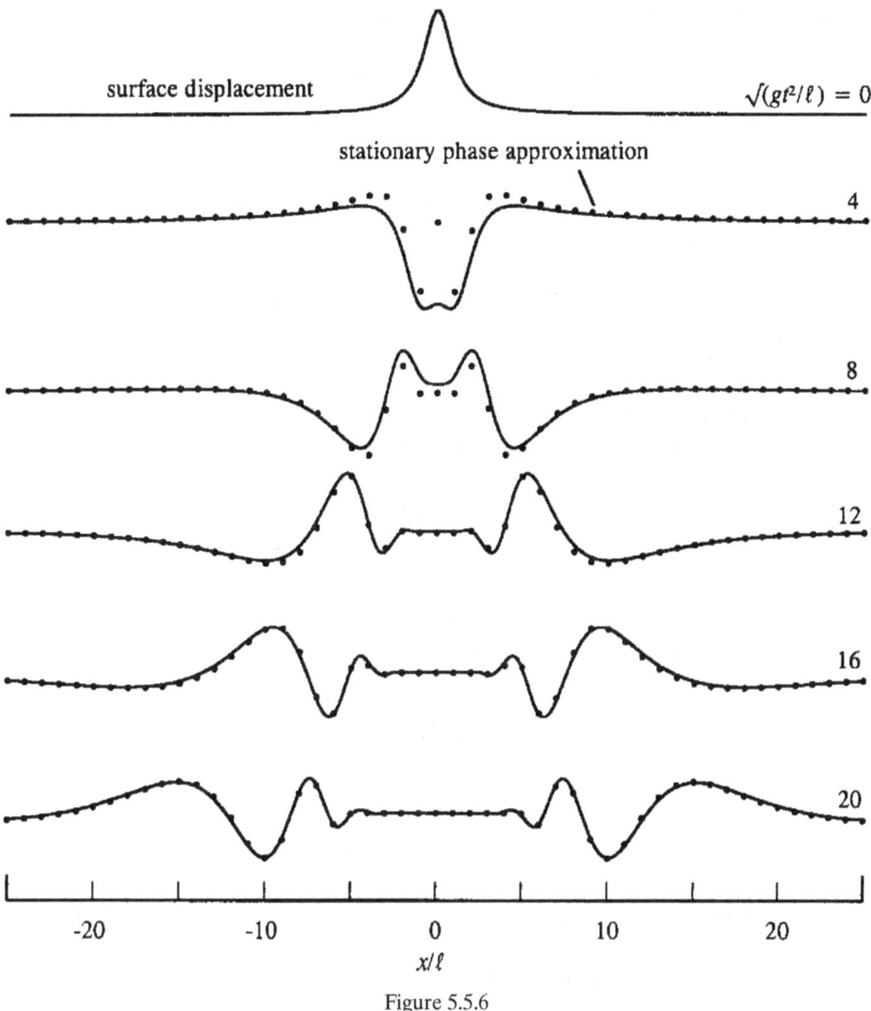

Figure 5.5.6

Equations (5.5.23) and (5.5.24) represent the surface displacement in terms of identical waveforms of opposite phase that correspond to the initial and final phases of the source motion. The phase difference depends on the nondimensional duration $\sqrt{g\tau^2/\ell}$ of the explosive bubble. The waveforms predicted by both formulae at small times are plotted in Figure 5.5.6 when $\sqrt{g\tau^2/\ell} = 3$, so that the wave generated in the contraction phase starts shortly before the second profile at $\sqrt{gt^2/\ell} = 4$. For incompressible flow the surface responds instantaneously at $t = 0$ to the beginning of the explosion, which produces the initial large elevation immediately above the source.

5.6 Initial-value problems in two surface dimensions

The application of the method of stationary phase to dispersive waves in two or in a higher number of dimensions will be discussed by reconsideration of the surface wave problem of §5.5.1, but with the initial disturbance confined to the neighbourhood of the origin $\mathbf{x} \equiv (x, y, z) = 0$.

5.6.1 Waves generated by a surface elevation symmetric about the origin

Let the motion start from rest with the initial surface elevation

$$\zeta = f_o(x, y) \equiv \frac{h_o}{[1 + (\varpi/\ell)^2]^{\frac{3}{2}}}, \quad 0 < \varpi = \sqrt{x^2 + y^2} < \infty, \quad t = 0. \quad (5.6.1)$$

It may be assumed that the uniform, initial value of the velocity potential $\varphi_o = 0$. Equations (5.5.4) become

$$\left. \begin{array}{l} \nabla^2(H\varphi) = 0, \quad z < 0 \\[2mm] \dfrac{\partial^2(H\varphi)}{\partial t^2} + g\dfrac{\partial(H\varphi)}{\partial z} = -g\delta(t)f_o(x, y), \quad z = 0 \end{array} \right\} -\infty < t < \infty, \quad (5.6.2)$$

where $H\varphi = \varphi$ when $t > 0$.

Write the solution of Laplace's equation in $z < 0$ as the Fourier integral

$$H\varphi = \int_{-\infty}^{\infty} \mathcal{A}(\mathbf{k}, \omega)e^{i(\mathbf{k}\cdot\mathbf{x}-\omega t)+zk} d^2\mathbf{k}d\omega, \quad \mathbf{k} = (k_1, k_2, 0), \quad k = |\mathbf{k}|,$$

where $\mathcal{A}(\mathbf{k}, \omega)$ is the Fourier transform of $H\varphi$ at $z = 0$. To avoid cumbersome notation it is convenient to adopt the convention that $x = x_1$ and $y = x_2$, so that scalar products such as $\mathbf{k} \cdot \mathbf{x}$ can also be written as $k_i x_i$. Surface condition (5.6.2) gives

$$\mathcal{A}(k, \omega) = \frac{g}{2\pi} \frac{f_o(\mathbf{k})}{(\omega^2 - \Omega^2)}, \quad \Omega(k) = \sqrt{gk},$$

with

$$f_o(\mathbf{k}) = \frac{1}{(2\pi)^2} \int_{-\infty}^{\infty} f_o(x, y)e^{-i\mathbf{k}\cdot\mathbf{x}} dxdy = \frac{h_o\ell^2}{2\pi}e^{-k\ell}. \quad (5.6.3)$$

Proceeding as in §5.5.1 [noting, in particular, that $f_o(\mathbf{k}) = f_o(-\mathbf{k})$], we find

$$\zeta = \int_{-\infty}^{\infty} f_o(\mathbf{k}) \cos[\mathbf{k} \cdot \mathbf{x} - \Omega(k)t] d^2\mathbf{k} \text{ for } t > 0. \quad (5.6.4)$$

The principle of stationary phase asserts that the main contribution to this integral at large times is from wavenumbers $\mathbf{k}$ clustered about the wavenumber $\mathbf{k}_o$ at which the phase

$$\Theta = \mathbf{k} \cdot \mathbf{x} - \Omega(k)t \quad (5.6.5)$$

is stationary. When $\mathbf{x}, t$ are given, $\mathbf{k}_o$ is determined by the condition

$$\frac{\partial\Omega}{\partial k_i} = \frac{x_i}{t}, \quad i = 1, 2, \quad (5.6.6)$$

that the vector group velocity $\partial\Omega/\partial\mathbf{k} \equiv (\partial\Omega/\partial k_1, \partial\Omega/\partial k_2)$ at $(\mathbf{x}, t)$ is just equal to $(x_1, x_2)/t$.

Expand the phase about the stationary point in powers of $\xi_i = k_i - k_{oi}$:

$$\mathbf{k} \cdot \mathbf{x} - \Omega(k)t \approx [\mathbf{k}_o \cdot \mathbf{x} - \Omega(k_o)t] + \xi_i\left[x_i - \frac{\partial\Omega}{\partial k_i}t\right] - \frac{\xi_i\xi_j}{2}t\frac{\partial^2\Omega}{\partial k_i\partial k_j}$$

$$\equiv [\mathbf{k}_o \cdot \mathbf{x} - \Omega(k_o)t] - \frac{\xi_i\xi_j}{2}t\frac{\partial^2\Omega}{\partial k_i\partial k_j},$$

where the derivatives are evaluated at $\mathbf{k} = \mathbf{k}_o$. Replace $f_o(\mathbf{k})$ in the integrand with its value $f_o(\mathbf{k}_o)$ at $\mathbf{k} = \mathbf{k}_o$. Then

$$\zeta \approx f_o(\mathbf{k}_o)\operatorname{Re}\left\{ e^{i[\mathbf{k}_o \cdot \mathbf{x} - \Omega(\mathbf{k}_o)t]} \int_{-\infty}^{\infty} e^{-i\frac{t}{2}\frac{\partial^2 \Omega}{\partial k_i \partial k_j}\xi_i\xi_j}\, d^2\xi \right\}.$$

The final integral can be transformed into a product of two integrals of the kind (5.5.8) by a simple rotation of axes. The precise value of the integral depends on the nature of the stationary point at $\mathbf{k}_o$, and is given generally by

$$\int_{-\infty}^{\infty} e^{-i\frac{t}{2}\frac{\partial^2 \Omega}{\partial k_i \partial k_j}\xi_i\xi_j}\, d^2\xi = \frac{2\pi\, e^{\chi\frac{i\pi}{4}}}{t\sqrt{|\det(\partial^2\Omega/\partial k_i \partial k_j)|}} \quad \text{where}\quad \chi = \begin{cases} -2, & \text{if } \Theta \text{ is a maximum} \\ +2, & \text{at a minimum} \\ 0, & \text{at a saddle point} \end{cases}.$$

$$(5.6.7)$$

For waves on deep water, $\Omega = \sqrt{gk}$ and the determinant

$$\det\left(\frac{\partial^2 \Omega}{\partial k_i \partial k_j}\right) = \frac{\partial^2 \Omega}{\partial k_1^2}\frac{\partial^2 \Omega}{\partial k_2^2} - \left(\frac{\partial^2 \Omega}{\partial k_1 \partial k_2}\right)^2 = -\frac{g}{8k^3} < 0.$$

Therefore $\mathbf{k}_o$ is a saddle point, $\chi = 0$, and the stationary phase approximation of the surface displacement becomes

$$\zeta = \frac{2\pi f_o(\mathbf{k}_o)}{t\sqrt{|\det(\partial^2\Omega/\partial k_i \partial k_j)|}}\cos[\mathbf{k}_o \cdot \mathbf{x} - \Omega(\mathbf{k}_o)t].$$

Now,

$$k_{oi} = \frac{gt^2}{4\varpi^3}\, x_i, \quad \Omega(k_o) = \frac{gt}{2\varpi};$$

hence, finally,

$$\frac{\zeta}{h_o} \approx \frac{1}{2\sqrt{2}}\left(\frac{\ell}{gt^2}\right)^{\frac{1}{2}}\left(\frac{gt^2\ell}{\varpi^2}\right)^{\frac{3}{2}} e^{-\frac{gt^2\ell}{4\varpi^2}}\cos\left(\frac{gt^2}{4\varpi}\right). \tag{5.6.8}$$

The overall amplitude of the surface disturbance decreases like $1/t$ (as opposed to $1/\sqrt{t}$ for one-dimensional propagation) because wave energy is now dispersed in two surface dimensions. This increased rate of decay is evident from Figure 5.6.1, where the solid curves are exact, numerically calculated profiles, and the dots are stationary phase approximation (5.6.8). Each surface profile at successive times is drawn to the same scale except for the first, which is drawn to a scale *one tenth* smaller, showing how surface spreading rapidly reduces the initial surface amplitude; contrast this with the corresponding plot for the one-dimensional case of Figure 5.5.3. The surface wave envelope has one maximum at

$$\frac{\varpi}{\ell} = \frac{\sqrt{gt^2/\ell}}{\sqrt{6}},$$

whose positions are indicated by the inverted arrows in the figure.

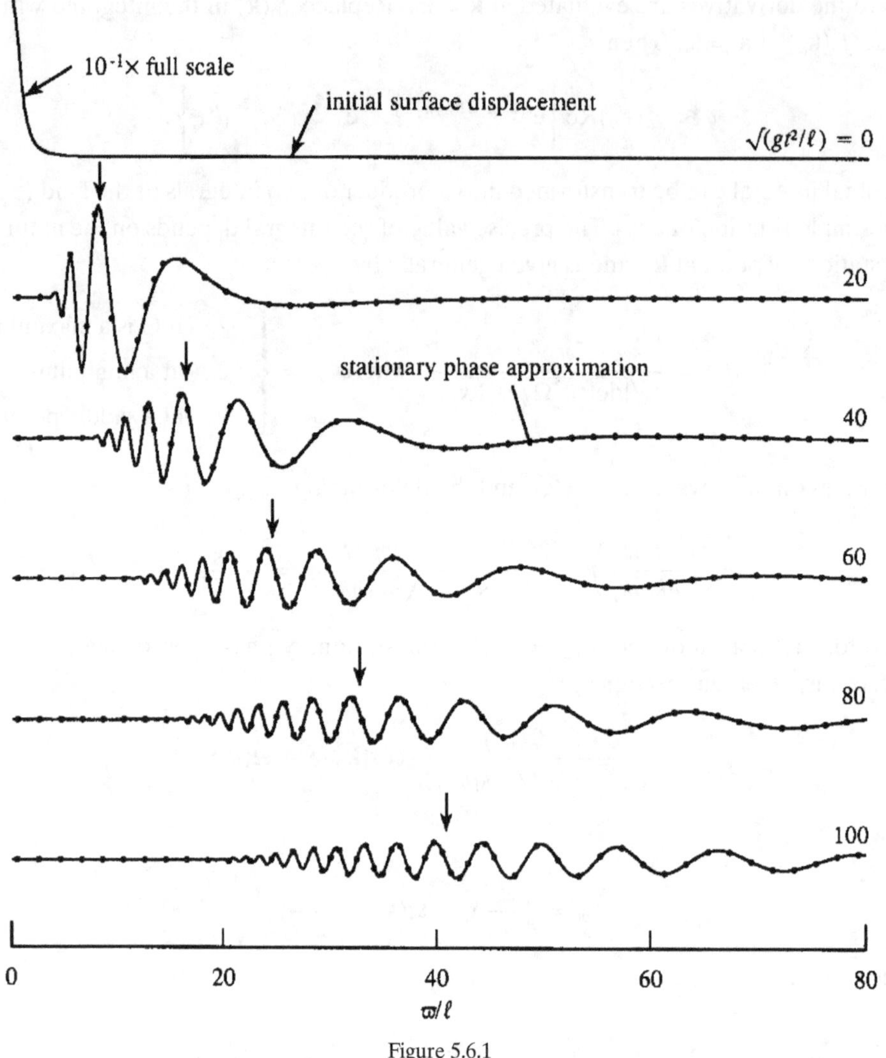

Figure 5.6.1

5.6.2 The energy equation in two dimensions

The radially spreading wave-energy distribution can also be calculated from the energy equation, as in §5.5.6. The group velocity at (ϖ, t) is x_i/t and the surface energy density E satisfies the conservation equation

$$\frac{\partial E}{\partial t} + \frac{\partial}{\partial x_j}\left(\frac{x_j}{t}\,E\right) = 0. \tag{5.6.9}$$

The general solution is

$$E = \frac{1}{t^2}\Psi\left(\frac{x_i}{t}\right),$$

where the factor $1/t^2$ is characteristic of the enhanced rate of decay of surface wave amplitude caused by two-dimensional spreading. The total wave energy

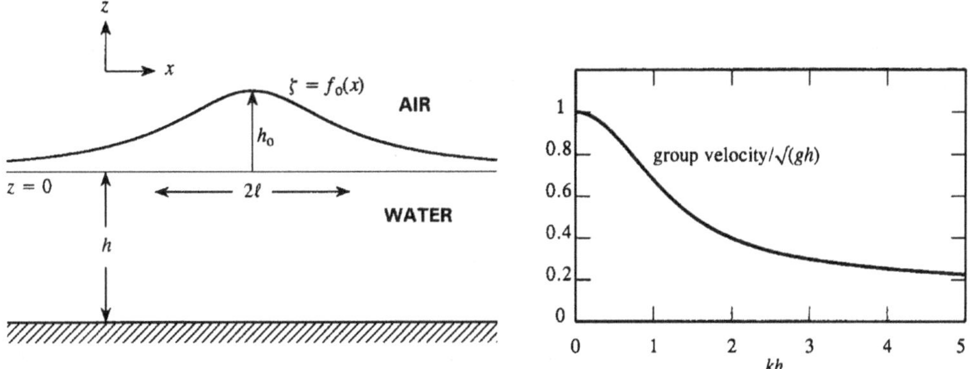

Figure 5.7.1

$\int_{-\infty}^{\infty} E(\mathbf{x}, t)\, d^2\mathbf{x}$ is equal to the potential energy determined by the initial elevation of the water $2\pi^2 \rho_o g \int_{-\infty}^{\infty} |f_o(\mathbf{k})|^2 d^2\mathbf{k}$. Group velocity equation (5.6.6) implies that $d^2\mathbf{x} = t^2 |\det(\partial^2 \Omega / \partial k_i \partial k_j)| d^2\mathbf{k}$; hence

$$E(x_i, t) = \frac{2\pi^2 \rho_o g |f_o(\mathbf{k})|^2}{t^2 |\det(\partial^2 \Omega / \partial k_i \partial k_j)|}, \quad \text{where } k_i = \frac{gt^2}{4\varpi^3} x_i.$$

In the fully dispersed region $E(x_i, t) = \frac{1}{2} \rho_o g |\zeta_o|^2$, where ζ_o is the local wave amplitude, and therefore

$$|\zeta_o| = \frac{2\pi |f_o(\mathbf{k})|}{t \sqrt{|\det(\partial^2 \Omega / \partial k_i \partial k_j)|}} \equiv \frac{h_o}{2\sqrt{2}} \left(\frac{\ell}{gt^2}\right)^{\frac{1}{2}} \left(\frac{gt^2 \ell}{\varpi^2}\right)^{\frac{3}{2}} e^{-\frac{gt^2 \ell}{4\varpi^2}},$$

which accords with the stationary phase result (5.6.8).

5.7 Surface motion near a wavefront

5.7.1 One-dimensional waves

The group velocity $\partial \Omega / \partial k = \frac{1}{2}\sqrt{g/k}$ for waves on deep water becomes infinitely large with increasing wavelength ($k \to 0$). An impulsive source at the origin therefore produces motion instantaneously at all points. Indeed, for the initial surface elevation problem of Figure 5.5.1, it follows from (5.5.6) by expanding in powers of the time that

$$\frac{\zeta}{h_o} \sim \frac{\ell}{x}\left(\frac{\ell}{x} + \frac{gt^2}{2x} + \cdots +\right), \quad x \to \infty.$$

The first term in the parentheses corresponds to the initial displacement of the free surface, the second to the leading effect of the waves for small values of gt^2/x. At any distant point x the surface motion therefore starts instantaneously, the amplitude initially increasing like $gt^2/2x$.

At first sight the conclusion is quite different when the fluid has finite depth (h in Figure 5.7.1), because the group velocity can never exceed $\sqrt{gh}$ (attained by long waves as $kh \to 0$) and the front of the surface waves generated by a disturbance at the origin would therefore be expected to be found at $x \sim t\sqrt{gh}$. However, this argument is based on the method of stationary phase, which actually becomes invalid at the wavefront where, for long waves (see §5.2.3),

$$\frac{\partial \Omega}{\partial k} \approx \sqrt{gh}\left[1 - \frac{(kh)^2}{2}\right] \quad \text{and} \quad \frac{\partial^2 \Omega}{\partial k^2} \to 0 \quad \text{as} \quad kh \to 0. \tag{5.7.1}$$

Let the motion be started from rest at $t = 0$ by an initial surface elevation $\zeta = f_o(x)$. By the method of §5.5.1 the surface displacement for $t > 0$ is found to be

$$\zeta = \int_{-\infty}^{\infty} f_o(k) \cos[kx - \Omega(k)t] \, dk, \tag{5.7.2}$$

where $\Omega = \sqrt{gk \tanh(kh)}$. We can determine the behaviour at large distances and small times as before by expanding in powers of the time and considering the behaviour when $x \gg h$. Taking $f_o(k)$ to be defined as in (5.5.5), we find

$$\frac{\zeta}{h_o} \sim \frac{\ell^2}{x^2}\left(1 + \frac{3gt^2h}{x^2} + \cdots +\right), \quad \frac{x}{h} \to \infty.$$

Thus, as in the case of waves on deep water, there is no delay at arbitrarily large distances before the motion starts, although it is very weak. Because the group velocity cannot exceed $\sqrt{gh}$, this infinitely fast response at large distances is merely a consequence of the assumption of incompressible flow, as the first intimations of fluid motion must actually be the arrival of acoustic disturbances that have infinite wave speed in an incompressible fluid.

To examine the motion near the wavefront at large x and t (where $x/t \sim \sqrt{gh}$), observe that integral (5.7.2) is then dominated by contributions from small values of kh, where $\Omega \approx k\sqrt{gh}[1 - (kh)^2/6]$ so that, for $x > 0$,

$$\frac{\zeta}{h_o} \approx \frac{\ell}{2}\int_0^{\infty} \cos\left[kh\left(\frac{x}{h} - \sqrt{\frac{gt^2}{h}}\right) + \frac{(kh)^3}{6}\sqrt{\frac{gt^2}{h}}\right] dk$$

$$= \frac{\pi\ell}{2h}\left[\frac{4h}{gt^2}\right]^{\frac{1}{6}} \mathrm{Ai}\left\{\left[\frac{4h}{gt^2}\right]^{\frac{1}{6}}\left(\frac{x}{h} - \sqrt{\frac{gt^2}{h}}\right)\right\}, \tag{5.7.3}$$

where $\mathrm{Ai}(x)$ is the Airy function (Figure 5.7.2) that satisfies

$$\int_0^{\infty} \cos(\sigma\xi^3 + x\xi)d\xi = [\pi/(3\sigma)^{\frac{1}{3}}]\,\mathrm{Ai}\left(x/(3\sigma)^{\frac{1}{3}}\right), \quad \sigma > 0,$$

and has the asymptotic properties

$$\mathrm{Ai}(x) \sim \begin{cases} \dfrac{e^{-\frac{2}{3}x^{\frac{3}{2}}}}{2\pi^{\frac{1}{2}}x^{\frac{1}{4}}}, & x \to +\infty \\[2ex] \dfrac{\sin\left(\frac{2}{3}|x|^{\frac{3}{2}} + \frac{\pi}{4}\right)}{\pi^{\frac{1}{2}}|x|^{\frac{1}{4}}}, & x \to -\infty \end{cases}.$$

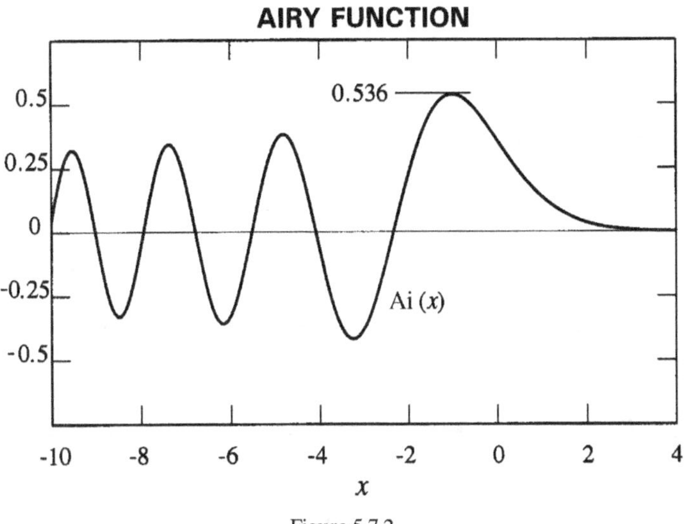

Figure 5.7.2

The surface displacement therefore decreases exponentially fast when $x/t >$ $\sqrt{gh}$. The wave amplitude rises to a maximum at the wavefront, where $\zeta/h_o \sim$ $0.536(\pi\ell/2h)(4h/gt^2)^{\frac{1}{6}}$, i.e. the peak amplitude decreases with time like $1/t^{\frac{1}{3}}$ as opposed to the marginally more rapid $1/t^{\frac{1}{2}}$ dependence predicted for points behind the wavefront by the stationary phase formula:

$$\frac{\zeta}{h_o} \approx \frac{\pi^{\frac{1}{2}}\ell e^{-k\ell}}{2^{\frac{1}{2}}\cosh(kh)\sqrt{t|\partial^2\Omega/\partial k^2|}} \cos\left(kx - \Omega t + \frac{i\pi}{4}\right), \quad x, \, t > 0,$$

where k is the positive root of $x/t = \partial\Omega(k)/\partial k$. Just to the rear of the wavefront,

$$kh \sim \sqrt{2\left(1 - \frac{x}{t\sqrt{gh}}\right)},$$

and the wave amplitude $|\zeta_o|$ is predicted by the stationary phase formula to be

$$\frac{|\zeta_o|}{h_o} \sim \frac{\ell}{h}\left(\frac{h}{gt^2}\right)^{\frac{1}{4}} \frac{\pi^{\frac{1}{2}}}{2^{\frac{3}{4}}|1 - x/t\sqrt{gh}|^{\frac{1}{4}}}$$

which is unbounded at the nominal wave front.

The evolution of the waveform with time is shown by the solid curves in Figure 5.7.3 for the case where $\ell = h$. The solid dots to the rear of the wavefront are the stationary phase approximation. The peak amplitude occurs at the wavefront and propagates at the speed $\sqrt{gh}$ of long waves. There is always motion ahead of the nominal wavefront, and the open circles in Figure 5.7.3 are the prediction of this decay according to Airy function approximation (5.7.3) – the motion decreases exponentially fast over a distance $\sim h(gt^2/4h)^{\frac{1}{6}}$ ($\sim 4h$ when $\sqrt{gt^2/h} = 100$). Although this increases slowly with time, the wave group to the rear of the front expands more rapidly in proportion to $\sim h(gt^2/h)^{\frac{1}{4}}$. When viewed on the scale of the characteristic wavelength at the front ($\sim 15h$ at $\sqrt{gt^2/h} = 100$) the wavefront may be said to be sharp.

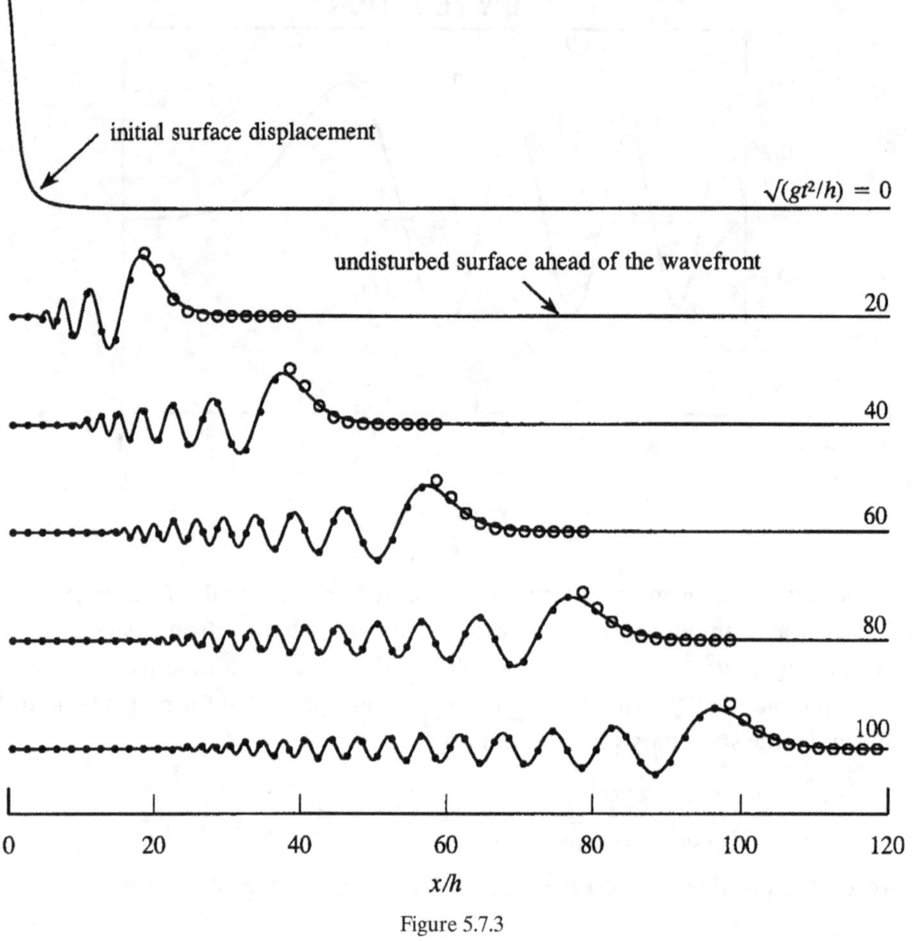

Figure 5.7.3

5.7.2 Waves generated by motion of the seabed

Let a section of the ocean floor $z = -h$ experience a sudden elevation $d(x, y)$ at time $t = 0$ (Figure 5.7.4). The velocity potential of the resulting motion satisfies

$$\nabla^2 \varphi = 0, \quad z < 0,$$

$$\frac{\partial^2 \varphi}{\partial t^2} + g \frac{\partial \varphi}{\partial z} = 0, \quad z = 0,$$

$$\frac{\partial \varphi}{\partial z} = f(x, y, t) \equiv \frac{\partial}{\partial t}[d(x, y)\mathrm{H}(t)], \quad z = -h. \tag{5.7.4}$$

Laplace's equation is satisfied by

$$\varphi = \int_{-\infty}^{\infty} \left[\mathcal{A}(\mathbf{k}, \omega)\mathrm{e}^{kz} + \mathcal{B}(\mathbf{k}, \omega)\mathrm{e}^{-kz} \right] \mathrm{e}^{i(\mathbf{k}\cdot\mathbf{x} - \omega t)} \, d^2\mathbf{k} d\omega,$$

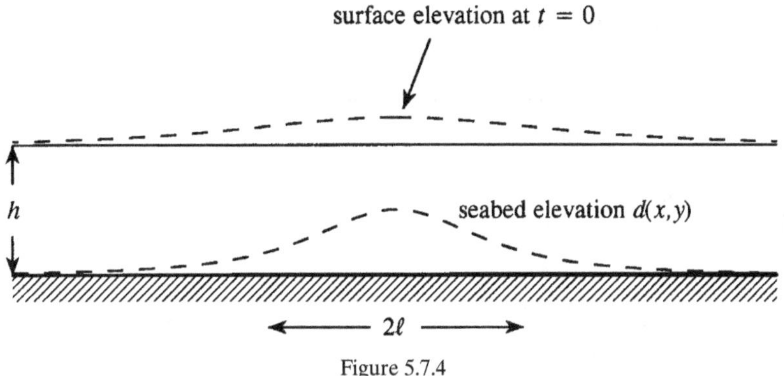

Figure 5.7.4

where the conditions at $z = 0$ and $z = -h$ respectively require

$$\mathcal{A}(\mathbf{k}, \omega)(\omega^2 - gk) + \mathcal{B}(\mathbf{k}, \omega)(\omega^2 + gk) = 0,$$

$$k\mathcal{A}(\mathbf{k}, \omega)\mathrm{e}^{-kh} - k\mathcal{B}(\mathbf{k}, \omega)\mathrm{e}^{kh} = f(\mathbf{k}, \omega).$$

These equations yield

$$\mathcal{A}(\mathbf{k}, \omega) = \frac{(gk + \omega^2)f(\mathbf{k}, \omega)}{2k\left[\omega^2 \cosh(kh) - gk\sinh(kh)\right]},$$

$$\mathcal{B}(\mathbf{k}, \omega) = \frac{(gk - \omega^2)f(\mathbf{k}, \omega)}{2k\left[\omega^2 \cosh(kh) - gk\sinh(kh)\right]}.$$

The free surface elevation $\zeta = -\partial\varphi/g\partial t$ at $z = 0$ is therefore

$$\zeta = \int_{-\infty}^{\infty} \frac{i\omega f(\mathbf{k}, \omega)\mathrm{e}^{i(\mathbf{k}\cdot\mathbf{x}-\omega t)} \, d^2k d\omega}{\omega^2 \cosh(kh) - gk\sinh(kh)}. \tag{5.7.5}$$

Consider the symmetric seabed disturbance

$$d(x, y) = \frac{d_o}{[1 + (\varpi/\ell)^2]^{\frac{3}{2}}},$$

for which

$$f(\mathbf{k}, \omega) = \frac{d_o\ell^2 \mathrm{e}^{-k\ell}}{(2\pi)^2}.$$

This does not depend on ω, so that [by residues, replacing ω in (5.7.5) with $\omega + i0$]

$$\zeta = \frac{d_o\ell^2 \mathrm{H}(t)}{2\pi} \int_{-\infty}^{\infty} \frac{\cos[\mathbf{k}\cdot\mathbf{x} - \Omega(k)t]}{\cosh(kh)} \mathrm{e}^{-k\ell} \, d^2\mathbf{k}, \tag{5.7.6}$$

where $\Omega(k) = \sqrt{gk\tanh(kh)}$. As $t \to +0$ this formula gives the initial elevation of the water. Figure 5.7.4 illustrates the relative displacements (drawn to an exaggerated scale) of the seabed and the free surface when $\ell = h$.

Integral (5.7.6) can be evaluated by stationary phase to give the surface motion for $t > 0$. However, the stationary points must be determined numerically, and it is therefore

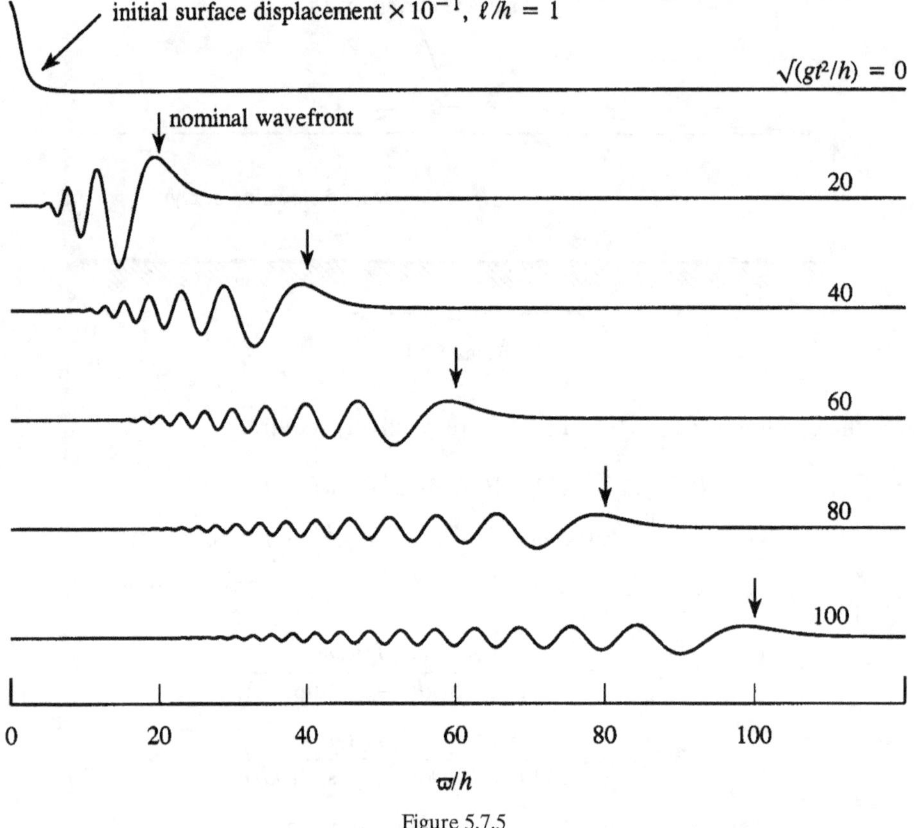

Figure 5.7.5

more convenient to perform the integration numerically. To do this we change to polar coordinates (k, θ), where θ is measured from the direction of $\mathbf{x} = (x, y)$, and evaluate the θ integral by using

$$\int_0^{2\pi} \cos[k\varpi \cos\theta - \Omega(k)t]\,d\theta = 2\int_0^{\pi} \cos(k\varpi \cos\theta)\cos[\Omega(k)t]\,d\theta = 2\pi\,J_0(k\varpi)\cos[\Omega(k)t].$$

Hence, (5.7.6) becomes

$$\frac{\zeta}{d_o} = \ell^2 \int_0^\infty \frac{\cos[\Omega(k)t]}{\cosh(kh)} k J_0(k\varpi) e^{-k\ell}\, dk$$

$$= \frac{\ell^2}{h^2} \int_0^\infty \cos\left[\left(\frac{gt^2}{h}\right)^{\frac{1}{2}} (\lambda \tanh \lambda)^{\frac{1}{2}}\right] J_0\left(\frac{\lambda\varpi}{h}\right) \frac{\lambda e^{-\frac{\lambda\ell}{h}}\, d\lambda}{\cosh \lambda}. \tag{5.7.7}$$

The surface wave plots in Figures 5.7.5 and 5.7.6 are respectively for $\ell/h = 1,\ 5$. In both cases the nominal wavefront of the disturbance is close to that determined by the shallow-water limit $\varpi = t\sqrt{gh}$ (indicated by the arrow heads). The waves decay

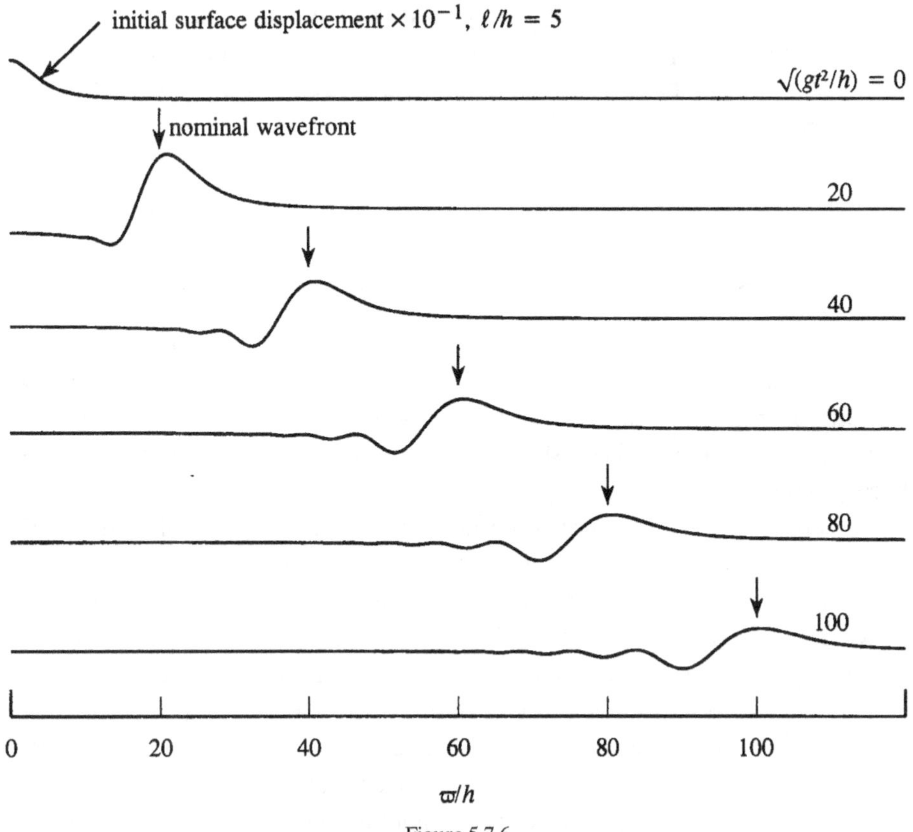

Figure 5.7.6

exponentially fast at larger distances, entirely in agreement with the Airy function theory of §5.7.1. Because of the long characteristic wavelength when $\ell/h = 5$, the wave
motion is then similar to predictions for the shallow-water problem of §5.4.3, the surface
displacement consisting of a large positive pulse that propagates essentially without dispersion (but with amplitude decreasing like $1/\sqrt{\varpi}$) followed by a region of depressed
water level. The pulse shape at $\sqrt{gt^2/h} = 20$ in Figure 5.7.6 is very similar to the wave
profile in Figure 5.4.2. Stationary phase theory implies that the rear end of the wave profile is associated with waves of minimal group velocity ($k \to \infty$). Such waves propagate
as on deep water, and [by use of the formula $\mathrm{J}_0(x) \sim (2/\pi x)^{\frac{1}{2}} \cos(x - \frac{\pi}{4})$, $x \to \infty$] the
dominant contribution to (5.7.7) is given by

$$\frac{\zeta}{d_o} \sim \frac{\ell^2}{h^2} \left(\frac{2h}{\pi \varpi}\right)^{\frac{1}{2}} \int_0^\infty \cos\left[\left(\frac{gt^2}{h}\right)^{\frac{1}{2}} \lambda^{\frac{1}{2}}\right] \cos\left(\frac{\lambda \varpi}{h} - \frac{\pi}{4}\right) \frac{\sqrt{\lambda}e^{-\frac{\lambda \ell}{h}} \, d\lambda}{\cosh \lambda}$$

$$\approx \frac{\ell^2}{\sqrt{2\pi} h^2} \left(\frac{h}{\varpi}\right)^{\frac{1}{2}} \int_0^\infty \cos\left[\frac{\lambda \varpi}{h} - \left(\frac{gt^2}{h}\right)^{\frac{1}{2}} \sqrt{\lambda} - \frac{\pi}{4}\right] \frac{\sqrt{\lambda}e^{-\frac{\lambda \ell}{h}} \, d\lambda}{\cosh \lambda}, \quad \frac{\varpi}{t} \to 0,\ t \to \infty.$$

The usual stationary phase argument applied to this integral would give a wave group whose amplitude is proportional to $\exp(-\frac{gt^2\ell}{4\varpi^2})$. Thus larger values of ℓ are associated with shorter wave groups because of the cut-off provided by this factor when t/ϖ becomes large. The truth of this assertion is made evident by a comparison of Figures 5.7.5 and 5.7.6.

5.7.3 Tsunami produced by an undersea earthquake

A surface gravity wave whose wavelength is much longer than the ocean depth is called a tsunami. In this 'shallow-water' limit the phase speed is constant and the wave propagates over great distances with little or no dispersion. The wave amplitude increases rapidly as it approaches a shore, causing wavefront steepening and the motion to become nonlinear.

An undersea earthquake causes both horizontal and vertical displacements of the seabed. However, only the vertical motions are effective in producing surface waves, and because the solid volume of the seabed is conserved the volumes of water displaced by upward and downward motions of the bed must be effectively equal and opposite. The seabed motion usually occurs over time scales that are negligible compared with that of the resulting tsunami. For a large earthquake the seabed displacement tends to be centred along a 'rupture' that may be several hundred kilometres long and about 100 km wide, the seabed rising on one side of the rupture and falling (by an amount consistent with conservation of volume) on the opposite side. We obtain a simple analytical model that preserves volume displacement by taking

$$d(x, y) = d_o \left\{ \frac{1}{[1 + (\varpi_+/\ell)^2]^{\frac{3}{2}}} - \frac{1}{[1 + (\varpi_-/\ell)^2]^{\frac{3}{2}}} \right\}, \quad \varpi_\pm = \sqrt{(x \pm L)^2 + y^2}.$$

in Equations (5.7.4). The seabed motion is equivalent to a volume source of dipole type (centred on the x axis with its axis parallel to the x direction) formed by a positive source at $x = -L$ and a sink at $x = +L$. The profile of the seabed in the plane of symmetry $y = 0$ is illustrated in Figure 5.7.7 when $\ell = L = h$, together with the initial asymmetric surface elevation.

We can plot the surface wave displacement at time $t > 0$ by combining the separate displacements produced by source and sink determined by formula (5.7.7) respectively for $\varpi = \varpi_\pm$. The resulting wave profile for $x > 0$, $y = 0$ is plotted in the lower part of Figure 5.7.7 for the case $\ell = L = 5h$. The principal disturbance travelling in the positive x direction is a large positive pulse of width $\sim 10h$ that propagates without change of form at the shallow-water wave speed $\sqrt{gh}$ (the 'nominal centroid' is at $x = t\sqrt{gh}$ and coincides with the undisturbed surface at the origin at $t = 0$). The motion represents an elementary tsunami of width $\sim 10h$ that greatly exceeds the depth of the ocean.

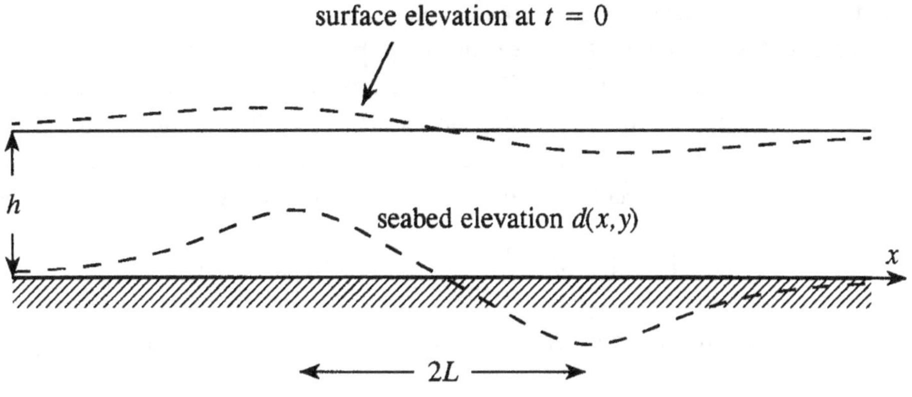

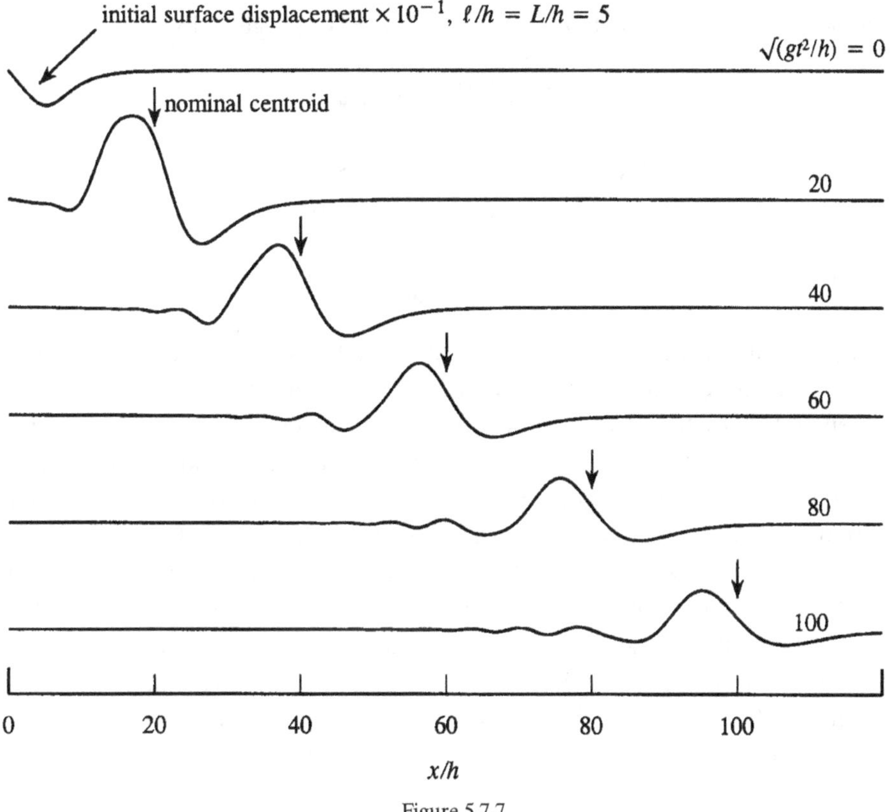

Figure 5.7.7

5.8 Periodic wave sources

Consider waves on water of arbitrary depth h generated by a time harmonic pressure distribution

$$p_o(x, y)e^{-i\omega_o t} = \int_{-\infty}^{\infty} p_o(\mathbf{k})e^{i(\mathbf{k}\cdot\mathbf{x} - \omega_o t)} \, d^2\mathbf{k}, \quad \omega_o > 0, \tag{5.8.1}$$

applied at the free surface $z = 0$. In applications we are of course concerned with real pressures, and it is understood that the solution of the wave problem will ultimately be combined with its complex conjugate. However, the exponential frequency dependence is convenient because in the linearised approximation all fluctuating quantities are proportional $e^{-i\omega_o t}$, and this factor may therefore be temporarily suppressed.

The velocity potential is accordingly defined by the Fourier integral [compare (5.1.11)]

$$\varphi = \int_{-\infty}^{\infty} \mathcal{A}(\mathbf{k}, \omega_o) \frac{\cosh[k(z+h)]}{\cosh(kh)} \, e^{i\mathbf{k}\cdot\mathbf{x}} d^2\mathbf{k}, \quad z \le 0,$$

where $\mathcal{A}(\mathbf{k}, \omega_o)$ is determined in terms of $p_o(\mathbf{k})$ by the surface condition of (5.1.5):

$$\mathcal{A}(\mathbf{k}, \omega_o) = -\frac{i\omega_o}{\rho_o} \frac{p_o(\mathbf{k})}{[\omega_o^2 - \Omega^2(k)]}, \quad \Omega^2(k) = gk \tanh(kh).$$

Hence the surface displacement $\zeta = (-1/i\omega_o)(\partial\varphi/\partial z)_{z=0}$ is given by

$$\zeta = \frac{1}{\rho_o g} \int_{-\infty}^{\infty} \frac{\Omega^2(k) p_o(\mathbf{k})}{[\omega_o^2 - \Omega^2(k)]} \, e^{i\mathbf{k}\cdot\mathbf{x}} \, d^2\mathbf{k}$$

$$= -\frac{p_o(x, y)}{\rho_o g} + \frac{\omega_o^2}{\rho_o g} \int_{-\infty}^{\infty} \frac{p_o(\mathbf{k})}{[\omega_o^2 - \Omega^2(k)]} \, e^{i\mathbf{k}\cdot\mathbf{x}} \, d^2\mathbf{k}. \tag{5.8.2}$$

5.8.1 One-dimensional waves

When ω_o is real the value of integral (5.8.2) is ambiguous because of the poles at $\Omega^2(k) = \omega_o^2$ on the real axis. Real poles were avoided in the initial-value problems of §5.5 effectively by the addition of a small positive imaginary part to the frequency, thereby ensuring a null solution for $t < 0$. We now require a solution satisfying the radiation condition of steady *outgoing* wave behaviour, in that wave energy must radiate away from the sources. This solution can also be obtained, however, by the same formal device of replacing ω_o with $\omega_o + i\epsilon$, where $\epsilon > 0$ and is subsequently allowed to vanish. This replacement implies that applied surface pressure (5.8.1) is proportional to $e^{\epsilon t}$ and therefore ultimately vanishes as $t \to -\infty$. A solution with the same time dependence must automatically comply with the radiation condition, as waves radiating towards the sources from infinity are absent when t is sufficiently large and negative.

Let us consider the simplest case of uniform forcing along the line $x = 0$, and put

$$p_o(x) = P_o \ell \delta(x),$$

where P_o may be regarded as a constant pressure applied over a narrow strip of width ℓ. For *deep* water ($h \to \infty$) $\Omega^2 = g|k|$, and (5.8.2) becomes

$$\zeta = -\frac{P_o \ell}{2\pi \rho_o g} \int_{-\infty}^{\infty} \frac{|k| e^{ikx} \, dk}{[|k| - (\omega_o + i\epsilon)^2/g]}$$

$$= -\frac{P_o \ell \delta(x)}{\rho_o g} - \frac{P_o \ell \omega_o^2}{2\pi \rho_o g^2} \int_{-\infty}^{\infty} \frac{e^{ikx} \, dk}{[|k| - (\omega_o + i\epsilon)^2/g]}, \quad \epsilon \to +0. \tag{5.8.3}$$

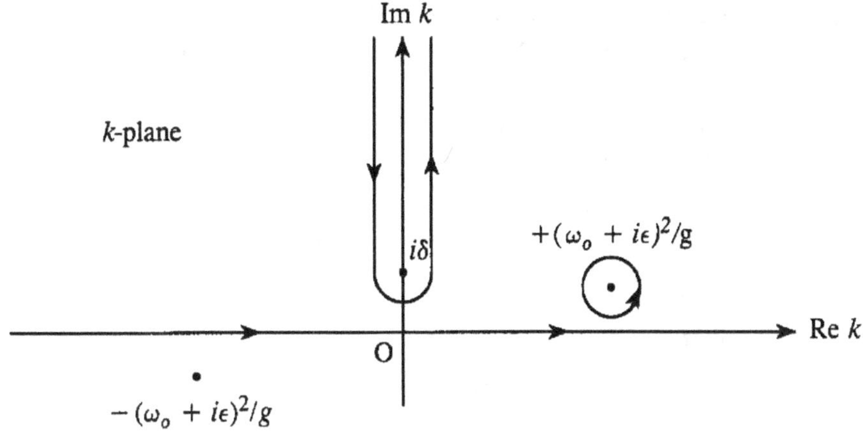

Figure 5.8.1

Because $\omega_o > 0$ there are simple poles at $k = \pm(\omega_o + i\epsilon)^2/g$ respectively above and below the real k axis (Figure 5.8.1). To evaluate the integral the non-analytic function $|k|$ is replaced with $\sqrt{k^2 + \delta^2}$, where $\delta > 0$ will ultimately vanish. This has branch cuts in the k plane extending from $\pm i\delta$ to $\pm i\infty$ on the imaginary axis, so that when $x > 0$,

$$\int_{-\infty}^{\infty} \frac{e^{ikx}\, dk}{[\sqrt{k^2 + \delta^2} - (\omega_o + i\epsilon)^2/g]} \;\to\; 2\pi i e^{i(\omega_o^2/g)x} + 2\int_0^{\infty} \frac{\lambda e^{-\lambda x}\, d\lambda}{\lambda^2 + (\omega_o^2/g)^2} \quad \text{as } \epsilon,\, \delta \;\to\; +0.$$

where the first term on the right-hand side is the residue contribution from the pole at $(\omega_o + i\epsilon)^2/g$ and the integral is taken along the upper branch cut. We evaluate the integral for $x > g/\omega_o^2$ by expanding the non-exponential part of the integrand in powers of λ, and in the first approximation is equal to $2(g/\omega_o^2 x)^2$. By restoring the time factor and taking real parts, we then find for $x > 0$

$$\frac{\zeta}{P_o/\rho_o g} \approx -\left[\ell\delta(x) + \frac{1}{\pi}\frac{g\ell}{\omega_o^2 x^2} + \cdots +\right]\cos\omega_o t + \frac{\omega_o^2 \ell}{g}\sin\left(\frac{\omega_o^2 x}{g} - \omega_o t\right). \tag{5.8.4}$$

The cosine term includes the local surface displacement beneath the applied pressure at $x = 0$ plus a *near-field* correction decreasing as $1/x^2$ and communicated instantaneously throughout the incompressible fluid. The sine term comes from the pole and represents an unattenuated surface wave that dominates the motion at large distances.

In general only pole contributions are significant in the far field and are the only component of the surface motion capable of withdrawing energy from the source and propagating it to infinity. Because the wave amplitude $\zeta_o = (P_o/\rho_o g)(\omega_o^2\ell/g)$, the rate at which energy radiates to $x = \pm\infty$ per unit span of the source is

$$2\left(\frac{\partial\Omega}{\partial k}\frac{1}{2}\rho_o g|\zeta_o|^2\right) = \frac{P_o\omega_o\ell}{2}\frac{P_o}{\rho_o g}\frac{\omega_o^2\ell}{g}. \tag{5.8.5}$$

EXAMPLE 1. Verify that (5.8.5) is equal to the rate of working of the applied surface pressure.

We have to evaluate $-\int_{-\infty}^{\infty} \langle p_o(x)\cos(\omega_o t)\partial\zeta/\partial t\rangle\, dx$, where $\langle\ \rangle$ denotes a time average, and $\partial\zeta/\partial t$ is given by the first line of (5.8.3) after restoration of the factor $e^{-i\omega_o t}$. Then

$$-\int_{-\infty}^{\infty}\left\langle p_o(x)\cos(\omega_o t)\frac{\partial\zeta}{\partial t}\right\rangle dx$$

$$= -\frac{P_o^2\ell^2}{8\pi\rho_o g}\left\langle\left[e^{-i\omega_o t}+e^{i\omega_o t}\right]\left[\int_{-\infty}^{\infty}\frac{i\omega_o|k|e^{-i\omega_o t}\,dk}{|k|-(\omega_o+i\epsilon)^2/g}+\text{c.c.}\right]\right\rangle$$

$$= \frac{P_o^2\ell^2}{4\rho_o g}\int_{-\infty}^{\infty}\frac{\omega_o|k|(2\omega_o\epsilon/g)}{\pi[(|k|-\omega_o^2/g)^2+(2\omega_o\epsilon/g)^2]}$$

$$\to \frac{P_o^2\ell^2}{4\rho_o g}\int_{-\infty}^{\infty}\omega_o|k|\delta\left(|k|-\omega_o^2/g\right)\,dk,\quad\text{as }\epsilon\to+0,$$

$$= \frac{P_o\omega_o\ell}{2}\frac{P_o}{\rho_o g}\frac{\omega_o^2\ell}{g},$$

where c.c. denotes the complex conjugate of the preceding quantity. In deriving this result ω_o has been replaced with $\omega_o+i\epsilon$ only in the denominator of the integrand in the first line. The reader can check that no terms are thereby lost from the final result as $\epsilon\to0$.

5.8.2 Periodic sources in two surface dimensions

Time harmonic waves radiating in two dimensions from a source localised at the origin are given more generally by

$$\zeta = \int_{-\infty}^{\infty}\frac{\mathcal{F}(\mathbf{k})}{D(\mathbf{k},\omega_o+i\epsilon)}\,e^{i[\mathbf{k}\cdot\mathbf{x}-(\omega_o+i\epsilon)t]}\,d^2\mathbf{k},\tag{5.8.6}$$

where $\mathcal{F}(\mathbf{k})$ represents the effective source and $D(\mathbf{k},\omega)=D(-\mathbf{k},-\omega)=\omega^2-\Omega^2(k)$ is called the 'dispersion function'. This notation is adopted to emphasise the generality of the following discussion. The positive imaginary part $i\epsilon$ added to the frequency causes the wavefield to grow with time, but at any finite time the total wave energy is finite.

Waves arriving at x_i in the fully dispersed far field are determined by wavenumbers $\mathbf{k}$ in the integral clustered around points $\mathbf{k}=\mathbf{k}_o$, say, at which

$$\frac{x_i}{\varpi}=\frac{\partial\Omega/\partial k_i}{|\partial\Omega/\partial k_i|},\quad \omega_o=\Omega(\mathbf{k}),\tag{5.8.7}$$

where $\omega_o=\Omega(\mathbf{k})$ is a branch of the dispersion curve $C_{\mathbf{k}}$: $D(\mathbf{k},\omega_o)=0$. Wave energy propagates at the group velocity $\partial\Omega/\partial k_i$, and this vector is normal to $C_{\mathbf{k}}$ at $\mathbf{k}=\mathbf{k}_o$ and parallel and in the same direction as x_i, so that $x_i\partial\Omega/\partial k_i>0$ at $\mathbf{k}_o$ (Figure 5.8.2).

The dominant far-field wave group at x_i is extracted from integral (5.8.6) by expansion of the integrand about all points $\mathbf{k}_o$ satisfying condition (5.8.7). To do this, let $\mathbf{n}\equiv\mathbf{n}(\mathbf{k}_o)$

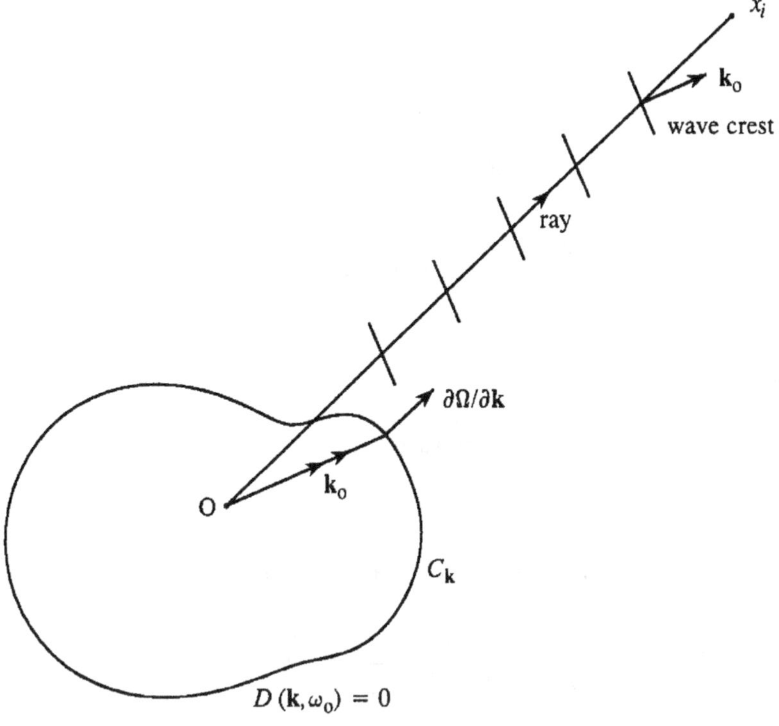

Figure 5.8.2

be the unit normal at $\mathbf{k}_o$ on $C_\mathbf{k}$, orientated in the direction of the group velocity $\partial\Omega/\partial\mathbf{k}$. In the neighbourhood of $\mathbf{k}_o$ (Figure 5.8.3) put

$$\mathbf{k} = \mathbf{k}_o + \mathbf{c}(s) + s_\perp\mathbf{n},$$

where $\mathbf{k}_o + \mathbf{c}(s)$ lies on $C_\mathbf{k}$, so that $D(\mathbf{k}_o + \mathbf{c}, \omega_o) = 0$, s is the arc length measured along $C_\mathbf{k}$ from $\mathbf{k}_o$, and $s_\perp$ is distance measured from $\mathbf{k}_o + \mathbf{c}$ parallel to $\mathbf{n}$. In the integrand of (5.8.6) $D(\mathbf{k}, \omega_o + i\epsilon)$ can be replaced for small values of $s_\perp$ and ϵ with

$$D(\mathbf{k}_o + \mathbf{c} + s_\perp\mathbf{n}, \omega_o + i\epsilon) = s_\perp D_{\mathbf{k}_n} + i\epsilon D_\omega,$$

where $D_{\mathbf{k}_n} = \mathbf{n} \cdot \partial D/\partial\mathbf{k}$, $D_\omega = \partial D/\partial\omega$ evaluated at $\mathbf{k}_o$. Then the dominant part of (5.8.6) is

$$\zeta \approx \int_{-\infty}^{\infty} \frac{\mathcal{F}(\mathbf{k}_o)}{s_\perp D_{\mathbf{k}_n} + i\epsilon D_\omega} \, e^{i[(\mathbf{k}_o+\mathbf{c})\cdot\mathbf{x}+s_\perp\varpi-(\omega_o+i\epsilon)t]} \, ds\,ds_\perp.$$

The $s_\perp$ integral is non-zero provided the pole lies in $\mathrm{Im}\, s_\perp > 0$, which it does because $s_\perp = -i\epsilon D_\omega/D_{\mathbf{k}_n} \equiv i\epsilon/(\mathbf{n} \cdot \partial\Omega/\partial\mathbf{k})$ is positive imaginary.

Next expand $\mathbf{c}(s)$ in the exponential in powers of s. Because $d\mathbf{c}/ds$ is the unit tangent on $C_\mathbf{k}$, and x_i and n_i are parallel, the leading term in the expansion is

$$\mathbf{c} \cdot \mathbf{x} \approx \frac{1}{2}s^2\mathbf{x} \cdot \left(\frac{d^2\mathbf{c}}{ds^2}\right)_{s=0} \equiv -\frac{1}{2}s^2\frac{\mathbf{x} \cdot \mathbf{n}}{\varpi_\mathbf{k}} = -\frac{s^2}{2}\frac{\varpi}{\varpi_\mathbf{k}},$$

where $\varpi_\mathbf{k}$ is the radius of curvature on $C_\mathbf{k}$ at $\mathbf{k}_o$.

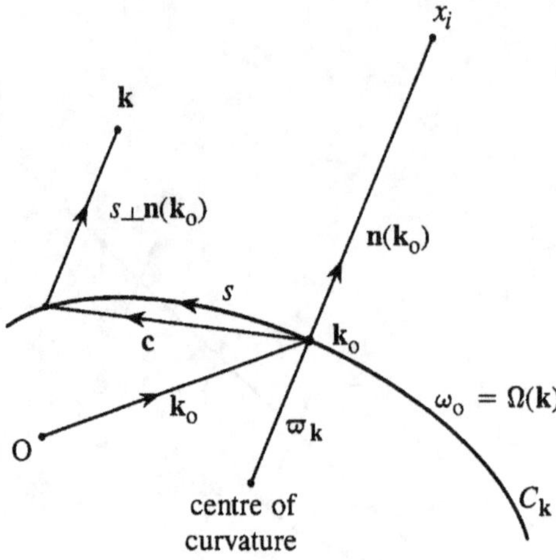

Figure 5.8.3

Both of the integrals with respect to $s_\perp$ and s can now be evaluated explicitly to yield

$$\zeta \approx (2\pi)^{\frac{3}{2}} \left(\frac{|\varpi_\mathbf{k}|}{\varpi}\right)^{\frac{1}{2}} \frac{i\mathcal{F}(\mathbf{k}_o)}{(\mathbf{n}\cdot\partial D/\partial\mathbf{k})_{\mathbf{k}=\mathbf{k}_o}}\, e^{i[\mathbf{k}_o\cdot\mathbf{x}-\omega_o t)-\frac{\pi}{4}\mathrm{sgn}(\varpi_\mathbf{k})]}, \quad \varpi \to \infty.$$

Now $\mathbf{n}\cdot(\partial D/\partial\mathbf{k})_{\mathbf{k}=\mathbf{k}_o} = -D_\omega(\mathbf{k}_o,\omega_o)|\partial\Omega/\partial\mathbf{k}|$; hence our final approximation to the far-field surface displacement becomes

$$\zeta \approx (2\pi)^{\frac{3}{2}} \sum_{\mathbf{k}_o} \left(\frac{|\varpi_\mathbf{k}|}{\varpi}\right)^{\frac{1}{2}} \frac{-i\mathcal{F}(\mathbf{k})}{D_\omega(\mathbf{k},\omega_o)|\partial\Omega/\partial\mathbf{k}|}\, e^{i[\mathbf{k}\cdot\mathbf{x}-\omega_o t)-\frac{\pi}{4}\mathrm{sgn}(\varpi_\mathbf{k})]}, \quad \varpi \to \infty, \quad (5.8.8)$$

where the summation is over all wavenumbers on $D(\mathbf{k},\omega_o) = 0$ satisfying

$$\frac{x_i}{\varpi} = \frac{\partial\Omega/\partial k_i}{|\partial\Omega/\partial k_i|} \quad \left(\text{so that } \mathbf{x}\cdot\frac{\partial\Omega}{\partial\mathbf{k}} > 0\right),$$

and $\mathrm{sgn}(\varpi_\mathbf{k}) = \pm 1$ according to whether $C_\mathbf{k}$ is convex or concave at $\mathbf{k}_o$ towards the far-field point x_i.

The fully dispersed waves determined by a given wavenumber $\mathbf{k}$ on the dispersion curve $D(\mathbf{k},\omega_o)$ radiate in the direction of the group velocity $\partial\Omega/\partial\mathbf{k}$, which is in the direction of one of the two normals. The following simple rule of thumb determines which of the normals: It is in the direction towards the neighbouring dispersion curve obtained when ω_o is *increased* by a small amount.

EXAMPLE 2. Evaluate far-field surface waves (5.8.2) produced on deep water by periodic surface pressure fluctuations near the origin.

For deep water $\Omega = \sqrt{gk}$, $C_\mathbf{k}$ is the circle $|\mathbf{k}| = \varpi_\mathbf{k} = k_o \equiv \omega_o^2/g$, $x_i/\varpi = k_i/|\mathbf{k}|$ and

$$\mathcal{F}(\mathbf{k}) = \frac{kp_o(\mathbf{k})}{\rho_o}, \quad D_\omega(\mathbf{k},\omega_o) = 2\omega_o, \quad \frac{\partial\Omega}{\partial k} = \frac{1}{2}\sqrt{\frac{g}{k}}.$$

Therefore, putting $(x, y) = \varpi(\cos\theta, \sin\theta)$, we have

$$\zeta \approx -\frac{(2\pi)^{\frac{3}{2}}}{\sqrt{k_o\varpi}}\frac{k_o^2}{\rho_o g}p_o(k_o\cos\theta, k_o\sin\theta)\,e^{i(k_o\varpi+\frac{\pi}{4})}, \quad \varpi\to\infty.$$

The angular distribution of the radiation is entirely dependent on the properties of the applied pressure, which is otherwise obvious because the surface wave properties are isotropic when the water is stationary.

5.8.3 The surface wave power

The complex frequency in the representation (5.8.6) causes the wave field to grow with time, but at any finite time the total wave energy is finite. To calculate the energy we must take the real part. Then the overall energy of the surface waves at time t is (Whitham 1961)

$$\int_{-\infty}^{\infty}\frac{1}{2}\rho_o g|\zeta_o|^2\,d^2\mathbf{x} \equiv \int_{-\infty}^{\infty}\rho_o g\langle\zeta^2\rangle\,d^2\mathbf{x}$$

$$= \rho_o g\int_{-\infty}^{\infty}\left\langle\left\{\frac{1}{2}\int_{-\infty}^{\infty}\frac{\mathcal{F}(\mathbf{k})e^{i[\mathbf{k}\cdot\mathbf{x}-i(\omega_o+i\epsilon)t]}}{D(\mathbf{k},\omega_o+i\epsilon)}\,d^2\mathbf{k}+\text{c.c.}\right\}^2\right\rangle d^2\mathbf{x}$$

$$= 2\pi^2\rho_o g\int_{-\infty}^{\infty}\frac{|\mathcal{F}(\mathbf{k})|^2 e^{2\epsilon t}\,d^2\mathbf{k}}{[D^2(\mathbf{k},\omega_o)+\epsilon^2 D_\omega^2(\mathbf{k},\omega_o)]}, \quad \text{for } \epsilon\ll 1,$$

where $D_\omega(\mathbf{k},\omega) = \partial D(\mathbf{k},\omega)/\partial\omega$, and ϵ is sufficiently small that the change in wave amplitude over a period $2\pi/\omega_o$ is negligible.

The net *power* $\Pi = \frac{\partial}{\partial t}\int_{-\infty}^{\infty}\rho_o g\langle\zeta^2\rangle\,d^2\mathbf{x}$ absorbed by the surface waves is therefore given by

$$\Pi = \lim_{\epsilon\to+0}4\pi^2\rho_o g\int_{-\infty}^{\infty}\frac{\epsilon|\mathcal{F}(\mathbf{k})|^2 e^{2\epsilon t}\,d^2\mathbf{k}}{[D^2(\mathbf{k},\omega_o)+\epsilon^2 D_\omega^2(\mathbf{k},\omega_o)]}$$

$$= 4\pi^3\rho_o g\int_{-\infty}^{\infty}\frac{|\mathcal{F}(\mathbf{k})|^2\,\delta[D(\mathbf{k},\omega_o)]\,d^2\mathbf{k}}{|D_\omega(\mathbf{k},\omega_o)|}. \tag{5.8.9}$$

The presence of the δ function confirms that only 'propagating' wavenumbers $\mathbf{k}$ lying on the wavenumber curve $C_\mathbf{k}$ participate in the radiation of surface wave energy. Putting $d^2\mathbf{k} = ds_\perp ds(\mathbf{k})$, where $ds(\mathbf{k})$ is the element of arc length on $C_\mathbf{k}$ and $ds_\perp$ is a local coordinate normal to $C_\mathbf{k}$, as in §5.8.2, we have

$$\Pi = 4\pi^3\rho_o g\int_{C_\mathbf{k}}\frac{|\mathcal{F}(\mathbf{k})|^2\,ds(\mathbf{k})}{|D_\omega(\mathbf{k},\omega_o)||D_\mathbf{k}(\mathbf{k},\omega_o)|},$$

where $D_\mathbf{k}(\mathbf{k},\omega_o) = \partial D(\mathbf{k},\omega_o)/\partial\mathbf{k}$. The dispersion function can be factorised into a product of terms of the type $\omega_o - \Omega(\mathbf{k})$ whose normal is parallel to the group velocity $\partial\Omega/\partial\mathbf{k}$, because $\partial\Omega/\partial\mathbf{k} = -D_\mathbf{k}(\mathbf{k},\omega)/D_\omega(\mathbf{k},\omega)$ [Figure 5.8.4(a)]. Hence the power supplied to the surface waves becomes

$$\Pi = 4\pi^3\rho_o g\int_{C_\mathbf{k}}\frac{|\mathcal{F}(\mathbf{k})|^2\,ds(\mathbf{k})}{|D_\omega(\mathbf{k},\omega_o)|^2|\partial\Omega/\partial\mathbf{k}|}. \tag{5.8.10}$$

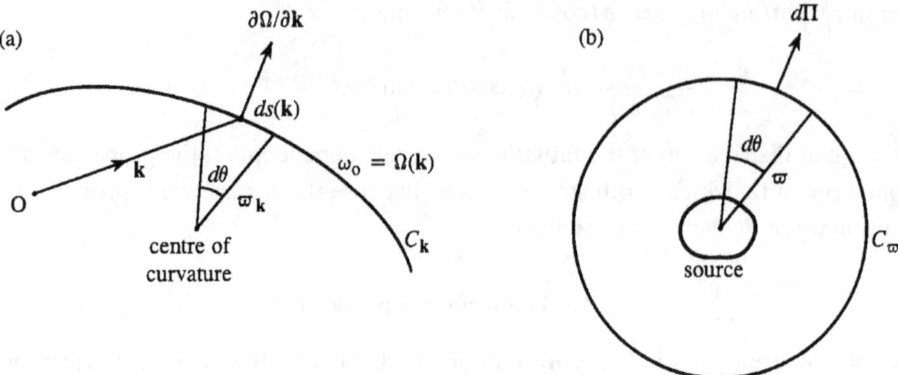

Figure 5.8.4

5.8.4 Surface wave amplitude

In the fully dispersed region the waves are locally of the form $\zeta = \zeta_o e^{i(\mathbf{k}\cdot\mathbf{x} - \omega_o t)}$, with slowly varying amplitude ζ_o and rapidly varying phase $\Theta = \mathbf{k}\cdot\mathbf{x} - \omega_o t$. For steady-state, time harmonic sources the energy density $E = \frac{1}{2}\rho_o g |\zeta_o|^2$ depends only on position, and the energy equation becomes

$$\frac{\partial}{\partial x_j}\left(\frac{x_j}{\varpi}\left|\frac{\partial\Omega}{\partial\mathbf{k}}\right| E\right) = 0.$$

Therefore

$$\Pi = \int_0^{2\pi}\left|\frac{\partial\Omega}{\partial\mathbf{k}}\right| E\,\varpi\,d\theta, \qquad (5.8.11)$$

where the integration is taken around the circumference of a large circle C_ϖ of radius ϖ in the far field, on which $(x, y) = \varpi(\cos\theta, \sin\theta)$ [Figure 5.8.4(b)].

Alternative representations (5.8.10) and (5.8.11) of Π are related by the identification of the unit normal $(\cos\theta, \sin\theta)$ on C_ϖ with the normal $\mathbf{n}(\mathbf{k})$ on $C_\mathbf{k}$, because wave energy radiating into $(\theta, d\theta)$ is precisely the energy radiating into $[\mathbf{n}(\mathbf{k}), ds(\mathbf{k})]$, where $d\theta$ and $ds(\mathbf{k})$ satisfy

$$d\theta = \frac{ds(\mathbf{k})}{|\varpi_\mathbf{k}|}, \qquad (5.8.12)$$

where, as before, $\varpi_\mathbf{k}$ is the radius of curvature on $C_\mathbf{k}$. Thus differential elements can be equated to obtain

$$E(x_i) = \frac{|\varpi_\mathbf{k}|}{\varpi}\frac{4\pi^3\rho_o g|\mathcal{F}(\mathbf{k})|^2}{|D_\omega(\mathbf{k}, \omega_o)|^2(\partial\Omega/\partial\mathbf{k})^2},$$

where $\mathbf{k}$ is expressed in terms of x_i/ϖ by means of Equations (5.8.7). However, $E = \frac{1}{2}\rho_o g|\zeta_o|^2$; therefore the wave amplitude is given in the far field by

$$|\zeta_o| \approx \sqrt{\frac{|\varpi_\mathbf{k}|}{\varpi}}\frac{(2\pi)^{\frac{3}{2}}|\mathcal{F}(\mathbf{k})|}{|D_\omega(\mathbf{k}, \omega_o)||\partial\Omega/\partial\mathbf{k}|}, \qquad \varpi \to \infty, \qquad (5.8.13)$$

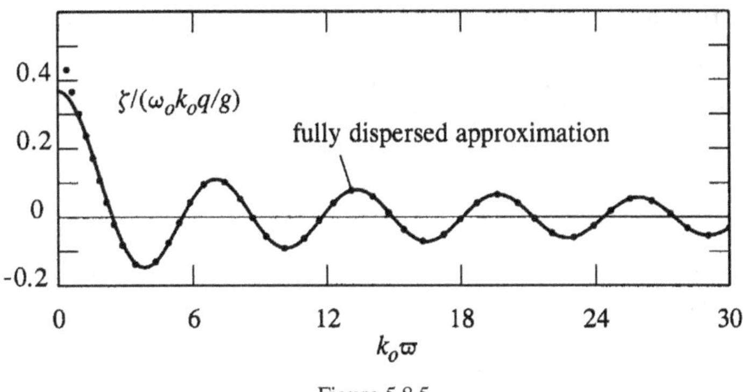

Figure 5.8.5

which agrees with asymptotic formula (5.8.8), with due account taken of the possibility of multiple branches of $C_{\mathbf{k}}$. Energy conservation demands that the amplitude of waves produced by a continuously radiating source should decay like $1/\sqrt{\varpi}$.

EXAMPLE 3. A point source of strength $qe^{-i\omega_o t}$ is situated at depth ℓ beneath the surface of an infinite expanse of deep water. Calculate the surface wave motion in the linearised approximation.

This is the periodic and three-dimensional analogue of the problem of Figure 5.5.5. The procedure described in §5.5.7 yields for the surface displacement

$$\zeta = \frac{i\omega_o q}{2\pi^2} \int_{-\infty}^{\infty} \frac{e^{-k\ell + i(\mathbf{k}\cdot\mathbf{x} - \omega_o t)}}{(\omega_o + i\epsilon)^2 - \Omega^2}\, d^2\mathbf{k}, \quad \Omega = \sqrt{gk}.$$

The far-field approximation supplies, in real form,

$$\frac{\zeta}{\omega_o k_o q/g} = \sqrt{\frac{2}{\pi k_o \varpi}}\, e^{-k_o \ell} \cos\left(k_o \varpi - \omega_o t - \frac{\pi}{4}\right), \quad \varpi \to \infty, \quad \text{where} \quad k_o = \frac{\omega_o^2}{g}.$$

Proceeding without approximation, we find

$$\frac{\zeta}{\omega_o k_o q/g} = -\frac{i}{\pi k_o} \int_0^{\infty} \frac{k J_0(k\varpi) e^{-k\ell - i\omega_o t}\, dk}{k - (\omega_o + i\epsilon)^2/g}.$$

The real part of this is plotted as a function $k_o \varpi$ as the solid curve in Figure 5.8.5 for $\omega_o t = 2n\pi$ and $k_o \ell = 1$. The dots are the far-field fully dispersed approximation.

5.9 Ship waves

The surface wave pattern of a moving source, such as that made by a ship, lacks the isotropy of many of the problems discussed previously. Let us consider a time harmonic surface pressure distribution translating at uniform speed V in the negative x direction over water of depth h, so that applied surface pressure (5.8.1) becomes

$$p_0(x + Vt, y)e^{-i\omega_o t} = \int_{-\infty}^{\infty} p_0(\mathbf{k}) e^{i[\mathbf{k}\cdot\mathbf{x} - (\omega_o - Vk_1)t]}\, d^2\mathbf{k}, \quad \omega_o,\ V > 0. \tag{5.9.1}$$

By the usual procedure, the resulting surface displacement is given by

$$\zeta = \frac{1}{\rho_o g} \int_{-\infty}^{\infty} \frac{\Omega^2(k) p_o(\mathbf{k})}{[(\omega_o - Vk_1)^2 - \Omega^2(k)]} \, e^{i[\mathbf{k}\cdot\mathbf{x} - (\omega_o - Vk_1)t]} \, d^2\mathbf{k}, \quad \Omega^2(k) = gk\tanh(kh)$$

$$= \frac{1}{\rho_o g} \int_{-\infty}^{\infty} \frac{\Omega^2(k) p_o(\mathbf{k})}{[(\omega_o - Vk_1)^2 - \Omega^2(k)]} \, e^{i[k_1(x+Vt) + k_2 y - \omega_o t]} \, d^2\mathbf{k}. \tag{5.9.2}$$

The phase $\Theta = \mathbf{k} \cdot \mathbf{x} - (\omega_o - Vk_1)t$, and the frequency of the surface displacement of wavenumber $\mathbf{k}$ is

$$-\frac{\partial\Theta}{\partial t} = \omega_o - Vk_1.$$

This is the frequency with respect to an observer fixed relative to the undisturbed water and evidently depends on the direction of propagation. Relative to the moving source, the frequency is

$$-\left(\frac{\partial}{\partial t} - V\frac{\partial}{\partial x}\right)\Theta = \omega_o.$$

It is therefore convenient to adopt a moving reference frame, by setting $\bar{x} = x + Vt$, $\bar{y} = y$, whereupon

$$\zeta = \frac{1}{\rho_o g} \int_{-\infty}^{\infty} \frac{\Omega^2(k) p_o(\mathbf{k})}{[(\omega_o - Vk_1)^2 - \Omega^2(k)]} \, e^{i(\mathbf{k}\cdot\bar{\mathbf{x}} - \omega_o t)} \, d^2\mathbf{k}. \tag{5.9.3}$$

5.9.1 Moving line pressure source

Let the moving pressure source be uniform in the y direction and be given by

$$p_o = \frac{P_o}{1 + (x + Vt)^2/\ell^2}, \quad \omega_o = 0.$$

Then $p_o(k) = (P_o\ell/2)e^{-|k|\ell}$, and conditions are steady in a frame moving with the source. On *deep water*, Equation (5.9.3) becomes

$$\zeta = \frac{P_o\ell}{2\rho_o g} \int_{-\infty}^{\infty} \frac{g|k|e^{-|k|\ell + ik\bar{x}} \, dk}{[(\omega_o + i\epsilon - Vk)^2 - g|k|]} \quad \text{as } \omega_o \to 0,$$

and the dispersion function

$$D(k, \omega_o) = \left[\omega_o - \left(Vk + \sqrt{g|k|}\right)\right]\left[\omega_o - \left(Vk - \sqrt{g|k|}\right)\right] \equiv [\omega_o - \Omega_+(k)][\omega_o - \Omega_-(k)].$$

These factors respectively vanish at $k = \mp g/V^2$ when $\omega_o = 0$. In either case $\partial\Omega_\pm/\partial k = \frac{1}{2}V$. Therefore the poles at $k = \pm g/V^2$ lie in the upper half-plane and contribute only to the radiation in $\bar{x} > 0$. In other words waves are found only to the rear of the moving pressure source. Evaluating the residue contributions, we find in the leading approximation

$$\frac{\zeta}{P_o\ell/\rho_o V^2} \approx -2\pi e^{-g\ell/V^2} \mathrm{H}(x + Vt) \sin\left[\frac{g}{V^2}(x + Vt)\right]. \tag{5.9.4}$$

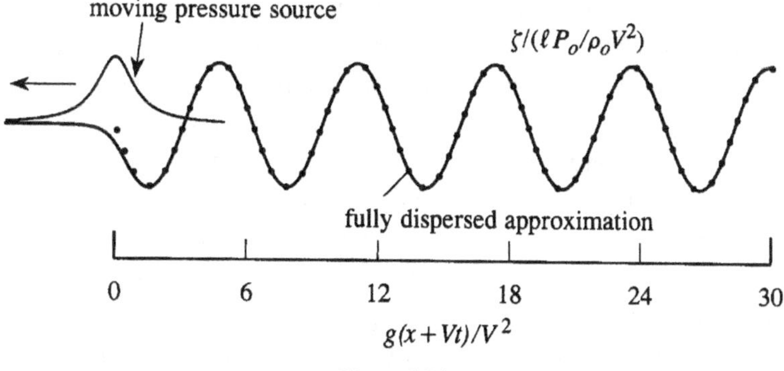

Figure 5.9.1

This sinusoidal wave pattern is stationary in a frame fixed with the moving source and consists of surface waves whose phase velocity relative to still water is just equal to the translation speed of the source. Wave energy is transmitted through the wavefield, away from the source, at the group velocity $\frac{1}{2}V$ relative to the source. No energy is radiated ahead of the source, as this would require the group velocity to exceed the phase speed.

The exact representation of the surface wave pattern involves in addition integrals respectively from the upper and lower branch cuts of $|k| = \lim_{\delta \to +0} \sqrt{k^2 + \delta^2}$ for $\bar{x} \gtrless 0$. In either case we find

$$\frac{\zeta}{P_o\ell/\rho_o V^2} = -2\pi e^{-g\ell/V^2} H(x + Vt) \sin\left[\frac{g}{V^2}(x + Vt)\right]$$

$$+ \int_0^\infty \frac{[\lambda \cos(\lambda\ell g/V^2) - \sin(\lambda\ell g/V^2)]}{\lambda^2 + 1} e^{-\lambda g|x+Vt|/V^2}\, d\lambda. \quad (5.9.5)$$

This is plotted as the solid curve in Figure 5.9.1. Fully dispersed wave approximation (5.9.4), shown dotted, is seen to differ from the exact solution only in the immediate vicinity of the moving pressure source.

When the water has finite depth h, we have $\Omega_\pm = Vk \pm \sqrt{gk\tanh(kh)}$. There are two poles at $k = \pm k_o$, where

$$\frac{V}{\sqrt{gh}} = \left[\frac{\tanh(k_o h)}{k_o h}\right]^{\frac{1}{2}}.$$

Because $\tanh(k_o h)/k_o h$ decreases monotonically from 1 to 0 over the range $0 \le k_o < \infty$, this equation has a real root only when the source speed $V < \sqrt{gh} =$ the maximum wave speed. When the source speed V exceeds $\sqrt{gh}$ no waves are formed, and the surface displacement consists of a local disturbance that travels along with the source and decays rapidly with distance from the source.

5.9.2 Wave-making resistance

The resistance experienced by a ship consists of various factors including frictional drag and wave-making. The wave drag experienced by a barge travelling along a narrow canal whose width is only marginally larger than that of the barge can be investigated by means of the one-dimensional moving pressure source of the previous section. In steady motion at speed V the surface motion in the wake of the barge consists of periodic waves of a wavelength whose phase velocity (relative to still water) is also V. The mean energy per unit area of water in the wake is $\frac{1}{2}\rho_0 g|\zeta_0|^2$. Consider the energy gain in the region between unit span of the barge and a fixed vertical plane $x = $ constant to the rear of the barge. The surface area of this region increases at rate V, and the rate of increase of wave energy is $\frac{1}{2}\rho_0 g|\zeta_0|^2 V$. However, there is also a flux of wave energy through the plane towards the barge at the group velocity (relative to still water) U, say, of the waves of phase velocity V. The rate at which energy is supplied by the barge is therefore $\frac{1}{2}\rho_0 g|\zeta_0|^2(V - U)$. This is just the rate of working RV of the wave-making resistance R per unit span. Hence the

$$\text{wave-making resistance } R = \frac{1}{2}\rho_0 g|\zeta_0|^2 \frac{(\text{phase velocity} - \text{still water group velocity})}{\text{phase velocity}}.$$

An alternative argument considers a wake of finite length. The rear end of the wake must advance at the group velocity, so that the net rate of increase in the length of the wake is $V - U$, and the corresponding net rate of increase of wake energy is again $\frac{1}{2}\rho_0 g|\zeta_0|^2(V - U)$. Or yet again, the wave power from the source is equal to $\frac{1}{2}\rho_0 g|\zeta_0|^2 \times$ (group velocity relative to the source), and this velocity is $(V - U)$.

In deep water $U = \frac{1}{2}V$, and therefore $R = \frac{1}{4}\rho_0 g|\zeta_0|^2$. For a barge spanning a canal of finite depth h, the wave speed and group velocity cannot exceed $\sqrt{gh}$, so that a steady wave pattern cannot be formed in the wake of a barge travelling faster than $V = \sqrt{gh}$. In this case the wave drag is dramatically reduced to levels controlled by that produced by two-dimensional waves at the side edges.

Formula (5.9.4) implies that for a surface disturbance of length $\sim 2\ell$ the wave resistance $R \sim e^{-2\ell g/V^2}$. It is therefore very small at low speeds and then rises to a maximum. This is typical of the measured dependence on Froude number $V^2/g\ell$ of the wave drag experienced by a ship, although details of the resistance curve depend very much on the shape of the hull. Dimensional arguments imply that, if the resistance depends only on the speed V, water density ρ_0, and gravity g, then

$$R = \rho_0 V^2 \ell^2 f\left(\frac{g\ell}{V^2}\right),$$

where ℓ is a linear dimension of the ship. The function $f(g\ell/V^2)$ of the Froude number depends on the detailed profile of the ship hull, but is the same for similarly shaped ships, and a knowledge of f obtained from model experiments can be used to predict the wave-making resistance at full scale.

A crude analytical model for estimating the wave-making resistance of a ship of length L is obtained by the placement of positive and negative pressure sources distance L apart respectively at the bow and stern to represent the corresponding high and low mean pressures at these points. The simple resistance model then predicts that

$$R \sim [\alpha - \beta\cos(gL/V^2)]e^{-2\ell g/V^2},$$

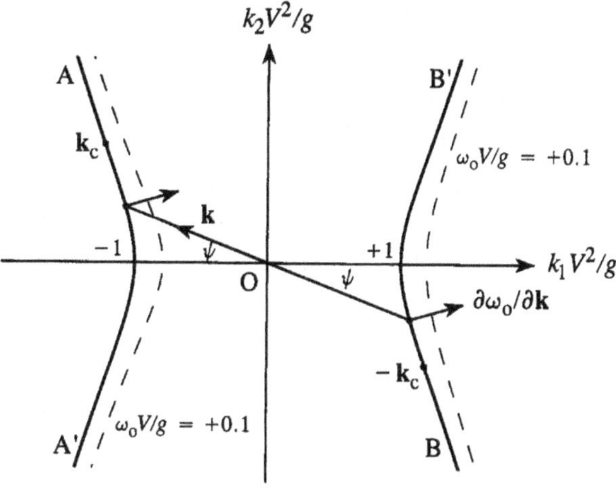

Figure 5.9.2

where the cosine factor represents effects of interference between the bow and stern waves. The constants α and β depend on the shape of the ship, and can often be chosen to give a good representation of the actual wave resistance.

5.9.3 Moving point-like pressure source

Consider the applied surface pressure distribution

$$p_o = \frac{P_o}{\{1 + [(x + Vt)^2 + y^2]/\ell^2\}^{\frac{3}{2}}}, \quad p_o(\mathbf{k}) \equiv \hat{p}_o(k) = \frac{P_o\ell^2}{2\pi}e^{-k\ell} \tag{5.9.6}$$

on water of infinite depth. The surface displacement is given by (5.9.2) with $\omega_o = 0$ and $\Omega^2 = gk$. According to the method of §5.8.2, the far-field surface waves at $\bar{\mathbf{x}}$ are associated with wavenumbers clustered around points $\mathbf{k}$ on the wavenumber curve

$$D(\mathbf{k}, \omega_o) \equiv (\omega_o - Vk_1)^2 - gk = 0$$

where condition (5.8.7) is satisfied – where the normal vector [parallel to $(\partial\omega_o/\partial\mathbf{k})_{\omega_o=0}$] is directed towards $\bar{\mathbf{x}}$. The curve has two branches (Figure 5.9.2) AA' and BB', the adjacent wavenumber curves for $\omega_o > 0$, are also plotted and indicate that the radiation is always in the direction $\bar{x}_1 > 0$ in the wake of the source.

Let us consider the radiation into $\bar{x}_1$, $\bar{x}_2 > 0$. This corresponds to the upper portion A ($k_2 > 0$) of AA' and the lower portion B of BB', for which the normals are directed into $\bar{x}_2 > 0$. It may be verified that, when $\omega_o = 0$, the pole of the integrand of (5.9.2) at

$$k_2 = -k_1\sqrt{\left(\frac{k_1 V^2}{g}\right)^2 - 1} \tag{5.9.7}$$

lies in $\operatorname{Im} k_2 > 0$ when $\epsilon > 0$ when $\mathbf{k}$ lies on A or B. Introduce the parametric representations

$$k_1 = \mp\frac{g}{V^2\cos\psi}, \quad k_2 = \pm\frac{g\sin\psi}{V^2\cos^2\psi}, \quad k = \frac{g}{V^2\cos^2\psi}, \quad 0 < \psi < \frac{\pi}{2}, \qquad (5.9.8)$$

where the upper/lower signs correspond respectively to $\mathbf{k}$ on A/B. For $\bar{x}_2 > 0$ the pole contribution to integral (5.9.2) may therefore be extracted, with the k_2 integral taken to be equal to the residues at poles (5.9.7) on A and B. The remaining integrations over A and B with respect to k_1 can then be expressed in terms of ψ in the combined form

$$\zeta = -\frac{4\pi g}{\rho_o V^4}\int_0^{\frac{\pi}{2}}\sec^4\psi\,\hat{p}_o\left(\frac{g\sec^2\psi}{V^2}\right)\sin\left[\frac{g}{V^2}\sec^2\psi\left(\bar{x}_1\cos\psi - \bar{x}_2\sin\psi\right)\right]d\psi \quad \text{for } \bar{x}_2 > 0,$$

$$= -\frac{4\pi g}{\rho_o V^4}\int_0^{\frac{\pi}{2}}\sec^4\psi\,\hat{p}_o\left(\frac{g\sec^2\psi}{V^2}\right)\sin\left[\frac{g\varpi}{V^2}\sec^2\psi\,\cos(\theta + \psi)\right]d\psi, \qquad (5.9.9)$$

where $\bar{\mathbf{x}} = \varpi(\cos\theta, \sin\theta)$.

The integration interval $0 < \psi < \psi_c = \tan^{-1}(1/\sqrt{2})$ corresponds to sections of the wavenumber curves A and B respectively between $k_2 = 0$ and the points labelled $\pm\mathbf{k}_c$, where

$$\frac{V^2\mathbf{k}_c}{g} = \left(-\sqrt{\frac{3}{2}}, \frac{\sqrt{3}}{2}\right).$$

These are singular points on the wavenumber curve where the curvature vanishes, i.e. where the group velocity is stationary. Now values of the wavenumber $\mathbf{k}$ are propagated without change along rays from the source in the direction of the group velocity. Because $\mathbf{k}$ is normal to the wave crests, those crests determined by 'small' wavenumbers $k < |\mathbf{k}_c|$ are more or less transverse to the direction of motion of the source (and similar to those generated by the one-dimensional source of §5.9.1). Those wavenumbers $\mathbf{k}$ 'above'/'below' $\pm\mathbf{k}_c$ on A and B [corresponding to the interval $\psi_c < \psi < \frac{\pi}{2}$ in (5.9.9)] represent shorter wavelength 'lateral' or diverging waves, with wave crests roughly aligned with the direction of motion of the source. However, because $\mathbf{k} = \pm\mathbf{k}_c$ are points of inflexion of the wavenumber curve, the group velocities at these points are inclined at the maximum possible angle θ_c to the $\bar{x}_1$ axis, called the *Kelvin ship wave angle*. On A, for example, $\omega_o = Vk_1 + \sqrt{gk}$ at $\omega_o = 0$, with group velocity

$$\frac{\partial\omega_o}{\partial\mathbf{k}} = \left(V + \frac{k_1\sqrt{g}}{2k^{\frac{3}{2}}}, \frac{k_2\sqrt{g}}{2k^{\frac{3}{2}}}\right) = V\left(\frac{2}{3}, \frac{1}{3\sqrt{2}}\right) \quad \text{at } \mathbf{k}_c,$$

so that the Kelvin angle $\theta_c = \sin^{-1}\frac{1}{3} \approx 19.47°$. Both families of waves are therefore confined to the same wedge-like region of angular width $\sim 38.94°$ in the wake of the source.

The remaining integration with respect to ψ in (5.9.9) is performed by the method of stationary phase. Before doing this, however, let us examine the expected shape of the transverse and lateral wave crests. The dominant contribution to integral

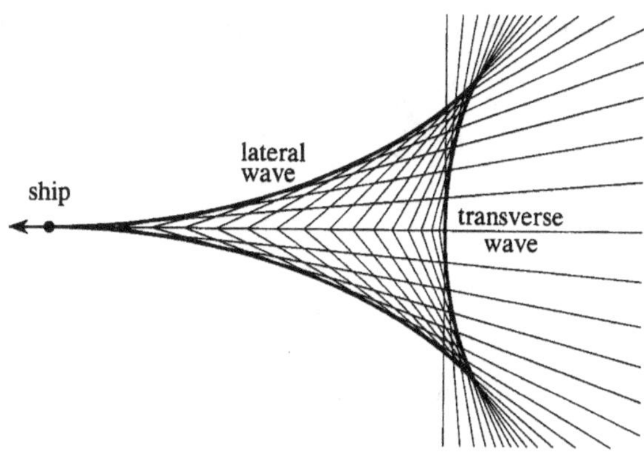

Figure 5.9.3

(5.9.9) is from values of ψ in the neighbourhood of the stationary points of the phase

$$\Theta = \frac{g}{V^2}\sec^2\psi\left(\bar{x}_1\cos\psi - \bar{x}_2\sin\psi\right).$$

The phase of each constituent of the integral is constant on a straight line. The stationary phase condition that $\partial\Theta/\partial\psi = 0$ implies that a wave crest is the *envelope* of this family of straight lines on which $\Theta = $ constant when ψ varies over its domain of definition. A typical family of constant phase lines is plotted in Figure 5.9.3. The interval $0 < \psi < \psi_c$ defines the upper half of the transverse wave crest; the reader can easily verify that the lower half of this wave corresponds to $-\psi_c < \psi < 0$. Similarly, the upper and lower lateral wave crests correspond to $\psi_c < |\psi| < \frac{\pi}{2}$. The transverse and lateral crests converge at the critical wavenumber $\mathbf{k}_c$. Because $\mathbf{k}_c$ is normal to the wavefront, the crests of the two wave families meet along a line inclined at an angle $\tan^{-1}(\sqrt{2}) = 54.74°$ to the wake axis.

Detailed confirmation of these predictions is obtained by evaluation of the stationary phase approximation to representation (5.9.9). When $\bar{x}_2 > 0$ the stationary values of $\Theta = \frac{g\varpi}{V^2}\sec^2\psi\,\cos(\theta + \psi)$ occur at the zeros ψ of

$$\Theta' = \sec^2\psi\,\cos(\theta + \psi)[2\tan\psi - \tan(\theta + \psi)] \tag{5.9.10}$$

in $[0, \frac{\pi}{2}]$, i.e., where

$$\tan(\theta + \psi) = 2\tan\psi,$$

$$\text{or}\quad \tan\psi = \frac{1 \pm \sqrt{1 - 8\tan^2\theta}}{4\tan\theta}, \quad 0 \le \theta < \sin^{-1}\frac{1}{3}, \tag{5.9.11}$$

where the upper/lower sign corresponds to the lateral/transverse surface wave. At these points the second and third derivatives with respect to ψ are given by

$$\Theta'' = \frac{g\varpi}{V^2}\sec^2\psi\,\cos(\theta + \psi)[1 + 6\tan^2\psi - 4\tan\psi\,\tan(\theta + \psi)],$$

$$\tag{5.9.12}$$

$$\Theta''' = \frac{g\varpi}{V^2}\sec^2\psi\,\cos(\theta + \psi)[10\tan\psi + 24\tan^3\psi - (5 + 18\tan^2\psi)\tan(\theta + \psi)].$$

In terms of these definitions the stationary phase approximation is given by

$$\zeta \approx \frac{-4\pi g}{\rho_o V^4} \sec^4\psi \, \hat{p}_o \left(\frac{g\sec^2\psi}{V^2}\right)\left(\frac{2\pi}{|\Theta''|}\right)^{\frac{1}{2}} \sin\left(\Theta + \frac{\pi}{4}\operatorname{sgn}\Theta''\right), \quad \frac{\varpi g}{V^2} \to \infty, \quad (5.9.13)$$

and in explicit form by

$$\zeta \approx \frac{-\pi^{\frac{1}{2}}}{2^{\frac{9}{4}}}\left(\frac{P_o}{\rho_o g}\right)\left(\frac{\ell g}{V^2}\right)^2\left(\frac{V^2}{\varpi g}\right)^{\frac{1}{2}} \frac{\sec^{\frac{1}{2}}\theta\left[3 \mp (1-8\tan^2\theta)^{\frac{1}{2}}\right]^{\frac{5}{2}}}{(1-8\tan^2\theta)^{\frac{1}{4}}\left[1-2\tan^2\theta \mp (1-8\tan^2\theta)^{\frac{1}{2}}\right]^{\frac{5}{4}}}$$

$$\times \exp\left\{-\frac{g\ell}{V^2}\frac{\left[3\mp(1-8\tan^2\theta)^{\frac{1}{2}}\right]^2}{8\left[1-2\tan^2\theta\mp(1-8\tan^2\theta)^{\frac{1}{2}}\right]}\right\}$$

$$\times \sin\left\{\frac{g\varpi}{V^2}\frac{\cos\theta\left[3\mp(1-8\tan^2\theta)^{\frac{1}{2}}\right]^2}{8\sqrt{2}\left[1-2\tan^2\theta\mp(1-8\tan^2\theta)^{\frac{1}{2}}\right]^{\frac{1}{2}}} \mp \frac{\pi}{4}\right\}, \quad \frac{\varpi g}{V^2} \to \infty. \quad (5.9.14)$$

The transverse waves ($+$ sign) dominate the inner wake region and have the longer wavelengths, the maximum wavelength being $2\pi V^2/g$ on the wake axis. The maximum wavelength of the shorter lateral waves (corresponding to $\mathbf{k} = \mathbf{k}_c$) is $4\pi V^2/3g$, this being equal also to the minimum wavelength of the transverse waves. At any point $\bar{\mathbf{x}} = \varpi(\cos\theta, \sin\theta)$ within a wedge of semi-angle of $\sim 19.47°$ two wave crests intersect – a lateral wave [upper sign in (5.9.14)] and a transverse wave (lower sign). The waves intersect at an angle χ equal to that between the corresponding wavenumbers $\mathbf{k}_L$ and $\mathbf{k}_T$, say, where

$$\frac{\mathbf{k}_L V^2}{g} = \sec^2\psi_-\,(\cos\psi_-, \sin\psi_-), \quad \frac{\mathbf{k}_T V^2}{g} = \sec^2\psi_+\,(\cos\psi_+, \sin\psi_+)$$

and [see (5.9.11)]
$$\tan\psi_\pm = \frac{1 \pm \sqrt{1-8\tan^2\theta}}{4\tan\theta}, \quad 0 \le \theta < \sin^{-1}\frac{1}{3}.$$

A simple calculation shows that

$$\cos\chi = \frac{\mathbf{k}_L \cdot \mathbf{k}_T}{|\mathbf{k}_L \cdot \mathbf{k}_T|} = 3\sin\theta, \quad 0 < \sin\theta < \frac{1}{3}, \quad (5.9.15)$$

so that the angle of intersection decreases from $90°$ to $0°$ as θ sweeps over the range $0° < \theta < 19.47°$. The relative amplitudes of these waves along any given ray θ can differ greatly, however. The wave crests are parallel and are inclined at $54.74°$ to the wake axis at the edge of the Kelvin wedge ($\theta \sim 19.47°$), where the converging constant phase curves of the two wave systems merge to form a *cusp*. However, wave crest does not merge with wave crest because there is a phase difference of $\frac{\pi}{2}$, although the wave amplitudes are equal.

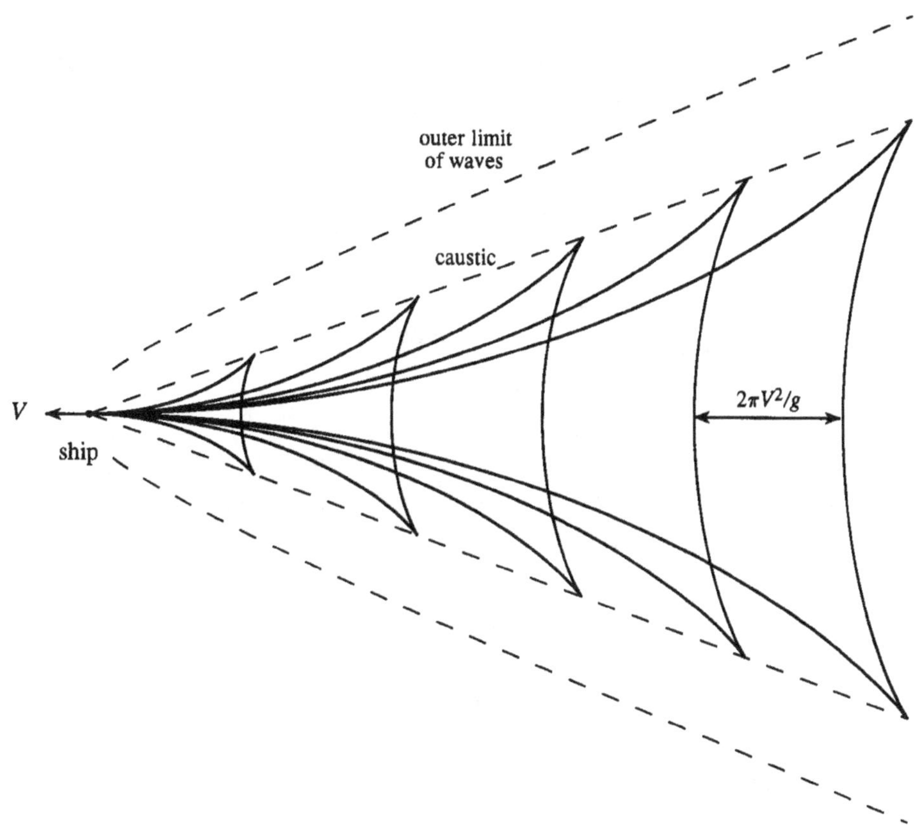

Figure 5.9.4

5.9.4 Plotting the wave crests

The pattern of wave crests (the solid curves in Figure 5.9.4) is defined by the family of
curves on which the phase of surface displacement (5.9.14) (the argument of the sine)
takes constant values differing by 2π, i.e. by the polar equation

$$\frac{g\varpi}{V^2} = \frac{8\sqrt{2}\left[1 - 2\tan^2\theta \mp (1 - 8\tan^2\theta)^{\frac{1}{2}}\right]^{\frac{1}{2}}}{\cos\theta\left[3 \mp (1 - 8\tan^2\theta)^{\frac{1}{2}}\right]^2}\left(2n \mp \frac{1}{4}\right)\pi, \quad n = 1, 2, 3, \ldots,$$

where the upper/lower sign is taken for the lateral/transverse wave. Waves of the maximum wavelength $2\pi V^2/g$ are transverse and follow in the wake of the source.

The wave crests are also defined by the equation

$$\mathbf{k} \cdot \bar{\mathbf{x}} = \bar{\Theta}, \tag{5.9.16}$$

where the right-hand side $\bar{\Theta}$ is constant. The normal to the wave crest at $\bar{\mathbf{x}}$ is parallel to
$\mathbf{k}$, which is a slowly varying function of $\bar{\mathbf{x}}$ and is constant along the ray through $\bar{\mathbf{x}}$ from
the source. Therefore, for a given value of $\mathbf{k}$,

$$\bar{\mathbf{x}} = |\bar{\mathbf{x}}|\frac{\partial\omega_o/\partial\mathbf{k}}{|\partial\omega_o/\partial\mathbf{k}|},$$

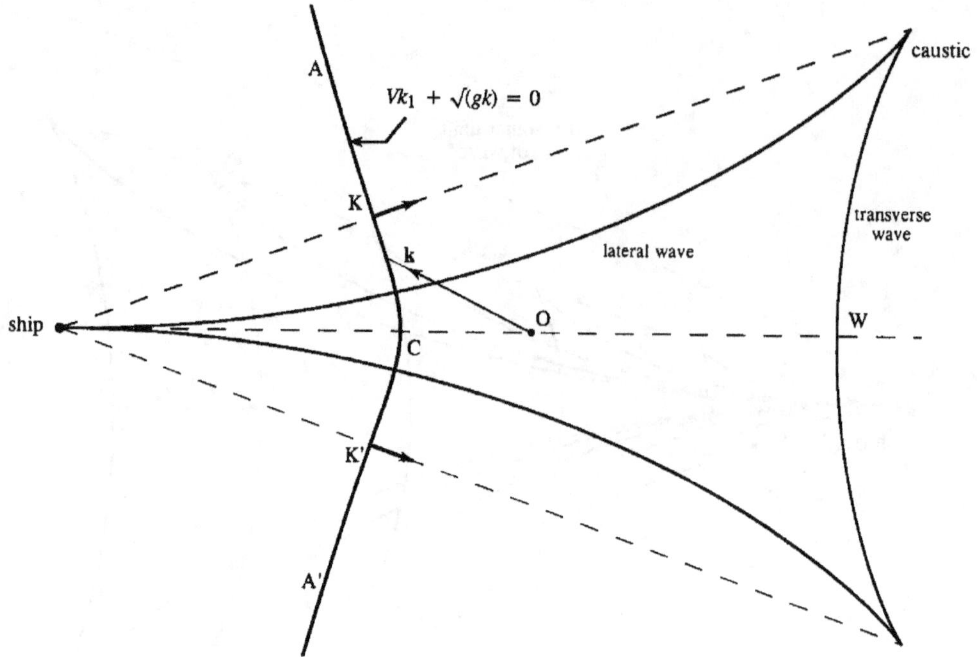

Figure 5.9.5

so that on the wave crest (5.9.16)

$$|\bar{\mathbf{x}}| = \frac{\bar{\Theta}\,|\partial\omega_o/\partial\mathbf{k}|}{\mathbf{k}\cdot\partial\omega_o/\partial\mathbf{k}} \quad\text{and}\quad \bar{\mathbf{x}} = \frac{\bar{\Theta}\,\partial\omega_o/\partial\mathbf{k}}{\mathbf{k}\cdot\partial\omega_o/\partial\mathbf{k}}. \tag{5.9.17}$$

When $\bar{\Theta}$ is prescribed, the second of these equations gives the position $\bar{\mathbf{x}}$ on the curve $\bar{\Theta} = $ constant that receives radiation from the point $\mathbf{k}$ on the wavenumber curve $D(\mathbf{k},\omega_o)=0$ ($\omega_o = 0$). In other words, it provides a parametric representation of the curve of constant phase in terms of admissible values of $\mathbf{k}$, which are restricted to lie on $D(\mathbf{k},\omega_o)=0$ for $\omega_o = 0$.

Thus on the branch AA' ($Vk_1 + \sqrt{gk} = 0$) of Figure 5.9.5,

$$\mathbf{k} = \frac{g}{V^2}\left(-\frac{1}{\cos\psi},\frac{\sin\psi}{\cos^2\psi}\right),\quad \frac{\partial\omega_o}{\partial\mathbf{k}} = V\left(1-\frac{\cos^2\psi}{2},\frac{\sin\psi\,\cos\psi}{2}\right),\quad -\frac{\pi}{2}<\psi<\frac{\pi}{2},$$

so that (5.9.17) becomes

$$\frac{g\bar{x}_1}{V^2} = -2\bar{\Theta}\cos\psi\left(1-\frac{\cos^2\psi}{2}\right),\quad \frac{g\bar{x}_2}{V^2} = -\bar{\Theta}\sin\psi\,\cos^2\psi. \tag{5.9.18}$$

These equations define a transverse wave crest when ψ varies within the range $|\psi| < \tan^{-1}(1/\sqrt{2}) = \psi_c = 35.26°$ (between K' and K in Figure 5.9.5). To obtain the plot shown it is necessary to choose a constant *negative* value of $\bar{\Theta}$; the reader can verify that this actually corresponds to a positive value of the argument of the sine in (5.9.14). The lateral wave is defined by the intervals $\psi_c < |\psi| < \frac{\pi}{2}$.

5.9.5 Behaviour at the caustic

The amplitudes of both the lateral and transverse waves are predicted by stationary phase formula (5.9.13) to be infinite along the line of cusps, which is called a *caustic*, where $\theta = \theta_c \equiv \sin^{-1}\frac{1}{3}$, $\Theta'' = 0$, and the stationary phase approximation fails. This corresponds to the critical wavenumber $\mathbf{k}_c$ at which the group velocity $\partial\omega_o/\partial\mathbf{k}$ is stationary. According to (5.9.13) the net surface displacement near $\theta = \theta_c$, produced by the lateral and transverse waves, is given approximately by

$$\zeta \sim -\frac{2^{\frac{1}{8}}3^{\frac{5}{4}}\pi^{\frac{1}{2}}}{|\theta_c - \theta|^{\frac{1}{4}}}\left(\frac{P_o}{\rho_o g}\right)\left(\frac{\ell g}{V^2}\right)^2\left(\frac{V^2}{\varpi g}\right)^{\frac{1}{2}}\exp\left(-\frac{3 g \ell}{2V^2}\right)\sin\left(\frac{3^{\frac{1}{2}}g\varpi}{2V^2}\right), \quad \frac{\varpi g}{V^2} \to \infty, \quad \theta \sim \theta_c.$$

$$(5.9.19)$$

To determine the behaviour of ζ on the cusp line $\theta = \theta_c$ we revert to integral representation (5.9.9) and approximate it for $\theta \sim \theta_c$ by expanding the phase to *third* order in $\psi - \psi_c$ [where $\psi_c = \tan^{-1}(1/\sqrt{2})$] and first order in $\theta - \theta_c$. Then

$$\Theta \equiv \frac{g\varpi}{V^2}\sec^2\psi\,\cos(\theta + \psi)$$

$$\approx \Theta_c + \left(\frac{\partial^2\Theta}{\partial\theta\,\partial\psi}\right)_c (\theta - \theta_c)\lambda + \frac{1}{6}\left(\frac{\partial^3\Theta}{\partial\psi^3}\right)_c \lambda^3$$

$$= \frac{g\varpi}{V^2}\frac{\sqrt{3}}{2}\left[1 - 3(\theta - \theta_c)\lambda - \frac{\lambda^3}{\sqrt{2}}\right], \quad \lambda = \psi - \psi_c.$$

Then (5.9.9) becomes

$$\zeta \approx -\frac{3^2\pi g}{\rho_o V^4}\hat{p}_o\left(\frac{3g}{2V^2}\right)\sin\left(\frac{\sqrt{3}}{2}\frac{g\varpi}{V^2}\right)\int_{-\infty}^{\infty}\cos\left\{\frac{g\varpi}{V^2}\frac{\sqrt{3}}{2\sqrt{2}}[\lambda^3 + 3\sqrt{2}(\theta - \theta_c)\lambda]\right\}d\lambda$$

$$= -\frac{3^{\frac{3}{2}}(2\pi)^2 g}{2^{\frac{1}{2}}\rho_o V^4}\left(\frac{V^2}{g\varpi}\right)^{\frac{1}{3}}\hat{p}_o\left(\frac{3g}{2V^2}\right)\sin\left(\frac{3^{\frac{1}{2}}}{2}\frac{g\varpi}{V^2}\right)\mathrm{Ai}\left[\left(\frac{g\varpi}{V^2}\right)^{\frac{2}{3}}\frac{3}{\sqrt{2}}(\theta - \theta_c)\right],$$

where Ai is the Airy function of §5.7. Substituting for $\hat{p}_o$, we obtain for the behaviour near the caustic

$$\zeta \approx -2^{\frac{1}{2}}3^{\frac{3}{2}}\pi\left(\frac{P_o}{\rho_o g}\right)\left(\frac{g\ell}{V^2}\right)^2\left(\frac{V^2}{g\varpi}\right)^{\frac{1}{3}}\exp\left(-\frac{3g\ell}{2V^2}\right)\sin\left(\frac{3^{\frac{1}{2}}}{2}\frac{g\varpi}{V^2}\right)$$

$$\times\mathrm{Ai}\left[\left(\frac{g\varpi}{V^2}\right)^{\frac{2}{3}}\frac{3}{\sqrt{2}}(\theta - \theta_c)\right]. \tag{5.9.20}$$

When θ is slightly smaller than the Kelvin angle $\theta_c \approx 19.47°$ the Airy function can be replaced with

$$\frac{2^{\frac{1}{8}}}{\pi^{\frac{1}{2}}3^{\frac{1}{4}}|\theta - \theta_c|^{\frac{1}{4}}}\left(\frac{V^2}{g\varpi}\right)^{\frac{1}{6}}\sin\left[2^{\frac{1}{4}}3^{\frac{1}{2}}\left(\frac{g\varpi}{V^2}\right)|\theta - \theta_c|^{\frac{3}{2}} + \frac{\pi}{4}\right] \sim \frac{2^{-\frac{3}{8}}}{\pi^{\frac{1}{2}}3^{\frac{1}{4}}|\theta - \theta_c|^{\frac{1}{4}}}\left(\frac{V^2}{g\varpi}\right)^{\frac{1}{6}},$$

in which case (5.9.20) reduces to stationary phase approximation (5.9.19).

The Airy function takes its maximum value ~ 0.536 very near $\theta = \theta_c$ when ϖ is large, so that the wave amplitude on the cusp line decreases with ϖ like $1/\varpi^{\frac{1}{3}}$, more slowly than for the transverse waves in the inner, long-wavelength region of the wake. Indeed on the nth transverse wave crest in the wake $g\varpi/V^2 \sim 2n\pi$, so that the ratio

$$\frac{\text{cusp line amplitude}}{\text{transverse wave amplitude}} \sim n^{\frac{1}{6}},$$

and therefore waves near the caustic tend to increase in prominence in the far wake.

Outside the Kelvin wedge $(\theta > \theta_c)$ the Airy function diminishes rapidly (Figure 5.7.2) into the region of undisturbed water. It becomes negligible when its argument exceeds about 2.5, i.e. at a perpendicular distance s from the line of cusps given by

$$\frac{gs}{V^2} \sim \frac{2.5\sqrt{2}}{3} \left(\frac{g\varpi}{V^2}\right)^{\frac{1}{3}}.$$

This distance defines the outer boundary of the wave motion, and actually represents substantial additional region of influence beyond the Kelvin wedge (Figure 5.9.4).

5.9.6 Wave-making power

The wave power Π generated by the moving ship can be evaluated by Whitham's method described in §5.8.3. For the source (5.9.6) the surface displacement is given by the real-valued expression

$$\zeta = \frac{1}{\rho_o} \int_{-\infty}^{\infty} \frac{k\hat{p}_o(k)}{[(i\epsilon - Vk_1)^2 - gk]} e^{i(\mathbf{k}\cdot\bar{\mathbf{x}}+\epsilon t)} d^2\mathbf{k}.$$

Therefore

$$\int_{-\infty}^{\infty} \langle \rho_o g \zeta^2 \rangle d^2\bar{\mathbf{x}} = \frac{(2\pi)^2 g}{\rho_o} \int_{-\infty}^{\infty} \frac{k^2 |\hat{p}_o(k)|^2 e^{2\epsilon t} d^2\mathbf{k}}{|(i\epsilon - Vk_1)^2 - gk|^2}.$$

Now, when $\epsilon \ll 1$ we can write $|(i\epsilon - Vk_1)^2 - gk|^2 = \{[(Vk_1)^2 - gk]^2 + 4\epsilon^2 V^2 k_1^2\}$, and

$$\lim_{\epsilon \to +0} \frac{2\epsilon|Vk_1|}{\{[(Vk_1)^2 - gk]^2 + 4\epsilon^2 V^2 k_1^2\}} = \pi\,\delta[(Vk_1)^2 - gk].$$

Hence the wave-making power $\Pi = \frac{\partial}{\partial t} \int_{-\infty}^{\infty} \langle \rho_o g \zeta^2 \rangle d^2\bar{\mathbf{x}}$ is given by

$$\Pi = \frac{4\pi^3 g}{\rho_o V} \int_{-\infty}^{\infty} \frac{k^2 |\hat{p}_o(k)|^2}{|k_1|} \delta[(Vk_1)^2 - gk]\, d^2\mathbf{k}. \tag{5.9.21}$$

The δ function confines the integration to the two branches of the wavenumber curve $D(\mathbf{k}, \omega_o)_{\omega_o=0} \equiv (Vk_1)^2 - gk = 0$. By integrating first with respect to k_2 and then using parametric representations of the type (5.9.8) on the wavenumber curve, we find

$$\Pi = \frac{8\pi^3 g^2}{\rho_o V^5} \int_{-\frac{\pi}{2}}^{\frac{\pi}{2}} \sec^5\psi \left|\hat{p}_o\left(\frac{g\sec^2\psi}{V^2}\right)\right|^2 d\psi. \tag{5.9.22}$$

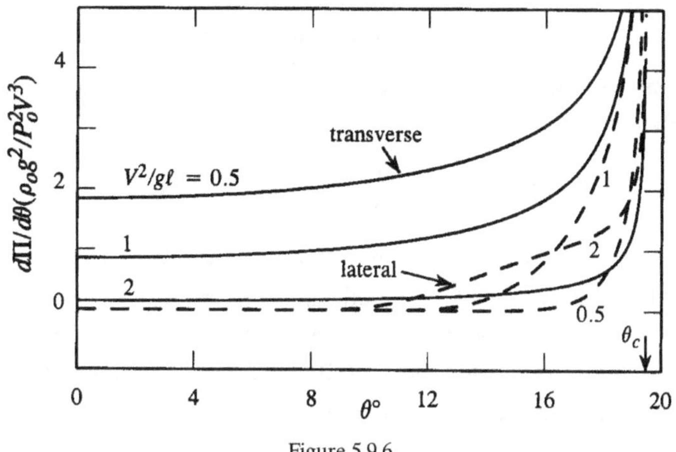

Figure 5.9.6

The integration ranges $|\psi| < \psi_c$ and $\psi_c < |\psi| < \frac{\pi}{2}$ $[\psi_c = \tan^{-1}(1/\sqrt{2})]$ correspond respectively to the power entering the transverse and lateral surface waves. By means of transformation (5.9.11) we can therefore write

$$\Pi = \frac{\pi^3 g^2}{2^{\frac{5}{2}} \rho_o V^5} \int_{-\theta_c}^{\theta_c} \frac{\sec^2\theta[3 \mp \sqrt{1 - 8\tan^2\theta}]^4}{\sqrt{1 - 8\tan^2\theta}(1 - 2\tan^2\theta \mp \sqrt{1 - 8\tan^2\theta})^{\frac{5}{2}}}$$

$$\times \left| \hat{p}_o \left[\frac{g}{V^2} \frac{[3 \mp \sqrt{1 - 8\tan^2\theta}]^2}{8(1 - 2\tan^2\theta \mp \sqrt{1 - 8\tan^2\theta})} \right] \right|^2 d\theta. \tag{5.9.23}$$

There are actually two integrals, the upper/lower sign corresponding respectively to the power delivered to the lateral and transverse surface waves; in each case the integrand represents the angular distribution of wave power within the Kelvin wedge. The inverse-square-root singularity at the caustic is integrable, so that the power is always finite.

When the ship source is defined as in (5.9.6), we can write

$$\Pi = \frac{\pi}{2^{\frac{9}{2}}} \frac{P_o^2 V^3}{\rho_o g^2} \left(\frac{g\ell}{V^2} \right)^4 \int_{-\theta_c}^{\theta_c} \frac{\sec^2\theta[3 \mp \sqrt{1 - 8\tan^2\theta}]^4}{\sqrt{1 - 8\tan^2\theta}(1 - 2\tan^2\theta \mp \sqrt{1 - 8\tan^2\theta})^{\frac{5}{2}}}$$

$$\times \exp\left[-\frac{g\ell}{V^2} \frac{[3 \mp \sqrt{1 - 8\tan^2\theta}]^2}{4(1 - 2\tan^2\theta \mp \sqrt{1 - 8\tan^2\theta})} \right] d\theta. \tag{5.9.24}$$

Figure 5.9.6 indicates the variation of $d\Pi/d\theta$ for each surface wave mode with the group velocity direction θ for three different values of the Froude number $V^2/g\ell$. A low Froude number favours the transverse modes; the lateral modes are more prominent when $V^2/g\ell$ becomes large.

5.9.7 Wave amplitude calculated from the power

In the fully dispersed far field at a radial distance ϖ from the ship, the wave power within the Kelvin wedge is also given by

$$\Pi = \sum \int_{-\theta_c}^{\theta_c} \frac{1}{2} \rho_o g |\zeta_o|^2 \left| \frac{\partial \omega_o}{\partial \mathbf{k}} \right| \varpi \, d\theta, \qquad (5.9.25)$$

where ζ_o is the local sinusoidal wave amplitude and the summation is over the two separate contributions from the lateral and transverse modes.

The group velocity is evaluated on the branches of the wavenumber curve in Figure 5.9.2, which have parametric representations (5.9.8), in terms of which $|\partial \omega_o / \partial \mathbf{k}| = V\sqrt{1 - \frac{3}{4}\cos^2\psi}$. Transformation (5.9.11) then yields

$$\left| \frac{\partial \omega_o}{\partial \mathbf{k}} \right| = \frac{2V \sec\theta}{3 \mp \sqrt{1 - 8\tan^2\theta}}$$

the upper/lower sign referring to the lateral/transverse mode. Hence, by equating the integrands of (5.9.25) and (5.9.24), we find

$$|\zeta_o|^2 = \frac{\pi}{2^{\frac{9}{2}}} \left(\frac{P_o}{\rho_o g} \right)^2 \left(\frac{g\ell}{V^2} \right)^4 \left(\frac{V^2}{g\varpi} \right) \frac{\sec\theta \, (3 \mp \sqrt{1 - 8\tan^2\theta})^5}{\sqrt{1 - 8\tan^2\theta}(1 - 2\tan^2\theta \mp \sqrt{1 - 8\tan^2\theta})^{\frac{5}{2}}}$$

$$\times \exp\left[-\frac{g\ell}{V^2} \frac{(3 \mp \sqrt{1 - 8\tan^2\theta})^2}{4(1 - 2\tan^2\theta \mp \sqrt{1 - 8\tan^2\theta})} \right],$$

which confirms result given by stationary phase approximation (5.9.14).

5.10 Ray theory

The propagation of waves over water of continuously variable depth cannot usually be treated analytically. However, when significant changes in the depth occur only over distances that are large compared with the surface wavelength it is possible to obtain an approximate representation of the wave motion by extension of the notion of group velocity to a medium with *slowly varying properties*. This is the basis of *ray theory*. The idea is quite general and applicable to waves of arbitrary type (sound, electromagnetic, elastic, etc.) in general, anisotropic wave-bearing media.

5.10.1 Kinematic theory of wave crests

Consider first a description of wave propagation through a uniform, wave-bearing medium in terms of a linear partial differential equation with constant coefficients. Solutions exist in the form of plane waves of the type $\zeta = \zeta_o e^{i(k_i x_i - \omega t)}$, where the amplitude ζ_o, the wavenumber $\mathbf{k}$, and frequency ω are each constant. The values of $\mathbf{k}$ and ω are related by the dispersion relation

$$D(\mathbf{k}, \omega) = 0. \qquad (5.10.1)$$

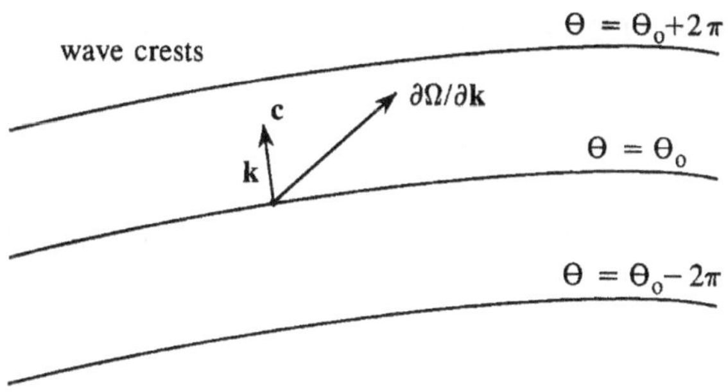

Figure 5.10.1

The phase $\Theta = k_i x_i - \omega t$ has the obvious properties

$$k_i = \frac{\partial\Theta}{\partial x_i}, \quad \omega = -\frac{\partial\Theta}{\partial t}. \tag{5.10.2}$$

We can also put $\zeta = \zeta_o e^{i(k_i x_i - \omega t)}$ in the *fully dispersed far field* of a localized source of waves where now, however, the amplitude $\zeta_o \equiv \zeta_o(x_i, t)$, wavenumber $\mathbf{k} \equiv \mathbf{k}(x_i, t)$, and frequency $\omega \equiv \omega(x_i, t)$ vary *slowly* with position and time on scales of the wavelength $2\pi/k$ and period $2\pi/\omega$. For example, in the case of a wave source near the coordinate origin that generates waves for a short time interval near $t = 0$, the stationary phase method of §§5.5–5.7 shows that the wavenumber and frequency present at $\mathbf{x}$ at time t in the fully dispersed far field are still related by dispersion formula (5.10.1) and are determined as a function of x_i/t by the group velocity condition

$$\frac{\partial\omega}{\partial k_i} = \frac{x_i}{t}, \quad \text{where} \quad \frac{\partial\omega}{\partial k_i} \equiv -\frac{\partial D(\mathbf{k}, \omega)/\partial k_i}{\partial D(\mathbf{k}, \omega)/\partial\omega}.$$

Moreover, the solutions of these equations continue to satisfy (5.10.2) because, for example,

$$\frac{\partial\Theta}{\partial x_i} \equiv \frac{\partial}{\partial x_i}(k_j x_j - \omega t) = k_i + t\frac{\partial k_j}{\partial x_i}\left(\frac{x_j}{t} - \frac{\partial\omega}{\partial k_j}\right) \equiv k_i.$$

These conclusions actually permit the wave problem in the fully dispersed far field to be reformulated in terms of phase function Θ. We now *define* the local values of the wavenumber and frequency by relations (5.10.2). They are required to satisfy dispersion relation (5.10.1), which now assumes the form of a nonlinear, first-order partial differential equation

$$D\left(\frac{\partial\Theta}{\partial x_i}, -\frac{\partial\Theta}{\partial t}\right) = 0 \tag{5.10.3}$$

called the *Hamilton–Jacobi* equation.

The phase Θ is constant on the wave crests and changes in value by $\pm2\pi$ between neighbouring crests (Figure 5.10.1). The normal to the wave crests is in the direction of $\mathbf{k} = \nabla\Theta$, and the frequency $\frac{\omega}{2\pi} = -\frac{1}{2\pi}\frac{\partial\Theta}{\partial t}$ is the rate at which crests advance past a fixed position in space. Surfaces of constant phase move at the *phase velocity* **c**, which satisfies

$$\frac{\partial \Theta}{\partial t} + c_j \frac{\partial \Theta}{\partial x_j} = 0,$$

and which is necessarily normal to the wavefronts, so that $c_j k_j / k = \omega / k$. The phase and its derivatives are kinematic variables, and Equations (5.10.2) imply that fully dispersed waves in an arbitrary medium satisfy

$$\frac{\partial k_i}{\partial t} + \frac{\partial \omega}{\partial x_i} = 0. \tag{5.10.4}$$

Different media are distinguished by their Hamilton–Jacobi equations (5.10.3), whose functional form in any particular case depends on the mechanical properties of the medium. This equation [and dispersion relation (5.10.1)] can actually represent several different wave types. For each wave type it becomes

$$\frac{\partial \Theta}{\partial t} + \Omega \left(\frac{\partial \Theta}{\partial x_i} \right) = 0, \quad \text{i.e.,} \quad \omega = \Omega(\mathbf{k}),$$

where $\omega - \Omega(\mathbf{k})$ is one of the factors of $D(\mathbf{k}, \omega)$. The equation is usually solved by the method of characteristics. The characteristics are more commonly called 'rays' and are solutions $\mathbf{x} = \mathbf{x}(t)$ of

$$\frac{d\mathbf{x}}{dt} = \frac{\partial \Omega}{\partial \mathbf{k}}.$$

This equation must be solved simultaneously with equations governing the variations of $\mathbf{k}$ and ω along the rays. Equations (5.10.2) imply that

$$\frac{\partial k_i}{\partial t} + \frac{\partial \Omega}{\partial k_j} \frac{\partial k_j}{\partial x_i} = 0.$$

However, because $\partial k_j / \partial x_i = \partial^2 \Theta / \partial x_i \partial x_j = \partial k_i / \partial x_j$, it follows that

$$\frac{\partial k_i}{\partial t} + \frac{\partial \Omega}{\partial k_j} \frac{\partial k_i}{\partial x_j} = 0,$$

$$\text{i.e.} \quad \frac{d\mathbf{k}}{dt} = 0 \quad \text{along a characteristic.}$$

Thus the wavenumber vector $\mathbf{k}$ propagates without change along a ray and so, also, does the frequency $\omega = \Omega(\mathbf{k})$. The rays are therefore straight lines parallel to the group velocity $\partial \Omega / \partial \mathbf{k}$ determined by the value of $\mathbf{k}$.

These conclusions are all restatements of results obtained previously from the principle of stationary phase. Once the rays have been found the distribution of wave energy is deduced from the energy equation

$$\frac{\partial E}{\partial t} + \frac{\partial}{\partial x_j} \left(\frac{\partial \Omega}{\partial k_j} E \right) = 0, \tag{5.10.5}$$

which states that the wave-energy density E propagates along rays at the group velocity.

5.10.2 Ray tracing in an inhomogeneous medium

The same procedure can be applied to study the propagation of fully dispersed waves in a medium that changes 'slowly' on a scale of wavelength. We put $\zeta = \zeta_o e^{i\Theta}$ and define the local wavenumber and frequency as in Equations (5.10.2). The dispersion relation $D(\mathbf{k}, \omega, \mathbf{x}) = 0$ now exhibits a slow dependence on position, and each possible wave type satisfies

$$\frac{\partial \Theta}{\partial t} + \Omega\left(\frac{\partial \Theta}{\partial x_i}, x_i\right) = 0, \quad \text{i.e.,} \quad \omega = \Omega(\mathbf{k}, \mathbf{x}), \tag{5.10.6}$$

Thus at different positions $\mathbf{x}$ in the medium waves of a given frequency will generally have different wavenumbers $\mathbf{k}$.

We now have the two relations

$$\frac{\partial \omega}{\partial x_i} = \frac{\partial \Omega}{\partial k_j}\frac{\partial k_j}{\partial x_i} + \frac{\partial \Omega}{\partial x_i} \quad \text{and} \quad \frac{\partial \omega}{\partial t} = \frac{\partial \Omega}{\partial k_j}\frac{\partial k_j}{\partial t}$$

from which Equations (5.10.2) and (5.10.4) imply respectively that

$$\frac{\partial k_i}{\partial t} + \frac{\partial \Omega}{\partial k_j}\frac{\partial k_i}{\partial x_j} = -\frac{\partial \Omega}{\partial x_i} \quad \text{and} \quad \frac{\partial \omega}{\partial t} + \frac{\partial \Omega}{\partial k_j}\frac{\partial \omega}{\partial x_j} = 0.$$

Hence the system of characteristic equations now becomes

$$\frac{d\mathbf{x}}{dt} = \frac{\partial \Omega}{\partial \mathbf{k}}, \quad \frac{d\mathbf{k}}{dt} = -\frac{\partial \Omega}{\partial \mathbf{x}}, \quad \frac{d\omega}{dt} = 0. \tag{5.10.7}$$

They determine the variation of wavenumber for an observer travelling along a ray at the group velocity $\partial\Omega/\partial\mathbf{k}$. For such an observer the frequency ω remains constant. The variation of the group velocity with position and time means that the waves are now *refracted* by the medium.

In a nondissipative, stationary medium governed by linear equations with time-independent coefficients, wave modes of different frequencies propagate independently, and therefore each mode conserves its wave energy that, however, now propagates along generally curvilinear rays at a variable group velocity. Equations (5.10.7) are analogous to Hamilton's equations of motion of a particle governed by Hamilton–Jacobi equation (5.10.6), in which $\mathbf{x}(t)$, $\mathbf{k}(t)$ are respectively the particle position and momentum at time t and $\omega = \Omega(\mathbf{k}, \mathbf{x})$ is its conserved total energy.

The conservation of energy by waves propagating along rays permits the wave amplitude to be calculated from energy equation (5.10.5) as soon as ray-tracing equations (5.10.7) have been solved.

5.10.3 Refraction of waves at a sloping beach

Surface gravity waves of frequency ω_o generated by a surface pressure distribution $p_o(x, y)\cos\omega_o t$ applied to water of depth h are determined by (§5.8)

$$\zeta = \frac{1}{\rho_o g}\text{Re}\int_{-\infty}^{\infty}\frac{\Omega^2(k)p_o(\mathbf{k})}{[\omega_o^2 - \Omega^2(k)]}\,e^{i(\mathbf{k}\cdot\mathbf{x}-\omega_o t)}\,d^2\mathbf{k}, \quad \text{where} \quad \Omega^2 = gk\tanh(kh).$$

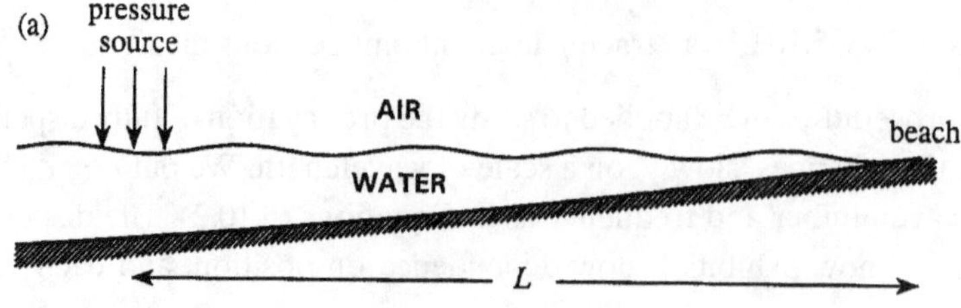

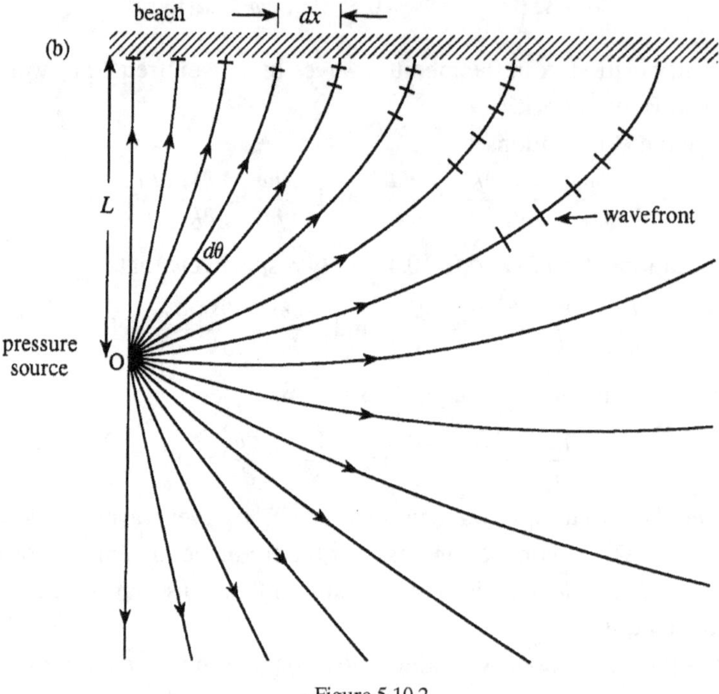

Figure 5.10.2

The corresponding surface wave power generated by the source is (see §5.8.3)

$$\Pi = \frac{2\pi^3|\omega_o|^3}{\rho_o g} \int_{-\infty}^{\infty} |p_o(\mathbf{k})|^2 \delta\left[\omega_o^2 - \Omega^2(k)\right] d^2\mathbf{k}. \tag{5.10.8}$$

Let the source be concentrated in the neighbourhood of the coordinate origin at distance L from a straight beach [at $y = L$, Figure 5.10.2(a)]. The water has 'slowly varying' depth equal to h_o at large distances from the shore and is given elsewhere by

$$h \equiv h(y) = h_o\left[1 - e^{-\alpha(L-y)}\right], \quad -\infty < y < L.$$

We express Equations (5.10.7) in a form suitable for numerical integration by putting

$$(X_1, X_2) = \left(\frac{x}{h_o}, \frac{y}{h_o}\right), \quad (K_1, K_2) = (k_1 h_o, k_2 h_o), \quad \bar{h} = \frac{h}{h_o}, \quad \bar{L} = \frac{L}{h_o}, \quad \bar{\alpha} = \alpha h_o, \quad s = \omega_o t.$$

Then the ray-tracing equations become

$$
\left.
\begin{aligned}
\left(\frac{dX_1}{ds}, \frac{dX_2}{ds}\right) &= \frac{(K_1, K_2)}{2K^2}\left[1 + \frac{2K\hbar}{\sinh(2K\hbar)}\right] \\[2mm]
\left(\frac{dK_1}{ds}, \frac{dK_2}{ds}\right) &= \left[0, \frac{\bar{\alpha}K^2 e^{-\bar{\alpha}(L - X_2)}}{2(\omega_o^2 h_o/g)\cosh^2(K\hbar)}\right] \\[2mm]
\frac{d\omega}{ds} &= 0
\end{aligned}
\right\}
\quad \text{where} \quad K = \sqrt{K_1^2 + K_2^2}. \qquad (5.10.9)
$$

These are to be solved subject to the initial conditions at the source, $s = 0$:

$$
X_1 = 0, \ \ X_2 = 0, \ \ K_1 = K\cos\theta, \ \ K_2 = K\sin\theta, \ \ \omega = \omega_o,
$$

where K is the positive zero of $\omega_o^2 h_o/g = K\tanh(K\hbar)$ at $X_2 = 0$. In the present case the group velocity and wavenumber vector are both tangential to the rays. The initial direction of a ray emanating from the source is defined by the angle θ ($0 < \theta < 2\pi$).

The solution of characteristic system (5.10.9) is a particular solution of the corresponding Hamilton–Jacobi equation,

$$
\frac{\partial\Theta}{\partial t} + \sqrt{g|\nabla\Theta|\tanh(|\nabla\Theta|h)} = 0,
$$

which, because $\mathbf{k} = \nabla\Theta$, $\omega_o = -\partial\Theta/\partial t$, is equivalent to the dispersion relation

$$
\omega_o = \sqrt{gk\tanh(kh)}.
$$

Because K_1 and ω_o do not vary along a ray, the dispersion relation could actually be used to calculate K_2 at any point along the ray and thereby reduce the the number of equations in system (5.10.9) to the two equations for X_1, X_2 only (see Examples 1 and 2 of this section). However, because the solution of the complete system (5.10.9) must automatically satisfy the dispersion relation, it is better to use the latter to provide a running check on the accuracy of the computed solution of the equations.

System (5.10.9) can be solved numerically (for example, by the Runge–Kutta procedure of §3.5.3). Rays are plotted in Figure 5.10.2(b) for the case $\omega_o^2 h_o/g = 1$. Refraction produced by the gradual decrease in the group velocity as the shore is approached causes rays to bend towards the shore, wavefronts (plotted at equal phase intervals in the figure) to line up with the shoreline, and the wavelength to decrease. To estimate the wave amplitude close to the shore, the steady-state energy equation is written as

$$
\frac{\partial}{\partial x_j}\left(\frac{\partial\Omega}{\partial k_j}\frac{1}{2}\rho_o g|\zeta_o|^2\right) = \Pi\delta(x)\delta(y), \qquad (5.10.10)
$$

where Π is wave power (5.10.8) generated by the source, which is localised at the origin. By setting $d^2\mathbf{k} = k\,dk\,d\theta$ in (5.10.8), we can write $\Pi = \int_0^{2\pi}\bar{\Pi}(\theta)\,d\theta$, where function $\bar{\Pi}(\theta)$ is determined by the source spectrum $p_o(\mathbf{k})$. If the energy equation is applied to the curvilinear angular region $d\theta$ of Figure 5.10.2(b), bounded by two rays whose distance apart at the shore is dx, we have

$$
\bar{\Pi}(\theta)\,d\theta = \left|\frac{\partial\Omega}{\partial k}\right|\frac{1}{2}\rho_o g|\zeta_o|^2\,dx,
$$

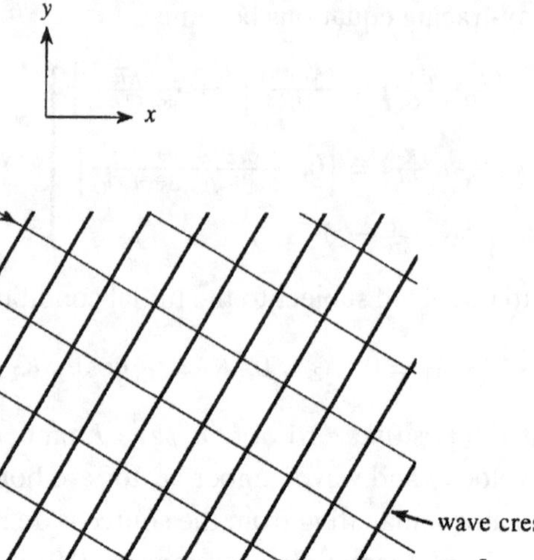

Figure 5.10.3

where $\partial\Omega/\partial k = \sqrt{gh(y)}$ in the shallow water near the shore. Hence, close to the shore,

$$|\zeta_o| \sim \left[\frac{2\bar{\Pi}(\theta)}{\rho_o g \sqrt{gh(y)}} \frac{d\theta}{dx} \right]^{\frac{1}{2}}.$$

EXAMPLE 1. STRAIGHT-CRESTED WAVES INCIDENT UPON A SLOPING BEACH
Straight-crested, time harmonic waves on shallow water of uniform depth h_o approach a straight shoreline $-\infty < x < \infty$, $y = 0$ from $y > 0$ (Figure 5.10.3). Calculate the ray paths and the pattern of the wave crests when the depth decreases linearly with y from h_o to zero in the region $0 < y < L$, where $h_o/L \ll 1$.

The depth

$$h = \begin{cases} h_o y/L, & 0 < y < L \\ h_o, & y > L \end{cases}$$

and the shallow-water dispersion relation is

$$\omega = \Omega(\mathbf{k}, y) = \begin{cases} k\sqrt{\alpha y}, & 0 < y < L, \\ k\sqrt{\alpha L}, & y > L, \end{cases} \qquad \text{where } \alpha = \frac{gh_o}{L}. \qquad (5.10.11)$$

The characteristic equations are

$$\frac{dx}{dt} = \frac{k_1}{k}\sqrt{\alpha L}, \quad \frac{dy}{dt} = \frac{k_2}{k}\sqrt{\alpha L}, \quad \frac{dk_1}{dt} = \frac{dk_2}{dt} = \frac{d\omega}{dt} = 0, \quad y > L,$$

$$\frac{dx}{dt} = \frac{k_1}{k}\sqrt{\alpha y}, \quad \frac{dy}{dt} = \frac{k_2}{k}\sqrt{\alpha y}, \quad \frac{dk_1}{dt} = 0, \quad \frac{dk_2}{dt} = -\frac{k}{2}\sqrt{\frac{\alpha}{y}}, \quad \frac{d\omega}{dt} = 0, \quad 0 < y < L.$$

Let ψ be the angle of incidence of waves in the deep-water region $(y > L)$, where $k = \sqrt{\omega^2/\alpha L}$. Then

$$k_1 = \sqrt{\frac{\omega^2}{\alpha L}}\, \sin \psi, \quad k_2 = -\sqrt{\frac{\omega^2}{\alpha L}}\, \cos \psi, \quad \text{for } y > L.$$

The frequency ω and k_1 take the same constant values in $y \gtrless L$. It follows, therefore, from dispersion relation (5.10.11) that

$$k_2 = -\sqrt{\frac{\omega^2}{\alpha L}\left(\frac{L}{y} - \sin^2 \psi\right)}^{\frac{1}{2}}, \quad 0 < y < L.$$

Thus the wavenumber and frequency are known everywhere, and the problem of solving the characteristic equations is reduced to solving the ray equations for x and y, which can be combined in the form

$$\frac{dy}{dx} = \begin{cases} -\frac{1}{\sin \psi}\left(\frac{L}{y} - \sin^2 \psi\right)^{\frac{1}{2}}, & 0 < y < L \\ -\frac{\cos \psi}{\sin \psi}, & y > L \end{cases}.$$

For the particular ray meeting the shoreline $(y = 0)$ at $x = x_r$ we find

$$\frac{x - x_r}{L} = \begin{cases} \frac{1}{\sin^2 \psi}\left[\sqrt{\frac{y}{L}}\sin \psi \sqrt{1 - \frac{y \sin^2 \psi}{L}} - \sin^{-1}\left(\sqrt{\frac{y}{L}}\sin \psi\right)\right], & 0 < y < L \\ -\frac{y \tan \psi}{L} + \frac{\tan \psi - \psi}{\sin^2 \psi}, & y > L \end{cases}.$$

Now $\sin^{-1} x \sim x + x^3/6 + \cdots +$ for small x. Therefore, near the beach $(y \to 0)$ the ray is given by

$$\frac{y}{L} \approx \left[\frac{3(x_r - x)}{2L \sin \psi}\right]^{\frac{2}{3}},$$

which cuts the x axis at right angles.

The phase $\Theta = \mathbf{k} \cdot \mathbf{x} - \omega t$ is constant on the wave crests, which are orthogonal to the rays and satisfy

$$k_1 dx + k_2 dy = 0,$$

$$\text{i.e.,} \quad \frac{dy}{dx} = \begin{cases} \sin \psi \left/ \left(\frac{L}{y} - \sin^2 \psi\right)^{\frac{1}{2}}\right., & 0 < y < L \\ \tan \psi, & y > L \end{cases}.$$

Hence the wave crest that touches the shoreline at $x = x_c$ is given by

$$\frac{x - x_c}{L} = \begin{cases} \frac{1}{\sin^2 \psi}\left[\sqrt{\frac{y}{L}}\sin \psi \sqrt{1 - \frac{y \sin^2 \psi}{L}} + \sin^{-1}\left(\sqrt{\frac{y}{L}}\sin \psi\right)\right], & 0 < y < L \\ \frac{y \cot \psi}{L} + \frac{\psi}{\sin^2 \psi}, & y > L \end{cases}.$$

At the shore the wave crest has the semi-parabolic form

$$\frac{y}{L} \approx \left[\frac{(x - x_c)\sin \psi}{2L}\right]^2, \quad \text{where } \operatorname{sgn}(x - x_c) = \operatorname{sgn}(\psi).$$

Therefore, when $\psi > 0$ (as in Figure 5.10.3), the crest meets the shore at x_c from $x > x_c$.

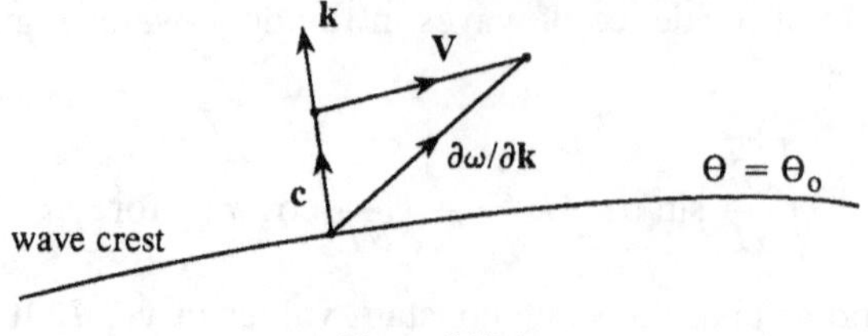

Figure 5.10.4

EXAMPLE 2. REFRACTION OF SOUND WAVES BY WIND The dispersion relation for sound waves of frequency ω (relative to the fixed reference frame of the ground) propagating in homogeneous air in nonuniform mean flow at velocity $\mathbf{V}(y)$ (Figures 5.10.4 and 5.10.5) is

$$\omega = \mathbf{k} \cdot \mathbf{V} + c_o k, \qquad (5.10.12)$$

where the 'speed of sound' c_o is assumed to be constant, and y is measured upwards from the ground (Figure 5.10.5).

The group velocity

$$\frac{\partial \omega}{\partial \mathbf{k}} = \mathbf{V} + c_o \mathbf{n}, \quad \text{where} \quad \mathbf{n} = \frac{\mathbf{k}}{k}$$

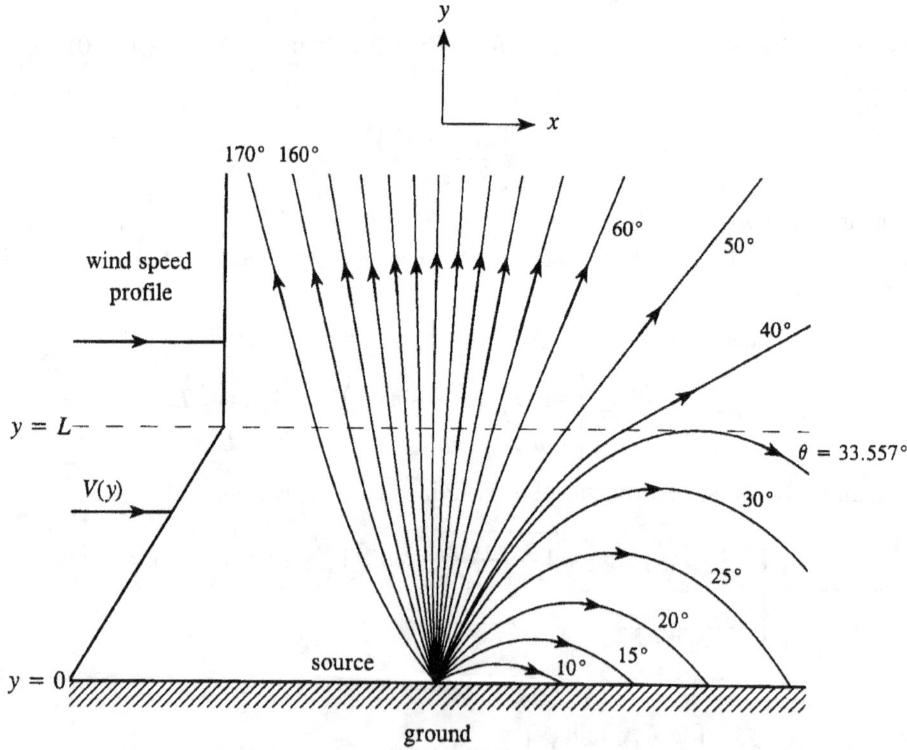

Figure 5.10.5

and the phase velocity

$$\mathbf{c} \equiv \frac{\omega}{k}\mathbf{n} = (\mathbf{V} \cdot \mathbf{n})\mathbf{n} + c_o\mathbf{n}$$

are not generally parallel, so that wave energy does not propagate in directions normal to the wave crests; also $|\partial\omega/\partial\mathbf{k}| \geq$ the phase speed. Actually, acoustic wave energy is not necessarily conserved during propagation along the characteristics ('rays') because it is possible for the sound to exchange energy with the background mean flow when the latter varies with position – although energy as a whole is conserved, it is the combined energies of the moving medium and the sound that is constant, and not just the acoustic energy (see §5.11).

When $\mathbf{V} = [V(y), 0, 0]$, so that $\omega = V(y)k_1 + c_o k$ and

$$\frac{\partial\omega}{\partial\mathbf{x}} = \left(0, k_1 \frac{\partial V}{\partial y}, 0\right),$$

the ray equations (for three-dimensional propagation) become

$$\frac{dx}{dt} = V + \frac{c_o k_1}{k}, \quad \frac{dy}{dt} = \frac{c_o k_2}{k}, \quad \frac{dz}{dt} = \frac{c_o k_3}{k}$$

$$\frac{dk_1}{dt} = 0, \quad \frac{dk_2}{dt} = -k_1 \frac{\partial V}{\partial y}, \quad \frac{dk_3}{dt} = 0, \quad \frac{d\omega}{dt} = 0. \tag{5.10.13}$$

Hence

$$k_1 = \text{constant}, \quad k_3 = \text{constant}, \quad \omega = \text{constant},$$

and from the dispersion relation,

$$k_2 = \pm\sqrt{(k_o - Mk_1)^2 - (k_1^2 + k_3^2)}, \quad \text{where} \quad k_o = \frac{\omega}{c_o}, \quad M = \frac{V(y)}{c_o}.$$

The problem now reduces to solving the ray path equations [first line of (5.10.13)]. Because the dispersion relation can be written as $k = k_o - Mk_1$, these equations reduce to

$$\frac{dx}{dy} = \frac{M(k_o - Mk_1) + k_1}{\pm\sqrt{(k_o - Mk_1)^2 - (k_1^2 + k_3^2)}},$$

$$\frac{dz}{dy} = \frac{k_3}{\pm\sqrt{(k_o - Mk_1)^2 - (k_1^2 + k_3^2)}}.$$

For the case illustrated in Figure 5.10.5, the wind speed $V(y)$ increases linearly from zero at the ground up to $y = L$ and remains constant thereafter, with

$$M = \begin{cases} 0.2\, y/L, & 0 < y < L \\ 0.2, & y > L \end{cases}.$$

Consider the radiation in the plane $z = 0$ produced by a point source on the ground at $\mathbf{x} = 0$. For the ray leaving the source inclined at angle θ to the x direction, we have $k_1 = k_o \cos\theta$, $k_3 = 0$, and

$$\frac{dx}{dy} = \frac{M(1 - M\cos\theta) + \cos\theta}{\pm\sqrt{(1 - M\cos\theta)^2 - \cos^2\theta}}.$$

Initially, at the source, we must have $k_2 > 0$, so the positive sign is to be taken. The sign must be changed when the ray reaches the critical height y_r of total reflection, where

$$1 - M(y_r)\cos\theta = \pm\cos\theta.$$

Only the '+' sign is relevant when $M < 1$, so that the critical height is determined by $\cos\theta = 1/(1 + M)$. This means that those rays for which $\cos\theta < 1/(1 + M_{\max})$ are *not* reflected (so that the positive sign is taken along the whole ray path). In the figure the critical angle for total reflection is $\theta = \cos^{-1}(1/1.2) \approx 33.6°$.

This example shows how the wind promotes propagation close to the ground in the downwind direction, but tends to produce an upwind 'shadow'. The critical ray (labelled $\theta = 33.557°$) in the figure is reflected at a 'caustic' at $y = L$ (§5.9).

5.11 Wave action

Ray theory is applicable in the fully dispersed region, where waves of different wavelengths and amplitudes are sufficiently spread out that wavelengths and amplitudes vary by negligible amounts over a wavelength or wave period. It is then convenient to speak of a local wave-energy density E, which is the wave energy per unit surface area (or per unit volume in more general systems), and is defined by averaging over a wavelength or wave period. We have relied on Rayleigh's proof that E in the fully dispersed region propagates at the group velocity to justify the calculation of wave amplitudes in ray theory. In Example 2 of §5.10, however, it was remarked that conservation of wave energy could not be ensured when waves propagate through a region that is itself in 'slowly varying' mean motion (*nonuniform* on a scale of many wavelengths). The energy of the whole system might be conserved, but there is no reason to suppose that there is no transfer of energy between the wave and the background flow.

This also raises the question of how wave energy should be defined in a medium in nonuniform motion. The simplest procedure is to use the same definition of E in both moving and stationary media – for a moving medium we must then adopt a local reference frame fixed relative to the local flow and define wave properties relative to this frame. The size of the region covered by the local reference frame must be large enough to contain many wavelengths, in other words, the background flow must vary significantly only over many wavelengths, the same approximation used in conventional ray theory. A transformation to the moving frame gives a trivial modification of previous ideas when the medium moves uniformly, such that a single moving reference frame can be used throughout the fluid. It is only in cases in which the background flow is nonuniform, such as when sound propagates through a variable wind (Example 2,

§5.10), that difficulties arise. Namely, how do we link or reconcile wave properties defined relative to the different local reference frames traversed by a propagating wave?

5.11.1 Variational description of a fully dispersed wave group

All of the general results of ray theory (§5.10) can actually be derived from a very simple application of Hamilton's principle of least action (Whitham 1965). The procedure is very general, but the details are first described for a fully dispersed wave on fluid of uniform depth h.

Consider the slowly varying wave train with surface displacement

$$\zeta = \mathrm{Re}\left[\zeta_o(x,t)e^{i\Theta(x,t)}\right] \tag{5.11.1}$$

where local values of the wavenumber and frequency are defined in the usual way by

$$k = \frac{\partial\Theta}{\partial x}, \quad \omega = -\frac{\partial\Theta}{\partial t},$$

and where the complex wave amplitude $\zeta_o(x,t)$ varies slowly over distances and times respectively of orders $1/k$, $1/\omega$.

When the surface elevation (5.11.1) is prescribed, the corresponding motion within the fluid can be calculated in terms of a velocity potential φ. If ζ_o, k, and ω are assumed to be *constant* we find

$$\varphi = \mathrm{Re}\left\{\frac{-i\omega\zeta_o\,\cosh[k(z+h)]e^{i\Theta(x,t)}}{k\,\sinh(kh)}\right\}, \tag{5.11.2}$$

which evidently satisfies the bottom boundary condition $\partial\varphi/\partial z = 0$ at $z = -h$. The formulae (5.11.1) and (5.11.2) satisfy the kinematic condition $\partial\zeta/\partial t = \partial\varphi/\partial z$ at $z = 0$, but the dynamic condition that the pressure vanishes at the free surface has *not* been imposed, so that the relationship between k and ω is so far undetermined.

Let us now calculate the potential and kinetic energies V, T of the motion per unit surface area by the usual process of averaging over a wavelength $\lambda = 2\pi/k$ (keeping k and ω fixed):

$$\left.\begin{aligned}
V &= \left\langle\int_0^\zeta \rho_o g z\,dz\right\rangle = \left\langle\frac{1}{2}\rho_o g\zeta^2\right\rangle = \frac{1}{4}\rho_o g a^2 \\
T &= \frac{1}{\lambda}\int_0^\lambda \frac{1}{2}\rho_o\left(\varphi\frac{\partial\varphi}{\partial z}\right)_{z=0} dx = \frac{\rho_o\omega^2 a^2}{4k\,\tanh(kh)}
\end{aligned}\right\} \quad \text{where} \quad a = |\zeta_o|.$$

These expressions are for a wave train of constant amplitude, wavenumber, and frequency, but in a first approximation they are also applicable for slowly varying values of a, k, and ω. The actual variations of a, k, and ω cannot be specified arbitrarily but must be compatible with the equations of motion.

At any instant, however, the motion is defined by the parameters a and Θ [$\arg(\zeta_o)$ being absorbed into Θ], which accordingly play the role of generalised coordinates for the local wave motion. Equations governing the evolution of the fully dispersed wave can therefore be derived by considering Lagrange's equations of motion determined by the *average Lagrangian*

$$\mathcal{L}(a, k, \omega) = T - V = \frac{1}{4}\rho_0 g a^2 \left[\frac{\omega^2}{gk\tanh(kh)} - 1\right], \quad k = \frac{\partial\Theta}{\partial x}, \quad \omega = -\frac{\partial\Theta}{\partial t}. \quad (5.11.3)$$

Hamilton's principle requires

$$\delta \int \mathcal{L}\left(a, \frac{\partial\Theta}{\partial x}, -\frac{\partial\Theta}{\partial t}\right) dxdt = 0$$

for changes δa, $\delta\Theta$ that vanish on the arbitrary boundaries of integration. Only the derivatives of the phase Θ appear explicitly in $\mathcal{L}$; it is a 'cyclic variable' whose 'generalised momentum' is an integral of the motion – in the present case it will be shown to correspond to conservation of wave energy.

The variation of a yields $\partial\mathcal{L}/\partial a = 0$. This is equivalent to $\mathcal{L} = 0$, and defines the dispersion relation or Hamilton–Jacobi equation of the wave motion:

$$\frac{\omega^2}{gk\tanh(kh)} - 1 = 0, \quad \text{where} \quad k = \frac{\partial\Theta}{\partial x}, \quad \omega = -\frac{\partial\Theta}{\partial t}. \quad (5.11.4)$$

The variation with respect to Θ gives the conservation equation

$$\frac{\partial\mathcal{L}_\omega}{\partial t} - \frac{\partial\mathcal{L}_k}{\partial x} = 0, \quad (5.11.5)$$

where $\mathcal{L}_\omega = \partial\mathcal{L}/\partial\omega$, etc. This can be transformed into the energy propagation equation. Indeed, because $\mathcal{L} = 0$, the group velocity is given by

$$\frac{\partial\omega}{\partial k} = -\frac{\mathcal{L}_k}{\mathcal{L}_\omega},$$

$$\therefore \quad \frac{\partial\mathcal{L}_\omega}{\partial t} + \frac{\partial}{\partial x}\left(\frac{\partial\omega}{\partial k}\mathcal{L}_\omega\right) = 0. \quad (5.11.6)$$

However, (5.11.3) and the dispersion relation reveal that

$$E = \frac{1}{2}\rho_0 g a^2 = \omega\mathcal{L}_\omega,$$

so that (5.11.6) becomes

$$\frac{\partial}{\partial t}\left(\frac{E}{\omega}\right) + \frac{\partial}{\partial x}\left(\frac{\partial\omega}{\partial k}\frac{E}{\omega}\right) = 0. \quad (5.11.7)$$

This is the required energy equation, because the frequency ω is constant along the characteristics ('rays') of the Hamilton–Jacobi equation (5.11.4). We have here an alternative to Rayleigh's proof that wave energy propagates at the group velocity.

5.11.2 Fully dispersed waves in a non-uniformly moving medium

To deal with the more general case of propagation through a medium of non-uniform properties, possibly also in non-uniform motion, we require, as before, that changes in the background properties of the medium must be slow on scales of the wavelength and wave period. We assume that locally the medium has a uniform velocity $\mathbf{V}(\mathbf{x}, t)$ and then adopt a local reference frame with respect to which it is permissible to assume the fluid is at rest. In the case of surface waves it will usually be necessary to assume also that $\mathbf{V}$ does not vary with depth from its local value at the surface, at least over distances comparable with the surface wavelength. The local fully dispersed wave is represented in the form of (5.11.1) relative to the local moving coordinate system, but with the following modifications:

(1) Denote by ω_R the wave frequency relative to an observer in the moving reference frame. Time derivatives relative to such an observer are equivalent to

$$\frac{D}{Dt} = \frac{\partial}{\partial t} + V_j \frac{\partial}{\partial x_j}.$$

Thus, $\omega_R = -D\Theta/Dt$, i.e.,

$$\omega_R = -\frac{\partial \Theta}{\partial t} - V_j \frac{\partial \Theta}{\partial x_j} \equiv \omega - \mathbf{k} \cdot \mathbf{V}, \tag{5.11.8}$$

where ω is the frequency relative to a fixed (inertial) reference frame.

(2) The Lagrangian $\mathcal{L}$ in the moving coordinate system will have the same form as for the medium at rest.

Thus, if we consider again surface waves on fluid of finite depth, we shall suppose that $h = h(x_i)$, $\mathbf{V} = \mathbf{V}(x_i, t)$ (where x_i are horizontal coordinates). Then in a local reference frame, (5.11.1) and (5.11.2) become

$$\zeta = \mathrm{Re}\left(\zeta_o e^{i\Theta}\right),$$

$$\varphi = \mathrm{Re}\left\{\frac{-i\omega_R\zeta_o \cosh[k(z+h)]e^{i\Theta}}{k \sinh(kh)}\right\},$$

and

$$\mathcal{L}(a, k_i, \omega_R) = \frac{1}{4}\rho_o g a^2 \left[\frac{\omega_R^2}{gk\tanh(kh)} - 1\right], \quad k_i = \frac{\partial\Theta}{\partial x_i}, \quad \omega_R = -\frac{D\Theta}{Dt}. \tag{5.11.9}$$

Hamilton's principle must be applied in the common, inertial reference frame, which requires

$$\delta \int \mathcal{L}\left(a, \frac{\partial\Theta}{\partial x_i}, -\frac{D\Theta}{Dt}\right) d^2\mathbf{x}dt = 0$$

for changes δa, $\delta\Theta$ that vanish on the arbitrary boundaries of integration. The variation of a yields the dispersion relation

$$\omega_R^2 \equiv (\omega - \mathbf{k} \cdot \mathbf{V})^2 = gk\tanh(kh), \quad \text{where} \quad k_i = \frac{\partial\Theta}{\partial x_i}, \quad \omega = -\frac{\partial\Theta}{\partial t}, \tag{5.11.10}$$

and the Θ variation supplies

$$\frac{\partial}{\partial t}\mathcal{L}_{\omega_R} + \frac{\partial}{\partial x_j}\left(V_j\mathcal{L}_{\omega_R} - \mathcal{L}_{k_j}\right) = 0. \tag{5.11.11}$$

Now

$$
\omega_R \mathcal{L}_{\omega_R} = \frac{1}{2}\rho_0 g a^2 \equiv E,
$$

$$
\omega_R \left(V_j \mathcal{L}_{\omega_R} - \mathcal{L}_{k_j} \right) = E \frac{\partial}{\partial k_j}\left(\mathbf{V} \cdot \mathbf{k} + \omega_R \right) \equiv E \frac{\partial \omega}{\partial k_j}.
$$

Hence, (5.11.11) becomes the following generalisation of (5.11.7):

$$
\frac{\partial}{\partial t}\left(\frac{E}{\omega_R} \right) + \frac{\partial}{\partial x_j}\left(\frac{\partial \omega}{\partial k_j}\,\frac{E}{\omega_R} \right) = 0.
$$

The quantity

$$
\frac{E}{\omega_R} \equiv \frac{E}{\omega - \mathbf{V}\cdot\mathbf{k}} \tag{5.11.12}
$$

is called the *wave-action* density. In a moving medium wave action is conserved along rays and not the local wave energy, because in general neither ω_R nor ω is constant along a ray.

Thus, for space–time-dependent moving water the ray equations (5.10.7) are replaced with

$$
\frac{d\mathbf{x}}{dt} = \frac{\partial \Omega}{\partial \mathbf{k}}, \quad \frac{d\mathbf{k}}{dt} = -\frac{\partial \Omega}{\partial \mathbf{x}}, \quad \frac{d\omega}{dt} = \frac{\partial \Omega}{\partial t}, \tag{5.11.13}
$$

where

$$
\Omega = \mathbf{V}\cdot\mathbf{k} \pm \sqrt{gk\,\tanh(kh)},
$$

coupled with the wave-action conservation equation:

$$
\frac{\partial}{\partial t}\left(\frac{E}{\omega - \mathbf{V}\cdot\mathbf{k}} \right) + \frac{\partial}{\partial x_j}\left(\frac{\partial \omega}{\partial k_j}\,\frac{E}{\omega - \mathbf{V}\cdot\mathbf{k}} \right) = 0. \tag{5.11.14}
$$

The preceding derivation for surface gravity waves is easily extended to other wave systems. In particular, for the problem of Example 2, §5.10, of sound propagation through a variable wind, we now see that it is the wave action E/ω_R that is conserved along rays. Because the wind speed $V(y)$ does not depend on time, the frequency ω relative to the fixed inertial frame is constant along a ray, but $\omega_R = \omega - V(y)k_1 \equiv \omega[1 - M(y)\cos\theta]$ varies with height, where θ is the initial inclination of the ray at the ground. Thus $E/[1 - M(y)\cos\theta]$ is conserved along rays, and this can be used to estimate changes in wave amplitude. Wave energy must increase (by extraction of energy from the background mean flow) when ω_R increases; the wave cedes energy to the mean flow when ω_R decreases. Water waves approaching a fast-flowing river outlet, for example (so that $\mathbf{k} \cdot \mathbf{V} < 0$), would increase in amplitude by extraction of energy from the mean stream.

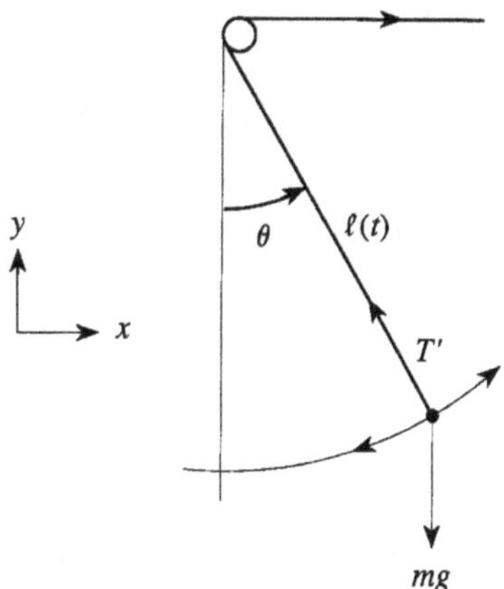

Figure 5.11.1

5.11.3 General wave-bearing media

These conclusions are easily generalised to arbitrary media for which wave propagation is described by linear equations. In all such cases the dispersion relation is

$$\mathcal{L}(a, \mathbf{k}, \omega_R) = 0,$$

with the obvious interpretation that the local mean values of the wave kinetic and potential energies are equal. Ray tracing is governed by

$$\frac{d\mathbf{x}}{dt} = \frac{\partial \Omega}{\partial \mathbf{k}}, \quad \frac{d\mathbf{k}}{dt} = -\frac{\partial \Omega}{\partial \mathbf{x}}, \quad \frac{d\omega}{dt} = \frac{\partial \Omega}{\partial t}, \quad \Omega = \mathbf{V} \cdot \mathbf{k} + \omega_R,$$

and wave energy (defined relative to the local reference frame) is determined by the wave-action conservation equation:

$$\frac{\partial}{\partial t}\left(\frac{E}{\omega_R}\right) + \frac{\partial}{\partial x_j}\left(\frac{\partial \Omega}{\partial k_j}\frac{E}{\omega_R}\right) = 0.$$

The mean potential energy will not normally depend explicitly on ω_R and the kinetic energy will be proportional to ω_R^2, so that

$$E = \omega_R \mathcal{L}_{\omega_R} - \mathcal{L} \equiv \omega_R \mathcal{L}_{\omega_R},$$

and in all cases

$$\text{wave-action density} = \frac{E}{\omega_R} = \mathcal{L}_{\omega_R}.$$

These equations conform to notions of *wave–particle duality*: The wavenumber $\mathbf{k}$ plays the same role as particle momentum in Hamiltonian mechanics, whereas the frequency ω_R plays the role of the Hamiltonian or energy – conservation of wave action implies that $E = \hbar \omega_R$ along a ray for a suitably valued constant $\hbar$.

EXAMPLE 1. ENERGY EQUATION FOR A SIMPLE PENDULUM OF VARYING LENGTH
Let the pendulum of Figure 5.11.1 execute small angular oscillations $\theta(t)$ about the

vertical and have an externally controlled length $\ell(t)$ whose time scale of variation is large compared with the period $\sim 2\pi\sqrt{\ell/g}$. The pendulum is not a closed mechanical system, and its energy E is therefore not conserved. However, because ℓ varies slowly the rate of change of E is also slow and proportional to $\dot\ell = d\ell/dt$. Thus $E = E(\ell)$, and there must be some combination of E and ℓ that remains constant during the oscillatory motions – this combination is the analogue of wave-action density (it is called an *adiabatic invariant* in particle mechanics).

The position of the bob in the xy plane at time t is

$$\mathbf{r} = \left[\ell(t)\sin\theta, \, -\ell(t)\cos\theta\right]$$

and the equations for small-amplitude oscillations are therefore

$$m\frac{d^2}{dt^2}(\ell\theta) = -T'\theta, \quad -m\frac{d^2\ell}{dt^2} = T' - mg.$$

Eliminating the tension T' and using the dot notation for differentiation, we obtain

$$\ddot\theta + \frac{2\dot\ell}{\ell}\dot\theta + \frac{g}{\ell}\theta = 0. \tag{5.11.15}$$

We can form an energy equation by multiplying by $\dot\theta$ and rearranging:

$$\frac{d}{dt}\left[\frac{1}{2}m(\ell\dot\theta)^2 + \frac{1}{2}mg\ell\theta^2\right] + \frac{1}{2}m\dot\theta^2\frac{d}{dt}\ell^2 - \frac{1}{2}mg\theta^2\frac{d\ell}{dt} = 0, \tag{5.11.16}$$

where $T = \frac{1}{2}m(\ell\dot\theta)^2$ is the kinetic energy of the bob relative to the instantaneous state of the pendulum, and $V = \frac{1}{2}mg\ell\theta^2$ is the potential energy.

The solution of (5.11.15) when the length variations are ignored can be taken in the form

$$\theta = a\cos(\omega t), \quad \text{where} \quad \omega = \sqrt{\frac{g}{\ell}}.$$

We now take this to be the solution when the length varies slowly on the time scale of the pendulum oscillations, so that $a(t)$ and $\omega(t)$ become slowly varying functions of time. We find the equation for a by replacing θ^2 and $\dot\theta^2$ in (5.11.16) with their values averaged over a wave period with a, ω, and ℓ held fixed:

$$\langle\theta^2\rangle = \frac{a^2}{2}, \quad \langle\dot\theta^2\rangle = \frac{\omega^2 a^2}{2} = \frac{ga^2}{2\ell}.$$

The averaged energy is then given by

$$E = \left\langle\frac{1}{2}m(\ell\dot\theta)^2 + \frac{1}{2}mg\ell\theta^2\right\rangle = \frac{mgla^2}{2},$$

and the energy equation becomes

$$\frac{dE}{dt} + \frac{E}{2\ell}\frac{d\ell}{dt} = 0, \quad \therefore \quad \frac{d}{dt}\left(E\sqrt{\ell}\right) = 0,$$

$$\text{i.e.} \quad \frac{d}{dt}\left(\frac{E}{\omega}\right) = 0,$$

which shows that the *action* E/ω is conserved as the pendulum changes in length. The pendulum energy increases as ω increases, i.e. as the length ℓ decreases, and vice versa.

EXAMPLE 2. AVERAGE LAGRANGIAN APPLIED TO A PENDULUM OF VARYING LENGTH The Lagrangian for the pendulum of Figure 5.11.1 when changes in $\ell(t)$ are ignored is

$$L(\theta, \dot\theta, \ell) = \frac{1}{2}m(\ell\dot\theta)^2 - \frac{1}{2}mg\ell\theta^2. \qquad (5.11.17)$$

However, the motion takes place in the presence of a slow change in the length ℓ, so that the pendulum regarded as a mechanical system is not 'closed' but is coupled to another system S, say, responsible for the changes in ℓ. We derive the equations of motion from Hamilton's principle by varying each of the generalised coordinates *independently* – in each variation the other coordinates are considered to be known. The θ equation of motion can therefore be found by use of the Lagrangian $\hat{L}$ of the whole system (which is assumed to be closed) with all other generalised coordinates regarded as specified functions of the time. Thus set

$$\hat{L} = L(\theta, \dot\theta, \ell) + L_S(\ell, \dot\ell, q, \dot q)$$

where $L_S(\ell, \dot\ell, q, \dot q)$ accounts for the system S and may involve generalised coordinates q that do not appear explicitly in L. By replacing ℓ and the coordinates q by given functions of the time, we have

$$\hat{L} = L[\theta, \dot\theta, \ell(t)] + L_S[\ell(t), \dot\ell(t), q(t), \dot q(t)],$$

where L_S can obviously make no contribution to the equation of motion for θ and may be discarded.

Thus in (5.11.17) we assume $\ell(t)$ is a prescribed, slowly varying function of the time and postulate a solution of the form

$$\theta = a\cos\Theta, \quad \text{where} \quad \omega = -\frac{\partial\Theta}{\partial t},$$

where the amplitude a and frequency ω are slowly varying functions of the time to be determined; a and the phase Θ play the roles of new generalised coordinates. This solution is substituted into $L[\theta, \dot\theta, \ell(t)]$ and the result is averaged over the period $2\pi/\omega$ of the pendulum with a, ω, and ℓ held fixed, yielding

$$\mathcal{L} = \frac{1}{4}ma^2\ell^2\left(\omega^2 - \frac{g}{\ell}\right).$$

Hamilton's principle $\delta\int\mathcal{L}\,dt = 0$ for arbitrary variations δa and $\delta\Theta$ in amplitude and phase now supplies:

$$\frac{\partial\mathcal{L}}{\partial a} = 0, \quad \frac{\partial}{\partial t}\mathcal{L}_\omega = 0.$$

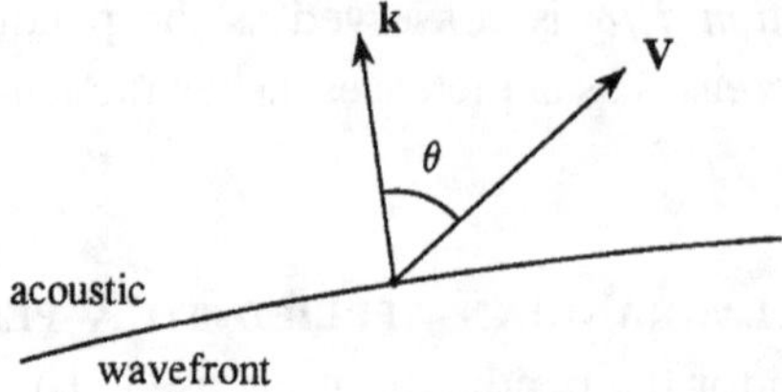

Figure 5.11.2

The first of these gives the frequency $\omega = \sqrt{g/\ell}$. Noting as before that the mean kinetic and potential energies are equal, we find that the energy of the pendulum $E = \omega \mathcal{L}_\omega$, and the second equation reduces to the equation of conservation of action

$$\frac{d}{dt}\left(\frac{E}{\omega}\right) = 0.$$

E/ω is evidently a constant of the motion because the phase Θ is a cyclic variable, which does not occur explicitly in the average Lagrangian $\mathcal{L}$.

EXAMPLE 3. ACOUSTIC WAVE ACTION Let sound propagate through an inviscid, isentropic medium in the presence of a non-uniform, background mean flow of velocity $\mathbf{V}(\mathbf{x})$ (Figure 5.11.2). When body forces are ignored, the perturbation energy equation is obtained from Equation (1.5.10) by subtracting out the corresponding equation in the absence of sound. The mean flow is assumed to vary over distances that are large compared with the wavelength of the sound, and it is assumed that the sound produces a local irrotational perturbation of the flow defined by a velocity potential φ and pressure and density perturbations p' and ρ'. The acoustic energy equation is then

$$\frac{\partial}{\partial t}\left[\frac{p'^2}{2\rho_o c_o^2} + \rho'\mathbf{V}\cdot\nabla\varphi + \frac{1}{2}\rho_o(\nabla\varphi)^2\right] + \mathrm{div}\left[-\frac{\partial\varphi}{\partial t}(\rho_o\nabla\varphi + \rho'\mathbf{V})\right] = 0, \qquad (5.11.18)$$

where ρ_o, and c_o are the mean density and sound speed (that may vary with position).

For time harmonic waves of frequency ω relative to the fixed frame and ω_R relative to a local moving frame, we have

$$\omega_R = c_o k,$$

$$\omega = c_o k + \mathbf{V}\cdot\mathbf{k}, = \omega_R\left(1 + \frac{\mathbf{V}\cdot\mathbf{k}}{c_o k}\right) \equiv \omega_R\left(1 + \mathbf{M}\cdot\hat{\mathbf{k}}\right),$$

$$\nabla\varphi = \frac{\hat{\mathbf{k}}p'}{\rho_o c_o},$$

$$-\frac{\partial\varphi}{\partial t} = i\omega\varphi = \left(\frac{\omega}{\omega_R}\right)i\omega_R\varphi = \frac{\omega p'}{\omega_R \rho_o},$$

$$\rho' = \frac{p'}{c_o^2},$$

where $\mathbf{M} = \mathbf{V}/c_o$, $\hat{\mathbf{k}} = \mathbf{k}/k$.

The energy density E averaged over a wave period is

$$E = \left\langle \frac{p'^2}{2\rho_o c_o^2} + \frac{1}{2}\rho_o(\nabla\varphi)^2 \right\rangle \equiv \left\langle \frac{p'^2}{\rho_o c_o^2} \right\rangle,$$

$$\therefore \quad \left\langle \frac{p'^2}{2\rho_o c_o^2} + \rho'\mathbf{V}\cdot\nabla\varphi + \frac{1}{2}\rho_o(\nabla\varphi)^2 \right\rangle = E(1 + \mathbf{M}\cdot\hat{\mathbf{k}}) \equiv \frac{\omega E}{\omega_R},$$

and

$$\left\langle -\frac{\partial\varphi}{\partial t}(\rho_o\nabla\varphi + \rho'\mathbf{V}) \right\rangle = \frac{\omega}{\omega_R}\left\langle \frac{p'^2}{\rho_o c_o^2} \right\rangle\left(c_o\frac{\mathbf{k}}{k} + \mathbf{V} \right) \equiv \frac{\omega}{\omega_R}E\frac{\partial\omega}{\partial\mathbf{k}}.$$

Hence, substituting into the averaged form of (5.11.18) and noting that $\omega = \text{constant}$ along a ray, we obtain the acoustic wave-action conservation equation in the form

$$\frac{\partial}{\partial t}\left(\frac{E}{\omega_R} \right) + \text{div}\left(\frac{\partial\omega}{\partial\mathbf{k}}\frac{E}{\omega_R} \right) = 0,$$

$$\text{where} \qquad \omega_R = \frac{\omega}{1 + M\cos\theta},$$

θ being the angle between the wave normal and $\mathbf{V}$.

5.12 Diffraction of surface waves by a breakwater

Docks and harbours are protected from waves arriving from the open sea by breakwaters. The simplest type of breakwater is a long vertical barrier projecting out from the shore and curving around the harbour. The harbour entrance frequently consists of a short channel through the gap between the ends of two such breakwaters, and effective harbour design requires an understanding of the interaction of incoming waves with the gap. Except for very long waves (of length comparable with the gap width) it is permissible to study surface waves interacting with the end of a single breakwater.

5.12.1 Diffraction by a long, straight breakwater

Consider waves in water of uniform depth h incident normally 'from the open sea' on a semi-infinite breakwater modelled by a thin, rigid barrier occupying $-\infty < x < 0$, $y = 0$, $z > -h$, where the 'top' of the barrier is above the mean water level $z = 0$ (Figure 5.12.1). Let the waves have frequency $\omega > 0$ and write the velocity potential $\varphi(x, y, z, t)$ in the form

$$\varphi = \Phi(x, y)\frac{\cosh[k_o(z + h)]}{\cosh[k_o h]}e^{-i\omega t}. \tag{5.12.1}$$

This automatically satisfies the rigid bottom condition $(\partial\varphi/\partial z)_{z=-h} = 0$. The free surface condition $(\partial^2\varphi/\partial t^2 + g\partial\varphi/\partial z)_{z=0} = 0$ implies that

$$(k_o h)\tanh(k_o h) = \frac{\omega^2 h}{g}. \tag{5.12.2}$$

This equation determines k_o when the frequency ω is specified.

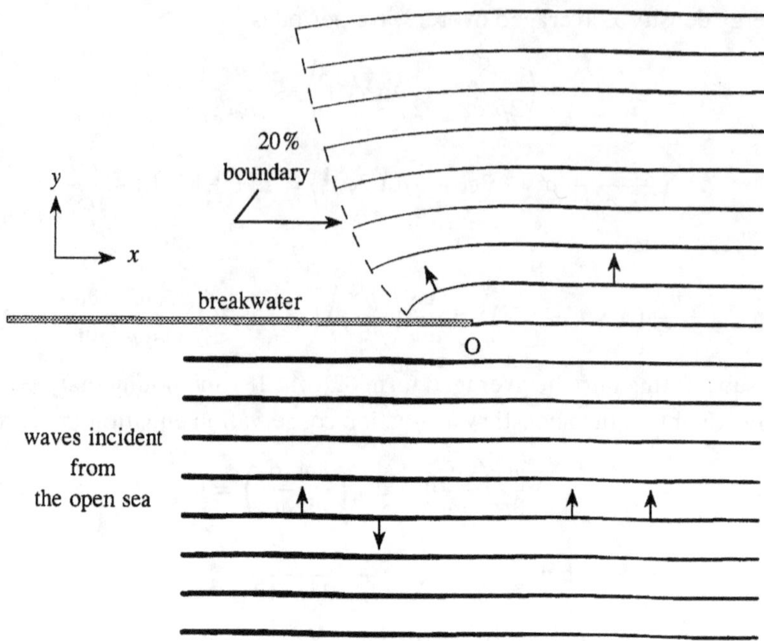

Figure 5.12.1

Laplace's equation $\nabla^2 \varphi = 0$ reduces to

$$\frac{\partial^2 \Phi}{\partial x^2} + \frac{\partial^2 \Phi}{\partial y^2} + k_o^2 \Phi = 0, \tag{5.12.3}$$

and the free surface elevation $\zeta = \dfrac{i\omega}{g} \Phi(x, y)e^{-i\omega t}$. $\tag{5.12.4}$

Dispersion equation (5.12.2) determines one positive real root k_o when $\omega > 0$. A wave incident normally upon the breakwater from the open sea ($y < 0$) may therefore be taken to be defined by $\Phi = \Phi_I \equiv e^{ik_o y}$. The solution of (5.12.3) includes this wave plus the wave diffracted by the breakwater. The combined field is then required to satisfy the condition that the normal component of velocity vanish on the breakwater:

$$\frac{\partial \Phi}{\partial y} = 0 \quad \text{for} \quad x < 0, \ y = 0. \tag{5.12.5}$$

5.12.2 Solution of the diffraction problem

By writing the diffracted component of the potential as a Fourier integral we can put

$$\Phi = e^{ik_o y} + \int_{-\infty}^{\infty} \hat{\Phi}(k, y)e^{ikx} \, dk, \tag{5.12.6}$$

where

$$\frac{\partial \hat{\Phi}}{\partial y^2} + (k_o^2 - k^2)\hat{\Phi} = 0.$$

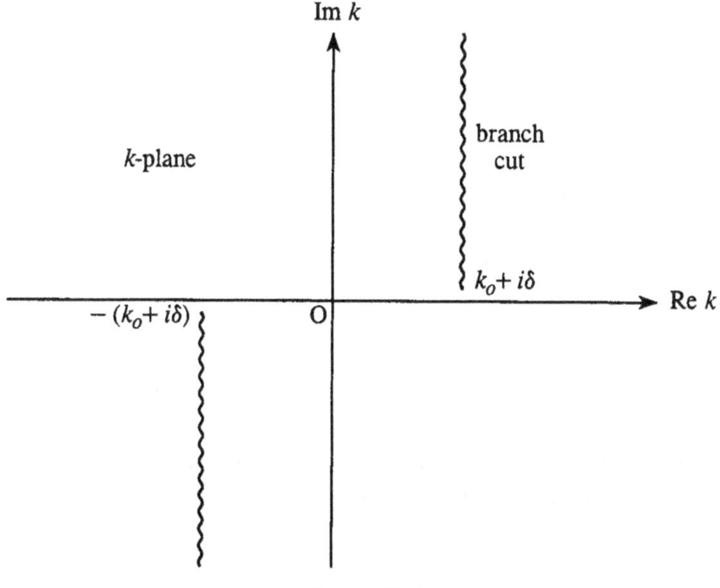

Figure 5.12.2

In accordance with our usual practice, we replace the frequency with $\omega + i\epsilon$, where $\omega > 0$ and ϵ is a small positive quantity that is ultimately allowed to vanish. Thus we enforce the radiation condition by seeking the particular solution that grows like $e^{\epsilon t}$.

The left-hand side of (5.12.2) is an increasing function of $k_o > 0$. Therefore the root of this equation also acquires a small *positive* imaginary part $i\delta$, say, so that $\sqrt{k_o^2 - k^2}$, regarded as a function of k, has branch points at $k = \pm(k_o + i\delta)$, respectively just *above* and *below* the real k axis. Let us define branch cuts for this function in the complex k plane to be straight rays in the upper and lower half-planes joining $k = \pm(k_o + i\delta)$ and $\pm i\infty$ (Figure 5.12.2), and consider only the branch of $\sqrt{k_o^2 - k^2}$ that is real and positive on the interval $-k_o < k < k_o$ of the real axis (when $\epsilon \to +0$). Then

$$\sqrt{k_o^2 - k^2} = +i|k_o^2 - k^2|^{\frac{1}{2}} \quad \text{for } |k| > k_o \text{ on the real axis,}$$

and the solution (5.12.6) that is *bounded* in $y \gtrless 0$ becomes

$$\Phi = e^{ik_o y} + \int_{-\infty}^{\infty} \mathcal{A}_{\pm}(k) e^{i\left(kx + |y|\sqrt{k_o^2 - k^2}\right)} \, dk$$

for suitable functions $\mathcal{A}_{\pm}(k)$.

Now $\partial\Phi/\partial y = 0$ for $x < 0$ and $y \to \pm 0$, and must be continuous in the water across the positive x axis. In other words, $\partial\Phi/\partial y$ is continuous across the whole of the x axis. This requires that $\mathcal{A}_+(k) = -\mathcal{A}_-(k) \equiv \mathcal{A}(k)$, say, and therefore that

$$\Phi = e^{ik_o y} + \text{sgn}(y) \int_{-\infty}^{\infty} \mathcal{A}(k) e^{i\left(kx + |y|\sqrt{k_o^2 - k^2}\right)} \, dk. \tag{5.12.7}$$

The function $A(k)$ must satisfy the following conditions:

(1) $\partial \Phi / \partial y = 0$ for $x < 0$, $y \to \pm 0$,

$$\therefore \quad k_o + \int_{-\infty}^{\infty} A(k)\sqrt{k_o^2 - k^2}\, e^{ikx}\, dk = 0, \quad x < 0,$$

which can be expressed in the more convenient form

$$\int_{-\infty}^{\infty} \left[A(k)\sqrt{k_o^2 - k^2} - \frac{k_o}{2\pi i(k+i0)} \right] e^{ikx}\, dk = 0, \quad x < 0; \tag{5.12.8}$$

(2) $\lim_{y \to +0} \Phi = \lim_{y \to -0} \Phi$ for $x > 0$,

$$\therefore \quad \int_{-\infty}^{\infty} A(k) e^{ikx}\, dk = 0, \quad x > 0. \tag{5.12.9}$$

Equations (5.12.8) and (5.12.9) constitute a pair of *Wiener–Hopf* dual integral equations for the function $A(k)$. They are satisfied if

$$A(k)\sqrt{k_o^2 - k^2} - \frac{k_o}{2\pi i(k+i0)} = L(k), \tag{5.12.10}$$

$$A(k) = U(k), \tag{5.12.11}$$

where $L(k)$ and $U(k)$ are 'lower' and 'upper' functions (i.e., regular in $\operatorname{Im} k \lessgtr 0$) that vanish as $|k| \to \infty$ respectively in $\operatorname{Im} k \lessgtr 0$. Eliminating $A(k)$, we obtain

$$U(k)\sqrt{k_o^2 - k^2} - \frac{k_o}{2\pi i(k+i0)} = L(k). \tag{5.12.12}$$

This is a standard Wiener–Hopf functional equation that we solve for $L(k)$ and $U(k)$ by rewriting it for real values of k in the form

$$\text{upper function} = \text{lower function.}$$

We accomplish this in two steps.

Step 1: Because the branch cuts are at $k = \pm(k_o + i\delta)$, $\delta \to +0$,

$$\sqrt{k_o^2 - k^2} = \sqrt{k_o + k} \times \sqrt{k_o - k} = \text{upper function} \times \text{lower function.}$$

Then (5.12.12) can be written as

$$U(k)\sqrt{k_o + k} - \frac{k_o}{2\pi i(k+i0)\sqrt{k_o - k}} = \frac{L(k)}{\sqrt{k_o - k}}, \tag{5.12.13}$$

i.e.   upper function $+$ mixed function $=$ lower function.

Step 2: The mixed function is now expressed as the sum of lower and upper functions:

$$\frac{k_o}{2\pi i(k+i0)\sqrt{k_o - k}}$$

$$= \frac{k_o}{2\pi i(k+i0)} \left(\frac{1}{\sqrt{k_o - k}} - \frac{1}{\sqrt{k_o + i0}} \right) + \frac{k_o}{2\pi i(k+i0)\sqrt{k_o + i0}}$$

$$\text{lower function} \qquad + \qquad \text{upper function}$$

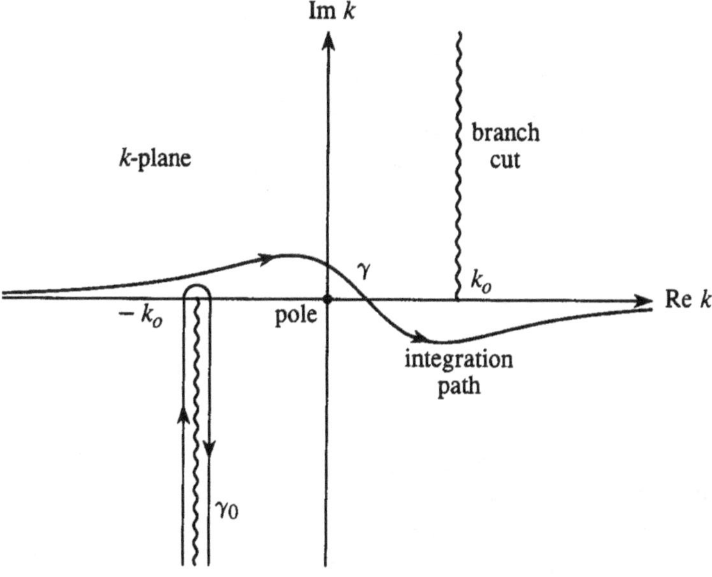

Figure 5.12.3

Hence (5.12.13) becomes

$$U(k)\sqrt{k_o+k} - \frac{\sqrt{k_o}}{2\pi i(k+i0)} = \frac{L(k)}{\sqrt{k_o-k}} + \frac{k_o}{2\pi i(k+i0)}\left(\frac{1}{\sqrt{k_o-k}} - \frac{1}{\sqrt{k_o+i0}}\right),$$

where the left-hand side is an 'upper' function and the right-hand side is a 'lower' function. Together they define a function that is regular and of algebraic growth throughout the whole of the k plane, and which must therefore be a polynomial. Thus

$$\mathcal{A}(k) \equiv U(k) = \frac{\sqrt{k_o}}{2\pi i\sqrt{k_o+k}}\left(\frac{1}{k+i0} + \sum_{n\geq 0}a_n k^n\right).$$

The coefficients a_n are chosen to ensure that the surface elevation ζ is continuous, in particular at the edge of the breakwater. Because of relation (5.12.4), this means that the integral in (5.12.7) must converge at the origin, which is possible only if all of the a_n vanish. Hence (5.12.7) becomes

$$\Phi = e^{ik_o y} + \frac{\mathrm{sgn}(y)\sqrt{k_o}}{2\pi i}\int_{-\infty}^{\infty} \frac{e^{i\left(kx+|y|\sqrt{k_o^2-k^2}\right)}}{\sqrt{k_o+k}(k+i0)}\,dk. \tag{5.12.14}$$

5.12.3 The surface wave pattern

The path of integration along the real k axis in (5.12.14) passes above the pole at $k=-i0$ (just below the real axis) and below/above the branch points at $k=\pm(k_o+i\delta)$. These branch points lie on the real axis in the limit $\epsilon \to +0$, in which case the integration contour must be deformed onto a path such as γ in Figure 5.12.3.

Consider first the waves in the immediate neighbourhood of the breakwater. When $y=0$ in (5.12.14) the integrand is regular in $\mathrm{Im}\,k > 0$. Therefore, for $x>0$ the path γ

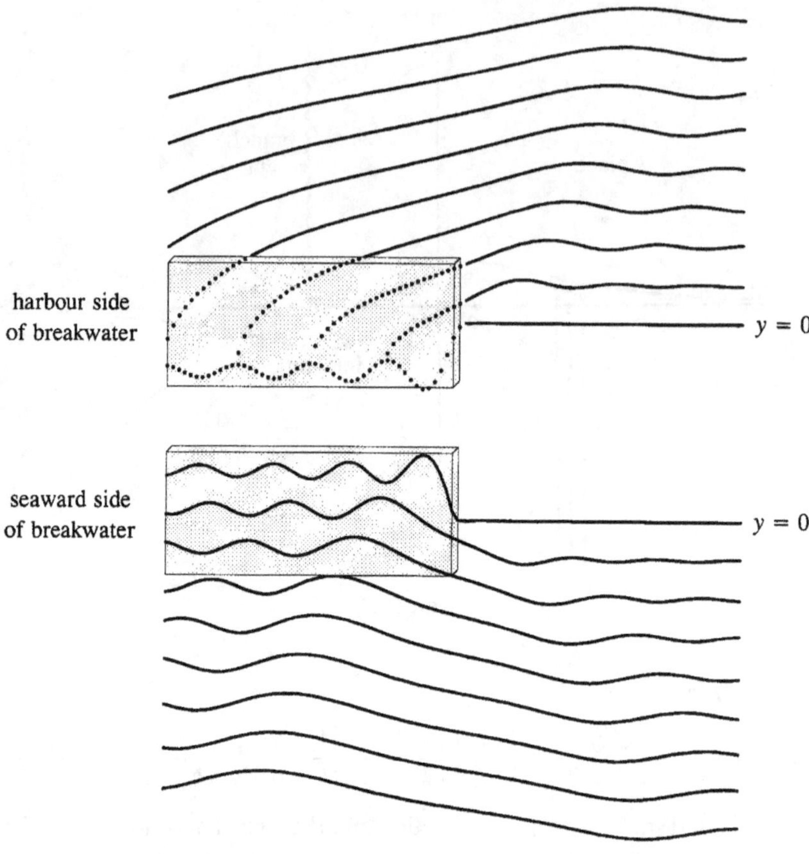

Figure 5.12.4

may be displaced to $+i\infty$ where the integrand is exponentially small without crossing any singularities. The diffracted component of Φ is therefore null along the positive x axis, that is, along the tangential extension of the breakwater. On this extension the fluctuations in the surface elevation are produced by the incident wave alone, so that (for normally incident waves) the amplitude of a wave crest is uniform along this extension. A schematic of the calculated wave crests is shown in Figure 5.12.4, the upper and lower parts of the diagram illustrate the waveforms on the harbour side ($y > 0$) and seaward side ($y < 0$) at the instant at which an incident crest (the lines labelled $y = 0$) is lined up with the extension of the breakwater.

We can calculate the surface elevation along the seaward and harbour faces of the breakwater ($x < 0,\ y = \mp 0$) by noting that the integrand of (5.12.14) is exponentially small at $k = -i\infty$ when $x < 0$ and $y = 0$. The integration path γ may therefore be displaced to $-i\infty$. A residue contribution from the pole at $k = 0$ is captured that cancels the incident wave $e^{ik_0 y}$ on the harbour side ($y = +0$) and doubles the wave-crest amplitude on the seaward side. Diffracted waves of opposite phase on $y = \pm 0$ radiate along the sides of the breakwater from its end, being the contributions from the integral around the contour γ_0 also captured during the displacement of γ to $k = -i\infty$. These waves decay like $1/|x|^{\frac{1}{2}}$ with distance from the end of the breakwater. The overall surface elevation on the seaward side is maximal at the instant shown in Figure 5.12.4; the surface displacement on the harbour side (the dotted continuation of the $y = 0$ curve) is produced entirely by the diffracted waves.

The wave profiles shown in Figure 5.12.4 near the breakwater in $y \gtrless 0$ can be calculated numerically from (5.12.14) by integration along a path such as γ (Figure 5.12.3) that avoids the pole and branch point singularities. The integrand decays exponentially fast as $k \to \pm\infty$ along the real axis, and convergence difficulties associated with a rapidly varying phase of the integrand are easily controlled for moderate values of $k_o|y|$. This has been done to plot the plan view of the wave crests in Figure 5.12.1, where the line thickness is proportional to the wave amplitude. The '20% boundary' may be taken to mark the edge of the region of calm water, or 'shadow zone', on the harbour side of the breakwater, beyond which the amplitude is less than 20% of its incidence value.

The residual surface motions within the shadow zone at large distances (many wavelengths) from the edge are a manifestation of waves 'diffracted' at the end of the breakwater, determined by the stationary phase approximation to integral (5.12.14). The stationary point is at $k = k_o \cos\theta$ and

$$\Phi_{\text{diffracted}} \sim -\frac{\sin(\theta/2)e^{i(k_o\varpi+\frac{\pi}{4})}}{(\pi k_o\varpi)^{\frac{1}{2}}\cos\theta} \quad \text{for} \quad k_o\varpi \to \infty, \quad \text{where} \quad (x, y) = \varpi(\cos\theta, \sin\theta).$$

$$(5.12.15)$$

This formula is applicable in all directions except within transition regions centred on $\theta = \pm\frac{\pi}{2}$, where $\cos\theta \to 0$, the first of which marks the boundary of the calm water on the harbour side of the breakwater. The lower transition region (centred on $\theta \sim -\frac{\pi}{2}$) divides surface areas of total reflection of waves incident upon the breakwater from the region of total transmission.

5.12.4 Uniform asymptotic approximation: Method of steepest descents

The utility of the method of stationary phase rests on Kelvin's argument that when $k_o\varpi \gg 1$ phase interference ensures that the main contribution to integral (5.12.14) is from the immediate neighbourhood of the point $k = k_o \cos\theta$ where

$$\Theta = kx + |y|\sqrt{k_o^2 - k^2} \tag{5.12.16}$$

is stationary. However, it is also necessary that other terms in the integrand should vary 'smoothly': the progressive failure of approximation (5.12.15) as $\theta \to \pm\frac{\pi}{2}$ is caused by the ultimate coincidence of the stationary point with the pole at $k = -i0$. The method of steepest descents (Debye 1909) can be used to improve the stationary phase approximation and to furnish a formula that is uniformly valid in θ in the presence of poles (and other singularities). The procedure is effected in two steps:

Step 1: Deform the integration contour γ of Figure 5.12.3 onto a new path γ_s passing through the stationary point ($k = k_o \cos\theta$, where $\Theta = k_o\varpi$) and on which

(i) $\text{Re}\,\Theta = k_o\varpi$,

(ii) $|e^{i\Theta}|$ decreases as k moves away from $k_o \cos\theta$ in either direction.

Such a path always exists for a regular function $\Theta(k)$ of the complex variable k.

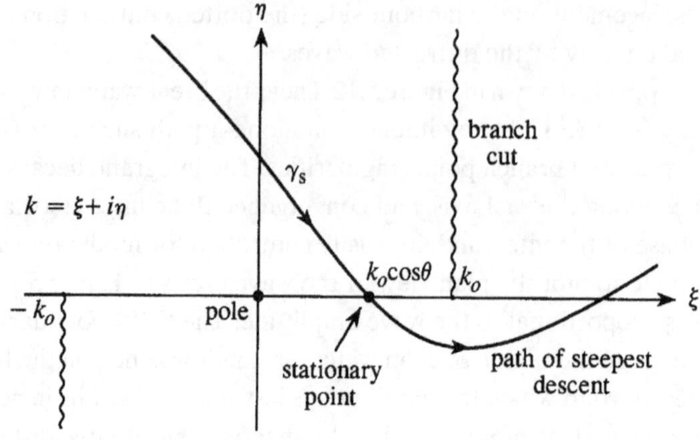

Figure 5.12.5

Put $k = \xi + i\eta$ on γ_s, where Equation (5.12.16) becomes

$$k_o\varpi + i\,\mathrm{Im}\,\Theta - (\xi + i\eta)x = |y|\sqrt{k_o^2 - (\xi + i\eta)^2}.$$

By eliminating $\mathrm{Im}\,\Theta$ between the two equations obtained by squaring both sides and equating real and imaginary parts, we find that $\mathrm{Re}\,\Theta = k_o\varpi$ along either of the curves

$$\eta = \frac{\pm(\xi - k_o\cos\theta)(k_o - \xi\cos\theta)}{|\sin\theta|\sqrt{(\xi - k_o\cos\theta)^2 + k_o^2\sin^2\theta}},$$

$$\text{on which}\quad \Theta = k_o\varpi \mp \frac{i\varpi(\xi - k_o\cos\theta)^2}{|\sin\theta|\sqrt{(\xi - k_o\cos\theta)^2 + k_o^2\sin^2\theta}}.$$

$$(5.12.17)$$

These curves intersect at right angles at the stationary point, where they cross the real axis in the directions $\pm\frac{\pi}{4}$.

We obtain the required contour γ_s by taking the lower signs in (5.12.17) (plotted in Figure 5.12.5 for $\cos\theta > 0$): $|e^{i\Theta}|$ decreases exponentially fast as ξ moves away from the stationary point in either direction along this path. This is actually the path of steepest descents from the *saddle point* of $|e^{i\Theta}|$ at $k = k_o\cos\theta$, along which $|e^{i\Theta}|$ decreases faster than along any other path through the stationary point. By evaluating the integral along γ_s we can be sure that the main contribution is from the neighbourhood of $k = k_o\cos\theta$ when $k_o\varpi \gg 1$.

If $\cos\theta < 0$ the pole at $k = -i0$ is crossed during the deformation of γ onto the steepest descents path γ_s. It is then necessary to augment the integral along γ_s by the residue contribution from the pole, i.e., by $-\mathrm{sgn}(y)e^{ik_o|y|}$. Because this occurs only for $x < 0$ we can put the resulting representation of (5.12.14) in the form

$$\Phi = e^{ik_o y} - \mathrm{sgn}(y)\mathrm{H}(-x)e^{ik_o|y|} + \frac{\mathrm{sgn}(y)\sqrt{k_o}}{2\pi i}\int_{\gamma_s}\frac{e^{i\Theta(k)}\,dk}{\sqrt{k_o + k}(k + i0)}, \qquad (5.12.18)$$

where Θ is given by (5.12.17) with the lower sign.

Step 2: Uniform approximation for the integral along γ_s. When $k_o \varpi \gg 1$ the main contribution to the integral is from a short segment of γ_s enclosing the stationary point, where Equations (5.12.17) with the lower signs may be replaced with

$$\eta = -(\xi - k_o \cos \theta), \quad \Theta(k) = k_o \varpi + \frac{i \varpi (\xi - k_o \cos \theta)^2}{k_o \sin^2 \theta}, \quad (k = \xi + i\eta).$$

By making the substitutions

$$\mu = (\xi - k_o \cos \theta)/k_o \quad \text{and} \quad dk \equiv d\xi + i d\eta = \sqrt{2} e^{-\frac{i\pi}{4}} k_o \, d\mu,$$

we find

$$\int_{\gamma_s} \frac{e^{i\Theta(k)} \, dk}{\sqrt{k_o + k}(k + i0)} \sim \sqrt{2} k_o e^{i(k_o \varpi - \frac{\pi}{4})} \int_{-\infty}^{\infty} \frac{e^{-k_o \varpi \mu^2 / \sin^2 \theta} \, d\mu}{\sqrt{k_o + k}(k + i0)} \quad \text{as} \quad k_o \varpi \to \infty, \quad (5.12.19)$$

where

$$k = \sqrt{2} e^{-\frac{i\pi}{4}} k_o \mu + k_o \cos \theta.$$

We find a uniform asymptotic approximation by splitting off the pole of the integrand as follows:

$$\frac{1}{\sqrt{k_o + k}(k + i0)} = \left[\frac{1}{\sqrt{k_o + k}(k + i0)} - \frac{1}{\sqrt{k_o}(k + i0)} \right] + \frac{1}{\sqrt{k_o}(k + i0)}$$

$$= \frac{-1}{\sqrt{k_o}\sqrt{k_o + k}\left(\sqrt{k_o} + \sqrt{k_o + k}\right)} + \frac{1}{\sqrt{k_o}(k + i0)}.$$

We obtain the asymptotic contribution to the integral from the first term in the usual way by replacing k with its value $k_o \cos \theta$ at the stationary point. For the second term (which is singular at $k = -i0$) we make the further substitution $\tau = \mu \sqrt{k_o \varpi}/|\sin \theta|$, and thereby deduce that

$$\int_{\gamma_s} \frac{e^{i\Theta(k)} \, dk}{\sqrt{k_o + k}(k + i0)} \sim \frac{-2|\sin \frac{\theta}{2}|}{\sqrt{k_o}(1 + \sqrt{2} \cos \frac{\theta}{2})} \sqrt{\frac{\pi}{k_o \varpi}} e^{i(k_o \varpi - \frac{\pi}{4})}$$

$$+ \frac{e^{ik_o \varpi}}{\sqrt{k_o}} \int_{-\infty}^{\infty} \frac{e^{-\tau^2} \, d\tau}{\left(\frac{\cos \theta}{|\sin \theta|} \sqrt{\frac{ik_o \varpi}{2}} - \tau \right)}.$$

The second term on the right-hand side is equal to

$$\frac{\pi \, \text{sgn}(x) e^{ik_o \varpi}}{i \sqrt{k_o}} \hat{w} \left(|\cot \theta| \sqrt{\frac{ik_o \varpi}{2}} \right)$$

where $\hat{w}(z)$ is a *Fresnel integral* [sometimes called the *plasma dispersion function* (Fried and Conte 1961)], defined for $\text{Im} \, z > 0$ by

$$\hat{w}(z) = \frac{i}{\pi} \int_{-\infty}^{\infty} \frac{e^{-\tau^2} \, d\tau}{z - \tau}. \tag{5.12.20}$$

The substitution of these results into (5.12.18) yields the required uniform approximation

$$\Phi \sim e^{ik_o y} - \mathrm{sgn}(y)\mathrm{H}(-x)e^{ik_o|y|} - \frac{\mathrm{sgn}(xy)}{2}\hat{w}\left(|\cot\theta|\sqrt{\frac{ik_o\varpi}{2}}\right)e^{ik_o\varpi}$$

$$+ \frac{\sin\frac{\theta}{2}\,e^{i(k_o\varpi+\frac{\pi}{4})}}{(1+\sqrt{2}\cos\frac{\theta}{2})\sqrt{\pi k_o\varpi}}, \qquad k_o\varpi \to \infty. \tag{5.12.21}$$

EXAMPLE 1. THE DIFFRACTED FAR FIELD Numerical work is facilitated by use of the Fresnel integral formulae (Abramowitz and Stegun 1970, §7)

$$\hat{w}\left(z\sqrt{\frac{i\pi}{2}}\right) = \sqrt{2}e^{-\frac{i\pi}{4}}\int_z^\infty e^{\frac{i\pi}{2}(\mu^2-z^2)}d\mu = \sqrt{2}e^{-\frac{i\pi}{4}}\left[\hat{g}(z)+i\,\hat{f}(z)\right],$$

$$\hat{g}(z) \sim \frac{1}{(\pi z)^3}, \quad \hat{f}(z) \sim \frac{1}{\pi z}, \quad |z|\to\infty, \quad |\arg z| < \frac{\pi}{2}.$$

When x is real and positive the functions $\hat{f}(x)$ and $\hat{g}(x)$ can be calculated from the rational fraction approximations:

$$\left.\begin{aligned}\hat{f}(x) &= \frac{1+0.926x}{2+1.792x+3.104x^2} \\[2ex] \hat{g}(x) &= \frac{1}{2+4.142x+3.492x^2+6.670x^3}\end{aligned}\right\} \quad 0 < x < +\infty.$$

Thus

$$-\frac{\mathrm{sgn}(xy)}{2}\hat{w}\left(|\cot\theta|\sqrt{\frac{ik_o\varpi}{2}}\right)e^{ik_o\varpi} \sim \frac{-\tan\theta\,e^{i(k_o\varpi+\frac{\pi}{4})}}{\sqrt{2\pi k_o\varpi}} \quad \text{when} \quad |\cot\theta|\sqrt{k_o\varpi} \to +\infty.$$

This approximation is applicable in radiation directions outside the parabolic 'shadow' transition regions of Figure 5.12.6, i.e. outside $|y| \sim k_o x^2$ where $|\cot\theta|\sqrt{k_o\varpi} > 1$. When combined with the final term on the right-hand side of (5.12.21) (which is valid in all directions) formula (5.12.15) for the overall diffracted far field is recovered.

EXAMPLE 2. THE TRANSITION REGIONS The 'geometrical optics' boundary of the shadow zone is the positive y axis. However, the transition is not sharp, but occurs over a finite region that increases with y in the manner indicated by the 20% boundary in Figure 5.12.1 (roughly corresponding to the surface region bounded by a parabola $y \sim k_o x^2$). The transition region on the seaward side of the breakwater is also evident in Figure 5.12.4.

An excellent picture of the surface waves in the transition region can be calculated from (5.12.21). The second and third terms in this formula are discontinuous across the geometric boundary $x = 0$; but $\hat{w}(0) = 1$, so that taken as a whole the formula predicts a continuous variation. The change in wave amplitude that occurs in crossing a transition region parabola, such as

$$\frac{y}{\lambda} = 2\pi\left(\frac{x}{\lambda}\right)^2, \quad \lambda = \frac{2\pi}{k_o} = \text{wavelength},$$

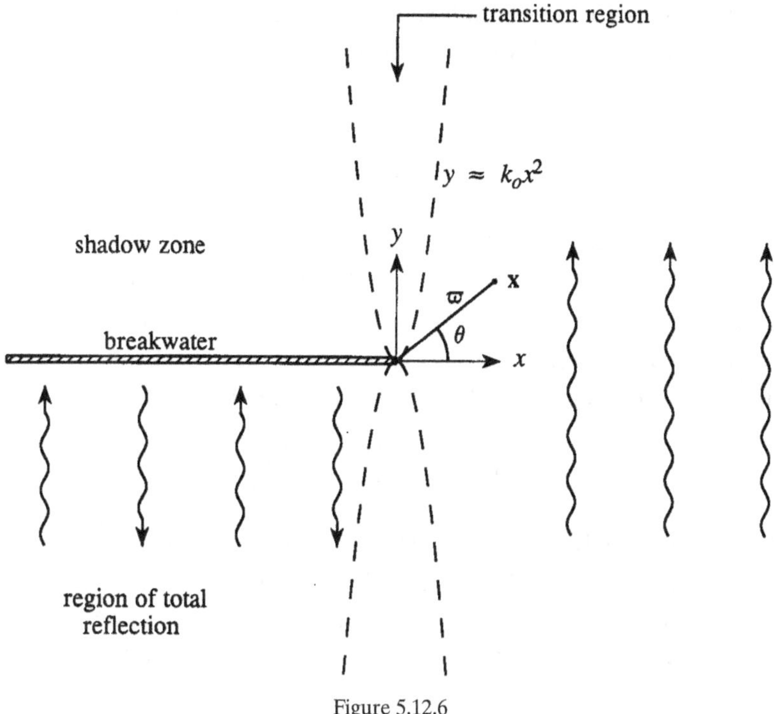

Figure 5.12.6

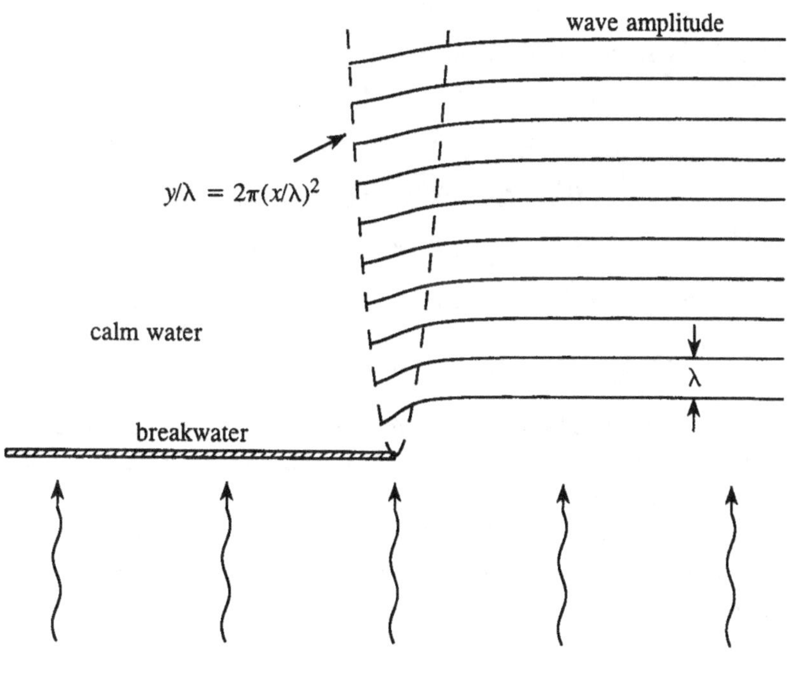

Figure 5.12.7

in the x direction is approximately independent of y. This is illustrated in the schematic amplitude plot in Figure 5.12.7, which is drawn by use of (5.12.21) and the formulae of Example 1.

PROBLEMS 5

1. A deep ocean wave of wavelength λ_o approaches a shoreline where the mean water depth is h. Show that for long waves the wavelength near the shore is $\lambda \approx \sqrt{2\pi h \lambda_o}$.

2. Show that when the shallow-water section of depth h_2 in Figure 5.4.3 terminates on the right-hand side at $x = L > 0$ at a vertical wall at which $v_x = 0$, the reflection coefficient for the wave $\zeta = \zeta_o R e^{-i(k_1 x_1 + \omega t)}$ reflected wave into $x < 0$ is given by

$$R = \frac{\cot(k_2 L) + i\sqrt{h_2/h_1}}{\cot(k_2 L) - i\sqrt{h_2/h_1}}.$$

Verify that wave energy is conserved during reflection.

3. Repeat the calculation of Example 3, §5.4, for a *floating dock* of mass m_o per unit span that is free to execute translational oscillations (without rotation) in the vertical direction. Show that if the vertical displacement of the dock from its equilibrium position is $\xi_o e^{-i\omega t}$ then the excess pressure on the underside of the dock is $p = i\rho_o \omega \varphi - \rho_o g \xi_o e^{-i\omega t}$, where

$$\varphi = \left[\frac{i\omega \xi_o x^2}{2h} - \frac{ig\zeta_o}{\omega}(\alpha x + \beta) \right] e^{-i\omega t}, \quad |x| < a.$$

Deduce that system (5.4.15), augmented by the equation of motion of the dock, becomes

$$e^{-ika} + R e^{ika} = -(\omega^2 a^2/2gh)\xi_o/\zeta_o - \alpha a + \beta,$$

$$ik\left(e^{-ika} - R e^{ika}\right) = (\omega^2 a/gh)\xi_o/\zeta_o + \alpha,$$

$$T e^{ika} = -(\omega^2 a^2/2gh)\xi_o/\zeta_o + \alpha a + \beta,$$

$$ikT e^{ika} = -(\omega^2 a/gh)\xi_o/\zeta_o + \alpha,$$

$$\left[\frac{\omega^2(m_o - \rho_o a^3/2h)}{2a\rho_o g} - 1 \right] \xi_o/\zeta_o + \beta = 0.$$

4. As for Question 3, but for a *freely floating dock* that can execute small-amplitude translational and rotational oscillations. Let the dock have moment of inertia I about the axis of symmetry (out of the plane of the paper in Figure 5.4.5), and rotate at angular velocity Ω. Show that the influence of the horizontal component (in the x direction) of the translational motion of the dock is second order and does not appear in the linear theory equations of motion.

5. The impulsive pressure

$$p_o = \frac{P_o \tau \delta(t)}{1 + (x/\ell)^2}, \quad P_o, \ \tau \text{ being constant,}$$

is applied at the surface $z = 0$ of water of density ρ_o and infinite depth. Show that the linearised velocity potential and surface displacement are given respectively by

$$\varphi = -\frac{\ell P_o \tau}{2\rho_o} H(t) \int_{-\infty}^{\infty} \cos[kx - \Omega(k)t] e^{(z-\ell)|k|}\, dk, \quad z < 0$$

$$\zeta = \frac{\ell P_o \tau}{2\rho_o g} \int_{-\infty}^{\infty} \Omega(k) \sin[kx - \Omega(k)t] e^{-\ell|k|}\, dk$$

where $\Omega(k) = \sqrt{g|k|}$.

Deduce the stationary phase approximation

$$\frac{\zeta}{(P_o/\rho_o g)\sqrt{g\tau^2/\ell}} = -\frac{\pi^{\frac{1}{2}}}{4} \left(\frac{\ell}{gt^2}\right)^{\frac{1}{4}} \left(\frac{gt^2\ell}{x^2}\right)^{\frac{5}{4}} e^{-\frac{gt^2\ell}{4x^2}} \sin\left(\frac{gt^2}{4|x|} - \frac{\pi}{4}\right),$$

and that the maximum of the main group is found at $x/t \approx \pm 0.447\sqrt{gl}$. The figure compares the exact linear surface displacement with this approximate result; the inverted arrows indicate the centre of the main group.

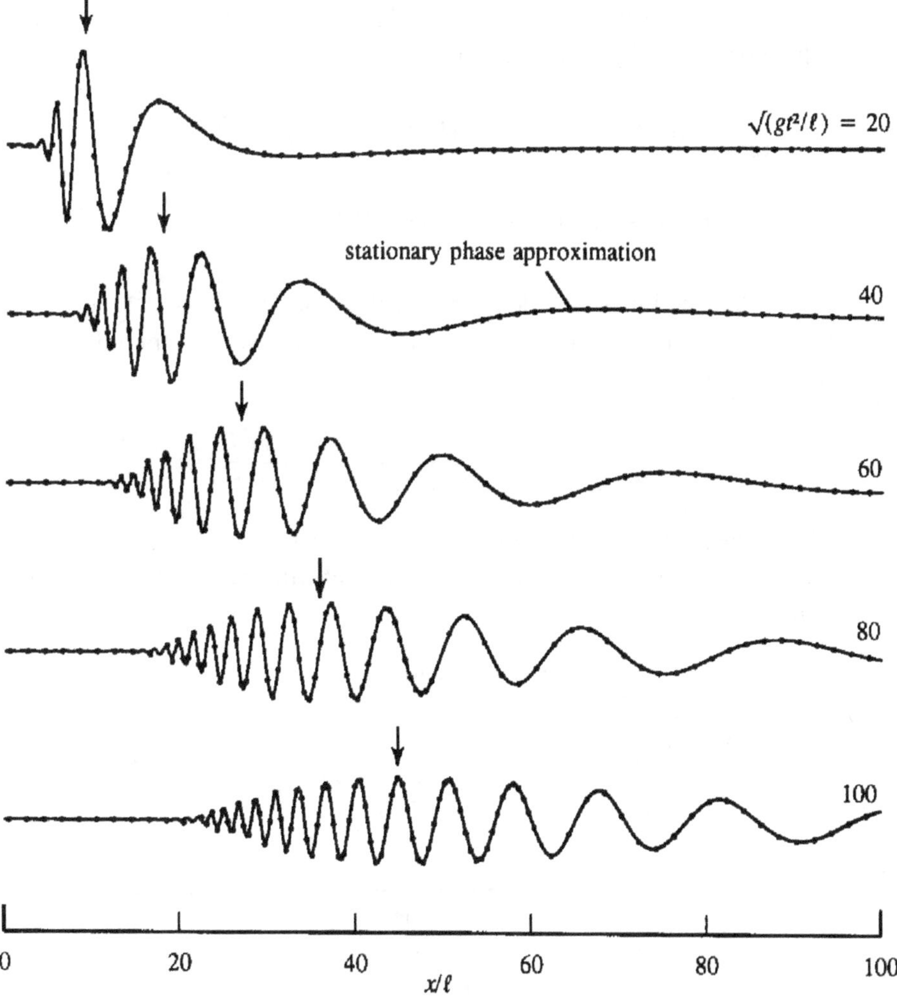

6. If conditions (5.5.1) for the initial-value wave problem of §5.5.1 are replaced with

$$\zeta = f_o(x), \quad \frac{\partial \zeta}{\partial t} = w_o(x) \quad \text{for} \quad -\infty < x < \infty \quad \text{at} \quad t = 0,$$

show that the velocity potential at $t = 0$ is given by

$$\varphi_o = \int_{-\infty}^{\infty} \frac{w_o(k)}{|k|} e^{ikx + z|k|} \, dk, \quad z \le 0,$$

and therefore that

$$H\varphi = \frac{1}{2\pi} \int_{-\infty}^{\infty} \frac{[i\omega w_o(k)/|k| + g f_o(k)]}{(\omega + i0)^2 - \Omega^2(k)} e^{i(kx - \omega t) + z|k|} \, dk \, d\omega, \quad z \le 0.$$

7. The impulsive pressure

$$p_o = \frac{P_o \tau \delta(t)}{[1 + (\varpi/\ell)^2]^{\frac{3}{2}}}, \quad \varpi = \sqrt{x^2 + y^2}, \quad P_o, \; \tau \text{ being constant,}$$

is applied at the surface $z = 0$ of water of density ρ_o and infinite depth. Show that the linearised velocity potential and surface displacement are given respectively by

$$\left.\begin{aligned}
\varphi &= -\frac{\ell^2 P_o \tau}{2\pi \rho_o} H(t) \int_{-\infty}^{\infty} \cos[\mathbf{k} \cdot \mathbf{x} - \Omega(k)t] e^{(z-\ell)k} \, d^2\mathbf{k}, \quad z < 0 \\[2mm]
\zeta &= \frac{\ell^2 P_o \tau}{2\pi \rho_o g} \int_{-\infty}^{\infty} \Omega(k) \sin[\mathbf{k} \cdot \mathbf{x} - \Omega(k)t] e^{-k\ell} \, d^2\mathbf{k}
\end{aligned}\right\} \quad \text{where } \Omega(k) = \sqrt{gk}.$$

Deduce that the initial kinetic energy of the motion is $\frac{\ell^4 P_o^2 \tau^2}{2\rho_o} \int_{-\infty}^{\infty} k e^{-2k\ell} \, d^2\mathbf{k}$, and use the energy method of §5.6.2 to show that the wave amplitude in the fully dispersed region is given by

$$\frac{|\zeta_o|}{(P_o/\rho_o g)\sqrt{g\tau^2/\ell}} = \frac{1}{4\sqrt{2}} \left(\frac{\ell}{gt^2}\right)^{\frac{1}{2}} \left(\frac{gt^2\ell}{\varpi^2}\right)^2 e^{-\frac{gt^2\ell}{4\varpi^2}}$$

and that the maximum of the main group is found at $\varpi/t \approx 0.354\sqrt{g\ell}$. The figure compares the exact linear surface displacement with the envelope determined by this formula.

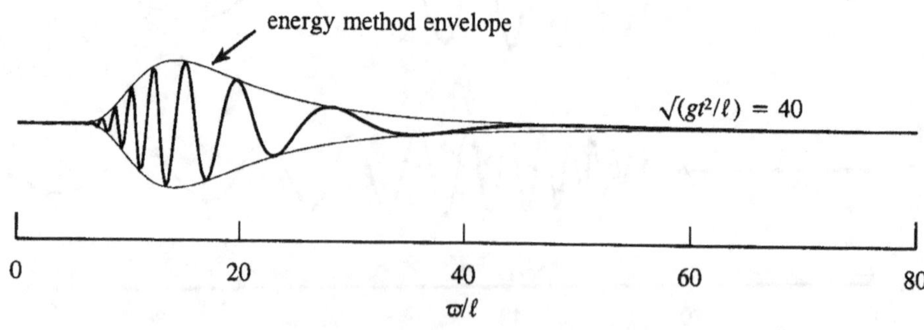

8. The velocity potential of the motion produced by a disturbance in the neighbourhood of the origin at time $t = 0$ in a three-dimensional dispersive medium is given by

$$\varphi = \iiint_{-\infty}^{\infty} f(\mathbf{k}) e^{i[\mathbf{k}\cdot\mathbf{x} - \Omega(\mathbf{k})t]}\, d^3\mathbf{k}, \quad t > 0.$$

Show that the stationary phase approximation to the potential at $(\mathbf{x}, t)$ is determined by wavenumbers clustered around the points $\mathbf{k} = \mathbf{k}_n$ at which $\mathbf{x}/t = (\partial\Omega/\partial\mathbf{k})_{\mathbf{k}=\mathbf{k}_n}$ and that as $t \to \infty$

$$\varphi \approx \sum_n \left(\frac{2\pi}{t}\right)^{\frac{3}{2}} \frac{f(\mathbf{k}_n)}{\sqrt{|\det(\partial^2\Omega/\partial k_i\,\partial k_j)_{\mathbf{k}=\mathbf{k}_n}|}}\, e^{i[\mathbf{k}_n\cdot\mathbf{x} - \Omega(\mathbf{k}_n)t + \frac{\chi_n\pi}{4}]},$$

where

$$\chi_n = \begin{cases} -3 & \text{when the phase } \Theta = \mathbf{k}\cdot\mathbf{x} - \Omega(\mathbf{k})t \text{ is a maximum at } \mathbf{k} = \mathbf{k}_n \\ +3 & \text{at a minimum} \\ +1 & \text{if } \Theta \text{ increases along 2 principal directions as } \mathbf{k} \text{ moves away from } \mathbf{k}_n \\ -1 & \text{if } \Theta \text{ increases along 1 principal direction from } \mathbf{k}_n \end{cases}$$

9. Derive representation (5.7.2) for the one-dimensional surface wave generated by an initial surface displacement on water of finite depth h.

10. The velocity potential of time harmonic motion produced by a disturbance in the neighbourhood of the origin in a three-dimensional dispersive medium is given by

$$\varphi = \int_{-\infty}^{\infty} \frac{\mathcal{F}(\mathbf{k})}{D(\mathbf{k}, \omega_o + i\epsilon)}\, e^{i[\mathbf{k}\cdot\mathbf{x} - (\omega_o + i\epsilon)t]}\, d^3\mathbf{k}, \quad \epsilon \to +0.$$

Derive Lighthill's (1960) asymptotic approximation

$$\varphi \approx 4\pi^2 i \sum_{k_j} \frac{\sqrt{|\varpi_1 \varpi_2|}}{|\mathbf{x}|} \frac{\mathcal{F}(\mathbf{k}_j)}{[\mathbf{n}(\mathbf{k})\cdot\partial D(\mathbf{k}, \omega_o)/\partial\mathbf{k}]_{\mathbf{k}=\mathbf{k}_j}}\, e^{i(\mathbf{k}_j\cdot\mathbf{x} - \omega_o t + \frac{\chi_j\pi}{4})}, \quad |\mathbf{x}| \to \infty,$$

where (in the notation of §5.8.2) the summation is over those wavenumbers $\mathbf{k}_j$ on $D(\mathbf{k}, \omega_o) = 0$ that satisfy $\mathbf{x}/|\mathbf{x}| = (\partial\Omega/\partial\mathbf{k})/|\partial\Omega/\partial\mathbf{k}| \equiv \mathbf{n}(\mathbf{k})$, at which ϖ_1, ϖ_2 are the principal radii of curvature, and

$$\chi_j = \begin{cases} -2 & \text{when } \varpi_1,\ \varpi_2 > 0, \text{ i.e. } D(\mathbf{k}, \omega_o) = 0 \text{ is convex towards } \mathbf{x} \text{ at } \mathbf{k} = \mathbf{k}_j \\ +2 & \text{when } \varpi_1,\ \varpi_2 < 0, \text{ when } D(\mathbf{k}, \omega_o) = 0 \text{ is concave towards } \mathbf{x} \\ 0 & \text{in the mixed case when } \varpi_1 \times \varpi_2 < 0 \end{cases}$$

11. The shallow-water phase speed in Example 1 of §5.10 is $c(y) = \sqrt{gh(y)}$. If $(k_1, k_2) = \frac{\omega}{c}(\sin\theta, -\cos\theta)$, derive Snell's law

$$\frac{\sin\theta}{c(y)} = \text{constant}.$$

12. Consider shallow-water waves in a region where the phase velocity $c = c_o(1 + \alpha y)$, $\alpha > 0$. Show that the equation of the ray that passes through $x = 0$, $y = 0$ at angle θ_o to the y axis is the circle

$$\left(x - \frac{1}{\alpha \tan \theta_o}\right)^2 + \left(y + \frac{1}{\alpha}\right)^2 = \frac{1}{\alpha^2 \sin^2 \theta_o}$$

13. Apply the average Lagrangian method of Example 2, §5.11, to the two-particle system in the figure. The masses M, m are connected by a light, inextensible string passing over frictionless pulleys. The mass $M > m$ falls in a vertical line, whereas m moves vertically and executes small amplitude oscillations about the vertical of variable energy E. Show that E/ω $(\omega = \sqrt{g/\ell})$ is a constant of the motion and the energy equation can be written as

$$\frac{1}{2}(M + m)\dot{\ell}^2 + (M - m)\ell g + E = \text{constant}, \quad \text{provided} \quad \frac{(M - m)}{(M + m)} \ll 1.$$

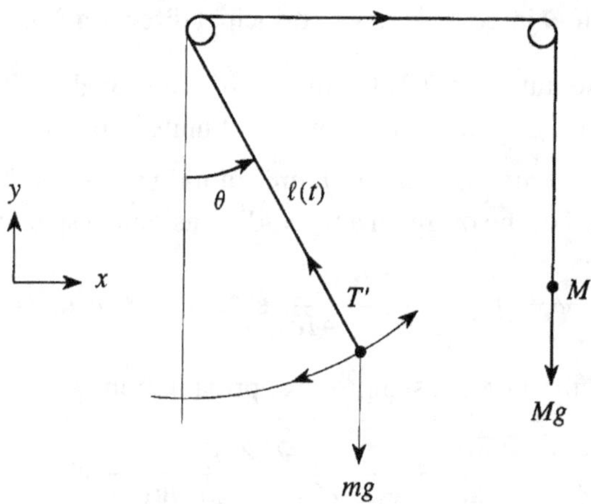

14. Fully dispersed waves propagate in a medium with dispersion relation $\omega = \Omega(\mathbf{k}, \mathbf{x}, t)$, which varies slowly on length and time scales respectively equal to the wavelength and wave period. The change in phase Θ between times $t = t_1$ and $t = t_2$ for an observer moving along a path $\mathbf{x} = \mathbf{x}(t)$ on which the wavenumber takes the value $\mathbf{k} = \mathbf{k}(t)$ is given by

$$\Theta = \int_{t_1}^{t_2} \left\{ \mathbf{k}(t) \cdot \frac{d\mathbf{x}}{dt}(t) - \Omega\left[\mathbf{k}(t), \mathbf{x}(t), t\right] \right\} \, dt.$$

Show that Θ is stationary for small changes $\delta\mathbf{x}(t)$, $\delta\mathbf{k}(t)$ that satisfy $\delta\mathbf{x}(t_1) = \delta\mathbf{x}(t_2) = 0$ if and only if the path is a ray (Lighthill 1978).

15. Consider the surface waves generated by an undersea earthquake that causes the seabed to rise by an amount $d(x, y)\mathcal{H}(t)$, where $\mathcal{H}(t) = \frac{1}{2} + \frac{1}{\pi} \tan^{-1}(t/\tau_o)$, $\tau_o > 0$.

Show that in the shallow-water approximation the surface elevation of the water satisfies

$$\frac{\partial^2 \zeta}{\partial t^2} - \frac{\partial}{\partial x_\alpha}\left(gh\frac{\partial \zeta}{\partial x_\alpha}\right) = \frac{\partial^2}{\partial t^2}\left[d(x, y)\mathcal{H}(t)\right].$$

Evaluate ζ using the shallow-water Green's function of §5.4.3 when

$$d(x, y) = \frac{d_o}{[1 + (\varpi/\ell)^2]^{\frac{3}{2}}}.$$

16. Consider the one-dimensional long-wavelength propagation of waves in a canal parallel to the x axis of slowly varying cross-sectional area A and width b *at the free surface*. Show that in the linearised approximation the surface elevation ζ satisfies

$$\frac{\partial^2 \zeta}{\partial t^2} = \frac{g}{b}\frac{\partial}{\partial x}\left(A\frac{\partial \zeta}{\partial x}\right).$$

17. An estuary extending from $x = 0$ to $x = a > 0$ has at x a rectangular cross section of uniform depth $h_o x$ and width $b_o x$, where h_o, b_o are constants. The estuary meets the open sea at $x = a$ where there exists a (low-frequency) tidal oscillation $\zeta = \zeta_o \cos \omega t$. Show that the surface elevation in the estuary is given by

$$\zeta = \zeta_o \cos(\omega t)\sqrt{\frac{a}{x}\frac{J_1(\alpha\sqrt{x})}{J_1(\alpha\sqrt{a})}}, \quad \text{where} \quad \alpha = \frac{2\omega}{\sqrt{gh_o}}.$$

18. Suppose the surface wave incident upon the breakwater of §5.12.1 propagates in the direction α $(0 < \alpha < \pi)$, so that $\Phi_I = e^{ik_o(x\cos\alpha + y\sin\alpha)}$. Show that

$$\Phi = e^{ik_o(x\cos\alpha + y\sin\alpha)} + \frac{\mathrm{sgn}(y)\sqrt{2k_o}\cos\frac{\alpha}{2}}{2\pi i}\int_{-\infty}^{\infty}\frac{e^{i\left(kx + |y|\sqrt{k_o^2 - k^2}\right)}\,dk}{\sqrt{k_o + k}(k - k_o\cos\alpha + i0)},$$

and determine a uniform approximation for Φ at large distances from the end of the breakwater.

6

Introduction to Acoustics

6.1 The wave equation

Vibrating bodies, regions of turbulent flow, the mixing of flows of different temperatures, and other unsteady phenomena produce fluctuations in pressure that propagate through a real fluid as sound, that is, as a succession of compressions and rarefactions. In applications we are usually concerned with wave amplitudes that are small relative to the mean undisturbed pressure. Therefore, in a first approximation the equations of motion may be linearised and combined to form the 'wave equation' for sound propagation. Sound waves are nondispersive in a homogeneous fluid over a broad range of frequencies for which viscous and thermal diffusion are small; in particular, this is true at audible frequencies, and it ensures that pressure fluctuations representing speech remain coherent over large distances.

The intensity of sound in air is usually measured on a decibel scale by the quantity

$$20 \times \log_{10}\left(\frac{|p|}{p_{\text{ref}}}\right),$$

where the p is a measure of the characteristic acoustic pressure and $p_{\text{ref}} = 2 \times 10^{-5}\,\text{N/m}^2$ is a standardised reference pressure. Thus $p = p_o \equiv 1$ atm $(= 10^5\,\text{N/m}^2)$ is equivalent to 194 dB. A very loud sound, $\sim$120 dB, corresponds to

$$\frac{p}{p_o} \approx \frac{2 \times 10^{-5}}{10^5} \times 10^{\left(\frac{120}{20}\right)} = 2 \times 10^{-4} \ll 1.$$

Nonlinear effects begin to be important for 'deafening' sounds of intensities $\sim$160 dB, for which $p/p_o \sim 0.02$ and $p \sim 0.3\,\text{lbs/in}^2$.

We shall see that sound waves are 'longitudinal disturbances', such that the passage of a wave in the form of a pressure fluctuation is accompanied by a back-and-forth motion of the fluid at the *acoustic particle velocity v* in the direction of propagation of the wave. In particular (§6.1.3)

$$\text{acoustic particle velocity} \approx \frac{\text{acoustic pressure}}{\rho_o \times \text{speed of sound}}.$$

390

In air the 'speed of sound' is about 340 m/s. Thus, at 120 dB $v \sim 5\,\mathrm{cm/s}$; at 160 dB $v \sim 5\,\mathrm{m/s}$.

6.1.1 The linear wave equation

In the simplest linearised approximation, sound is assumed to propagate through a *stationary* ideal fluid of mean pressure p_o and density ρ_o. Let the departures of the pressure and density from these mean values be denoted by p', ρ' where $p'/p_o \ll 1$, $\rho'/\rho_o \ll 1$. Linearised, inviscid momentum equation (1.6.5) becomes

$$\rho_o \frac{\partial \mathbf{v}}{\partial t} + \nabla p' = \mathbf{F}. \tag{6.1.1}$$

Take the continuity equation in the generalised form (2.5.1), involving the volume source distribution $q(\mathbf{x}, t)$, where (see Chapter 2) q is the rate of increase of fluid volume per unit volume produced, for example, by volume pulsations of a small body, of a bubble, or by an externally applied heat source. Linearisation yields

$$\frac{1}{\rho_o}\frac{\partial \rho'}{\partial t} + \operatorname{div}\mathbf{v} = q, \tag{6.1.2}$$

and by eliminating $\mathbf{v}$ between (6.1.1) and (6.1.2) we find

$$\frac{\partial^2 \rho'}{\partial t^2} - \nabla^2 p' = \rho_o \frac{\partial q}{\partial t} - \operatorname{div}\mathbf{F}. \tag{6.1.3}$$

To obtain a single equation governing the acoustic the pressure p' in terms of q and $\mathbf{F}$, we make use of the ideal fluid approximation (§2.1) of homentropic motion: $\rho = \rho(p, s_o)$, where s_o is the uniform and constant value of the specific entropy of the fluid. Thus energy losses that are due to viscous and thermal diffusion between neighbouring fluid particles are explicitly neglected during the passage of the sound wave (i.e. each fluid particle is assumed to expand and contract *adiabatically*). Then, because we can also put $p = p(\rho, s_o)$ and $p_o = p(\rho_o, s_o)$, small linearised disturbances must satisfy

$$p_o + p' = p(\rho_o + \rho', s_o) \approx p(\rho_o, s_o) + \left[\frac{\partial p}{\partial \rho}(\rho, s_o)\right]\rho', \tag{6.1.4}$$

where the derivative is evaluated at the undisturbed values of the pressure and density $(p_o,\ \rho_o)$. The quantity $c^2 = (\partial p/\partial \rho)_s$ has the dimensions of velocity squared, and its square root evaluated at p_o and ρ_o (and s_o) is called the *speed of sound*:

$$c_o = \sqrt{\left(\frac{\partial p}{\partial \rho}\right)_s}. \tag{6.1.5}$$

Hence $\rho' = p'/c_o^2$, and Equation (6.1.3) reduces to the inhomogeneous, acoustic wave equation for the pressure:

$$\left(\frac{1}{c_o^2}\frac{\partial^2}{\partial t^2} - \nabla^2\right)p = \rho_o \frac{\partial q}{\partial t} - \operatorname{div}\mathbf{F}, \tag{6.1.6}$$

where the prime $'$ on p' has been discarded. This describes the production of sound by the volume source q and the force distribution $\mathbf{F}$. When $q \equiv 0$ and $\mathbf{F} \equiv 0$ the equation governs sound propagation from sources on the boundaries of the fluid, such as the vibrating cone of a loudspeaker.

The unsteady motion can be described by a velocity potential φ provided $\mathbf{F} = 0$ or $\mathbf{F}$ can be expressed in terms of potential function, as in §2.3.1. The most important case in practice is $\mathbf{F} = 0$ and this will be assumed below. Then $\mathbf{v} = \nabla\varphi$, and the acoustic pressure can be written as

$$p = -\rho_o \frac{\partial \varphi}{\partial t}. \tag{6.1.7}$$

It follows from (6.1.6) (with $\mathbf{F} = 0$) that the velocity potential satisfies

$$\left(\frac{1}{c_o^2} \frac{\partial^2}{\partial t^2} - \nabla^2 \right) \varphi = -q(\mathbf{x}, t). \tag{6.1.8}$$

Causality can be invoked to justify the neglect of any time-independent constants of integration.

In the 'propagation zone', where the source terms $q = 0$, $\mathbf{F} = 0$, fluctuations in $\mathbf{v}$ and p, ρ (and in thermodynamic quantities such as the temperature T, internal energy e, and enthalpy w, but *not* the specific entropy s, which remains constant) propagate as sound governed by the homogeneous form of (6.1.8). The velocity fluctuation $\mathbf{v}$ produced by the passage of the wave is called the **acoustic particle velocity**.

6.1.2 Plane waves

A plane acoustic wave propagating in the x direction satisfies

$$\left(\frac{1}{c_o^2} \frac{\partial^2}{\partial t^2} - \frac{\partial^2}{\partial x^2} \right) \varphi = 0, \tag{6.1.9}$$

which has the general solution

$$\varphi = \Phi\left(t - \frac{x}{c_o} \right) + \Psi\left(t + \frac{x}{c_o} \right),$$

where Φ, Ψ are arbitrary functions that respectively represent waves travelling at speed c_o without change of form in the positive and negative x directions. The relation $\mathbf{v} = \nabla\varphi$ shows that the acoustic particle velocity is parallel to the propagation direction (the waves are 'longitudinal').

It follows from this result that fluctuations in v, p, ρ, T, and w in a plane wave propagating parallel to the x axis are related by

$$v = \pm\frac{p}{\rho_o c_o}, \quad \rho' = \frac{p}{c_o^2}, \quad T' = \frac{p}{\rho_o c_p}, \quad w' = \frac{p}{\rho_o}, \tag{6.1.10}$$

where p is the acoustic pressure, c_p is the specific heat at constant pressure, and the $\pm$ sign is taken according as the wave propagates in the positive or negative x direction.

Table 6.1.1 *Speed of sound and acoustic wavelength*

	c_o				λ at 1 kHz	
	m/s	f/s	km/h	mph	m	ft
Air	340	1100	1225	750	0.3	1
Water	1500	5000	5400	3400	1.5	5

6.1.3 Speed of sound

In an ideal gas of uniform mean temperature T_o,

$$c_o = \sqrt{\gamma p_o/\rho_o} = \sqrt{\gamma R T_o}, \tag{6.1.11}$$

where $\gamma = c_p/c_v$ is the ratio of specific heats (c_v being the specific heat at constant volume) and R is the gas constant. For dry air at $T_o = 10°C$, $c_o \approx 337\,\text{m/s}$.

Table 6.1.1 lists the approximate speeds of sound in air and in water and the corresponding acoustic wavelength λ at a frequency of 1 kHz (sound of frequency f Hz has wavelength $\lambda = c_o/f$).

EXAMPLE 1. WAVES IN A UNIFORM TUBE GENERATED BY AN OSCILLATING PISTON
The end $x = 0$ of an infinitely long, uniform tube is closed by a smoothly sliding piston executing small-amplitude normal oscillations at velocity $u_o(t)$ [Figure 6.1.1(a)]. If x increases along the tube, linear acoustic theory and the radiation condition require that $p = \Phi(t - x/c_o)$. At $x = 0$ the velocities of the fluid and piston are the same, so that

$$u_o(t) \equiv \frac{\Phi(t)}{\rho_o c_o}, \qquad \therefore \quad p = \rho_o c_o u_o(t - x/c_o) \ \text{ for } \ x > 0.$$

In practice, a solution of this kind, in which energy is confined by the tube to propagate in waves of constant cross section, becomes progressively invalid as x increases, because of the accumulation of small effects of flow nonlinearity. Nonlinear analysis reveals that at a point in the wave where the particle velocity is v the wave actually propagates at speed $c_o + v$, so that wave elements where v is large and positive produce 'wave steepening', resulting ultimately in the formation of 'shock waves'. This type of behaviour is important, for example, for waves generated in a long railway tunnel by the piston effect of an entering high-speed train.

EXAMPLE 2. REFLECTION AT A CLOSED END [Figure 6.1.1(b)] Let the plane wave $p = p_I(t - x/c_o)$ approach from $x < 0$ the closed, rigid end at $x = 0$ of a uniform, semi-infinite tube. The reflected pressure $p_R(t + x/c_o)$ is determined by the condition that the (normal) fluid velocity must vanish at $x = 0$. Therefore $p_I(t)/\rho_o c_o - p_R(t)/\rho_o c_o = 0$, and the overall pressure within the tube is given by

$$p = p_I(t - x/c_o) + p_I(t + x/c_o), \quad x < 0.$$

Reflection at the rigid end causes 'pressure doubling at the wall' where $p = 2p_I(t)$.

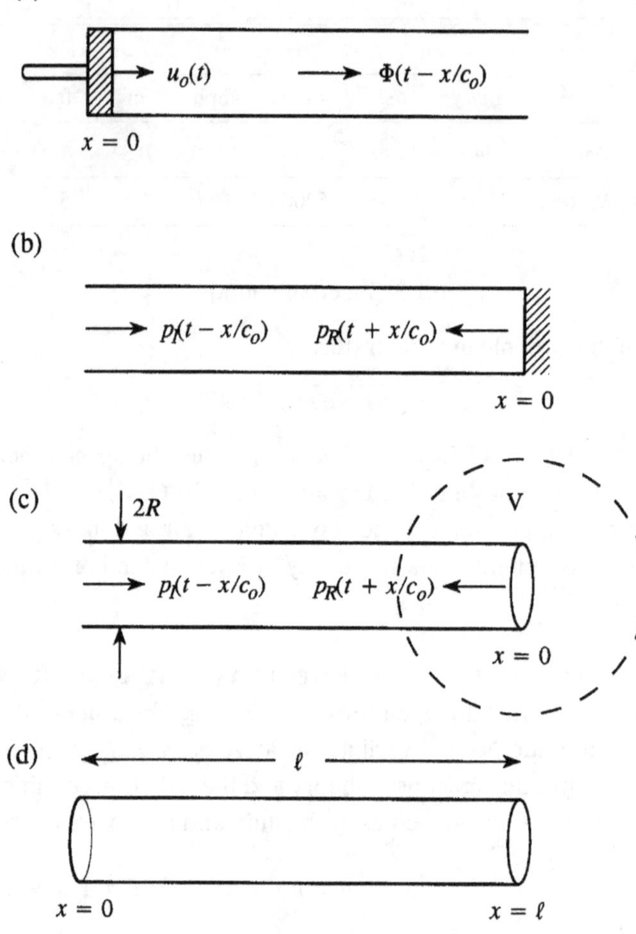

Figure 6.1.1

EXAMPLE 3. REFLECTION AT AN OPEN END [Figure 6.1.1(c)] When the wavelength of the sound is large compared with the radius R of an open-ended circular cylindrical tube, the first approximation to the condition satisfied by the acoustic pressure at the open end ($x = 0$) is that the overall pressure $p = 0$. Indeed, because of the free expansion of the fluid outside the tube, the pressure outside may be assumed to vanish compared with the incident pressure $p_I(t - x/c_o) \sim p_I(t)$. Then integration of the linearised momentum equation over the fluid contained in a spherical region V of radius R_s, where $R \ll R_s \ll \lambda = $ the acoustic wavelength, reveals that

$$\pi R^2 p \approx \frac{\partial}{\partial t} \int_V \rho_o v_1 \, d^3\mathbf{x} \sim \frac{f V p_I(t)}{c_o},$$

where $v_1 \sim O[p_I(t)/\rho_o c_o]$ is the particle velocity parallel to the tube, p is the net pressure within the tube close to the open end, and f is the frequency of the sound.

Therefore, near the open end,

$$p \sim \frac{fV}{\pi R^2 c_o} p_I(t) \sim \frac{R_s^3}{R^2 \lambda} p_I(t) \ll p_I(t) \quad \text{as} \quad \lambda = \frac{c_o}{f} \to \infty.$$

Thus, relative to the incident pressure, the pressure at the open end when $\lambda \gg R$ may be assumed to vanish. The pressure wave reflected back into the tube at the end is therefore approximately $-p_I(t + x/c_o)$,

$$p \approx p_I(t - x/c_o) - p_I(t + x/c_o), \quad x < 0,$$

and the acoustic particle velocity in the mouth of the tube $\approx 2p_I(t)/\rho_o c_o$, twice that attributable to the incident wave alone.

EXAMPLE 4. LOW-FREQUENCY RESONANT OSCILLATIONS IN A PIPE WITH OPEN ENDS [Figure 6.1.1(d)] A pressure wave of complex amplitude p' and radian frequency ω propagating in the $\pm x$ direction has the representation $p = \text{Re}[p'e^{-i\omega(t \mp x/c_o)}]$. Therefore the combination

$$p = \text{Re}\left[p'e^{-i\omega(t - x/c_o)} - p'e^{-i\omega(t + x/c_o)} \right]$$

vanishes at $x = 0$, and vanishes also at $x = \ell$ for those frequencies satisfying

$$e^{i\omega\ell/c_o} - e^{-i\omega\ell/c_o} = 0, \quad \text{i.e.,} \quad \sin(k_o \ell) = 0,$$

where $k_o = \omega/c_o$ is called the *acoustic wavenumber*. This equation accordingly determines the 'resonance frequencies' of an open-ended tube of length ℓ in the low-frequency approximation in which the pressure is assumed to vanish at the ends, *viz.*,

$$f = \frac{\omega}{2\pi} \approx \frac{n c_o}{2\ell}, \quad n = 1, 2, 3, \ldots.$$

Thus the gravest resonance frequency of a pipe of length $\ell = 1\,\text{m}$ is about $170\,\text{Hz}$ ($c_o \sim 340\,\text{m/s}$ in air), whose wavelength $\sim 2\ell = 2\,\text{m}$ does not depend on the speed of sound.

The simple one-dimensional theory neglects energy losses from the ends of the tube by radiation into the ambient atmosphere, and neglects also viscous and thermal losses in 'acoustic boundary layers' at the walls of the tube. These cause the wave amplitude to decay after several wave periods, so that resonant oscillations within the tube actually persist for only a finite time after the source of excitation is removed.

6.2 Acoustic Green's function

6.2.1 The impulsive point source

The impulsive point source of strength $\delta(\mathbf{x})\delta(t)$ is non-zero for an infinitesimal time at $t = 0$ and generalises the time-independent volume source of §§2.5, 2.6 in the inhomogeneous Laplace equation. The usual convention, however, is to reverse the sign of the source and to consider the corresponding inhomogeneous wave equation in the form

$$\left(\frac{1}{c_o^2}\frac{\partial^2}{\partial t^2} - \nabla^2\right)\varphi = \delta(\mathbf{x})\delta(t). \tag{6.2.1}$$

Because the source vanishes for $t < 0$ we are interested in only the *causal* solution, which is non-zero only for $t > 0$.

The solution must be radially symmetric. Putting $r = |\mathbf{x}|$ and observing that radial symmetry implies that

$$\nabla^2\varphi \equiv \frac{1}{r^2}\frac{\partial}{\partial r}\left(r^2\frac{\partial}{\partial r}\right)\varphi \equiv \frac{1}{r}\frac{\partial^2}{\partial r^2}(r\varphi).$$

it follows that (6.2.1) reduces to the *one-dimensional* wave equation for $r\varphi$:

$$\frac{1}{c_o^2}\frac{\partial^2}{\partial t^2}(r\varphi) - \frac{\partial^2}{\partial r^2}(r\varphi) = 0, \quad \text{when} \quad r > 0. \tag{6.2.2}$$

The general solution $r\varphi = \Phi(t - r/c_o) + \Psi(t + r/c_o)$, for arbitrary functions Φ and Ψ, yields

$$\varphi = \frac{\Phi\left(t - \frac{r}{c_o}\right)}{r} + \frac{\Psi\left(t + \frac{r}{c_o}\right)}{r}, \quad r > 0. \tag{6.2.3}$$

The terms on the right-hand side respectively represent spherically symmetric disturbances propagating in the directions of increasing and decreasing values of r at the speed of sound. Causality requires the incoming wave to be omitted by setting $\Psi = 0$, as it represents a disturbance arriving from $r = \infty$ and must necessarily be non-zero for $t < 0$. This statement of the causality principle is equivalent to imposing a *radiation condition* that the wave must radiate *away* from the source.

We can determine the functional form of Φ by the method of §2.5 or more simply by noting that, when $r \to 0$, $\varphi \sim \Phi(t)/r$ and therefore that temporal derivatives $\partial/\partial t$ become negligible compared with spatial derivatives $\partial/\partial r$. In other words the solution must resemble that for an incompressible fluid very close to the source. Hence we must have $\Phi(t) = \delta(t)/4\pi$, and causal solution of (6.2.1) becomes

$$\varphi(\mathbf{x}, t) = \frac{1}{4\pi r}\delta\left(t - \frac{r}{c_o}\right) \equiv \frac{1}{4\pi|\mathbf{x}|}\delta\left(t - \frac{|\mathbf{x}|}{c_o}\right). \tag{6.2.4}$$

The sound wave consists of a spherical pulse that is non-zero only on the surface of the sphere $r = c_o t > 0$ expanding at the speed of sound; it vanishes *everywhere* for $t < 0$.

6.2.2 Green's function

The free-space Green's function $G(\mathbf{x}, \mathbf{y}, t - \tau)$ is the *causal* solution of the wave equation generated by the impulsive point source $\delta(\mathbf{x} - \mathbf{y})\delta(t - \tau)$, located at the point $\mathbf{x} = \mathbf{y}$ at time $t = \tau$. We obtain G from solution (6.2.4) by replacing $\mathbf{x}$ with $\mathbf{x} - \mathbf{y}$ and t with $t - \tau$ (compare §2.6). Thus, if

$$\left(\frac{1}{c_o^2}\frac{\partial^2}{\partial t^2} - \nabla^2\right) G = \delta(\mathbf{x} - \mathbf{y})\delta(t - \tau), \quad \text{where} \quad G = 0 \text{ for } t < \tau, \tag{6.2.5}$$

then

$$G(\mathbf{x}, \mathbf{y}, t - \tau) = \frac{1}{4\pi|\mathbf{x} - \mathbf{y}|}\delta\left(t - \tau - \frac{|\mathbf{x} - \mathbf{y}|}{c_o}\right), \tag{6.2.6}$$

a spherically symmetric pulse radiating from the source at $\mathbf{y}$ for times $t > \tau$. The pulse amplitude decreases inversely with distance $|\mathbf{x} - \mathbf{y}|$ from the source.

6.2.3 Retarded potential

Write general acoustic equation (6.1.6) in the form

$$\left(\frac{1}{c_o^2}\frac{\partial^2}{\partial t^2} - \nabla^2\right) p = \mathcal{F}(\mathbf{x}, t), \tag{6.2.7}$$

where the generalised source $\mathcal{F}(\mathbf{x}, t)$ generates sound waves that propagate away from the source region, in accordance with the radiation condition.

This source distribution is equivalent to the following distribution of point sources (see §2.6):

$$\iint_{-\infty}^{\infty} \mathcal{F}(\mathbf{y}, \tau)\delta(\mathbf{x} - \mathbf{y})\delta(t - \tau)d^3y d\tau.$$

Point-source solution (6.2.6) and the principle of superposition therefore supply the outgoing wave solution of (6.2.7) in the form

$$p(\mathbf{x}, t) = \iint_{-\infty}^{\infty} \mathcal{F}(\mathbf{y}, \tau)G(\mathbf{x}, \mathbf{y}, t - \tau)d^3y d\tau \tag{6.2.8}$$

$$= \frac{1}{4\pi}\iint_{-\infty}^{\infty}\frac{\mathcal{F}(\mathbf{y}, \tau)}{|\mathbf{x} - \mathbf{y}|}\delta\left(t - \tau - \frac{|\mathbf{x} - \mathbf{y}|}{c_o}\right)d^3y d\tau, \tag{6.2.9}$$

$$\text{i.e. } p(\mathbf{x}, t) = \frac{1}{4\pi}\int_{-\infty}^{\infty}\frac{\mathcal{F}\left(\mathbf{y}, t - \frac{|\mathbf{x}-\mathbf{y}|}{c_o}\right)}{|\mathbf{x} - \mathbf{y}|}d^3y. \tag{6.2.10}$$

Formula (6.2.10) defines a **retarded potential** solution of the wave equation, representing the pressure at $\mathbf{x}$ at time t as a linear combination of contributions from sources at points $\mathbf{y}$ emitting sound at the earlier times $t - |\mathbf{x} - \mathbf{y}|/c_o$, where $|\mathbf{x} - \mathbf{y}|/c_o$ is the time of travel of sound from $\mathbf{y}$ to $\mathbf{x}$.

6.2.4 Sound from a vibrating sphere

The small rigid sphere of Figure 2.7.1, oscillating at speed $U(t)$ in the x_1 direction about its centre at the origin, is equivalent to a point-volume dipole source of strength

$2\pi a^3 U(t)$ when its radius a is *small* relative to the characteristic acoustic wavelength. The velocity potential is the causal solution of

$$\left(\frac{1}{c_o^2}\frac{\partial^2}{\partial t^2} - \nabla^2\right)\varphi = \frac{\partial}{\partial x_j}[f_j(t)\delta(\mathbf{x})], \quad \text{where} \quad f = [2\pi a^3 U(t), 0, 0].$$

By using the identity

$$\frac{\partial}{\partial y_j}\frac{\delta\left(t - \tau - \frac{|\mathbf{x}-\mathbf{y}|}{c_o}\right)}{|\mathbf{x}-\mathbf{y}|} = -\frac{\partial}{\partial x_j}\frac{\delta\left(t - \tau - \frac{|\mathbf{x}-\mathbf{y}|}{c_o}\right)}{|\mathbf{x}-\mathbf{y}|} \tag{6.2.11}$$

and representation (6.2.9), we find by integration by parts (compare §2.7)

$$\varphi(\mathbf{x}, t) = \frac{1}{4\pi}\iint_{-\infty}^{\infty}\frac{\partial}{\partial y_j}\left[f_j(\tau)\delta(\mathbf{y})\right]\frac{\delta\left(t - \tau - \frac{|\mathbf{x}-\mathbf{y}|}{c_o}\right)}{|\mathbf{x}-\mathbf{y}|}d^3\mathbf{y}d\tau$$

$$= \frac{1}{4\pi}\iint_{-\infty}^{\infty}f_j(\tau)\delta(\mathbf{y})\frac{\partial}{\partial x_j}\left[\frac{\delta\left(t - \tau - \frac{|\mathbf{x}-\mathbf{y}|}{c_o}\right)}{|\mathbf{x}-\mathbf{y}|}\right]d^3\mathbf{y}d\tau$$

$$= \frac{1}{4\pi}\frac{\partial}{\partial x_j}\iint_{-\infty}^{\infty}f_j(\tau)\delta(\mathbf{y})\left[\frac{\delta\left(t - \tau - \frac{|\mathbf{x}-\mathbf{y}|}{c_o}\right)}{|\mathbf{x}-\mathbf{y}|}\right]d^3\mathbf{y}d\tau$$

$$\equiv \frac{\partial}{\partial x_1}\left[\frac{2\pi a^3 U\left(t - \frac{|\mathbf{x}|}{c_o}\right)}{4\pi|\mathbf{x}|}\right],$$

i.e. (putting $r = |\mathbf{x}|$ and $x_1 = r\cos\theta$),

$$\varphi = -\frac{a^3\cos\theta}{2r^2}U\left(t - \frac{r}{c_o}\right) - \frac{a^3\cos\theta}{2c_o r}\frac{\partial U}{\partial t}\left(t - \frac{r}{c_o}\right).$$

$$\underbrace{\qquad\qquad}_{\text{near-field}} \qquad\qquad \underbrace{\qquad\qquad}_{\text{far-field}}$$

The 'near-field' component coincides with the 'hydrodynamic' velocity potential (2.7.10) for incompressible flow when the time retardation r/c_o in the argument of U is ignored; indeed the motion becomes incompressible when $c_o \to \infty$. The near field is dominant when

$$\frac{1}{r} \gg \frac{1}{c_o}\frac{1}{U}\frac{\partial U}{\partial t} \sim \frac{f}{c_o},$$

where f is the characteristic frequency of the oscillations. Because sound of frequency f travels a distance

$$c_o/f = \lambda \equiv \text{one acoustic wavelength}$$

in one period of oscillation $1/f$, the near field is dominant when $r \ll \lambda$.

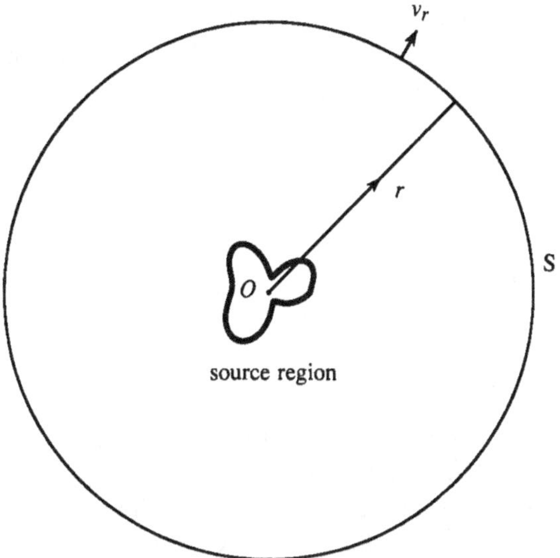

Figure 6.2.1

When the motion is irrotational and incompressible there is no net loss of energy by the oscillating sphere (§2.10). However, when c_o is finite the sphere possesses a 'far-field' or 'acoustic' region consisting of propagating sound waves, which takes over from the near field when $r \gg \lambda$. These waves carry energy away from the sphere. Our calculation has made the implicit assumption that the motion close to the sphere is roughly the same as in incompressible flow, and therefore that $a \ll \lambda$. The sphere is therefore **acoustically compact**, i.e. it is small compared with the wavelengths of the sound waves it is producing (or, more generally, with which it interacts).

6.2.5 Acoustic energy flux

At large distances r from a localised, three-dimensional acoustic source the amplitudes of all radiating quantities $\sim O(1/r)$. A knowledge of the acoustic field to this order is sufficient to permit the calculation of the *acoustic energy flux*. When r is large the wavefronts are approximately plane, and the acoustic particle velocity is radial with $v_r = p/\rho_o c_o$ (Figure 6.2.1). The quantity

$$I = pv_r = \frac{p^2}{\rho_o c_o} \tag{6.2.12}$$

is called the **acoustic intensity**. It is the rate of transmission of acoustic energy per unit area of wavefront.

The **acoustic power** Π radiated by the source is given by

$$\Pi = \oint_S pv_r \, dS = \oint_S \frac{p^2}{\rho_o c_o} \, dS, \tag{6.2.13}$$

where the integration is over the surface S of a large sphere of radius r centred on the source region. Because the surface area $= 4\pi r^2$ it is necessary to know p and v_r

only correct to order $1/r$ on S in order to evaluate the integral. Smaller contributions (determined, for example, by the near field of the vibrating sphere in §6.2.4) decrease too quickly to make a finite contribution to the integral as $r \to \infty$.

6.2.6 Green's function in one space dimension: Method of descent

The Green's function for *plane* waves that propagate in one dimension, say parallel to the x_1 axis, is the causal solution $G(x_1, y_1, t - \tau)$ of

$$\left(\frac{1}{c_o^2} \frac{\partial^2}{\partial t^2} - \frac{\partial^2}{\partial x_1^2} \right) G = \delta(x_1 - y_1)\delta(t - \tau), \quad \text{where} \ \ G = 0 \ \text{for} \ t < \tau. \quad (6.2.14)$$

Evidently, we can obtain this equation formally by integrating the corresponding three-dimensional equation (6.2.5) over the plane of uniformity $-\infty < y_2, \ y_3 < \infty$. This is Hadamard's (1952) *method of descent* to a lower space dimension. Using formula (6.2.6) for G in three dimensions, we find that (6.2.14) is satisfied by

$$G(x_1, y_1, t - \tau) = \iint_{-\infty}^{\infty} \frac{1}{4\pi |\mathbf{x} - \mathbf{y}|} \delta\left(t - \tau - \frac{|\mathbf{x} - \mathbf{y}|}{c_o} \right) dy_2 dy_3$$

$$= \int_0^{\infty} \delta\left(t - \tau - \frac{\sqrt{|x_1 - y_1|^2 + \mu^2}}{c_o} \right) \frac{\mu \, d\mu}{2\sqrt{|x_1 - y_1|^2 + \mu^2}}$$

$$= \frac{c_o}{2} \mathrm{H}\left(t - \tau - \frac{|x_1 - y_1|}{c_o} \right). \quad (6.2.15)$$

Thus, whereas in three dimensions G consists of a spherically radiating singular pulse that vanishes everywhere except at the wavefront, the corresponding one-dimensional wave is finite and consists of two simple discontinuities propagating in both directions from the source at the speed of sound to the rear of which the amplitude is constant and equal to $c_o/2$.

EXAMPLE 1. ALTERNATIVE DERIVATION OF GREEN'S FUNCTION IN ONE DIMENSION
Symmetry and the radiation condition imply that the causal solution of

$$\left(\frac{1}{c_o^2} \frac{\partial^2}{\partial t^2} - \frac{\partial^2}{\partial x^2} \right) \varphi = \delta(x)\delta(t), \quad \text{where} \ \ \varphi = 0 \ \text{for} \ t < 0,$$

has the form $\varphi = \Phi(t - |x|/c_o)$ when $x \neq 0$. By substituting this trial solution into the equation, noting that

$$\frac{\partial^2 \Phi}{\partial x^2}(t - |x|/c_o) \equiv \frac{1}{c_o^2} \Phi''(t - |x|/c_o) - \frac{2}{c_o}\delta(x)\Phi'(t),$$

where the prime denotes differentiation with respect to the argument, we find that $\Phi'(t) = \frac{1}{2}c_o\delta(t)$, and therefore that $\Phi(t) = \frac{1}{2}c_o\mathrm{H}(t)$, where the condition $\varphi = 0$, $t < 0$ implies that the constant of integration is zero. Hence $\varphi = \frac{1}{2}c_o\mathrm{H}(t - |x|/c_o)$, and this leads directly to formula (6.2.15) for Green's function.

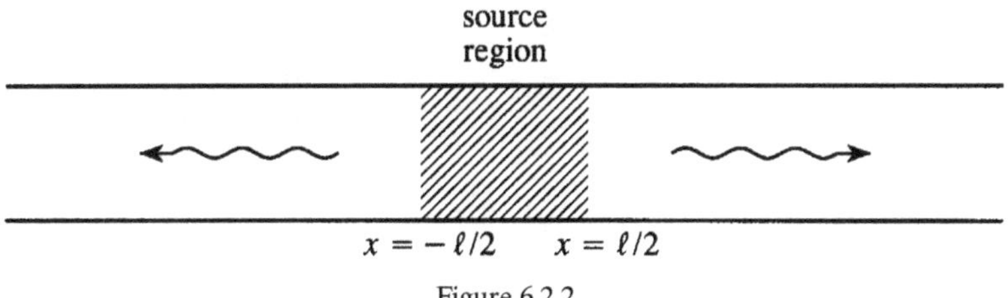

Figure 6.2.2

6.2.7 Waves generated by a one-dimensional volume source

Consider the causal solution of

$$\left(\frac{1}{c_o^2}\frac{\partial^2}{\partial t^2} - \frac{\partial^2}{\partial x^2}\right) p = \rho_o \frac{\partial q}{\partial t}(x, t), \quad -\infty < x < \infty.$$

Using the Green's function (6.2.15) and assuming that $q(x, t) \to 0$ as $t \to -\infty$, we obtain

$$p(x, t) = \frac{\rho_o c_o}{2} \iint_{-\infty}^{\infty} \frac{\partial q}{\partial \tau}(y, \tau) \mathrm{H}\left(t - \tau - \frac{|x - y|}{c_o}\right) dy d\tau$$

$$= \frac{\rho_o c_o}{2} \iint_{-\infty}^{\infty} q(y, \tau) \delta\left(t - \tau - \frac{|x - y|}{c_o}\right) dy d\tau$$

$$= \frac{\rho_o c_o}{2} \int_{-\infty}^{\infty} q\left(y, t - \frac{|x - y|}{c_o}\right) dy, \tag{6.2.16}$$

which is the acoustic pressure in 'retarded potential' form.

EXAMPLE 2. Let the source $q(x, t) = q_o \cos \Omega t$, where the source strength $q_o = \text{constant}$, be confined to the interval $-\frac{\ell}{2} < x < \frac{\ell}{2}$. Then (6.2.16) supplies the outgoing waves for $|x| > \ell/2$ (see Figure 6.2.2):

$$p(x, t) = \frac{\rho_o c_o q_o \lambda}{2\pi} \sin\left(\frac{\pi \ell}{\lambda}\right) \cos\left[\Omega\left(t - \frac{|x|}{c_o}\right)\right], \tag{6.2.17}$$

where $\lambda = 2\pi c_o/\Omega$ is the acoustic wavelength.

The source is *compact* when $\ell \ll \lambda$, so that

$$p(x, t) \approx \frac{\rho_o c_o q_o \ell}{\pi} \cos\left[\Omega\left(t - \frac{|x|}{c_o}\right)\right],$$

and all parts of the source region are radiating in phase. Otherwise, sound received at x from different points of the source (i.e., from different points in $-\ell/2 < x < \ell/2$) can interfere destructively and can result in complete 'silence' when ℓ is a multiple of the acoustic wavelength.

6.3 Kirchhoff's formula

Green's formula (2.8.7) for the inhomogeneous Laplace equation is easily extended to provide a formal representation of the solution of acoustic equation (6.1.6) in a region V

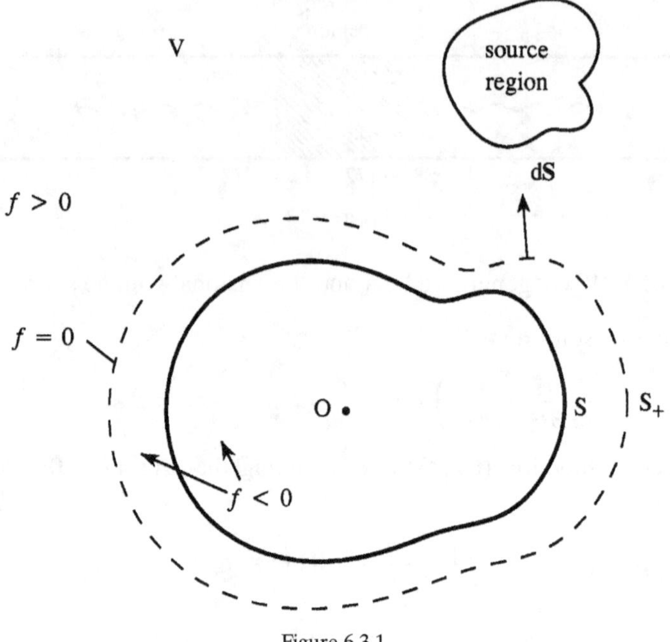

Figure 6.3.1

contained within a system of arbitrary *fixed* boundaries (or control surfaces). Consider a single closed surface S_+ defined by $f(\mathbf{x}) = 0$, with $f(\mathbf{x}) > 0$ in the region V outside S_+ (Figure 6.3.1). Assume that any fixed, solid boundaries S lie within S_+. As in §4.7.3, S_+ may subsequently be permitted to shrink down to coincide with S.

Multiply equation (6.1.6) by $H \equiv H(f)$ and form the inhomogeneous equation for the new variable Hp by use of the identities

$$H\nabla^2 p \equiv \nabla^2(Hp) - \nabla \cdot (p\nabla H) - \nabla H \cdot \nabla p,$$

$$H\frac{\partial^2 p}{\partial t^2} \equiv \frac{\partial^2(Hp)}{\partial t^2},$$

$$H\left(\rho_o\frac{\partial q}{\partial t} - \operatorname{div}\mathbf{F}\right) \equiv \frac{\partial(Hq)}{\partial t} - \operatorname{div}(H\mathbf{F}) + \nabla H \cdot \mathbf{F}.$$

We find, using the linearised momentum equation (6.1.1),

$$\left(\frac{1}{c_o^2}\frac{\partial^2}{\partial t^2} - \nabla^2\right)(Hp) = \frac{\partial(Hq)}{\partial t} - \operatorname{div}(H\mathbf{F}) - \operatorname{div}(p\nabla H) + \nabla H \cdot (-\nabla p + \mathbf{F})$$

$$= \frac{\partial(Hq)}{\partial t} - \operatorname{div}(H\mathbf{F}) - \operatorname{div}(p\nabla H) + \nabla H \cdot \rho_o\frac{\partial\mathbf{v}}{\partial t}. \qquad (6.3.1)$$

The third and last terms on the right-hand side are respectively equivalent to *dipole* and *monopole* sources on S_+. These surface sources and the prescribed volume sources and force distribution in V formally determine Hp *everywhere*, including the region within S_+ where $H(f) = 0$.

The formal, causal solution valid throughout all space can therefore be obtained by analogy with general solution (6.2.8). The result is called *Kirchhoff's formula*, which, by use of relation (6.2.11) and transformations (2.8.2), (2.8.3), can be cast in the form

$$(\mathrm{H}p)(\mathbf{x}, t) = \frac{\rho_o}{4\pi} \frac{\partial}{\partial t} \oint_S \frac{v_n(\mathbf{y}, t - |\mathbf{x} - \mathbf{y}|/c_o)}{|\mathbf{x} - \mathbf{y}|} \, d\mathrm{S}(\mathbf{y})$$

$$- \frac{1}{4\pi} \frac{\partial}{\partial x_j} \oint_S \frac{p(\mathbf{y}, t - |\mathbf{x} - \mathbf{y}|/c_o)}{|\mathbf{x} - \mathbf{y}|} n_j \, d\mathrm{S}(\mathbf{y})$$

$$+ \frac{1}{4\pi} \frac{\partial}{\partial t} \int_V \frac{q(\mathbf{y}, t - |\mathbf{x} - \mathbf{y}|/c_o)}{|\mathbf{x} - \mathbf{y}|} \, d^3\mathbf{y}$$

$$- \frac{1}{4\pi} \frac{\partial}{\partial x_j} \int_V \frac{F_j(\mathbf{y}, t - |\mathbf{x} - \mathbf{y}|/c_o)}{|\mathbf{x} - \mathbf{y}|} \, d^3\mathbf{y}, \tag{6.3.2}$$

where the surface S_+ has been permitted to shrink down onto S. As in Green's formula (2.8.7), the surface terms in (6.3.2) cannot be prescribed independently.

6.4 Compact Green's function

6.4.1 Generalised Kirchhoff formula

The discussion of §4.7.3 indicates that a formal representation of the solution of Equation (6.3.1) *within the fluid* can be expressed in terms of *any* causal Green's function $G(\mathbf{x}, \mathbf{y}, t - \tau)$ that satisfies (6.2.5) for $\mathbf{x}$ and $\mathbf{y}$ within the fluid region (outside S in Figure 6.3.1). Then, within the fluid

$$(\mathrm{H}p)(\mathbf{x}, t) = \iint G(\mathbf{x}, \mathbf{y}, t - \tau)$$

$$\times \left[\frac{\partial(\mathrm{H}q)}{\partial \tau} - \mathrm{div}\,(\mathrm{H}\mathbf{F}) - \mathrm{div}\,(p\nabla\mathrm{H}) + \nabla\mathrm{H} \cdot \rho_o \frac{\partial \mathbf{v}}{\partial \tau} \right] (\mathbf{y}, \tau) \, d^3\mathbf{y}d\tau,$$

where the integrations are over the fluid region and over all retarded times τ. Proceeding as in §4.7.3 and letting S_+ shrink down to S, we find for $\mathbf{x}$ within the fluid

$$p(\mathbf{x}, t) = \iint_V \left[G(\mathbf{x}, \mathbf{y}, t - \tau) \frac{\partial q}{\partial \tau}(\mathbf{y}, \tau) + \frac{\partial G}{\partial \mathbf{y}}(\mathbf{x}, \mathbf{y}, t - \tau) \cdot \mathbf{F}(\mathbf{y}, \tau) \right] d^3\mathbf{y}d\tau$$

$$+ \iint_S \left[G(\mathbf{x}, \mathbf{y}, t - \tau) \rho_o \frac{\partial v_n}{\partial \tau}(\mathbf{y}, \tau) + \frac{\partial G}{\partial y_n}(\mathbf{x}, \mathbf{y}, t - \tau) p(\mathbf{y}, \tau) \right] d\mathrm{S}(\mathbf{y})d\tau.$$

$$\tag{6.4.1}$$

The first integral is the contribution from volume sources and forces within the fluid; the surface integral gives similarly the contribution from the surface volume sources of strength v_n per unit area and the normal surface forces of strength p per unit area.

6.4.2 The time harmonic wave equation

If the surface S is executing small normal oscillations at a known velocity $v_n(\mathbf{x}, t)$ it would be possible to determine the radiated sound from (6.3.2) or (6.4.1) provided the surface pressure p is also known. An alternative procedure is to *eliminate* the surface pressure term on S by using generalised Kirchhoff formula (6.4.1) with G chosen to have vanishing normal derivative. It is possible to do this in a straightforward manner when the surface S is *acoustically compact* by extension of the method of far-field Green's functions of §2.16.

In the notation of §5.4.2, put

$$G(\mathbf{x}, \mathbf{y}, t - \tau) = -\frac{1}{2\pi} \int_{-\infty}^{\infty} G(\mathbf{x}, \mathbf{y}, \omega) e^{-i\omega(t-\tau)} \, d\omega. \tag{6.4.2}$$

Then $G(\mathbf{x}, \mathbf{y}, \omega)$ is the Green's function for the time harmonic (*Helmholtz*) equation:

$$(\nabla^2 + k_o^2)G(\mathbf{x}, \mathbf{y}, \omega) = \delta(\mathbf{x} - \mathbf{y}), \quad k_o = \frac{\omega}{c_o}. \tag{6.4.3}$$

We shall make use of two important results for this equation.

First, solutions satisfying the same radiation condition and surface conditions on S also satisfy the *reciprocal theorem*

$$G(\mathbf{x}, \mathbf{y}, \omega) = G(\mathbf{y}, \mathbf{x}, \omega). \tag{6.4.4}$$

The proof is similar to that given in §2.15 for incompressible flow.

Second, the free-space solution of (6.4.3), representing time harmonic waves radiating to infinity from a source at $\mathbf{y}$, is proportional to the Fourier time transform of time-dependent Green's function (6.2.6):

$$G(\mathbf{x}, \mathbf{y}, \omega) = -\int_{-\infty}^{\infty} G(\mathbf{x}, \mathbf{y}, t - \tau) e^{i\omega(t-\tau)} \, dt$$

$$= -\int_{-\infty}^{\infty} \frac{\delta(t - \tau - |\mathbf{x} - \mathbf{y}|/c_o)}{4\pi |\mathbf{x} - \mathbf{y}|} e^{i\omega(t-\tau)} \, dt$$

$$= -\frac{e^{ik_o|\mathbf{x}-\mathbf{y}|}}{4\pi |\mathbf{x} - \mathbf{y}|}. \tag{6.4.5}$$

6.4.3 The compact approximation

Let the Green's function $G(\mathbf{x}, \mathbf{y}, \omega)$ have a *vanishing* normal derivative on S, and suppose that S is acoustically compact and of characteristic dimension ℓ, so that $k_o\ell \ll 1$. The reciprocal theorem ensures that $\partial G/\partial y_n$ and $\partial G/\partial x_n$ both vanish on S. Let the observer at $\mathbf{x}$ be in the acoustic far field. We shall determine the approximate form of $G(\mathbf{x}, \mathbf{y}, \omega)$ by solving as a function of $\mathbf{y}$ the reciprocal problem in which the source is placed at the far-field point $\mathbf{x}$. This is just the method of §2.16.

Then $G(\mathbf{x}, \mathbf{y}, \omega)$ is the solution of a scattering problem in which spherical wave (6.4.5) produced by the point source at $\mathbf{x}$ is incident upon S. Take the coordinate origin inside S (at O in Figure 6.3.1) and let $\mathbf{y} \sim O(\ell)$ be in the neighbourhood of S. The compactness condition $k_o\ell \ll 1$ permits $G(\mathbf{x}, \mathbf{y}, \omega)$ to be expanded in the following form, analogous to (2.16.6):

$$G(\mathbf{x}, \mathbf{y}, \omega) = \frac{-e^{ik_o|\mathbf{x}|}}{4\pi|\mathbf{x}|}\left\{1 - \frac{ik_ox_i}{|\mathbf{x}|}\left[y_i - \varphi_i^*(\mathbf{y})\right] + \cdots + \right\}, \quad \mathbf{y} \sim O(\ell),\ |\mathbf{x}| \to \infty. \quad (6.4.6)$$

The first term in the large braces represents incident wave (6.4.5) evaluated at $\mathbf{y} = 0$. The next term is $O(k_o\ell)$ and includes a component $-ik_ox_iy_i/|\mathbf{x}|$ from the incident wave plus a correction $-ik_ox_i\varphi_i^*(\mathbf{y})/|\mathbf{x}|$ that is due to S. To this order of approximation $Y_i(\mathbf{y}) \equiv y_i - \varphi_i^*(\mathbf{y})$ satisfies Laplace's equation with $\partial Y_i/\partial y_n = 0$ on S, i.e., $\mathbf{Y}(\mathbf{y})$ is the Kirchhoff vector introduced in §2.16.1. The terms omitted in (6.4.6) are of the order of $(k_o\ell)^2$ or smaller. When they are ignored the resulting approximation for G is called the *compact Green's function*, and it can be used to determine the monopole and dipole components of the radiation.

The time-domain form of the compact Green's function is found by substitution into (6.4.2):

$$\begin{aligned}
G(\mathbf{x}, \mathbf{y}, t - \tau) &= \frac{1}{2\pi}\int_{-\infty}^{\infty}\frac{e^{ik_o|\mathbf{x}|}}{4\pi|\mathbf{x}|}\left(1 - \frac{ik_ox_iY_i}{|\mathbf{x}|} + \cdots + \right)e^{-i\omega(t-\tau)}\,d\omega \\
&\approx \frac{1}{4\pi|\mathbf{x}|}\left[\delta(t - \tau - |\mathbf{x}|/c_o) + \frac{\mathbf{x}\cdot\mathbf{Y}}{c_o|\mathbf{x}|}\delta'(t - \tau - |\mathbf{x}|/c_o)\right] \\
&\approx \frac{1}{4\pi|\mathbf{x}|}\delta\left[t - \tau - (|\mathbf{x}| - \mathbf{x}\cdot\mathbf{Y}/|\mathbf{x}|)/c_o\right] \\
&\approx \frac{1}{4\pi|\mathbf{x} - \mathbf{Y}|}\delta\left(t - \tau - |\mathbf{x} - \mathbf{Y}|/c_o\right), \quad |\mathbf{x}| \to \infty. \quad (6.4.7)
\end{aligned}$$

We can make this formula symmetric, in accordance with reciprocity, by replacing $\mathbf{x}$ with $\mathbf{X} \equiv \mathbf{x} - \varphi^*(\mathbf{x})$, after which we define the **compact Green's function for a body bounded by a surface S**:

$$G(\mathbf{x}, \mathbf{y}, t - \tau) = \frac{1}{4\pi|\mathbf{X} - \mathbf{Y}|}\delta\left(t - \tau - \frac{|\mathbf{X} - \mathbf{Y}|}{c_o}\right), \quad (6.4.8)$$

where the Kirchhoff vectors $X_i = x_i - \varphi_i^*(\mathbf{x})$, $Y_i = y_i - \varphi_i^*(\mathbf{y})$, and φ_i^* is the velocity potential of the incompressible flow that would be produced by rigid-body motion of S at unit speed in the i direction.

The frequency domain approximation

$$G(\mathbf{x}, \mathbf{y}; \omega) = \frac{-e^{ik_o|\mathbf{X}-\mathbf{Y}|}}{4\pi|\mathbf{X} - \mathbf{Y}|}, \quad (6.4.9)$$

is the natural generalisation of far-field Green's function (2.18.3) for incompressible flow.

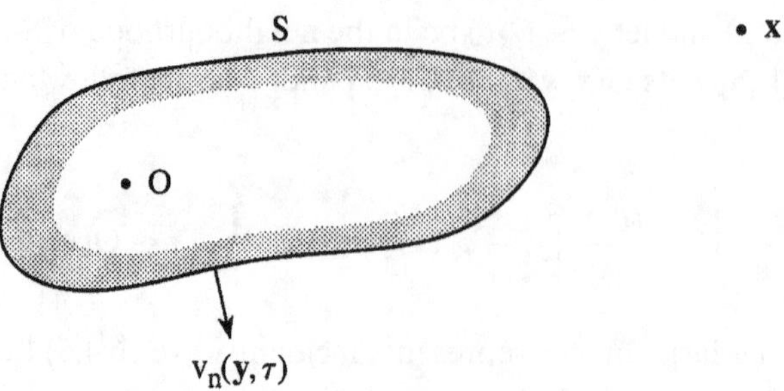

Figure 6.4.1

EXAMPLE 1. LOW-FREQUENCY RADIATION FROM A VIBRATING BODY The closed surface S of a body in an acoustic medium vibrates with small amplitude normal velocity $v_n(\mathbf{x}, t)$ at sufficiently low frequency that the body is acoustically compact (Figure 6.4.1). In the absence of sources and body forces and when G is chosen such that $\partial G/\partial y_n = 0$ on S, generalised Kirchhoff formula (6.4.1) supplies

$$p(\mathbf{x}, t) = \rho_o \frac{\partial}{\partial t} \iint_S G(\mathbf{x}, \mathbf{y}, t - \tau) v_n(\mathbf{y}, \tau)\, dS(\mathbf{y}) d\tau.$$

To determine the approximate solution in the far field, the compact Green's function (6.4.8) is introduced and the integrand is expanded correct to *dipole order*:

$$p(\mathbf{x}, t) \approx \rho_o \frac{\partial}{\partial t} \int_{-\infty}^{\infty} \oint_S \frac{v_n(\mathbf{y}, \tau)}{4\pi |\mathbf{X} - \mathbf{Y}|} \delta\left(t - \tau - \frac{|\mathbf{X} - \mathbf{Y}|}{c_o}\right) dS(\mathbf{y}) d\tau,$$

$$= \frac{\rho_o}{4\pi |\mathbf{x}|} \frac{\partial}{\partial t} \int_{-\infty}^{\infty} \oint_S v_n(\mathbf{y}, \tau) \delta\left(t - \tau - \frac{|\mathbf{x}|}{c_o} + \frac{\mathbf{x} \cdot \mathbf{Y}}{c_o |\mathbf{x}|}\right) dS(\mathbf{y}) d\tau \left(\mathbf{X} \sim \mathbf{x} \text{ as } |\mathbf{x}| \to \infty\right)$$

$$= \frac{\rho_o}{4\pi |\mathbf{x}|} \frac{\partial}{\partial t} \int_{-\infty}^{\infty} \oint_S v_n(\mathbf{y}, \tau) \left[\delta\left(t - \tau - \frac{|\mathbf{x}|}{c_o}\right) + \delta'\left(t - \tau - \frac{|\mathbf{x}|}{c_o}\right) \frac{x_j Y_j}{c_o |\mathbf{x}|}\right] dS(\mathbf{y}) d\tau,$$

where the prime denotes differentiation with respect to time. Performing the integration with respect to τ, we obtain

$$p(\mathbf{x}, t) \approx \frac{\rho_o}{4\pi |\mathbf{x}|} \frac{\partial}{\partial t} \oint_S v_n\left(\mathbf{y}, t - \frac{|\mathbf{x}|}{c_o}\right) dS(\mathbf{y}) + \frac{\rho_o x_j}{4\pi c_o |\mathbf{x}|^2} \frac{\partial^2}{\partial t^2} \oint_S v_n\left(\mathbf{y}, t - \frac{|\mathbf{x}|}{c_o}\right) Y_j(\mathbf{y}) dS(\mathbf{y}).$$

The first integral is a *monopole* sound wave that is non-zero only if the volume enclosed by S is *pulsating*. It is then the most important component of the far field – the second integral is smaller by a factor $\sim O(\omega \ell/c_o) \ll 1$ (because $\partial/\partial t \sim \omega$ and $Y_j \sim \ell$).

The monopole vanishes for a *rigid body* executing small-amplitude translational oscillations at velocity $\mathbf{U}(t)$, so that $v_n(\mathbf{y}, \tau) = n_i(\mathbf{y}) U_i(\tau)$, where $\mathbf{n}(\mathbf{y})$ is the surface normal

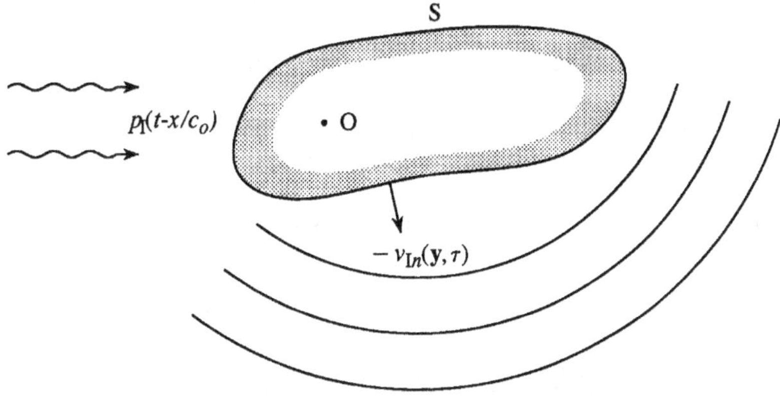

Figure 6.4.2

directed into the fluid. The substitution $Y_j = y_j - \varphi_j^*(\mathbf{y})$ in the second integral then gives the *dipole* radiation:

$$p(\mathbf{x}, t) \approx \frac{\rho_o x_j}{4\pi c_o |\mathbf{x}|^2} \frac{\partial^2 U_i}{\partial t^2} \left(t - \frac{|\mathbf{x}|}{c_o}\right) \oint_S (n_i y_j - n_i \varphi_j^*) dS, \quad |\mathbf{x}| \to \infty,$$

$$= \frac{x_i}{4\pi c_o |\mathbf{x}|^2} \left(m_o \delta_{ij} + \mathrm{M}_{ij}\right) \frac{\partial^2 U_j}{\partial t^2} \left(t - \frac{|\mathbf{x}|}{c_o}\right)$$

$$= \frac{x_i}{4\pi c_o |\mathbf{x}|^2} \left(m_o \frac{\partial^2 U_i}{\partial t^2} - \frac{\partial F_i}{\partial t}\right) \left(t - \frac{|\mathbf{x}|}{c_o}\right),$$

where m_o is the mass of fluid displaced by the body, M_{ij} is the added-mass tensor of the body, and $\mathbf{F}$ is the force exerted on the body by the fluid (§2.14).

6.4.4 Rayleigh scattering: Scattering by a compact body

The scattered radiation produced when sound impinges on a rigid, acoustically compact body (Figure 6.4.2) is equivalent in the far field to that generated by a monopole and a dipole source. Let the incident wave be propagating in the x direction and be given by

$$p = p_I(t - x/c_o),$$

and let $p_s(\mathbf{x}, t)$ denote the scattered pressure. Then p_s satisfies the homogeneous wave equation

$$\left(\frac{1}{c_o^2} \frac{\partial^2}{\partial t^2} - \nabla^2\right) p_s = 0,$$

and the solution in the far field is given by generalised Kirchhoff formula (6.4.1) and the compact Green's function in the form

$$p_s(\mathbf{x}, t) = \rho_o \frac{\partial}{\partial t} \iint_S G(\mathbf{x}, \mathbf{y}, t - \tau) v_{sn}(\mathbf{y}, \tau) \, dS(\mathbf{y}) d\tau,$$

where $\mathbf{v}_s$ is the *scattered* component of the acoustic particle velocity.

Now

$$v_{sn} + v_{In} = 0 \quad \text{on the rigid surface S,}$$

where $\mathbf{v}_I = (v_I, 0, 0)$ is the particle velocity in the incident wave. Therefore, expanding the compact Green's function to dipole order, we obtain

$$p_s(\mathbf{x}, t) \approx -\frac{\rho_o}{4\pi |\mathbf{x}|} \frac{\partial}{\partial t} \oint_S v_{In}([t] - y_1/c_o)\, dS(\mathbf{y})$$

$$-\frac{\rho_o x_j}{4\pi c_o |\mathbf{x}|^2} \frac{\partial^2}{\partial t^2} \oint_S Y_j(\mathbf{y}) v_{In}([t] - y_1/c_o)\, dS(\mathbf{y}), \quad |\mathbf{x}| \to \infty, \quad (6.4.10)$$

where $[t] = t - |\mathbf{x}|/c_o$ is the retarded time.

The continuity equation applied to the incident field implies that

$$\oint_S v_{In}([t] - y_1/c_o)\, dS(\mathbf{y}) = -\int_{V_S} \frac{1}{\rho_o c_o^2} \frac{\partial p_I}{\partial t}([t] - y_1/c_o)\, d^3\mathbf{y} \sim -\frac{V_S}{\rho_o c_o^2} \frac{\partial p_I}{\partial t}([t]),$$

where the volume integration is over the interior of S (of volume V_S) within which variations of the retarded time can be ignored. Similarly, setting $v_{In} = (p_I/\rho_o c_o)n_1$ in the second integral of (6.4.10), we find

$$\oint_S Y_j(\mathbf{y}) v_{In}([t] - y_1/c_o)\, dS(\mathbf{y}) = \frac{1}{\rho_o c_o} \oint_S Y_j(\mathbf{y}) n_1(\mathbf{y}) p_I([t] - y_1/c_o)\, dS(\mathbf{y}).$$

When retarded time variations in this integral are neglected, the resulting approximations for the two integrals in (6.4.10) are both second order in the 'compactness parameter' $\omega \ell/c_o \ll 1$, where ℓ is the effective diameter of the body and ω is a typical frequency;

$$\therefore \quad \oint_S Y_j(\mathbf{y}) v_{In}([t] - y_1/c_o)\, dS(\mathbf{y}) \sim \frac{p_I([t])}{\rho_o c_o} \left(V_S \delta_{1j} + \frac{M_{1j}}{\rho_o} \right).$$

Hence (6.4.10) becomes

$$p_s \approx \frac{1}{4\pi \rho_o c_o^2 |\mathbf{x}|} \left\{ m_o - \left(m_o \cos\theta + \frac{x_j M_{1j}}{|\mathbf{x}|} \right) \right\} \left[\frac{\partial^2 p_I}{\partial t^2} \right], \quad |\mathbf{x}| \to \infty, \quad (6.4.11)$$

where $\cos\theta = x_1/|\mathbf{x}|$, the square brackets denote evaluation at the retarded time $t - |\mathbf{x}|/c_o$, and the two terms in the braces represent respectively radiation of monopole and dipole type. Both of these components are of the same nominal magnitude $\sim O(k_o^2 \ell^2) \ll 1$ relative to the incident wave, where $k_o = \omega/c_o$.

EXAMPLE 2. The acoustic pressure p_s scattered when a sound wave $p_I(\mathbf{x}, t)$ is incident upon a compact, stationary rigid body of volume V_S is determined at large distances by the outgoing solution of

$$\left(\frac{1}{c_o^2} \frac{\partial^2}{\partial t^2} - \nabla^2 \right) p_s = \rho_o \frac{\partial}{\partial t} [q\delta(\mathbf{x})] - \text{div}[\mathbf{f}\delta(\mathbf{x})], \quad (6.4.12)$$

$$q = \frac{V_S}{\rho_o c_o^2} \frac{\partial p_I}{\partial t}; \quad f_i = (m_o \delta_{ij} + M_{ij}) \frac{1}{\rho_o} \frac{\partial p_I}{\partial x_j} \equiv -(m_o \delta_{ij} + M_{ij}) \frac{\partial v_{Ij}}{\partial t}, \quad (6.4.13)$$

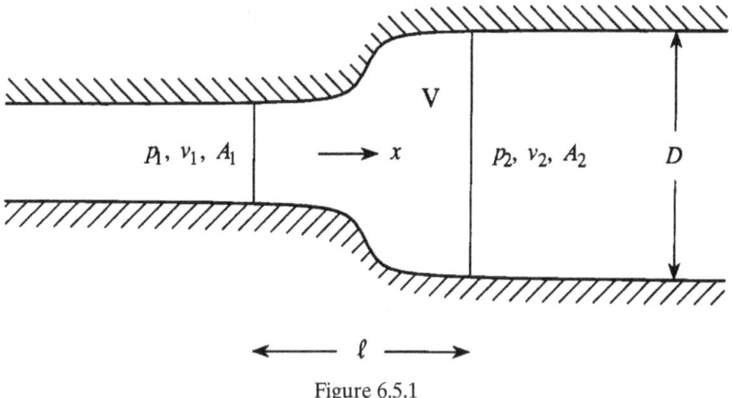

Figure 6.5.1

where the coordinate origin is within the body, $\mathbf{v}_I$ is the incident wave acoustic particle velocity, and m_o is the mass of fluid displaced by the body. The monopole strength q is equal and opposite to the volume flux into a surface that just encloses the body produced by the incident sound alone; $\mathbf{f}$ is the net force exerted on the fluid by the body.

6.5 One-dimensional propagation through junctions

The propagation of low-frequency sound along gas and oil pipelines, through air conditioning systems, through the branching airways in the lungs, etc., constitutes a complex, but essentially one-dimensional, problem. The fundamental question that must first be resolved concerns conditions to be applied at a junction across which there is a change in, say, the cross-sectional area A of the 'wave guide'. In fact, junctions across which the mean fluid density and sound speed change (with or without changes in A) are also important. In all cases we shall assume that the frequency is sufficiently small that the acoustic wavelength is much longer than the duct diameter D and that transitions in the duct and fluid properties occur over distances that are much smaller than the wavelength $\sim \lambda$ of the sound on either side of the junction.

Thus in Figure 6.5.1 both of the lengths ℓ and D are taken to be *acoustically compact*, so that λ and $k_o \sim 2\pi/\lambda$ satisfy

$$\lambda \gg \ell, \; D, \quad k_o\ell, \; k_oD \ll 1.$$

In these circumstances we can show that the pressure and the *volume velocity* are continuous across the junction. That is, if p_1, v_1, A_1 and p_2, v_2, A_2 are the respective values of the pressure, particle velocity (in the x direction), and cross-section area in the uniform sections just to the left and right of the transition region, then

$$\left. \begin{array}{c} p_1 = p_2 \\[2mm] A_1 v_1 = A_2 v_2 \end{array} \right\} \quad \text{at the junction.} \qquad (6.5.1)$$

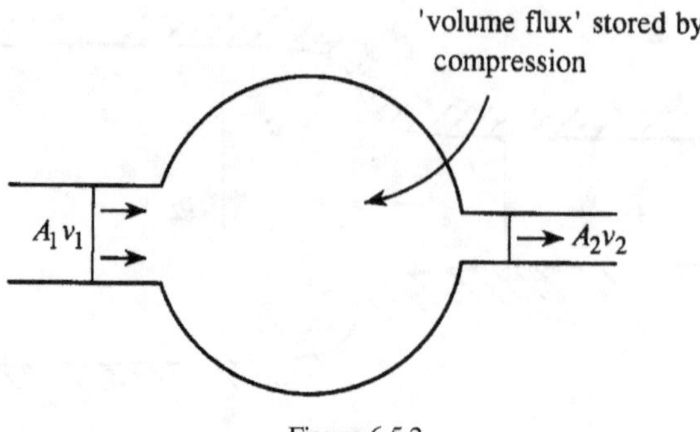

Figure 6.5.2

6.5.1 Continuity of volume velocity

We establish the continuity of volume velocity by integrating the linearised continuity equation $[1/\rho_o(\mathbf{x})]\partial\rho/\partial t + \mathrm{div}\,\mathbf{v} = 0$ over the volume $\ell\bar{A}(A_1 < \bar{A} < A_2)$ of the region V of variable cross section separating the uniform parts of the duct. If an overbar denotes a representative mean value within V, then for sound of frequency ω

$$\frac{\omega\bar{p}}{\bar{\rho}_o\bar{c}_o^2}\ell\bar{A} + A_2 v_2 - A_1 v_1 = 0.$$

This can be written as

$$A_1 v_1 - A_2 v_2 \sim (\bar{k}_o\ell)\bar{A}\frac{\bar{p}}{\bar{\rho}_o\bar{c}_o}.$$

Now $\bar{p}/\bar{\rho}_o\bar{c}_o \sim \bar{v}$, where the velocity $\bar{v}$ is of the same order as v_1, v_2. Therefore, because $\bar{k}_o\ell \ll 1$, it follows that $A_1 v_1 = A_2 v_2$ except in cases where $\bar{A}$ is very large compared with A_1 or A_2. The latter restriction would apply, for example, to a junction with a 'bulb' that can temporarily store acoustic energy (Figure 6.5.2).

6.5.2 Continuity of pressure

To establish continuity of pressure the linearised momentum equation $\rho_o(\mathbf{x})\partial\mathbf{v}/\partial t + \nabla p = 0$ is integrated over the axial width ℓ of the region V of Figure 6.5.1. Then

$$\omega\ell\bar{\rho}_o\bar{v} + p_2 - p_1 = 0,$$

where $\bar{v}$ is a suitable mean value of the axial particle velocity. Now, in order of magnitude,

$$\bar{A}\bar{v} \sim \frac{A_1 p_1}{\rho_1 c_1} \sim \frac{A_2 p_2}{\rho_2 c_2},$$

$$\therefore \quad p_1 = p_2 + O\left(\bar{k}_o\ell\frac{A_{1,2}}{\bar{A}}\bar{p}\right).$$

The condition $\bar{k}_o\ell \ll 1$ accordingly ensures continuity of pressure except possibly in cases in which one of $A_1/\bar{A}$, $A_2/\bar{A}$ is large; for example (Figure 6.5.3), when the two uniform duct sections are connected by a narrow 'neck' in which a large acceleration of the flow is accompanied by a large pressure gradient.

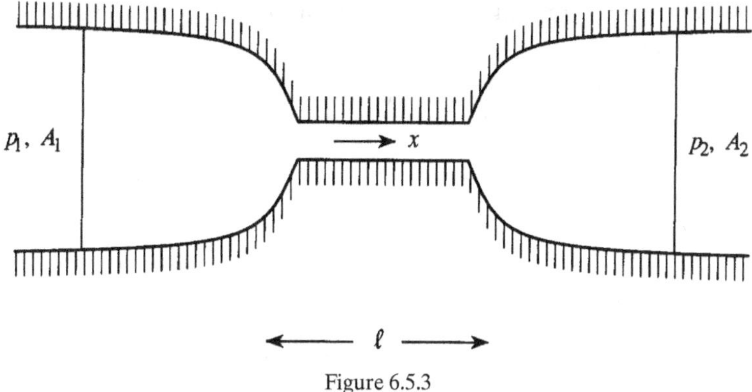

Figure 6.5.3

6.5.3 Reflection and transmission at a junction

Henceforth we shall assume that a general pressure disturbance has been decomposed with respect to frequency ω into a set of sinusoidal components proportional to $e^{-i\omega t}$, and consider the propagation of a typical member of the set through the one-dimensional system. For the junction of Figure 6.5.4 we take the coordinate origin at the centroid of the junction with x measured along the common axis from left to right, and consider a pressure wave $p = p' e^{i(k_1 x - \omega t)}$ of complex amplitude p' incident from $x < 0$. It will be convenient to suppress the exponential time factor, so that the pressure waves reflected and transmitted at the junction may be denoted respectively by $p' R e^{-ik_1 x}$ and $p' T e^{ik_2 x}$, where R and T are suitable reflection and transmission coefficients. The overall pressure and volume velocities to the left and right of the junction can accordingly be taken in the forms

$$p_1 = p'\left(e^{ik_1 x} + R e^{-ik_1 x}\right), \quad A_1 v_1 = \frac{p' A_1}{\rho_1 c_1}\left(e^{ik_1 x} - R e^{-ik_1 x}\right), \quad x < 0,$$

$$p_2 = p' T e^{ik_2 x}, \qquad\qquad A_2 v_2 = \frac{p' A_2}{\rho_2 c_2} T e^{ik_2 x}, \quad x > 0.$$

Continuity of pressure and volume velocity at the junction ($x = 0$) therefore yield

$$1 + R = T, \tag{6.5.2}$$

$$\frac{A_1}{\rho_1 c_1}(1 - R) = \frac{A_2}{\rho_2 c_2} T. \tag{6.5.3}$$

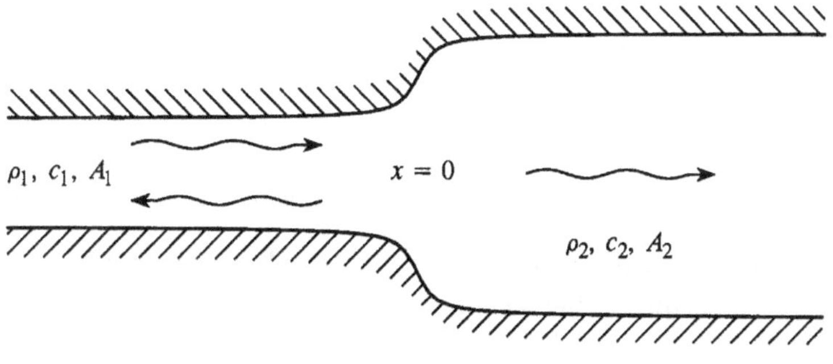

Figure 6.5.4

DEFINITION For a wave propagating in a uniform duct with the mean properties A_j, ρ_j, c_j, the quantity

$$\bar{Y}_j = \frac{A_j}{\rho_j c_j} \equiv \frac{\text{volume flux in the propagation direction}}{\text{pressure}} \qquad (6.5.4)$$

is called the *bare* acoustic admittance of the duct. This notation permits the volume velocity equation (6.5.3) to be written as

$$\bar{Y}_1(1 - R) = \bar{Y}_2\, T. \qquad (6.5.5)$$

The reflection and transmission coefficients are then deduced from (6.5.2) and (6.5.5) to be

$$R = \frac{\bar{Y}_1 - \bar{Y}_2}{\bar{Y}_1 + \bar{Y}_2}, \quad T = \frac{2\bar{Y}_1}{\bar{Y}_1 + \bar{Y}_2}. \qquad (6.5.6)$$

EXAMPLE 1. UNIFORM FLUID PROPERTIES When the mean density and sound speed are constant across the junction, the reflection and transmission coefficients are given by the simpler expressions

$$R = \frac{A_1 - A_2}{A_1 + A_2}, \quad T = \frac{2A_1}{A_1 + A_2}. \qquad (6.5.7)$$

The phase of the reflected wave relative to the incident pressure is therefore positive/negative according as $A_1 \gtrless A_2$. The junction behaves like an 'open end' when $A_2 \gg A_1$, in which case $R \sim -1$, $T \sim 0$, i.e. an equal and opposite 'expansion' wave is reflected back into the left-hand duct.

EXAMPLE 2. NON-REFLECTING JUNCTION A wave is transmitted without reflection (or phase shift) provided $\bar{Y}_1 = \bar{Y}_2$. Wave reflection at an area change can therefore be avoided by adjustment of the 'acoustic impedance' $\rho_2 c_2$ to equal $\rho_1 c_1 A_2 / A_1$. In many physiological applications, involving propagation along ducts with distensible walls, the effective sound speed is strongly dependent on the elastic properties of the walls. In such cases $\bar{Y}$ could remain constant across a junction even if the properties of the *fluid* are unchanged.

EXAMPLE 3. REFLECTION AT A BRANCHING JUNCTION Let the wave $p = p'e^{ik_0x}$ be incident upon the branching junction of Figure 6.5.5 from $x < 0$ in the duct with bare admittance $\bar{Y}_0 = A_0/\rho_0 c_0$. Using the notation of the figure, we have

$$p_0 = p'\left(e^{ik_0x} + Re^{-ik_0x}\right), \quad A_0 v_0 = p'\bar{Y}_0\left(e^{ik_0x} - Re^{-ik_0x}\right), \quad x < 0,$$

$$p_1 = p'\,Te^{ik_1x_1}, \qquad\qquad A_1 v_1 = p'\bar{Y}_1\,Te^{ik_1x_1}, \quad x_1 > 0,$$

$$p_2 = p'\,Te^{ik_2x_2}, \qquad\qquad A_2 v_2 = p'\bar{Y}_2\,Te^{ik_2x_2}, \quad x_2 > 0.$$

The transmission coefficient T is evidently the same for both transmitted waves because all pressures must be equal at the junction.

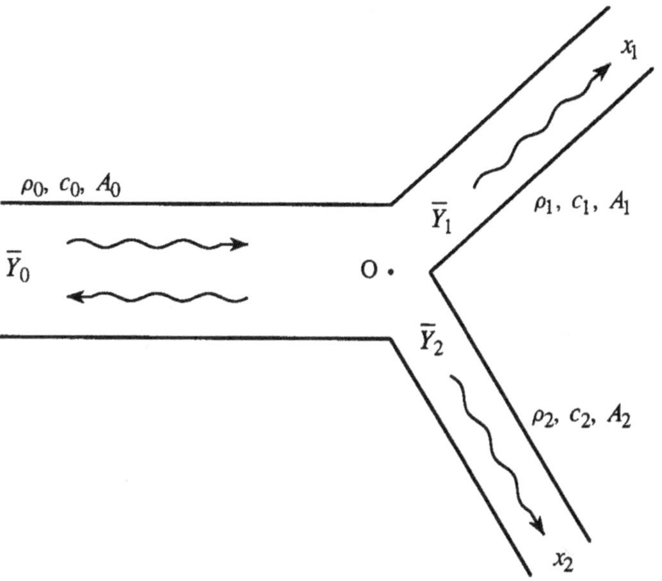

Figure 6.5.5

Then the conditions of continuity of pressure and volume flow rates at $x_0 = x_1 = x_2 = 0$ supply

$$1 + R = T, \quad \bar{Y}_0(1 - R) = (\bar{Y}_1 + \bar{Y}_2)\,T,$$

$$\therefore \quad R = \frac{\bar{Y}_0 - (\bar{Y}_1 + \bar{Y}_2)}{\bar{Y}_0 + (\bar{Y}_1 + \bar{Y}_2)}, \quad T = \frac{2\bar{Y}_0}{\bar{Y}_0 + (\bar{Y}_1 + \bar{Y}_2)}.$$

More generally, if the left-hand duct of Figure 6.5.5 splits into N separate branches at the junction, where the nth branch has admittance $\bar{Y}_n$, then

$$R = \frac{\bar{Y}_0 - \sum_{n=1}^{N} \bar{Y}_n}{\bar{Y}_0 + \sum_{n=1}^{N} \bar{Y}_n}, \quad T = \frac{2\bar{Y}_0}{\bar{Y}_0 + \sum_{n=1}^{N} \bar{Y}_n}. \tag{6.5.8}$$

6.6 Branching systems

The concept of 'bare acoustic admittance' will now be generalised to permit the simultaneous consideration of waves propagating in both directions. We first select a *reference* direction (the *positive* direction of a local x axis) with respect to which the local fluid particle and volume velocities are defined. The general acoustic admittance is then defined by

$$Y = \frac{\text{volume flux in the reference direction}}{\text{pressure}}. \tag{6.6.1}$$

This is generally a function of position, except for unidirectional propagation in an infinite uniform duct, when $Y = \bar{Y}$.

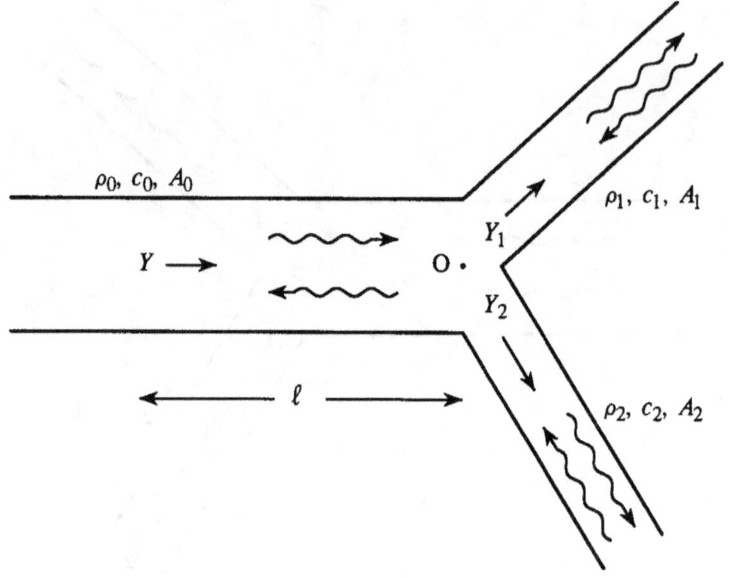

Figure 6.6.1

The determination of Y in complex branching systems is greatly facilitated by application of a fundamental formula now to be derived.

6.6.1 Fundamental formula

A duct of mean properties ρ_0, c_0, A_0 splits into N branches at $x = 0$, as illustrated in Figure 6.6.1 for $N = 2$. Let Y be the admittance at $x = -\ell$ directed towards the branch (in the positive x direction), and let Y_n denote the admittance at the entrance to the nth branch, the reference direction being *into* the branch, indicated by the arrow in the figure. Then

$$Y = \bar{Y}_0 \left[\frac{\sum_{n=1}^{N} Y_n - i\bar{Y}_0 \tan(k_0\ell)}{\bar{Y}_0 - i\left(\sum_{n=1}^{N} Y_n\right)\tan(k_0\ell)} \right], \quad \text{where} \quad \bar{Y}_0 = \frac{A_0}{\rho_0 c_0}, \quad k_0 = \frac{\omega}{c_0}. \tag{6.6.2}$$

PROOF In the primary duct $(A = A_0)$ we put

$$p = p'\left(e^{ik_0 x} + Re^{-ik_0 x}\right).$$

At $x = -\ell$ the definition of Y implies that

$$Y = \frac{\bar{Y}_0\left(e^{-ik_0\ell} - Re^{ik_0\ell}\right)}{\left(e^{-ik_0\ell} + Re^{ik_0\ell}\right)}, \tag{6.6.3}$$

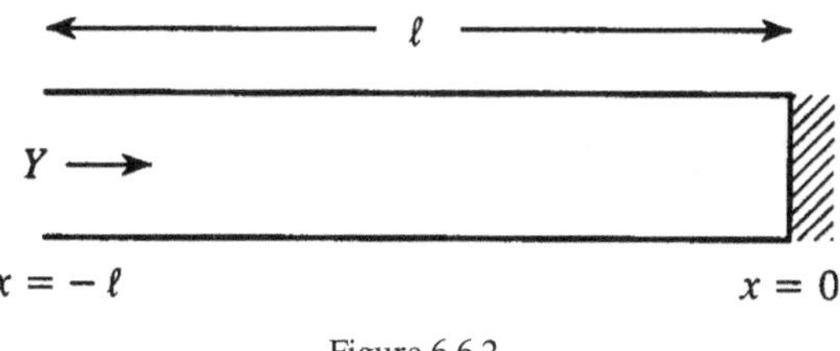

Figure 6.6.2

The pressure is continuous at the junction ($x = 0$) and equal to $p'(1 + R)$. Therefore, continuity of volume flux implies that

$$\bar{Y}_0(1 - R) = (1 + R) \sum_{n=1}^{N} Y_n,$$

$$\therefore \quad R = \frac{\bar{Y}_0 - \sum_{n=1}^{N} Y_n}{\bar{Y}_0 + \sum_{n=1}^{N} Y_n},$$

and we now obtain formula (6.6.2) by substituting for R in (6.6.3).

Observe that in (6.6.2) propagation in each of the N branches is not required to be unidirectional.

EXAMPLE 1. QUARTER-WAVE RESONATOR The resonance frequencies of long waves in a rigid tube of length ℓ ($-\ell < x < 0$) open at one end only (Figure 6.6.2) are (in a first approximation) the solutions of

$$\cos\left(\frac{\omega\ell}{c_0}\right) = 0. \tag{6.6.4}$$

The wavelength λ of the gravest mode $\sim 4\ell$.

The *input admittance* Y at the open end is given by (6.6.2) when the closed end $x = 0$ is regarded as a junction at which the admittance vanishes (so that $\sum_n Y_n \equiv 0$):

$$\therefore \quad Y = -i\bar{Y}_0 \tan\left(\frac{\omega\ell}{c_0}\right). \tag{6.6.5}$$

An arbitrarily small pressure fluctuation at the open end produces an infinitely large volume flux in the opening at the resonance frequencies satisfying (6.6.4), i.e. the resonance condition corresponds to

$$Y = \infty \quad \text{or} \quad \frac{1}{Y} = 0. \tag{6.6.6}$$

6.6.2 Energy transmission

The mean acoustic power transmitted along a duct of area A in the reference direction is $A\langle I \rangle$, where $I = pv$ is the acoustic intensity (§6.2.5) and the angle brackets denote a time average. The acoustic pressure and velocity must be taken in *real* form after

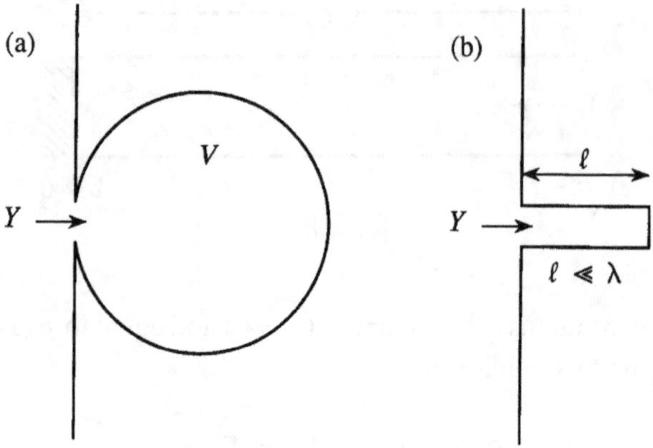

Figure 6.6.3

restoration of the time factor $\mathrm{e}^{-i\omega t}$. Thus, at a point in the duct where the complex amplitude of the pressure is p' and the admittance is Y, we write

$$p = \mathrm{Re}\left(p'\mathrm{e}^{-i\omega t}\right) \quad \text{and} \quad Av = \mathrm{Re}\left(Yp'\mathrm{e}^{-i\omega t}\right).$$

Then

$$\text{acoustic power} \equiv \Pi = \frac{1}{2}|p'|^2\mathrm{Re}\,Y. \tag{6.6.7}$$

For unidirectional propagation in an infinite duct $\Pi = \frac{1}{2}|p'|^2\mathrm{Re}\,\bar{Y}$, where $\bar{Y} = A/\rho_o c_o$ can be a *complex*-valued quantity when account is taken of wave damping in the duct (in thin-wall boundary layers, for example), which causes the effective sound speed c_o to possess a small negative imaginary part. On the other hand, for the closed end duct of Figure 6.6.2 (and in the absence of damping) the 'input admittance' (6.6.5) is pure imaginary, and on average no energy is absorbed by the resonator.

6.6.3 Acoustically compact cavity

Small, acoustically compact cavities [Figure 6.6.3(a)] can temporarily 'store' fluid volume, and this is reflected in the form of the input admittance Y. Apart from small corrections dependent on the possible damping of acoustic energy within the cavity, the value of Y does not depend on cavity *shape* but only on its volume or *capacity*, and can therefore be determined by consideration of a cavity in the form of a short, uniform duct with a closed end [Figure 6.6.3(b)].

We apply fundamental formula (6.6.2) to the duct of Figure 6.6.3(b) by taking $\sum_n Y_n \equiv 0$ for the 'branch' at the closed end and considering the compact limit $k_0\ell \ll 1$:

$$Y = -i\bar{Y}_0\tan(k_0\ell) \sim \frac{-i\omega(A_0\ell)}{\rho_0 c_0^2},$$

$$\therefore \quad \text{cavity input admittance} \quad Y = \frac{-i\omega V}{\rho_0 c_0^2}, \quad V = \text{cavity volume}. \tag{6.6.8}$$

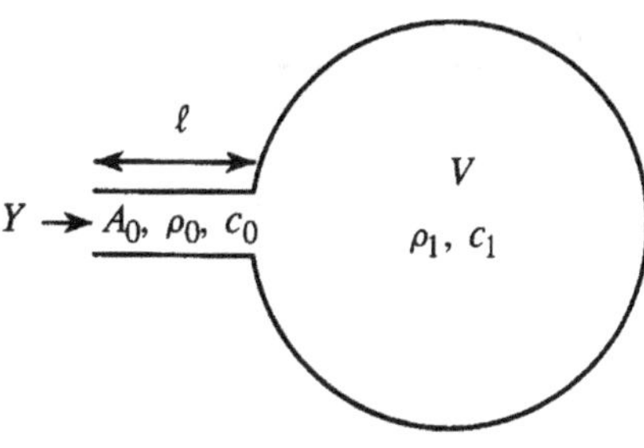

Figure 6.6.4

This is pure imaginary, so that the cavity does not absorb energy when the effective interior sound speed is real.

EXAMPLE 2. DIRECT CALCULATION OF CAVITY ADMITTANCE In a compact cavity the pressure and density are approximately uniform. Therefore integration of continuity equation (6.1.2) (with $q = 0$) over the interior volume V of the cavity of Figure 6.6.3(a) yields

$$\text{volume flux into cavity} = \frac{V}{\rho_0}\frac{\partial \rho}{\partial t} = \frac{V}{\rho_0 c_0^2}\frac{\partial p}{\partial t} = \frac{-i\omega V p}{\rho_0 c_0^2},$$

$$\therefore \quad Y = \frac{\text{volume flux into cavity}}{p}$$

$$\equiv \frac{-i\omega V}{\rho_0 c_0^2}.$$

6.6.4 The Helmholtz resonator

The Helmholtz resonator consists of an acoustically compact cavity, or 'bulb', of interior volume V that communicates with the atmosphere by means of a short, narrow-necked channel of cross-section A_0 and length $\ell \ll V^{\frac{1}{3}}$ (see Figure 6.6.4).

The system behaves as a simple harmonic oscillator with a reciprocating flow into and out of the cavity through the neck. Velocity fluctuations in the cavity are negligible, but the pressure variations provide the necessary 'spring' force pA_0 required for maintaining the periodic acceleration of the fluid slug of mass $\sim \rho_0 A_0 \ell$ in the neck (in practice the effective length of this slug exceeds ℓ because fluid outside the channel is also set into finite oscillatory motion; this is important only if ℓ is small compared with the channel diameter and can be handled by the introduction of suitable 'end corrections').

In general it may be assumed that the fluids in the cavity and neck have different mean densities and sound speeds, as implied in Figure 6.6.4. The input impedance of the neck opening is determined by fundamental formula (6.6.2) to be

$$Y = \bar{Y}_0 \left[\frac{Y_1 - i\bar{Y}_0 \tan(k_0\ell)}{\bar{Y}_0 - i Y_1 \tan(k_0\ell)} \right], \quad \text{where} \quad \bar{Y}_0 = \frac{A_0}{\rho_0 c_0}, \quad Y_1 = \frac{-i\omega V}{\rho_1 c_1^2}.$$

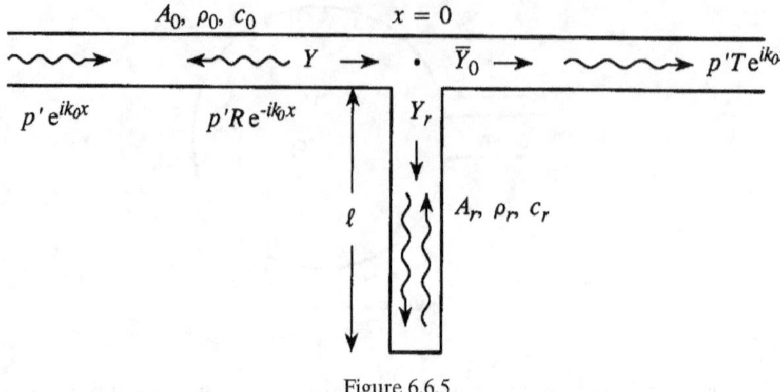

Figure 6.6.5

However, $\tan(k_0\ell) \sim k_0\ell$ when the neck is compact. Therefore, substituting for $\bar{Y}_0$ and Y_1, we find

$$Y = \frac{-i\omega\left(V/\rho_1 c_1^2 + \ell A_0/\rho_0 c_0^2\right)}{\left[1 - (\rho_0/\rho_1)\omega^2 V\ell/A_0 c_1^2\right]}.$$

The resonance condition $Y = \infty$ accordingly supplies the following formula for the Helmholtz resonance frequency:

$$\omega^2 = \left(\frac{\rho_1}{\rho_0}\right)\frac{A_0 c_1^2}{V\ell} \tag{6.6.9}$$

EXAMPLE 3. For a standard wine bottle

$$V = 750\,\text{cm}^3, \quad A_0 = 1\,\text{cm}^2, \quad \ell = 5\,\text{cm}.$$

Hence (taking $\rho_0 = \rho_1 = 1.23\,\text{kg/m}^3$, $c_0 = c_1 = 340\,\text{m/s}$) the Helmholtz resonance frequency of the sound excited by blowing over the mouth of the bottle is approximately $f = \omega/2\pi \approx 115\,\text{Hz}$.

6.6.5 Acoustic filter

A side-branch resonator can be used to block the transmission of sound of a given frequency across the branch. In the arrangement shown in Figure 6.6.5 the closed side branch forms a resonator of input admittance $Y_r = -i\bar{Y}_r \tan(k_r\ell)$, $(\bar{Y}_r = A_r/\rho_r c_r)$. In the primary wave guide $\bar{Y}_0 = A_0/\rho_0 c_0$, and the pressure on either side of the junction is given by

$$p = p'\left(e^{ik_0 x} + Re^{-ik_0 x}\right), \quad x < 0,$$
$$p = p'Te^{ik_0 x}, \quad x > 0,$$

where

$$R = \frac{-Y_r}{2\bar{Y}_0 + Y_r}, \quad T = \frac{2\bar{Y}_0}{2\bar{Y}_0 + Y_r}.$$

At the resonance frequencies Y_r becomes infinite and $T \to 0$, $R \to -1$. Therefore no acoustic energy is transmitted beyond the junction when ω is equal to one of the frequencies $\omega_n = (2n-1)\pi c_r / 2\ell$, $n = 1, 2, 3, \ldots$. A Helmholtz resonator with admittance

$$Y_r = \frac{i\omega A_r}{\rho_r \ell} \frac{(1 + \ell A_r / V)}{(\omega^2 - \omega_o^2)} \quad \text{and resonance frequency} \quad \omega_o = c_r \sqrt{\frac{A_r}{V\ell}}$$

can be used to block a single frequency ($V = $ resonator volume, $\ell = $ neck length, $A_r = $ neck cross-sectional area).

Acoustic energy must be conserved in the absence of damping. By direct calculation the net acoustic powers in the x direction on the left and to the right of the junction are given by

$$\Pi_L = \frac{1}{2} |p'|^2 \left(1 - |R|^2\right) \bar{Y}_0, \quad \Pi_R = \frac{1}{2} |p'|^2 |T|^2 \bar{Y}_0.$$

These are equal provided Y_r is pure imaginary (no dissipation in the resonator), i.e. energy conservation implies that $|R|^2 + |T|^2 = 1$ in a uniform duct. Furthermore, it is clear that

$$\Pi_L = \Pi_R = \frac{1}{2} |p_0|^2 \, \text{Re} \, Y,$$

where $p_0 = p'T$ is the pressure at $x = 0$.

At resonance the motion in the side-branch resonator remains finite:

$$p_0 \sim \frac{2\bar{Y}_0 p'}{Y_r}, \quad v_r = \frac{Y_r p_0}{A_r} \sim \frac{2\bar{Y}_0 p'}{A_r} \quad \text{as} \quad |Y_r| \to \infty.$$

We can examine the effect of damping in the resonator by replacing the resonator wavenumber $k_r \equiv \omega/c_r$ with $k_r + i\delta$, $\delta \ll k_r$, where the damping factor $\delta > 0$ when $\omega > 0$. Then, at resonance

$$Y_r \approx \frac{-i\bar{Y}_r \sin(k_r \ell)}{\cos(k_r \ell + i\delta\ell)} \sim \frac{\bar{Y}_r}{\delta\ell}.$$

Therefore, the power dissipated in the resonator is

$$\Pi_D = \frac{1}{2} |p_0|^2 \text{Re} \, Y_r \sim \frac{2\delta\ell \bar{Y}_0^2}{\bar{Y}_r} |p'|^2,$$

so that a fraction

$$\frac{\Pi_D}{\Pi_I} = \frac{4\delta\ell \bar{Y}_0}{\bar{Y}_r}$$

of the incident wave power $\Pi_I = \frac{1}{2} |p'|^2 \bar{Y}_0$ is dissipated in the resonator.

6.6.6 Admittance of a narrow constriction

A narrow, acoustically compact duct of length ℓ and bare admittance $\bar{Y}_0$ enters a very much larger wave guide of admittance Y_{out}, where $\bar{Y}_0 \ll Y_{\text{out}}$ (Figure 6.6.6).

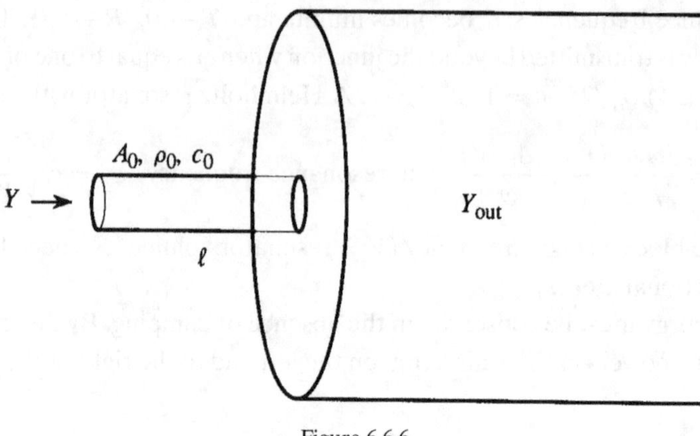

Figure 6.6.6

The fundamental formula (6.6.2) gives the input admittance of the duct in the form

$$Y = \bar{Y}_0 \left[\frac{Y_{\text{out}} - i\bar{Y}_0 \tan(k_0\ell)}{\bar{Y}_0 - iY_{\text{out}} \tan(k_0\ell)} \right].$$

The conditions $\bar{Y}_0 \ll Y_{\text{out}}$ and $k_0\ell \ll 1$ permit the simplification $Y = \bar{Y}_0 Y_{\text{out}}/(\bar{Y}_0 - iY_{\text{out}}k_0\ell)$,

$$\therefore \quad \frac{1}{Y} = \frac{1}{Y_{\text{out}}} + \frac{1}{Y_c}, \quad \text{where} \quad Y_c = \frac{iA_0}{\omega\ell\rho_0},$$

$$\text{i.e.} \quad \begin{array}{c} \text{admittance of a} \\ \text{narrow constriction} \end{array} \quad Y_c = \frac{iA_0}{\rho_0\omega\ell}. \tag{6.6.10}$$

This result provides a correction to the usual condition of continuity of pressure at a junction (§6.5.2). The inertia of fluid in the narrow section (necessary to maintain continuity of volume velocity) must be overcome by a finite drop in pressure across a narrow constriction. The inverse $1/Y$ of the admittance constitutes a complex resistance or 'impedance', and the constriction produces an additional 'in-series' resistance in the acoustic circuit such that (the volume flux being constant)

$$\text{input pressure} = \text{constriction pressure} + \text{output pressure},$$

$$\text{i.e.,} \quad \frac{1}{Y_{\text{in}}} = \frac{1}{Y_c} + \frac{1}{Y_{\text{out}}}. \tag{6.6.11}$$

EXAMPLE 4. BULBOUS TERMINATION A Helmholtz resonator attached to the end of an acoustic wave guide produces total reflection at the resonance frequency without change of phase. Let the resonator be formed by a cavity of volume V having a narrow neck of length ℓ and cross section A_0 (Figure 6.6.7). Take the mean density and sound speed to be uniform and equal respectively to ρ_0, c_0. The volume flux into the resonator

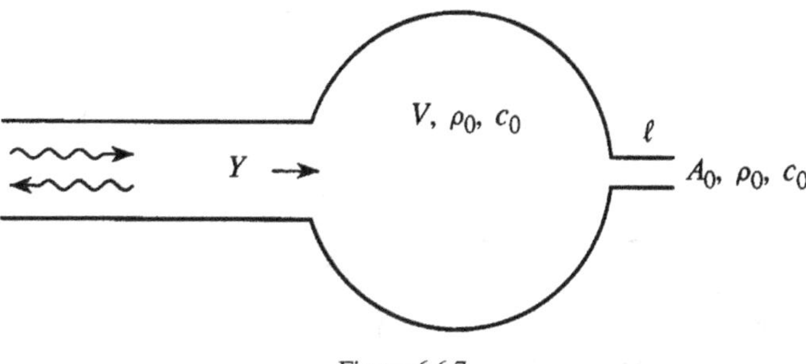

Figure 6.6.7

from the wave guide must be equal to that absorbed by the resonator by compression *plus* that leaving via the neck:

$$\therefore \quad Y = Y_{\text{cavity}} + Y_{\text{neck}} = \frac{-i\omega V}{\rho_0 c_0^2} + \frac{i A_0}{\omega \ell \rho_0},$$

$$\text{i.e.} \quad Y = \frac{-i V}{\omega \rho_0 c_0^2}\left(\omega^2 - \omega_0^2\right), \quad \omega_0 = c_0 \sqrt{\frac{A_0}{V\ell}}.$$

At the resonance frequency ω_0 of the resonator the volume velocity into the cavity from the wave guide vanishes, so that no acoustic energy can escape from the wave guide, but is totally reflected with reflection coefficient $R = +1$. This is because at resonance the normal velocity on the interior walls of the cavity must vanish everywhere *except* at the entrance to the neck.

6.7 Radiation from an open end

The assumption (§6.1.3, Example 3) that pressure fluctuations vanish at the open end of a duct leads to the anomalous prediction that none of the energy of an impinging sound wave incident from within the duct is radiated from the open end. The linear acoustic problem can be solved explicitly for sound of arbitrary frequency (by the Wiener–Hopf method, §5.12.2) in the case of a hard-walled semi-infinite circular-cylindrical duct. For long waves the problem can be treated by extension of a method originally due to Rayleigh (1945).

6.7.1 Rayleigh's method for low-frequency sound

Let the time harmonic sound wave $p = p' e^{ik_o x}$ be incident from within on the open end of a thin-walled, circular-cylindrical duct of radius R (Figure 6.7.1). The coordinate origin is taken at the centre O of the open end, and the axis of the duct coincides with the negative x axis. It is required to determine the pressure wave $p' \alpha e^{-ik_o x}$, say, reflected back into the duct, the sound radiated from the open end, and the motion in the neighbourhood of the open end in the acoustically compact limit $k_o R \ll 1$.

It is convenient to perform the calculations in terms of the velocity potential φ rather than the pressure. At low frequencies the motion in the region A of the duct mouth

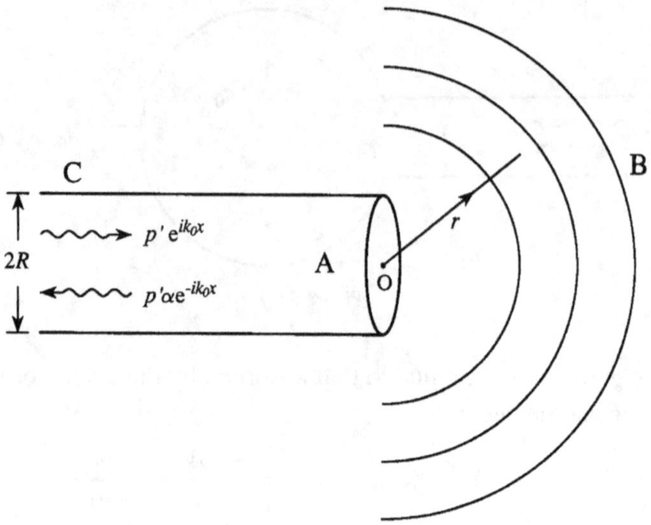

Figure 6.7.1

can be regarded as incompressible and can be expressed in terms of the known velocity potential (§2.12.6) of axisymmetric flow from a circular-cylindrical duct. On the other hand, outside the mouth at B, at distances $r \gg R$ the potential φ must resemble that for a point source $\propto e^{ik_o r}/4\pi r$ (§6.4.2). Therefore we can put

$$\varphi = e^{ik_o x} + \alpha e^{-ik_o x} \quad \text{for } |x| \gg R \text{ within the duct at C} \tag{6.7.1}$$

$$= \beta + \gamma \varphi^*(\mathbf{x}) \quad \text{at A} \tag{6.7.2}$$

$$= -\frac{\delta A e^{ik_o r}}{4\pi r} \quad \text{at B} \quad (r = |\mathbf{x}|), \tag{6.7.3}$$

where $\alpha, \beta, \gamma, \delta$ are frequency-dependent constants to be determined, $A = \pi R^2$ is the duct cross-sectional area, and $\varphi^*(\mathbf{x})$ is the velocity potential of incompressible axisymmetric flow from the mouth, normalised such that

$$\varphi^*(\mathbf{x}) \approx x - \ell_E \quad \text{as } x \to -\infty \text{ within the duct,}$$
$$\approx -A/4\pi r \text{ as } |\mathbf{x}| \to \infty \text{ outside the duct.} \tag{6.7.4}$$

The length $\ell_E \sim 0.61 R$ is known as the *end correction* (see Question 17 of Problems 2).

We find the values of the constants in (6.7.1)–(6.7.3) by 'matching' the different expressions for φ in regions of overlap. Thus, within the duct in the interval $R \ll |x| \ll \lambda$ (where $\lambda = 2\pi/k_o$ is the acoustic wavelength), formulae (6.7.1) and (6.7.2) yield

$$\varphi = 1 + \alpha + ik_o x(1 - \alpha) = \beta + \gamma(x - \ell_E).$$

Therefore

$$1 + \alpha = \beta - \gamma \ell_E,$$
$$ik_o(1 - \alpha) = \gamma.$$

Similarly, in the interval $R \ll r \ll \lambda$ outside the mouth, we find from (6.7.2) and (6.7.3) that

$$\beta = -\frac{i\delta k_o A}{4\pi},$$

$$\gamma = \delta.$$

These simultaneous equations yield [correct to $O(k_o^2 R^2)$]

$$\alpha = -\left(\frac{1 + ik_o(\ell_E + ik_o A/4\pi)}{1 - ik_o(\ell_E + ik_o A/4\pi)}\right), \quad \beta = \frac{k_o^2 A}{2\pi},$$

$$\gamma = \delta = \frac{2ik_o}{1 - ik_o(\ell_E + ik_o A/4\pi)}. \tag{6.7.5}$$

6.7.2 The reflection coefficient

In a more conventional notation we put

$$p = p'\left(e^{ik_o x} + \mathcal{R}e^{-ik_o x}\right), \quad x \ll 0,$$

where the reflection coefficient $\mathcal{R}$ is identified with the coefficient α of (6.7.5). To the same order of approximation (which retains the same approximation to the modulus and phase of $\mathcal{R}$) we can write

$$\mathcal{R} = -e^{2ik_o(\ell_E + ik_o A/4\pi)}. \tag{6.7.6}$$

The acoustic power incident upon the open end from within is $\Pi_I = \frac{1}{2}|p'|^2 \bar{Y}_o$, and the reflected power is $|\mathcal{R}|^2 \Pi_I$. Therefore

$$\begin{array}{c}\text{fraction of the incident acoustic power}\\ \text{radiated from the open end}\end{array} = 1 - e^{-k_o^2 A/\pi} = (k_o R)^2,$$

where $A = \pi R^2$.

When radiation damping is ignored ($k_o^2 R^2 \ll 1$) the pressure within the duct becomes

$$p = p'\left(e^{ik_o x} - e^{-ik_o(x - 2\ell_E)}\right) \equiv 2ip'e^{ik_o \ell_E} \sin\left[k_o(x - \ell_E)\right],$$

which formally vanishes at $x = \ell_E$ *outside* the duct. In other words, to this order of approximation the solution coincides with that we obtain by requiring that $p = 0$ at $x = \ell_E$, so that the classical low-frequency theory remains valid provided the duct is increased in length by ℓ_E. This increase takes account of the inertia of fluid outside the duct that is set in motion by the sound.

6.7.3 Admittance of the open end

The pressure and volume velocity in the duct are given by

$$p = p'\left(e^{ik_o x} + \mathcal{R}e^{-ik_o x}\right), \quad Av = \bar{Y}p'\left(e^{ik_o x} - \mathcal{R}e^{-ik_o x}\right),$$

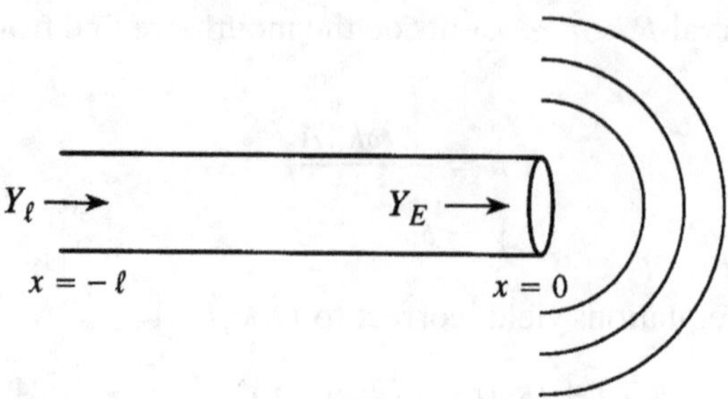

Figure 6.7.2

where $\bar{Y} = A/\rho_o c_o$ and

$$R = -\left[\frac{1 + ik_o(\ell_E + ik_o A/4\pi)}{1 - ik_o(\ell_E + ik_o A/4\pi)}\right].$$

Thus the open-end admittance Y_E (at $x = 0$ in Figure 6.7.2) is

$$Y_E = \bar{Y}\frac{(1 - R)}{(1 + R)} = \frac{iA}{\rho_o \omega(\ell_E + ik_o A/4\pi)}. \tag{6.7.7}$$

This is the same as for a short constriction of length ℓ_E (see §6.6.6) with a small imaginary correction $+ik_o A/4\pi$ representing the effect of dissipation due to radiation damping.

The admittance Y_ℓ at $x = -\ell$ is easily evaluated by use of fundamental formula (6.6.2):

$$Y_\ell = \bar{Y}\left[\frac{Y_E - i\bar{Y}\tan(k_o\ell)}{\bar{Y} - iY_E\tan(k_o\ell)}\right]. \tag{6.7.8}$$

When $k_o\ell \ll 1$,

$$Y_\ell = \bar{Y}\left(\frac{Y_E}{\bar{Y} - ik_o\ell Y_E}\right),$$

$$\therefore \quad \frac{1}{Y_\ell} = \frac{1}{Y_E} + \frac{k_o\ell}{i\bar{Y}} = \frac{\rho_o\omega(\ell_E + ik_o A/4\pi)}{iA} + \frac{\rho_o\omega\ell}{iA},$$

i.e., $\qquad Y_\ell = \dfrac{iA}{\rho_o\omega(\ell + \ell_E + ik_o A/4\pi)}, \tag{6.7.9}$

gives the effect of the end correction and radiation damping on the admittance (6.6.10) of a short constriction of length ℓ and cross section A.

6.7.4 Open-end input admittance

Sound waves are generated within the duct when the open end is 'immersed' within a region of fluctuating pressure. The wave amplitude depends on the duct 'input' admittance, which is identical with the open-end admittance Y_E given by (6.7.7). This conclusion is a consequence of the reciprocal theorem, but may be established directly by Rayleigh's method, as follows:

$$\text{Figure 6.7.3}$$

Let φ_I denote the uniform velocity potential outside the duct that corresponds to a uniform applied external pressure $p = i\rho_o\omega\varphi_I$. The appropriate local representations of the potential φ are given in Figure 6.7.3 for the regions A, B, C of §6.7.1. The interior admittance Y_1 depends on boundary conditions far within the duct, and satisfies

$$Y_1 = \frac{UA}{i\rho_o c_o \omega \varphi_o}.$$

The input admittance is defined by

$$Y = \frac{UA}{p} = \frac{UA}{i\rho_o\omega\varphi_I}.$$

By means of the matching procedure described in §6.7.1 we find

$$\varphi_I = \varphi_o - U(\ell_E + ik_o A/4\pi),$$

$$\therefore \quad Y = \frac{Y_1}{1 - Y_1(i\rho_o\omega/A)(\ell_E + ik_o A/4\pi)} \equiv \frac{Y_1}{1 + Y_1/Y_E},$$

$$\text{i.e.} \quad \frac{1}{Y} = \frac{1}{Y_1} + \frac{1}{Y_E}, \tag{6.7.10}$$

$$\therefore \quad \text{input admittance} = Y_E \equiv \frac{iA}{\rho_o\omega(\ell_E + ik_o A/4\pi)}. \tag{6.7.11}$$

EXAMPLE 1. RESONANCES OF AN OPEN-ENDED PIPE Let the end $x = -\ell$ of the pipe in Figure 6.7.2 be open. Inside the pipe at $x = -\ell$ we have [from (6.7.8)]

$$Y_\ell = \bar{Y}\left[\frac{Y_E - i\bar{Y}\tan(k_o\ell)}{\bar{Y} - iY_E\tan(k_o\ell)}\right].$$

By (6.7.10), with $Y_1 = Y_\ell$, the input impedance Y at the end $x = -\ell$ is given by

$$\frac{1}{Y} = \frac{1}{Y_\ell} + \frac{1}{Y_E} \equiv \frac{Y_E + Y_\ell}{Y_E Y_\ell}.$$

Therefore the resonance condition is $Y_E + Y_\ell = 0$. To leading order this yields the equation

$$\tan(k_o\ell) + \frac{2i\bar{Y}}{Y_E} = 0.$$

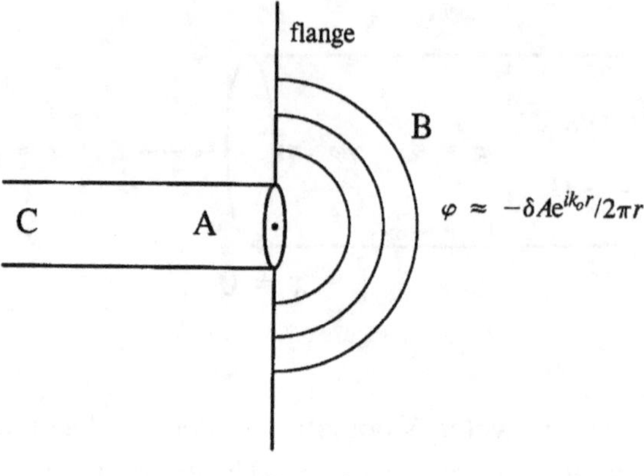

Figure 6.7.4

This can be written, by use of formula (6.7.7) for Y_E,

$$\sin\left[k_o(\ell + 2\ell_E + ik_o A/2\pi)\right] = 0,$$

$$\therefore \quad \omega \approx \frac{n\pi c_o}{\ell + 2\ell_E}\left[1 - \frac{in\pi A}{2(\ell + 2\ell_E)^2}\right], \quad n = 1,\, 2,\, 3,\, \dots.$$

Hence the effective length of the pipe determining the resonance frequencies is increased from ℓ to $\ell + 2\ell_E$, and the resonance frequencies have small *negative* imaginary parts (for the low-order modes n) that account for damping by radiation from the ends.

6.7.5 Flanged opening

The application of Rayleigh's method to determine the admittance of an open end with an *infinite* flange (Figure 6.7.4) proceeds exactly as in §§6.7.1, 6.7.3. However, the solid angle of spread of the sound radiated from the opening is now restricted to 2π, and the potential at B must therefore be taken in the form given in the figure. Similarly, (6.7.4) becomes

$$\varphi^*(\mathbf{x}) \approx x - \ell_E \quad \text{as } x \to -\infty \text{ within the duct,}$$
$$\approx -A/2\pi r \quad \text{as } r \to \infty \text{ outside the duct.} \tag{6.7.12}$$

The end correction for a flanged opening must be determined by numerical solution of Laplace's equation. This yields $\ell_E \approx 0.82R$ for a circular cylindrical duct [a result also obtained by Rayleigh (1945), who used an approximate analytical solution of Laplace's equation – see Example 3]. The admittance of the flanged opening is then given by the following modification of (6.7.7):

$$Y_E = \frac{iA}{\rho_o\omega(\ell_E + ik_o A/2\pi)}, \quad \ell_E \approx 0.82R. \tag{6.7.13}$$

In cases in which the flange is large compared with the duct diameter yet *small* compared with the acoustic wavelength, the end correction takes the modified value

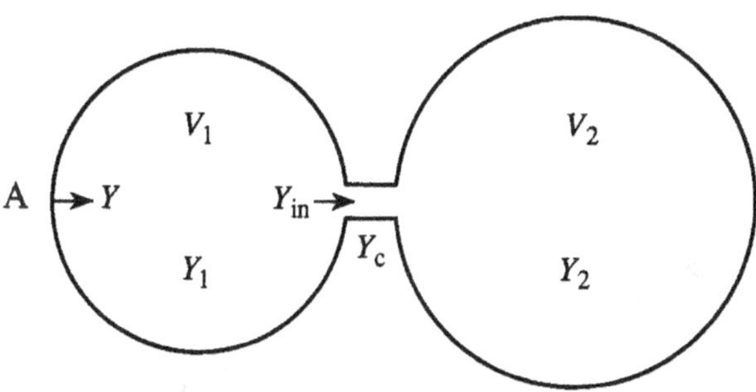

Figure 6.7.5

$\ell_E = 0.82R$, but the damping factor $ik_oA/2\pi$ in (6.7.13) should be replaced with its value $ik_oA/4\pi$ for an unflanged opening, because on the scale of the acoustic wavelength the 'flange is absent' and cannot prevent the spherical spreading of the sound.

NOTE When the open end of the duct is enclosed in an acoustic chamber or wave guide whose cross section is small relative to the acoustic wavelength, but greatly exceeds the duct cross section (as in Figure 6.6.6, for example), the assumed form in Figure 6.7.4 for the acoustic potential is applicable only close to the duct exit, when strictly the compressibility factor $e^{ik_o r}$, which is correct only for radiation into free space, must be replaced with 1. Then we must take

$$Y_E = \frac{iA}{\rho_o \omega \ell_E} \tag{6.7.14}$$

for both flanged and unflanged duct terminations.

A special and important case arises when the duct in Figure 6.7.4 is *absent*, so that the fluctuating flow is through a small aperture in a wall of infinitesimal thickness. The constriction then has zero length, and the end correction (for 'each end' of the aperture) can be determined analytically to be $\ell_E = \frac{\pi}{4}R \sim 0.78R$, which is a little smaller than the value (6.7.13) in the presence of a duct (Example 4). The aperture as a whole is therefore equivalent to a constriction of length $2\ell_E = \frac{\pi}{2}R$.

EXAMPLE 2. Find the resonance frequency of acoustic oscillations of a gas of mean density ρ_o and sound speed c_o filling the closed system formed by two rigid, acoustically compact cavities of volumes V_1, V_2 connected by a short, narrow cylindrical neck of length ℓ and cross-sectional area A_o (Figure 6.7.5).

Using the notation shown in the figure, denote by Y_1, Y_2, Y_c the respective admittances of the cavities V_1, V_2 and the neck:

$$Y_1 = \frac{-i\omega V_1}{\rho_o c_o^2}, \quad Y_2 = \frac{-i\omega V_2}{\rho_o c_o^2}, \quad Y_c = \frac{iA_o}{\rho_o \omega(\ell + 2\ell_E)},$$

where the end corrections at each end of the neck are assumed to be equal. The 'input admittance' Y_{in} satisfies

$$\frac{1}{Y_{\text{in}}} = \frac{1}{Y_c} + \frac{1}{Y_2}.$$

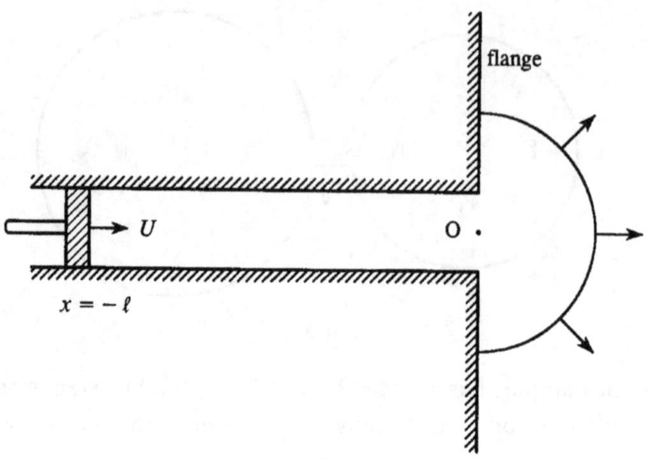

Figure 6.7.6

If we temporarily assume the existence of a small opening at a remote point A of V_1, with input admittance Y, then

$$Y = Y_1 + Y_{in}, \quad \text{so that} \quad \frac{1}{Y - Y_1} = \frac{1}{Y_c} + \frac{1}{Y_2}.$$

However, $Y = 0$ at a rigid boundary; therefore the resonance frequency must satisfy

$$\frac{1}{Y_1} + \frac{1}{Y_c} + \frac{1}{Y_2} = 0,$$

$$\therefore \quad \omega = c_o \sqrt{\frac{A_o}{(\ell + 2\ell_E)} \left(\frac{1}{V_1} + \frac{1}{V_2} \right)}.$$

6.7.6 Physical significance of the end correction

Consider a duct of cross-sectional area A with a freely moving piston of equal area that advances towards the open end at speed U (Figure 6.7.6). Assume the fluid motion is ideal and *incompressible*. The motion is also irrotational with velocity potential $U\varphi^*(\mathbf{x})$, where φ^* is defined as in (6.7.12). Then the energy T of the incompressible motion is entirely kinetic, and

$$T \equiv \frac{\rho_o U^2}{2} \int (\nabla \varphi^*)^2 \, d^3\mathbf{x} = \frac{\rho_o U^2}{2} A(\ell + \ell_E), \tag{6.7.15}$$

where the integration is over the fluid volume, outside the duct and within the duct between the open end and the piston, and $\ell \gg R$ is the instantaneous distance of the piston from the open end.

This formula is a consequence of the divergence theorem and asymptotic formulae (6.7.12):

$$\int (\nabla \varphi^*)^2 \, d^3\mathbf{x} = - \int_{\text{piston}} \varphi^* \frac{\partial \varphi^*}{\partial x} \, dy dz \equiv A(\ell + \ell_E),$$

where the second integration is over the wetted face of the piston at $x = -\ell$, where the fluid motion is sensibly uniform and $\varphi^* \sim x - \ell_E$ when $|x| \gg R$.

The motion produced by the piston is uniform and of speed U in the duct except very close to the open end. Thus the kinetic energy of the whole motion $\sim \frac{1}{2}\rho_o U^2 A\ell$ plus a correction to account for the nonuniform motion near the end and outside the duct. According to (6.7.15) this correction is $\frac{1}{2}\rho_o U^2 A\ell_E$, so that the end correction ℓ_E is the equivalent length of uniform flow in the duct whose kinetic energy is just equal to that of the nonuniform 'exit flow'.

Rayleigh (1870) obtained estimates of the value of ℓ_E by application of Kelvin's theorem (§2.10) that the irrotational kinetic energy T is the minimum for all possible motions produced by the piston. For any other motion with kinetic energy T', say, we must have

$$\ell + \ell_E \leq \frac{2T'}{\rho_o U^2 A}. \tag{6.7.16}$$

The procedure involves the calculation of T' from a solution of the piston problem containing one or more disposable parameters, whose values are then chosen to make T' a minimum (Example 3).

EXAMPLE 3. END CORRECTION OF A FLANGED DUCT (Rayleigh 1870) Consider a flanged, semi-infinite circular cylindrical duct of radius R (Figure 6.7.6), and use the following expression for the normal derivative $\partial\varphi^*/\partial x$ in the entrance plane:

$$\frac{\partial\varphi^*}{\partial x} = \alpha\left(1 + \frac{\mu r^2}{R^2} + \frac{\mu' r^4}{R^4}\right), \quad x = 0, \ r = \sqrt{y^2 + z^2} < R, \tag{6.7.17}$$

where μ, μ' are constants. Volume flow continuity requires that

$$\alpha = \frac{1}{1 + \mu/2 + \mu'/3}.$$

Representation (6.7.17) does not have the usual potential flow singularity at the edge $r = R$ of the opening [where potential theory predicts $\partial\varphi^*/\partial x \sim 1/(1 - r/R)^{\frac{1}{3}}$], but its use should still supply a good approximation for ℓ_E provided μ and μ' are chosen to make the kinetic energy T' a minimum. Indeed, the much simpler 'piston' approximation, which we obtain by setting $\alpha = 1$ and $\mu = \mu' = 0$, yields the estimate $\ell_E \approx 0.85R$, which is only about 4% in excess of the true value.

Within the duct,

$$\varphi^*(\mathbf{x}) = x - \ell_E + \sum_{n=1}^{\infty} \alpha_n J_0\left(\lambda_n \frac{r}{R}\right) e^{\lambda_n x/R}, \quad x < 0, \ r < R,$$

where J_ν is the Bessel function of order ν. This expansion satisfies the axisymmetric form of Laplace's equation and has vanishing normal derivative on the wall $r = R$

Table 6.7.1

μ	μ'	ℓ_E/R
-1.0120	1.9515	0.8242
0.0	1.1030	0.8254
1.1520	0.0	0.8281

provided λ_n is the nth *positive* root of $dJ_0(\lambda)/d\lambda \equiv -J_1(\lambda) = 0$. The coefficients α_n are chosen to satisfy (6.7.17), which yields

$$\alpha_n = \frac{4\alpha R}{\lambda_n^3 J_0(\lambda_n)}\left[\mu + 2\mu'\left(1 - \frac{8}{\lambda_n^2}\right)\right].$$

The kinetic energy T_I of the motion within the section V_I, say, of the duct contained in the interval $-\ell < x < 0$ (where $\ell \gg R$) is given by

$$\frac{2T_I}{\rho_o U^2} = \int_{V_I} (\nabla\varphi^*)^2 d^3\mathbf{x} \equiv 2\pi \int_0^R \left[\left(\varphi^*\frac{\partial\varphi^*}{\partial x}\right)_{x=0} - \left(\varphi^*\frac{\partial\varphi^*}{\partial x}\right)_{x=-\ell}\right] r\,dr$$

$$= \pi R^2 \left\{\ell + 16\alpha^2 R \sum_{n=1}^{\infty} \frac{\left[\mu + 2\mu'\left(1 - \frac{8}{\lambda_n^2}\right)\right]^2}{\lambda_n^5}\right\}.$$

In the region $x > 0$ the flow spreads hemispherically, and

$$\varphi^*(\mathbf{x}) = -\frac{1}{2\pi}\int_{S_0} \frac{u(y', z')dy'dz'}{\sqrt{x^2 + (y - y')^2 + (z - z')^2}}, \quad x > 0,$$

where the integration is over the face of the open end S_0 of the duct and $u = \partial\varphi^*/\partial x$ is normal velocity (6.7.17). The kinetic energy T_E of the exterior motion is calculated from the formula

$$\frac{2T_E}{\rho_o U^2} = -\int_{S_0} \varphi^*\frac{\partial\varphi^*}{\partial x}dydz.$$

whereupon the relation (6.7.16) satisfied by $T' = T_I + T_E$ becomes

$$\ell + \ell_E \le \frac{2}{\rho_o U^2 A}(T_I + T_E) = \ell + 16\alpha^2 R \sum_{n=1}^{\infty} \frac{\left[\mu + 2\mu'\left(1 - \frac{8}{\lambda_n^2}\right)\right]^2}{\lambda_n^5}$$

$$+ \frac{8\alpha^2 R}{3\pi}\left(1 + \frac{14\mu}{15} + \frac{314\mu'}{525} + \frac{5\mu^2}{21} + \frac{214\mu\mu'}{675} + \frac{89\mu'^2}{825}\right). \quad (6.7.18)$$

The values of μ, μ' minimizing the right-hand side of this expression are shown in the first row of Table 6.7.1. The corresponding estimate $\ell_E = 0.8242\,R$ for the end correction [which we obtain by replacing '$\le$' with '$=$' in (6.7.18)] is given in column three.

The table also lists alternative (larger and therefore less accurate) estimates for ℓ_E obtained when one of μ, μ' is required to vanish.

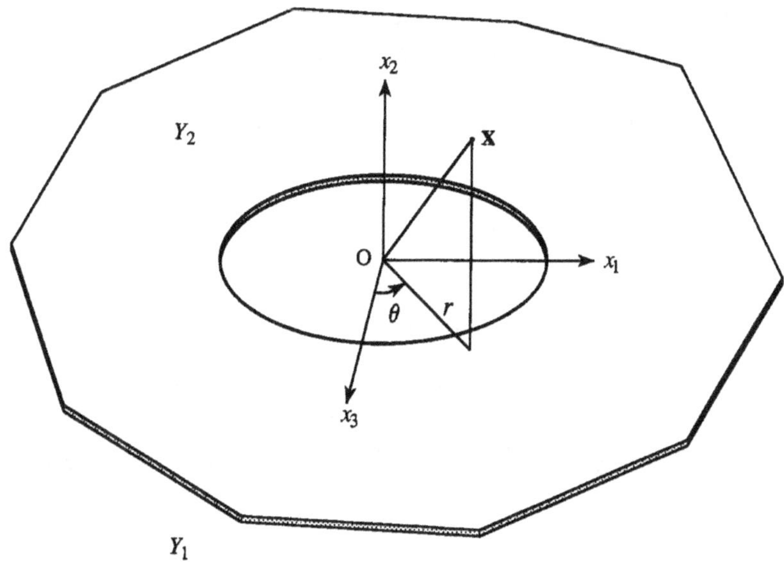

Figure 6.7.7

6.7.7 Admittance of a circular aperture

Consider time harmonic flow ($\propto e^{-i\omega t}$) through an acoustically compact circular aperture of radius R in a rigid wall of infinitesimal thickness produced by a uniform pressure load $p_1 - p_2$. The 'upper' and 'lower' faces of the wall coincide respectively with $x_2 = \pm 0$ in Figure 6.7.7, and p_2, p_1 are the pressures above and below the wall. The motion near the aperture is regarded as incompressible, with volume flux Q (in the x_2 direction), so that $Y_1 = Q/p_1$, $Y_2 = Q/p_2$.

Put $p_1 = i\rho_o\omega\varphi_1$, $p_2 = i\rho_o\omega\varphi_2$. Then the flow corresponding to the uniform potential increase $\varphi_2 - \varphi_1$ can be described by a velocity potential Φ such that the total potential is $\varphi = \varphi_{2,1} + \Phi$, respectively above and below the wall. Φ satisfies Laplace's equation, and $\Phi \sim \mp Q/2\pi|\mathbf{x}|$ as $x_2 \to \pm\infty$. Symmetry demands that the motion in the plane of the aperture must be in the normal direction, so that Φ must be constant there.

Φ is determined by use of plane wall solution (2.8.10) or generalised Kirchhoff representation (6.4.1) applied to incompressible flow, where for Laplace's equation $G(\mathbf{x}, \mathbf{y}, t - \tau)$ is replaced by $-G(\mathbf{x}, \mathbf{y})\delta(t - \tau)$, with

$$G(\mathbf{x}, \mathbf{y}) = \frac{-1}{4\pi|\mathbf{x} - \mathbf{y}|} + \frac{-1}{4\pi|\mathbf{x} - \bar{\mathbf{y}}|}, \quad \bar{\mathbf{y}} = (y_1, -y_2, y_3), \tag{6.7.19}$$

which has vanishing normal derivatives $\partial G/\partial x_2$, $\partial G/\partial y_2$ respectively on x_2 and $y_2 = 0$. Then

$$\varphi(\mathbf{x}) = \varphi_{1,2} + \frac{\text{sgn}(x_2)}{2\pi} \int_{-\infty}^{\infty} \frac{v_2(y_1, 0, y_3)}{|\mathbf{x} - \mathbf{y}|} \, dy_1 dy_3, \quad (y_2 = 0),$$

where the integration is over the upper/lower surfaces $y_2 = \pm 0$ respectively for $x_2 \gtrless 0$. Because $v_2 \equiv 0$ on the rigid portions of the wall, the integration may be restricted to the region S occupied by the aperture. The condition that φ (and the pressure) must be continuous across the aperture yields the integral equation

$$\oint_S \frac{v_2(y_1, 0, y_3)}{|\mathbf{x} - \mathbf{y}|}\, dy_1 dy_3 = \pi(\varphi_2 - \varphi_1), \quad \text{for } x_2 = y_2 = 0, \quad r = \sqrt{x_1^2 + x_3^2} < R \qquad (6.7.20)$$

for the aperture velocity v_2, which is clearly a function of r alone because $\varphi_2 - \varphi_1$ is constant. The solution is (Copson 1947)

$$v_2(r) = \frac{(\varphi_2 - \varphi_1)}{\pi\sqrt{R^2 - r^2}}, \quad r < R.$$

Thus $Q = \oint_S v_2(x_1, 0, x_3) dx_1 dx_3 = 2R(\varphi_2 - \varphi_1)$, i.e. $\varphi_2/Q - \varphi_1/Q = 1/2R$,

$$\therefore \quad \frac{1}{Y_1} = \frac{1}{Y_2} + \frac{\rho_o \omega}{2i R} \equiv \frac{1}{Y_2} + \frac{\rho_o \omega(\pi R/2)}{i A}, \quad \text{where} \quad A = \pi R^2,$$

$$\therefore \quad Y_{\text{aperture}} \equiv \frac{i A}{\rho_o \omega \ell_{\text{aperture}}}, \quad \text{where} \quad \ell_{\text{aperture}} = \frac{\pi R}{2}.$$

The end correction for 'one side' of the aperture is therefore $\ell_E = \frac{\pi}{4} R$.

6.8 Webster's equation

Consider the propagation of low-frequency sound of wavelength λ in a duct of 'slowly varying' cross-sectional area $S(x)$, where x is measured along the centreline of the duct. It is convenient to regard the duct cross section as circular, with variable radius $R = R(x) \equiv \sqrt{S(x)/\pi} \ll \lambda$, but this is not a necessary limitation.

When the divergence of the duct

$$\sim \frac{R}{S}\frac{dS}{dx} \ll 1,$$

the acoustic motion may be regarded as one dimensional, all quantities varying as functions of the axial variable x and the time. Then $\rho_o v S$ is mass flux through the cross section $S(x)$ at x in the x direction, where $v(x, t)$ is the acoustic particle velocity, and the net mass *outflow* from a thin slice of the duct at x of thickness δx is $\delta x\, \partial(\rho_o v S)/\partial x$ (Figure 6.8.1). Continuity requires this to be the same as $-\delta x S\, \partial\rho/\partial t$, so that when the mean density ρ_o is constant the continuity equation can be written as

$$\frac{1}{\rho_o}\frac{\partial\rho}{\partial t} + \frac{1}{S}\frac{\partial}{\partial x}(Sv) = 0 \qquad (6.8.1)$$

In the absence of applied body forces this may be combined with the axial momentum equation $\rho_o \partial v/\partial t + \partial p/\partial x = 0$ and the isentropic relation $dp = c_o^2 d\rho$ to yield **Webster's equation**:

$$\frac{1}{c_o^2}\frac{\partial^2 p}{\partial t^2} - \frac{1}{S}\frac{\partial}{\partial x}\left(S\frac{\partial p}{\partial x}\right) = 0. \qquad (6.8.2)$$

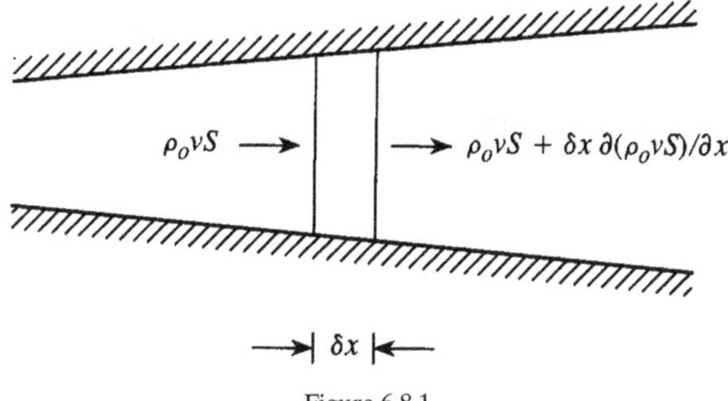

Figure 6.8.1

The same equation is satisfied by the velocity potential φ (where $p = -\rho_o \partial\varphi/\partial t$). In particular, for time harmonic waves ($\propto e^{-i\omega t}$) we have

$$\frac{1}{S}\frac{\partial}{\partial x}\left(S\frac{\partial\varphi}{\partial x}\right) + k_o^2\varphi = 0, \quad k_o = \frac{\omega}{c_o}. \tag{6.8.3}$$

EXAMPLE 1. THE EXPONENTIAL HORN Let $S = A_o e^{x/\ell_h}$, and let the horn extend between $x = 0$ and $x = \ell > 0$ (Figure 6.8.2). The solution of Equation (6.8.3) is

$$\varphi = e^{-x/2\ell_h}\left(a e^{i\kappa x} + b e^{-i\kappa x}\right), \quad \kappa = \sqrt{k_o^2 - \frac{1}{4\ell_h^2}}, \tag{6.8.4}$$

where $a,\ b$ are arbitrary constants.

Take the open-end condition at $x = \ell$ to be $p = 0$, i.e. $\varphi = 0$, and let the complex pressure amplitude at $x = 0$ be $p \equiv i\rho_o\omega\varphi = p'$. Then

$$a = \frac{p'}{i\rho_o\omega\left(1 - e^{2i\kappa\ell}\right)}, \quad b = \frac{-p' e^{2i\kappa\ell}}{i\rho_o\omega\left(1 - e^{2i\kappa\ell}\right)}.$$

Therefore the input admittance Y at $x = 0$ (where $S = A_o$) is given by

$$Y = i\bar{Y}_o\left[\sqrt{1 - \frac{1}{4k_o^2\ell_h^2}}\,\cot\left(k_o\ell\sqrt{1 - \frac{1}{4k_o^2\ell_h^2}}\right) + \frac{1}{2\ell_h}\right], \tag{6.8.5}$$

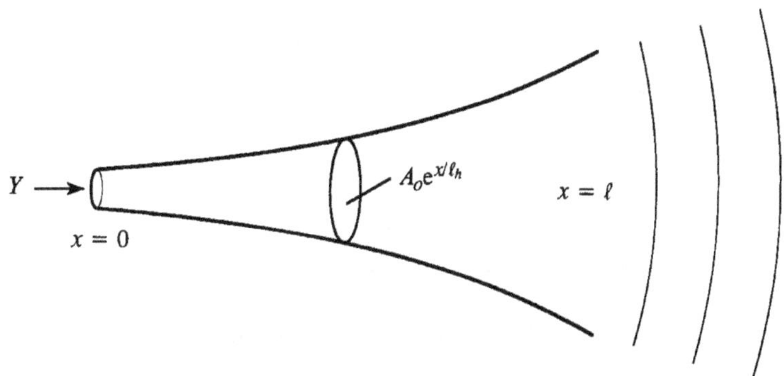

Figure 6.8.2

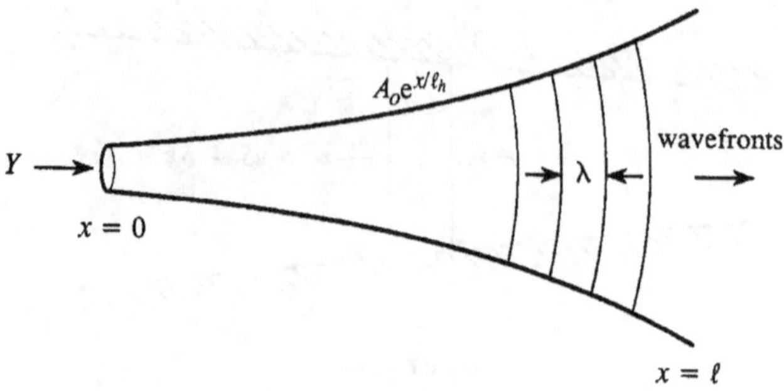

Figure 6.8.3

where $\bar{Y}_o = A_o/\rho_o c_o$ is the bare admittance of a uniform duct of area A_o. Y is pure imaginary, so that no energy escapes from the end of the duct. This is the same as for a uniform duct subject to the simple open-end condition $p = 0$. When $\ell_h \to \infty$, Equation (6.8.5) reduces to the formula $Y = \bar{Y}_o \cot(k_o \ell)$ for a duct of uniform cross section.

EXAMPLE 2. LARGE-DIAMETER HORN The reflection of sound at the open end of the horn can be neglected when the wavelength is small compared with the horn exit diameter (Figure 6.8.3). For an exponential horn, solution (6.8.4) then reduces to

$$\varphi = ae^{-x/2\ell_h + i\kappa x}, \quad \kappa = \sqrt{k_o^2 - \frac{1}{4\ell_h^2}}, \quad 0 < x < \ell.$$

The input admittance now becomes

$$Y = \bar{Y}_o \left(\sqrt{1 - \frac{1}{4k_o^2 \ell_h^2}} + \frac{i}{2k_o \ell_h} \right),$$

and the acoustic power radiated from the horn is

$$\Pi = \frac{|p'|^2}{2} \operatorname{Re} Y = \frac{|p'|^2}{2} \bar{Y}_o \sqrt{1 - \frac{c_o^2}{4\omega^2 \ell_h^2}}, \quad \text{provided} \quad \omega > \frac{c_o}{2\ell_h}.$$

Thus radiation occurs only at sufficiently high frequency that the acoustic wavelength $\lambda < 4\pi \ell_h \sim 12\ell_h$, i.e. the wavelength has to be more than an order of magnitude smaller than the length scale ℓ_h of the cross-sectional area variations.

EXAMPLE 3. LARGE-DIAMETER CONICAL HORN For a conical horn of length ℓ (Figure 6.8.4) we set

$$S = \alpha x^2 \quad \text{for} \quad 0 < x_1 < x < x_2, \quad \text{where} \quad \ell = x_2 - x_1,$$

and $\tan^{-1} \sqrt{\alpha/\pi}$ is the semi-angle of the cone. Time harmonic solutions of (6.8.3) are

$$\varphi = \frac{1}{x} \left(a\, e^{ik_o x} + b\, e^{-ik_o x} \right), \quad x_1 < x < x_2. \tag{6.8.6}$$

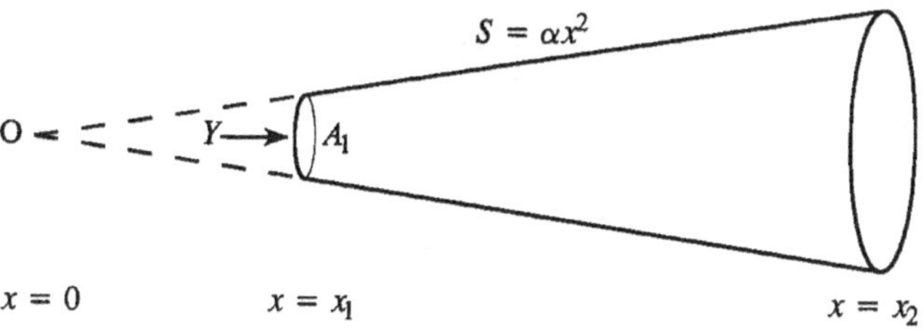

Figure 6.8.4

We can set $b = 0$ when the frequency is large enough that there is no reflection from the open end $x = x_2$. Then the input admittance at $x = x_1$ is found to be

$$Y = \bar{Y}_1 \left(1 + \frac{i}{k_o x_1}\right),$$

where $\bar{Y}_1 = A_1/\rho_o c_o \equiv \alpha x_1^2/\rho_o c_o$ is the bare admittance of a uniform duct of cross-sectional area A_1.

6.9 Radiation into a semi-infinite duct

Acoustic sources situated near the open end of a long duct generate sound waves within the duct. When wavelengths are large compared with the duct diameter and the sources are several diameters from the duct entrance, the pressure field incident upon the entrance may be regarded as uniform, and the sound radiated into the duct can be calculated by use of input admittance (6.7.13) or (6.7.11) respectively for flanged and unflanged entrances. However, the efficiency with which sound is generated is strongly influenced by the duct when the entrance lies in the hydrodynamic near field of the source. We must then use a compact Green's function to determine the radiation into the duct.

6.9.1 The compact Green's function

It is required to determine the compact approximation to the solution of

$$\left(\frac{1}{c_o^2}\frac{\partial^2}{\partial t^2} - \nabla^2\right) G = \delta(\mathbf{x} - \mathbf{y})\delta(t - \tau), \quad \text{where} \quad G = 0 \text{ for } t < \tau, \tag{6.9.1}$$

for $\mathbf{x}$ far within the duct and source positions $\mathbf{y}$ in the neighbourhood of the duct entrance (Howe 1998). Figure 6.9.1 illustrates the situation for an unflanged circular cylindrical duct, but the result to be subsequently obtained is applicable to ducts with arbitrary exit geometry provided only that the duct cross-sectional area A is uniform in the 'propagation zone' $|x_1| \gg 2R$ within the duct, where $R = \sqrt{A/\pi}$ is the effective interior duct radius.

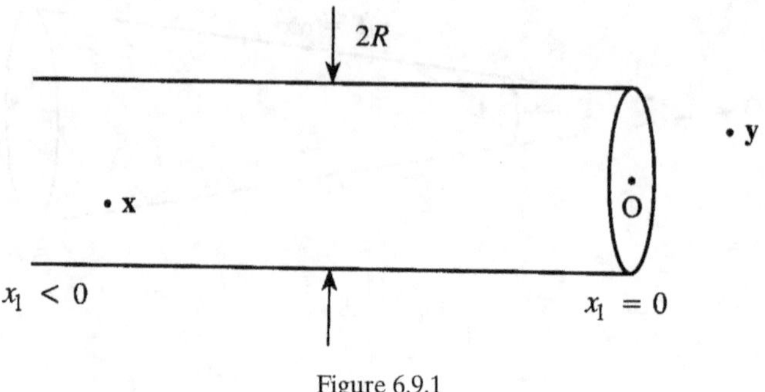

Figure 6.9.1

In the usual way (§6.4.2) we put

$$G(\mathbf{x}, \mathbf{y}, t - \tau) = -\frac{1}{2\pi} \int_{-\infty}^{\infty} G(\mathbf{x}, \mathbf{y}, \omega) e^{-i\omega(t-\tau)} \, d\omega, \qquad (6.9.2)$$

where $(\nabla^2 + k_o^2) G(\mathbf{x}, \mathbf{y}, \omega) = \delta(\mathbf{x} - \mathbf{y}), \quad k_o = \omega/c_o.$ $\qquad (6.9.3)$

The solution is required at points $\mathbf{x}$ within the duct at distances $|x_1| \gg R$ from the open end and for source points $\mathbf{y}$ near the entrance. In the compact approximation ($k_o R \ll 1$) only plane sound waves can propagate in the duct (Example 1), and this permits the problem to be solved easily by the reciprocal method of §6.4.3.

In the reciprocal problem the source is regarded as placed at $\mathbf{x}$ far within the duct and the solution $G(\mathbf{y}, \mathbf{x}, \omega) \equiv G(\mathbf{x}, \mathbf{y}, \omega)$ is sought as a function of $\mathbf{y}$ near the entrance. The first step involves the determination the wave incident upon the entrance produced by this source. Then Rayleigh's method (§6.7.1) is used to couple this wave to the motion at the duct entrance.

This incident wave is plane ($\propto e^{ik_o y_1}$) when $k_o R$ is small, and its governing equation can be found by the integration of (6.9.3) (regarded as an equation in which $\mathbf{y}$ is the independent variable) across the uniform duct cross section (with respect to the transverse variables y_2, y_3). The incident plane wave G_I is therefore the outgoing solution of

$$\left(\frac{\partial^2}{\partial y_1^2} + k_o^2 \right) G_\mathrm{I} = \frac{1}{A} \delta(x_1 - y_1).$$

This is just the equation satisfied by the zeroth-order component of a double Fourier series expansion of $G(\mathbf{y}, \mathbf{x}, \omega)$ in terms of the transverse variables. Solving the equation, we obtain the wave incident upon the entrance $y_1 = 0$ in the form

$$G_\mathrm{I} = \frac{e^{ik_o(y_1 - x_1)}}{2ik_o A}, \qquad x_1 < y_1.$$

Thus, to the *right* of the source in Figure 6.9.1 ($y_1 > x_1$), but at distances $\gg R$ from the entrance, we can put

$$G(\mathbf{y}, \mathbf{x}, \omega) = \frac{e^{-ik_o x_1}}{2ik_o A} \left(e^{ik_o y_1} + \mathcal{R} e^{-ik_o y_1} \right), \qquad (6.9.4)$$

where the reflection coefficient $\mathcal{R}$ accounts for conditions at the open end. Near the end, where $k_o y_1 \ll 1$, we have

$$G(\mathbf{y}, \mathbf{x}, \omega) \approx \frac{e^{-ik_o x_1}}{2ik_o A}\left[1 + \mathcal{R} + ik_o y_1(1 - \mathcal{R})\right], \quad R \ll |y_1| \ll 1/k_o. \qquad (6.9.5)$$

When $k_o R \ll 1$ the motion in the entrance is the same as for incompressible flow, and if we ignore the small correction associated with sound radiation from the open end (§6.7), we can write [for $\mathbf{y}$ outside the entrance and $|\mathbf{y}| \sim O(R)$ inside the duct]

$$G(\mathbf{y}, \mathbf{x}, \omega) \approx \alpha \varphi^*(\mathbf{y}),$$

where $\varphi^*(\mathbf{y})$ is defined as in §6.7.1, so that

$$G(\mathbf{y}, \mathbf{x}, \omega) \sim \alpha(y_1 - \ell_E) \quad \text{for} \quad |y_1| \gg R \quad \text{within the duct}.$$

By matching this asymptotic formula with (6.9.5) we find [neglecting small quantities $\sim O(k_o^2 R^2)$ relative to unity]

$$\mathcal{R} \approx -e^{2ik_o \ell_E}, \quad \alpha \approx \frac{1}{A} e^{-ik_o(x_1 - \ell_E)}.$$

This approximation for the reflection coefficient agrees with (6.7.6) when the contribution from radiation damping is discarded. Hence

$$G(\mathbf{y}, \mathbf{x}, \omega) \approx \begin{cases} \dfrac{\varphi^*(\mathbf{y})}{A} e^{-ik_o(x_1 - \ell_E)}, & \text{near the entrance and outside} \\[2mm] \dfrac{e^{-ik_o(x_1 - \ell_E)}}{2ik_o A}\left[e^{ik_o(y_1 - \ell_E)} - e^{-ik_o(y_1 - \ell_E)}\right], & |y_1| \gg R \text{ in the duct}. \end{cases} \qquad (6.9.6)$$

Now $\varphi^*(\mathbf{y}) \sim O(R)$ in the vicinity of the duct entrance and decreases to zero like $-A/4\pi|\mathbf{y}|$ with distance from the entrance outside the duct. Hence, when $k_o R \ll 1$ the following representation provides a uniform approximation to both of the formulae in (6.9.6):

$$G(\mathbf{y}, \mathbf{x}, \omega) \equiv G(\mathbf{x}, \mathbf{y}, \omega) \approx \frac{e^{-ik_o(x_1 - \ell_E)}}{2ik_o A}\left[e^{ik_o \varphi^*(\mathbf{y})} - e^{-ik_o \varphi^*(\mathbf{y})}\right].$$

This is applicable provided the point $\mathbf{x}$ is within the duct far from the entrance. However, we may generalise it by replacing the factor $(x_1 - \ell_E)$ in the first exponent with $\varphi^*(\mathbf{x})$ and rewriting the result in the form

$$G(\mathbf{x}, \mathbf{y}, \omega) \approx \frac{1}{2ik_o A}\left\{e^{ik_o|\varphi^*(\mathbf{x}) - \varphi^*(\mathbf{y})|} - e^{-ik_o[\varphi^*(\mathbf{x}) + \varphi^*(\mathbf{y})]}\right\}. \qquad (6.9.7)$$

The functional dependence of this expression on $\mathbf{x}$ and $\mathbf{y}$ is consistent with reciprocity, and it is valid for arbitrary source and observer locations provided at least one of them is within the duct at a distance from the mouth greatly exceeding R.

By substituting this expression for $G(\mathbf{x}, \mathbf{y}, \omega)$ into (6.9.2) and imposing the causality condition by indenting the integration contour to pass above the pole at $\omega = 0$, we obtain the **open-end compact Green's function**:

$$G(\mathbf{x}, \mathbf{y}, t - \tau) = \frac{c_o}{2A} \left(\mathrm{H}\left[t - \tau - \frac{|\varphi^*(\mathbf{x}) - \varphi^*(\mathbf{y})|}{c_o} \right] - \mathrm{H}\left\{ t - \tau + \frac{[\varphi^*(\mathbf{x}) + \varphi^*(\mathbf{y})]}{c_o} \right\} \right).$$

$$(6.9.8)$$

When $\mathbf{x}$ and $\mathbf{y}$ both lie within the duct at distances $\gg R$ from the open end, we can put $\varphi^*(\mathbf{x}) = x_1 - \ell_E$ and $\varphi^*(\mathbf{y}) = y_1 - \ell_E$, so that at low frequencies the Green's function becomes

$$G(\mathbf{x}, \mathbf{y}, t - \tau) = \frac{c_o}{2A} \left\{ \mathrm{H}\left(t - \tau - \frac{|x_1 - y_1|}{c_o} \right) - \mathrm{H}\left[t - \tau + \frac{(x_1 + y_1 - 2\ell_E)}{c_o} \right] \right\}.$$

$$(6.9.9)$$

The corresponding result for a duct infinite in both directions is

$$G(\mathbf{x}, \mathbf{y}, t - \tau) = \frac{c_o}{2A} \mathrm{H}\left(t - \tau - \frac{|x_1 - y_1|}{c_o} \right), \qquad (6.9.10)$$

so that the second term in the braces of (6.9.9) represents the influence of reflection and phase reversal at the open end, including the phase shift associated with the real part of the end correction, but neglecting radiation from the open end.

EXAMPLE 1. GREEN'S FUNCTION FOR THE INTERIOR OF AN INFINITE, HARD-WALLED CYLINDRICAL DUCT $G(\mathbf{x}, \mathbf{y}, \omega)$ satisfies (6.9.3) for $r < R$. Use cylindrical polar coordinates $\mathbf{x} = (r, \theta, x)$ and set $\mathbf{y} = (r', \theta', x')$. Then G has the expansion

$$G(\mathbf{x}, \mathbf{y}, \omega) = -\frac{i}{\pi R^2} \sum_{n,m=0}^{\infty} \frac{\sigma_m J_m(\lambda_{mn} r'/R) J_m(\lambda_{mn} r/R) \cos[m(\theta - \theta')] e^{i\gamma_{mn}|x - x'|}}{\gamma_{mn} \left(1 - \frac{m^2}{\lambda_{mn}^2} \right) J_m^2(\lambda_{mn})},$$

where $\sigma_0 = \frac{1}{2}$, $\sigma_m = 1$ $(m > 0)$, λ_{mn} is the nth non-negative zero of $\partial J_m(x)/\partial x$, and

$$\gamma_{mn} = \mathrm{sgn}(k_o)|k_o^2 - \lambda_{mn}^2/R^2|^{1/2} \quad \text{for } |k_o| > \lambda_{mn}/R,$$
$$= i|k_o^2 - \lambda_{mn}^2/R^2|^{1/2} \quad \text{for } |k_o| < \lambda_{mn}/R.$$

The nm mode decays exponentially with axial distance from the source when $|k_o| < \lambda_{mn}/R$, i.e., at frequencies ω smaller than $c_o\lambda_{mn}/R$ in absolute value. The smallest positive value of λ_{mn} is $\lambda_{1,0} \approx 1.841$, so that only the axially propagating *plane-wave* mode $m = n = 0$ can propagate to $x = \pm\infty$ when the frequency is below the *cut-off frequency* $\sim 0.29 c_o/R$ Hz.

EXAMPLE 2. Determine the low-frequency pressure waves radiated into a semi-infinite duct of radius R from an acoustically compact volume source $q(x_1, t)$ located within the duct at distance ℓ from the open end $(x_1 = 0)$ when $R \ll \ell \ll \lambda =$ the acoustic wavelength.

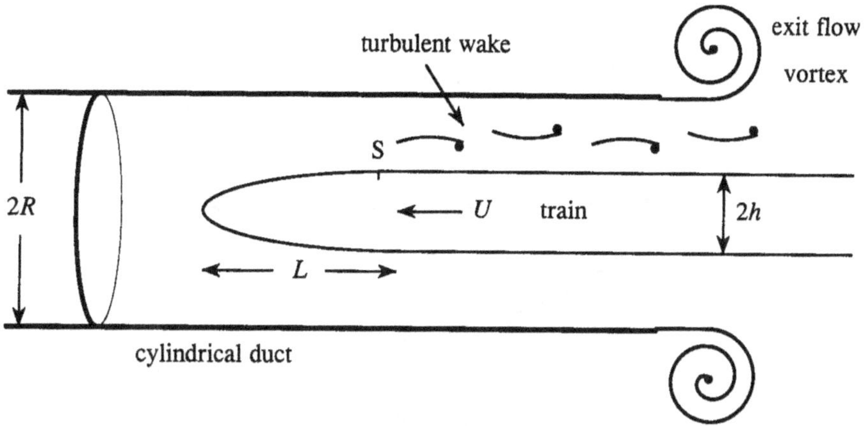

Figure 6.9.2

Take Green's function in the form (6.9.9). When $|y_1| \ll \lambda$ and $x_1 \to -\infty$,

$$G(\mathbf{x}, \mathbf{y}, t - \tau) \sim -(y_1 - \ell_E)\delta\left[t - \tau + \frac{(x_1 - \ell_E)}{c_o}\right].$$

The pressure waves radiated into the interior of the duct are therefore given by

$$p(x, t) = -\rho_o \frac{\partial}{\partial t} \int (y_1 - \ell_E) q\left[y_1, t + \frac{(x_1 - \ell_E)}{c_o}\right] dy_1.$$

This is smaller by a factor of $\sim k_o \ell \sim \ell/\lambda \ll 1$ than the corresponding pressure in an infinite duct [Equation (6.2.16)]:

$$\frac{\rho_o c_o}{2} \int q\left(y_1, t - \frac{|x_1 - y_1|}{c_o}\right) dy_1.$$

6.9.2 Wave generation by a train entering a tunnel

A train entering a tunnel pushes aside the stationary air, most of which flows over the train and out of the tunnel portal, but a residual 'piston effect' produces a compression wave that propagates ahead of the train into the tunnel at the speed of sound. For high-speed trains (of speeds exceeding about 200 km/h) the wavefront often steepens into a 'shock wave' during propagation in a long tunnel (because nonlinear convection causes higher pressures to propagate progressively faster than the speed of sound) and emerges from the distant exit as a loud bang or 'crack' called a 'micro-pressure' wave.

The formation of the compression wave is studied experimentally at model scale by projecting a 'wire-guided' *axisymmetric* 'train' into a circular-cylindrical duct about 7 m long and of internal radius $R \sim 5$ cm (Figure 6.9.2). The train travels at constant speed U in the negative x_1 direction, where the origin O is at the centre of the tunnel entrance plane with the x_1 axis coincident with the tunnel axis. The train cross section becomes uniform with constant area $A_o = \pi h^2$ at a distance L from the train nose, where h is the

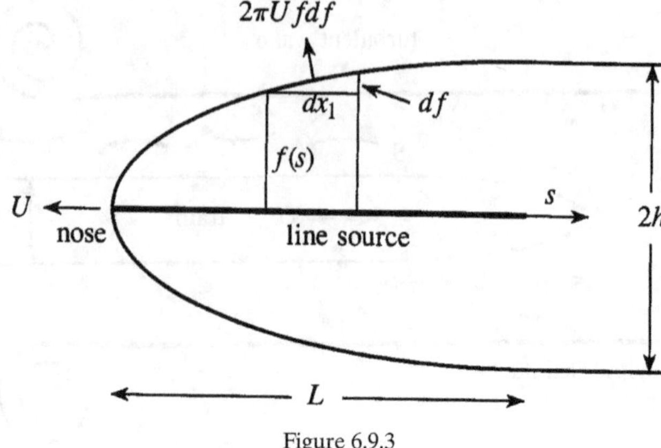

Figure 6.9.3

maximum train radius. The Mach number $M = U/c_o$ does not exceed about 0.4, and the *blockage* $A_o/A \le 0.2$, where $A = \pi R^2$ is the tunnel cross-sectional area.

The main pressure rise p_e across the wavefront is produced in a time interval $\sim R/U$ during which the train nose may be assumed to enter the tunnel. The Reynolds number $UR/\nu \sim 10^5$ ($\sim 10^7$ at full scale), and the initial form of p_e may therefore be calculated with the neglect of viscous diffusion. At later times, however, flow separation near the point labelled S (Figure 6.9.2) just to the rear of the nose produces a turbulent wake that is a source of an additional very low-frequency pressure wave p_w, say, that forms an extensive 'tail' behind the wavefront where the pressure slowly increases. The net acoustic pressure radiated into the tunnel is therefore

$$p = p_e + p_w. \tag{6.9.11}$$

A slow-moving 'exit flow vortex' ejected from the tunnel entrance as the train enters also makes a small contribution to the compression wave, but its effect is small compared with p_w and will be ignored.

In the simplest experimental configuration the axes of symmetry of the tunnel and train coincide (the train slides axisymmetrically along a wire stretched along the tunnel axis). When the blockage A_o/A is small, the moving train is acoustically equivalent to a *monopole* line source translating at the speed U of the train and distributed along the axis of symmetry of the nose (Figure 6.9.3). To calculate the source strength let $\varpi = \sqrt{x_2^2 + x_3^2}$ denote the perpendicular distance from the train axis, and let the axisymmetric profile of the train be

$$\varpi = f(x_1 + Ut),$$

where $A_T(s) = \pi f^2(s)$ is the cross-sectional area of the train at distance s from the nose tip, which crosses the entrance plane $x_1 = 0$ at $t = 0$. The rate at which air is displaced by a section of train of length dx_1 is therefore

$$2\pi U f(x_1 + Ut)df(x_1 + Ut) = 2\pi U f(x_1 + Ut)\frac{\partial f}{\partial x_1}(x_1 + Ut)dx_1 \equiv U\frac{\partial A_T}{\partial x_1}(x_1 + Ut)dx_1.$$

Thus the

$$\text{monopole source strength per unit length of the train} = U\frac{\partial A_{\mathrm{T}}}{\partial x_1}(x_1 + Ut).$$

This is collapsed onto a *line source* $q(\mathbf{x}, t)$ on the train axis by putting

$$q = U\frac{\partial A_{\mathrm{T}}}{\partial x_1}(x_1 + Ut)\delta(x_2)\delta(x_3). \tag{6.9.12}$$

Equation (6.1.6) for the main pressure rise p_e produced by the *inviscid* interaction of the tunnel and train therefore becomes

$$\left(\frac{1}{c_o^2}\frac{\partial^2}{\partial t^2} - \nabla^2\right)p_e = \frac{\partial}{\partial t}\left[\rho_o U\frac{\partial A_{\mathrm{T}}}{\partial x_1}(x_1 + Ut)\delta(x_2)\delta(x_3)\right]. \tag{6.9.13}$$

The monopole does not depend on time when viewed in a reference frame moving with the train and therefore generates only a *near field* when travelling within the tunnel or in free space. The compression wave is produced when the source interacts with the tunnel portal as the train enters the tunnel, over a time $\sim R/U$. The characteristic thickness of the wavefront therefore $\sim R/M \gg R$, and Equation (6.9.13) can therefore be solved by use of the compact Green's function (6.9.8).

This yields at position $\mathbf{x}$ in the tunnel ahead of the train:

$$p_e \approx \rho_o\frac{\partial}{\partial t}\iint_{-\infty}^{\infty} U\frac{\partial A_{\mathrm{T}}}{\partial y_1}(y_1 + U\tau)G(\mathbf{x}, y_1, 0, 0; t - \tau)dy_1 d\tau$$

$$= \frac{\rho_o U c_o}{2A}\int_{-\infty}^{\infty}\left\{A_{\mathrm{T}}'\Big(y_1 - M\varphi^*(y_1, 0, 0) + U[t]\Big) - A_{\mathrm{T}}'\Big(y_1 + M\varphi^*(y_1, 0, 0) + U[t]\Big)\right\}dy_1,$$

$$\tag{6.9.14}$$

where the prime on A_{T} denotes differentiation with respect to the argument, and $[t] = t + (x_1 - \ell_E)/c_o$ is the effective retarded time. This *linear theory* approximation determines the *initial* form of the compression wave profile, before the onset of nonlinear steepening, and is applicable within the region several tunnel diameters ahead of the train, during and just after tunnel entry.

The main contributions to integral (6.9.14) are from the vicinities of the nose and tail of the train, where the cross-sectional area A_{T} is changing. However, the compression wave is generated when the nose enters the tunnel and we may calculate it by temporarily considering a train of semi-infinite length. During the formation of the wave, and in the particular case in which the Mach number is small enough that terms $\sim O(M^2)$ are negligible, the term $M\varphi^*$ in the arguments of A_{T}' in (6.9.14) is small. By expanding to first order in $M\varphi^*$ and integrating by parts, we then find

$$p_e \approx \frac{\rho_o U^2}{A}\int_{-\infty}^{\infty}\frac{\partial A_{\mathrm{T}}}{\partial y_1}(y_1 + U[t])\frac{\partial\varphi^*}{\partial y_1}(y_1, 0, 0)dy_1, \quad M^2 \ll 1. \tag{6.9.15}$$

The validity of this formula can be extended in two ways:

First observe that, once the nose has passed into the tunnel, $\partial\varphi^*/\partial y_1 = 1$ in the region occupied by the nose, and (6.9.15) predicts the overall pressure rise across the wavefront to be $\Delta p_e \approx \rho_o U^2 A_o / A$. However, the asymptotic pressure rise can be calculated *exactly*, with no restriction on Mach number, to be

$$\Delta p_e = \frac{\rho_o U^2 A_o}{A(1 - M^2)},$$

because this is attained when $\varphi^*(y_1, 0, 0) \approx y_1 - \ell_E$ in (6.9.14). This suggests that approximation (6.9.15) can be extrapolated to finite Mach numbers by dividing by $(1 - M^2)$.

Second, a more detailed treatment of the train problem reveals that the right-hand side of Equation (6.9.13) for p_e should also include a dipole source, whose strength is proportional to the excess drag on the nose produced by the compression wave. In a first approximation the effect of this dipole can be accommodated by the multiplication of prediction (6.9.15) by

$$\left(1 + \frac{A_o}{A}\right).$$

Hence, incorporating both of these corrections, we obtain

$$p_e \equiv p_e\left(t + \frac{x_1}{c_o}\right) \approx \frac{\rho_o U^2}{A(1 - M^2)}\left(1 + \frac{A_o}{A}\right)\int_{-\infty}^{\infty}\frac{\partial A_T}{\partial y_1}(y_1 + U[t])\frac{\partial\varphi^*}{\partial y_1}(y_1, 0, 0)dy_1.$$

$$(6.9.16)$$

This extrapolation of the linear theory to finite values of M turns out to be applicable for $M < 0.4$ (Howe *et al.* 2000).

The subjective influence of the compression wave in a tunnel depends on the rate of change of the acoustic pressure. Similarly, the amplitude of the micro-pressure wave radiated from the distant open end is proportional to the 'pressure gradient' $\partial p_e/\partial t$ at the wavefront (Problems 6, Question 18). This can be calculated from (6.9.16) by use of integration by parts (observing that $\partial A_T/\partial x = 0$ at $x = \pm\infty$) as follows:

$$\frac{\partial p_e}{\partial t} = \frac{\rho_o U^2}{A(1 - M^2)}\left(1 + \frac{A_o}{A}\right)\int_{-\infty}^{\infty}\frac{\partial^2 A_T}{\partial t \partial y_1}(y_1 + U[t])\frac{\partial\varphi^*}{\partial y_1}(y_1, 0, 0)dy_1$$

$$= \frac{\rho_o U^3}{A(1 - M^2)}\left(1 + \frac{A_o}{A}\right)\int_{-\infty}^{\infty}\frac{\partial^2 A_T}{\partial y_1^2}(y_1 + U[t])\frac{\partial\varphi^*}{\partial y_1}(y_1, 0, 0)dy_1$$

$$= -\frac{\rho_o U^3}{A(1 - M^2)}\left(1 + \frac{A_o}{A}\right)\int_{-\infty}^{\infty}\frac{\partial A_T}{\partial y_1}(y_1 + U[t])\frac{\partial^2\varphi^*}{\partial y_1^2}(y_1, 0, 0)dy_1. \quad (6.9.17)$$

This formula shows that the fine details of the profile of the compression wavefront depend critically on the shape of the open end, because $\partial^2\varphi^*/\partial y_1^2 \sim 0$ except in the vicinity of the portal. Practical attempts to modify the profile therefore involve modifications in the geometry of the tunnel portal.

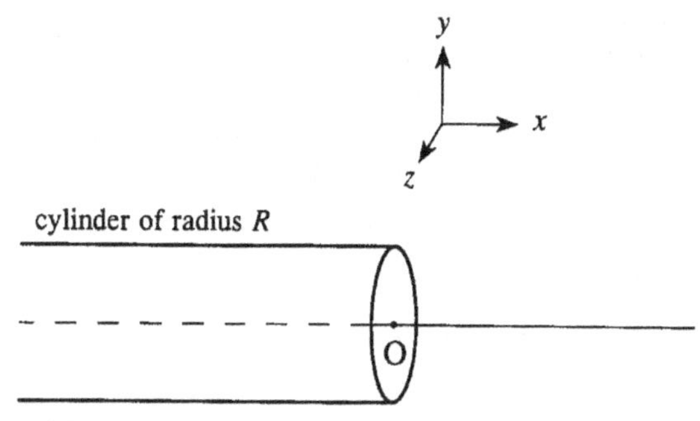

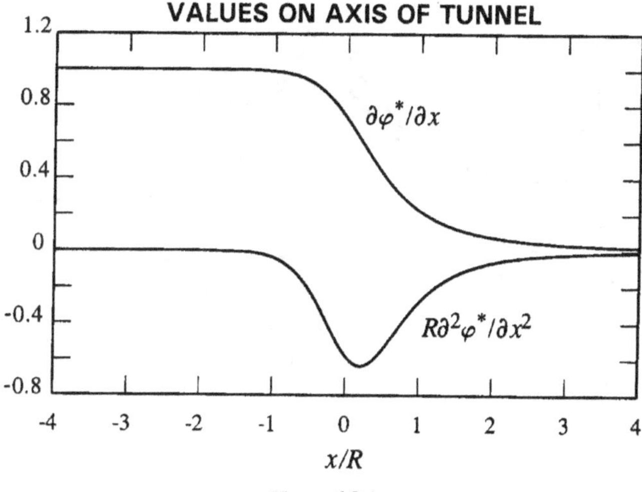

Figure 6.9.4

EXAMPLE 3. EVALUATION OF $\partial\varphi^*/\partial X$. The streamlines of the ideal flow from a circular cylinder of radius R defined by the potential $\varphi^*(\mathbf{x})$ are shown in Figure 2.12.8. The quantities appearing in (6.9.16) and (6.9.17) can be evaluated from

$$\frac{\partial\varphi^*}{\partial x}(\mathbf{x}) = \frac{1}{2} - \frac{1}{2\pi}\int_0^\infty I_0\left(\frac{\lambda\varpi}{R}\right)\left[\frac{2K_1(\lambda)}{I_1(\lambda)}\right]^{\frac{1}{2}}\sin\left\{\lambda\left[\frac{x}{R} + \mathcal{Z}(\lambda)\right]\right\}d\lambda,$$

$$\mathcal{Z}(\lambda) = \frac{1}{\pi}\int_0^\infty \ln\left[\frac{K_1(\mu)I_1(\mu)}{K_1(\lambda)I_1(\lambda)}\right]\frac{d\mu}{\mu^2 - \lambda^2},$$

where (Figure 6.9.4) $\mathbf{x} = (x, y, z)$, the origin is at the centre of the duct exit, $\varpi = \sqrt{y^2 + z^2} < R$, and I_0, I_1, K_1 are modified Bessel functions. The values of $\partial\varphi^*/\partial x$ and $\partial^2\varphi^*/\partial x^2$ on the axis of the duct are plotted in the figure.

EXAMPLE 4. SNUB-NOSED TRAIN When the 'aspect ratio' h/L of the train nose is very large (Figure 6.9.5), the formal limit $L \to 0$ transforms the distributed line source (6.9.12) into the point source

$$q = UA_o\delta(x_1 + Ut)\delta(x_2)\delta(x_3).$$

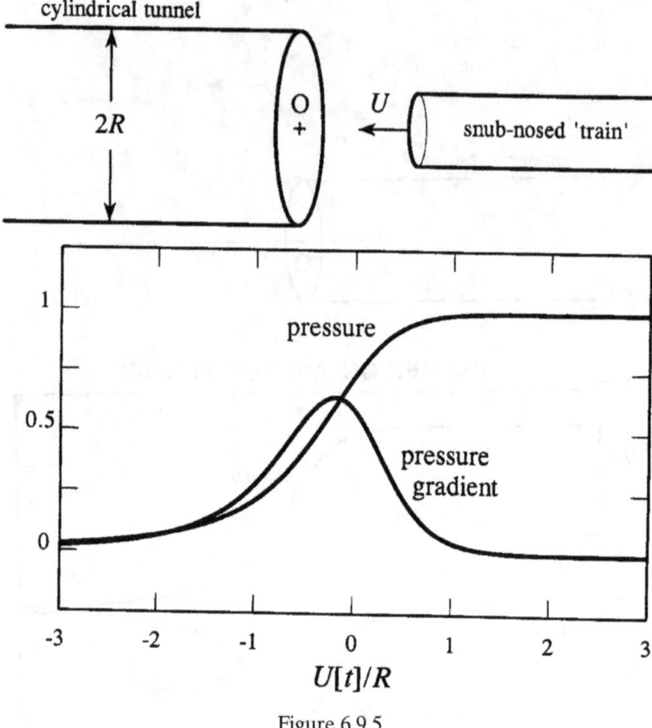

$$U[t]/R$$

Figure 6.9.5

Then equations (6.9.16) and (6.9.17) yield

$$p_e \left/ \frac{\rho_o U^2}{(1 - M^2)} \frac{A_o}{A} \left(1 + \frac{A_o}{A}\right) = \frac{\partial \varphi^*}{\partial y_1}(-U[t], 0, 0), \right.$$

$$\frac{\partial p_e}{\partial t} \left/ \frac{\rho_o U^3}{R(1 - M^2)} \frac{A_o}{A} \left(1 + \frac{A_o}{A}\right) = - R\frac{\partial^2 \varphi^*}{\partial y_1^2}(-U[t], 0, 0). \right.$$

These dimensionless measures of the compression wave pressure and pressure gradient are plotted against $U[t]/R$ in Figure 6.9.5. The plots illustrate that the order of magnitude of the *rise time* $\sim 2R/U$, a conclusion that is valid when the blockage A_o/A is less than about 0.25.

EXAMPLE 5. TRAIN WITH AN ELLIPTIC NOSE PROFILE (Howe and Iida 2003) Laboratory scale measurements of the compression wave are shown in Figure 6.9.6 (△). The tunnel consisted of a 6.5-m-long, thin-walled, circular-cylindrical tube of internal radius $R = 0.05$ m. The train nose had an ellipsoidal profile defined by

$$\varpi = h\sqrt{\frac{x}{L}\left(2 - \frac{x}{L}\right)}, \quad 0 < x < L \quad \left(\varpi = \sqrt{y^2 + z^2}\right),$$

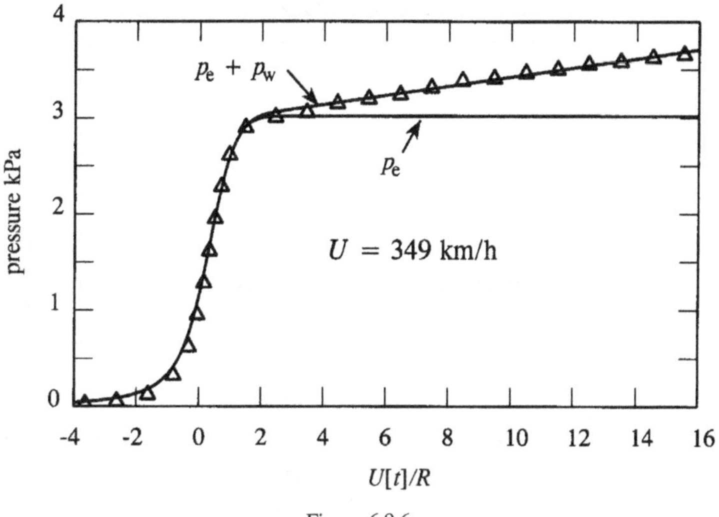

Figure 6.9.6

where $h = 2.235\,\mathrm{cm}$, $L = 6.705\,\mathrm{cm}$, so that (ignoring the rear end of the train)

$$\frac{A_\mathrm{T}(s)}{A_o} = \begin{cases} \frac{s}{L}\left(2 - \frac{s}{L}\right), & 0 < s < L \\ 1, & s > L \end{cases},$$

and

$$\frac{L}{h} = 3, \quad \frac{A_o}{A} = 0.2.$$

The train was projected at $U = 349\,\mathrm{km/h}$ ($M = 0.285$) along a wire stretched along the tunnel axis and passing smoothly through a coaxial bore hole in the train. The tail of the model train had an identical ellipsoidal shape, and the overall length of the train was 123.9 cm.

The front of the nose entered the tunnel at $t = 0$, and the pressure was measured within the tunnel at a distance $\ell_m = 1.5\,\mathrm{m}$ from the entrance. In the figure the pressure at $x = -\ell_m$ is plotted against $U[t]/R$, where $[t] = t - (\ell_m + \ell_E)/c_o$. The pressure rise begins just before the nose enters the tunnel at $U[t]/R = 0$, the main pressure rise being complete by $U[t]/R = 2$; the subsequent slow linear growth in the measured pressure represents the contribution of the pressure p_w generated by the turbulent wake of Figure 6.9.2. The figure shows that the inviscid pressure rise p_e [calculated with (6.9.16)] gives an excellent representation of the wavefront profile.

6.10 Damping of sound in a smooth-walled duct

Fluid motion very close to a solid boundary is controlled by viscous stresses in a thin 'boundary layer' across which there is a rapid adjustment in velocity to satisfy the no-slip condition; similarly, a temperature gradient across a 'thermal' boundary layer brings the temperatures of the solid and fluid into equality. A consideration of dimensions indicates that for acoustic waves of frequency ω the viscous boundary-layer 'thickness'

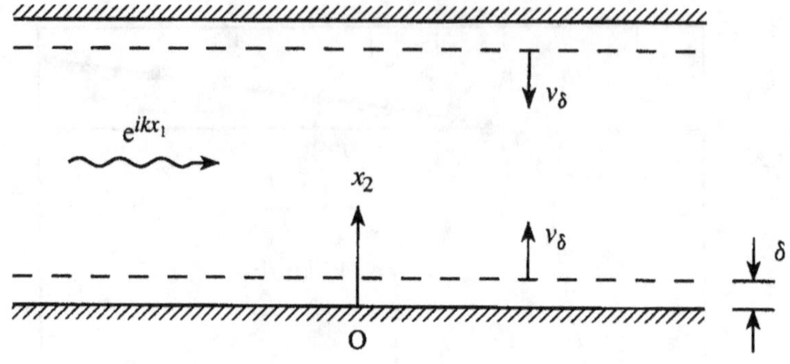

Figure 6.10.1

$\delta \sim \sqrt{\nu/\omega}$ (§4.4, Example 1). For propagation in a smooth-walled duct of radius R, the boundary-layer thickness is very much smaller than R when the *acoustic* Reynolds number $\omega R^2/\nu \gg 1$, a condition that is satisfied both in air and in water for most practical values of R and ω. The same conclusion holds for the thermal boundary layer, whose thickness $\sim \sqrt{\chi/\omega}$, where $\chi \sim O(\nu)$ is the thermometric conductivity. Thus the thicknesses of both boundary layers are roughly the same, and it will be convenient to denote either one by δ according to the context.

6.10.1 Time harmonic propagation in a duct

Take the x_1 axis parallel to the duct and consider a plane wave $\propto e^{ikx_1}$ of frequency $\omega > 0$ with *complex* wavenumber k (Figure 6.10.1). The imaginary part of k determines the damping of sound by the viscous and thermal boundary layers. Because the boundary layers are very thin, motion in the normal ('radial') direction in the boundary layers is constrained by the duct wall, so that the pressure must be approximately uniform across the boundary layer. Outside the boundary layers, the motion is irrotational and governed by

$$\left(\nabla^2 + k_o^2\right)\varphi = 0, \quad \text{where} \quad k_o = \frac{\omega}{c_o}.$$

Integrate this equation over the irrotational, 'inner' cross section of the duct (outside the boundary layers). To a first approximation the condition $\delta \ll R$ and the divergence theorem yield

$$\frac{\partial^2 \bar{\varphi}}{\partial x_1^2} + k_o^2 \bar{\varphi} - \frac{\ell_p}{A} v_\delta = 0 \tag{6.10.1}$$

where v_δ is the *inward* normal component of velocity at the outer edge of the boundary layer directed into the irrotational region, ℓ_p is perimeter of the duct, A its cross-sectional area, and

$$\bar{\varphi}(x_1) = \frac{1}{A} \int_A \varphi(\mathbf{x})\, dx_2 dx_3$$

is the mean acoustic potential at x_1.

The value of v_δ is made up of separate contributions from the viscous and thermal boundary layers. Take the coordinate origin at a point O on the wall, with the x_2 axis radially *inwards*. Then within the boundary layers the motion in an $x_1 x_2$ plane is locally the same as for two-dimensional grazing propagation of sound over a *plane* wall, and $v_\delta = v_2(x_1, \delta)$.

To calculate v_δ let us temporarily denote by p', ρ', $\mathbf{v}'$ the perturbation pressure, density, and velocity in the *acoustic* field, which satisfy the ideal fluid equations in $x_2 > \delta$. However, these equations may be regarded as satisfied by the acoustic variables within the whole of $x_2 > 0$, including the boundary layers. Because both the exact and acoustic fields satisfy the linearised continuity equation, and must vary like e^{ikx_1}, we can then form the difference

$$-\frac{i\omega}{\rho_o}(\rho - \rho') + ik(v_1 - v_1') + \frac{\partial}{\partial x_2}(v_2 - v_2') = 0,$$

and integrate over $0 < x_2 < \delta$. However, $v_2 \equiv v_2'$ at $x_2 = \delta$, and $v_2 = 0$ and $v_2' = v_\delta$ at $x_2 = 0$ (because $k_o\delta \ll 1$),

$$\therefore \quad v_\delta = v_{\text{viscous}} + v_{\text{thermal}},$$

$$v_{\text{viscous}} = -ik \int_0^\infty (v_1 - v_1')dx_2 \tag{6.10.2}$$

$$v_{\text{thermal}} = \frac{i\omega}{\rho_o} \int_0^\infty (\rho - \rho')dx_2$$

where the upper limits of integration have been extended to ∞ because the integrands vanish for $x_2 > \delta$.

The motion very close to the wall is retarded by viscous shear stresses, causing fluid to be displaced outwards relative to an ideal fluid at a rate equal to v_{viscous}. The component v_{thermal} is the outward velocity produced by volumetric expansion occurring when a temperature gradient is established in the boundary layer to bring the wall and fluid into equilibrium following adiabatic heating or cooling by the incident sound.

6.10.2 The viscous contribution

We calculate the viscous contribution to v_δ by using the *boundary-layer approximation* (cf. §4.4.7). The boundary layer is driven by the acoustic pressure gradient $\partial p'/\partial x_1$, which is opposed by viscous *shear* stresses in accordance with the linearised, tangential component of momentum equation (1.4.10) (with body forces omitted):

$$-i\omega(v_1 - v_1') = v\frac{\partial^2 v_1}{\partial x_2^2}, \quad x_2 < \delta, \tag{6.10.3}$$

where $v_1' = (-i/\rho_o\omega)\partial p'/\partial x_1 = kp'/\rho_o\omega$ is the acoustic particle 'slip' velocity along the wall, which is uniform across the boundary layer. The shear stress term on the right-hand side exceeds all other discarded viscous terms by a factor of $\sim 1/(k_o\delta)^2 \gg 1$, and the

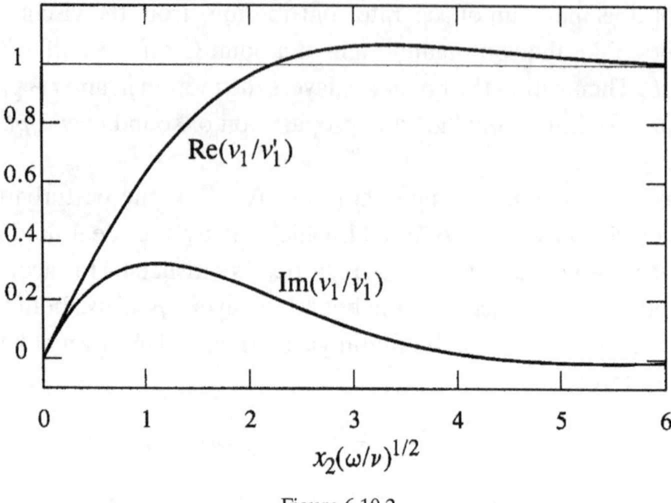

Figure 6.10.2

omitted compressive stresses (involving div $\mathbf{v}$) are smaller than the acoustic pressure gradient by a factor of the order of $k_o \delta/(\delta c_o/\nu) \ll 1$.

The solution must remain finite at $x_2 = +\infty$ and satisfy the no-slip condition at the wall, and is given by the *Kirchhoff–Stokes* formula

$$v_1 = v_1' \left(1 - e^{ix_2\sqrt{i\omega/\nu}}\right), \tag{6.10.4}$$

where $\mathrm{Im}\sqrt{i\omega} > 0$. The real and imaginary parts of v_1/v_1' are plotted against $x_2\,(\omega/\nu)^{\frac{1}{2}}$ in Figure 6.10.2. These plots indicate that the viscous boundary-layer thickness $\delta \sim \delta_\nu$, where

$$\delta_\nu = 2\sqrt{\frac{\nu}{\omega}}.$$

Typical values of δ_ν for propagation in air and water are given in the following table:

f, Hz	10^2	10^3	10^4	
Air	3×10^{-4}	10^{-4}	3×10^{-5}	$\left.\vphantom{\begin{array}{c}a\\b\end{array}}\right\}\ \delta_\nu$ m
Water	10^{-4}	3×10^{-5}	10^{-5}	

Hence, the first of (6.10.2) supplies

$$v_{\text{viscous}} = \int_0^\infty \frac{k\nu}{\omega}\frac{\partial^2 v_1}{\partial x_2^2}\,dx_2 = -\left(\frac{k\nu}{\omega}\frac{\partial v_1}{\partial x_2}\right)_{x_2=0} = -v_1'k\sqrt{\frac{\nu}{\omega}}e^{-\frac{i\pi}{4}}$$

$$= -\bar{\varphi}k^2\sqrt{\frac{\nu}{\omega}}e^{\frac{i\pi}{4}} \approx -\bar{\varphi}k_o^2\sqrt{\frac{\nu}{\omega}}e^{\frac{i\pi}{4}}, \tag{6.10.5}$$

where to a sufficient approximation k has been replaced with the undisturbed wavenumber k_o because (except at very high frequencies) the damping produces only a small change in the acoustic wavenumber.

6.10.3 The thermal contribution

The thermal contribution to v_δ results from the volumetric changes in the boundary layer produced by the conduction of heat to and from the wall over that produced by the adiabatic expansions and compressions caused by the sound. Let T_o denote the mean temperature of the fluid and wall. Then

$$d\rho = \frac{dp}{c_o^2} + \left(\frac{\partial \rho}{\partial s}\right)_p ds \equiv \frac{dp}{c_o^2} - \frac{\rho_o \beta T_o ds}{c_p},$$

where

$$\beta = -\frac{1}{\rho}\left(\frac{\partial \rho}{\partial T}\right)_p$$

is the coefficient of expansion at constant pressure. The density and pressure fluctuations in the sound wave satisfy $d\rho' = dp'/c_o^2$,

$$\therefore \quad \rho - \rho' \approx -\frac{\rho_o \beta T_o}{c_p} s, \tag{6.10.6}$$

where s is the change in the specific entropy from its uniform mean value in the absence of the sound.

The variation of s across the thermal boundary layer is governed by the linearised form of energy equation (1.6.6). The equation is simplified by use of the relation

$$T ds = dw - \frac{dp}{\rho}, \quad w = \text{enthalpy}. \tag{6.10.7}$$

Because $ds = 0$ in the acoustic wave,

$$0 = w' - \frac{p'}{\rho_o}.$$

The pressure fluctuations within the boundary layer are produced by the sound and equal p', but there is a local temperature difference $\delta T = T - T'$ between the actual temperature perturbation $T(x_2)$ and the adiabatic temperature fluctuation $T' \equiv p'/\rho_o c_p$ in the sound wave. Hence there is a corresponding additional change in the enthalpy equal to

$$\delta w = \left(\frac{\partial w}{\partial T}\right)_p (T - T').$$

Now $(\partial w/\partial T)_p = c_p$; therefore Equation (6.10.7) supplies the linearised relation

$$T_o s = c_p(T - T'),$$
$$\therefore \quad (6.10.6) \text{ becomes} \quad \rho - \rho' = -\rho_o \beta (T - T'). \tag{6.10.8}$$

Thus, by substituting from (6.10.8) for s into linearised energy equation (1.6.6) and applying the boundary-layer approximation, we find

$$-i\omega(T - T') = \chi \frac{\partial^2 T}{\partial x_2^2}, \quad 0 < x_2 < \delta, \quad \text{where} \quad \chi = \frac{\kappa}{\rho_o c_p}.$$

The heat capacity of the duct wall is assumed to be sufficiently large that the temperature fluctuations vanish at the wall, so that the appropriate solution is

$$T = T'\left(1 - e^{ix_2\sqrt{i\omega/\chi}}\right).$$

Using this and the second of (6.10.8) we find that the velocity v_{thermal} of (6.10.2) is

$$v_{\text{thermal}} = -\beta\chi\left(\frac{\partial T}{\partial x_2}\right)_{x_2=0} = -\frac{\omega^2\beta\bar{\varphi}}{c_p}\sqrt{\frac{\chi}{\omega}}e^{\frac{i\pi}{4}}. \tag{6.10.9}$$

6.10.4 The thermo-viscous damping coefficient

Substituting from (6.10.5) and (6.10.9) into the first of Eqs. (6.10.2), and thence into (6.10.1), we find that plane-wave propagation in the duct satisfies

$$\frac{\partial^2\bar{\varphi}}{\partial x^2} + k_o^2\left[1 + \frac{\ell_p}{A\sqrt{\omega}}\left(\sqrt{v} + \frac{\beta c_o^2}{c_p}\sqrt{\chi}\right)e^{\frac{i\pi}{4}}\right]\bar{\varphi} = 0, \tag{6.10.10}$$

where the second term in the brackets accounts for dissipation in the wall boundary layers. Solutions proportional to e^{ikx} are possible provided that

$$k = \pm(k_o + i\mu),$$

$$\mu \approx \frac{k_o\ell_p}{2A\sqrt{2\omega}}\left(\sqrt{v} + \frac{\beta c_o^2}{c_p}\sqrt{\chi}\right), \tag{6.10.11}$$

where only the imaginary part μ of the boundary-layer term has been retained.

In the special case of propagation in an ideal gas of temperature T_o, the coefficient of expansion $\beta = 1/T_o$, and $\beta c_o^2/c_p = \gamma - 1$, where γ is the ratio of the specific heats. The following table gives the approximate propagation distance x_d for air in a circular-cylindrical duct of radius R over which the wave amplitude decreases by a factor of $e^{-1} \sim 0.37$, so that the wave power is decreased to about 13% ($\sim -9\,\text{dB}$) of its initial value.

f Hz	10^2	10^3	10^4
x_d/R	3000	1000	300

PROBLEMS 6

1. In the absence of mean body forces, show that the sound generated by a volume source q and force $\mathbf{F}$ in a stationary fluid of uniform mean pressure p_o but variable mean density $\rho_o = \rho_o(\mathbf{x})$ is governed by

$$\frac{1}{\rho_o c_o^2}\frac{\partial^2 p}{\partial t^2} - \frac{\partial}{\partial x_j}\left(\frac{1}{\rho_o}\frac{\partial p}{\partial x_j}\right) = \frac{\partial q}{\partial t} - \operatorname{div}\left(\frac{1}{\rho_o}\mathbf{F}\right).$$

2. When the influence of gravity is included, the force $\mathbf{F}$ in (6.1.6) involves a component $\rho_o\mathbf{g}$, where $\mathbf{g}$ is the acceleration that is due to gravity. The mean pressure and density

vary with position, such that $\nabla p_o = \rho_o \mathbf{g}$, and in the absence of time-dependent forces (6.1.1) becomes

$$\rho_o \partial \mathbf{v}/\partial t + \nabla p' = \rho' \mathbf{g}.$$

Deduce that gravity can be neglected when the acoustic frequency $\omega \gg g/c_o$ (~ 0.03 s^{-1} in air at sea level).

3. If the tube in Figure 6.1.1(d) is closed at $x = 0$, show that the lowest-order resonance frequencies satisfy $\cos(k_o \ell) = 0$, and therefore that the wavelength of the gravest mode $\lambda \sim 4\ell$.

4. If the source strength $\mathcal{F}(\mathbf{x}, t)$ in retarded potential (6.2.10) is non-zero only in the neighbourhood of the origin, show that

$$p(\mathbf{x}, t) \approx \frac{1}{4\pi |\mathbf{x}|} \int_{-\infty}^{\infty} \mathcal{F}\left(\mathbf{y}, t - \frac{|\mathbf{x}|}{c_o} + \frac{\mathbf{x} \cdot \mathbf{y}}{c_o |\mathbf{x}|}\right) d^3 \mathbf{y}, \quad |\mathbf{x}| \to \infty.$$

5. For a localised *quadrupole* source

$$\mathcal{F}(\mathbf{x}, t) = \frac{\partial^2 T_{ij}}{\partial x_i \partial x_j}(\mathbf{x}, t),$$

show for retarded potential (6.2.10) that

$$p(\mathbf{x}, t) = \frac{1}{4\pi} \frac{\partial^2}{\partial x_i \partial x_j} \int_{-\infty}^{\infty} \frac{T_{ij}(\mathbf{y}, t - |\mathbf{x} - \mathbf{y}|/c_o)}{|\mathbf{x} - \mathbf{y}|} d^3 \mathbf{y}$$

$$\approx \frac{x_i x_j}{4\pi c_o^2 |\mathbf{x}|^3} \frac{\partial^2}{\partial t^2} \int_{-\infty}^{\infty} T_{ij}\left(\mathbf{y}, t - \frac{|\mathbf{x}|}{c_o} + \frac{\mathbf{x} \cdot \mathbf{y}}{c_o |\mathbf{x}|}\right) d^3 \mathbf{y}, \quad |\mathbf{x}| \to \infty.$$

6. Use Hadamard's method of descent to show that the acoustic Green's function [the *causal* solution of Equation (6.2.5)] in two space dimensions is

$$G(\mathbf{x}, \mathbf{y}, t - \tau) = \frac{H(t - \tau - |\mathbf{x} - \mathbf{y}|/c_o)}{2\pi \sqrt{(t - \tau)^2 - |\mathbf{x} - \mathbf{y}|^2/c_o^2}}.$$

7. If the dipole strength $f(x, t)$ vanishes at $x = \pm\infty$, show that the causal solution of

$$\left(\frac{1}{c_o^2} \frac{\partial^2}{\partial t^2} - \frac{\partial^2}{\partial x^2}\right) p = -\frac{\partial f}{\partial x}(x, t), \quad -\infty < x < \infty,$$

is

$$p(x, t) = \frac{1}{2} \int_{-\infty}^{\infty} \operatorname{sgn}(x - y) f\left(y, t - \frac{|x - y|}{c_o}\right) dy.$$

8. A rigid sphere of radius a is centred at the origin and oscillates in the x_1 direction at velocity $U(t)$. If the sphere is acoustically compact, show that the acoustic pressure

$$p(\mathbf{x}, t) \approx \frac{\rho_o a^3 \cos\theta}{2 c_o |\mathbf{x}|} \frac{\partial^2 U}{\partial t^2}(t - |\mathbf{x}|/c_o), \quad |\mathbf{x}| \to \infty, \quad \text{where } \frac{x_1}{|\mathbf{x}|} = \cos\theta.$$

9. A compact rigid disc of radius a executes small-amplitude vibrations at velocity $U(t)$ normal to itself. In the undisturbed state it lies in the plane $x_1 = 0$ with its centre

at the origin. If $\varphi_1^*(\mathbf{x}) = \mp(2/\pi)\sqrt{a^2 - x_2^2 - x_3^2}$ on the faces $x_1 = \pm 0$, $\sqrt{x_2^2 + x_3^2} < a$ of the disc, show that the acoustic pressure generated by the motion is given by

$$p(\mathbf{x}, t) \approx \frac{2\rho_o a^3 \cos\theta}{3\pi c_o |\mathbf{x}|} \frac{\partial^2 U}{\partial t^2} (t - |\mathbf{x}|/c_o), \quad |\mathbf{x}| \to \infty, \quad \cos\theta = \frac{x_1}{|\mathbf{x}|}.$$

10. Use Kirchhoff's formula (6.3.2) to derive the formula

$$p(\mathbf{x}, t) \approx \frac{x_i}{4\pi c_o |\mathbf{x}|^2} \left(m_o \frac{\partial^2 U_i}{\partial t^2} - \frac{\partial F_i}{\partial t} \right) \left(t - \frac{|\mathbf{x}|}{c_o} \right), \quad |\mathbf{x}| \to \infty$$

for the sound radiated by a compact vibrating rigid body, where m_o is the mass of fluid displaced by the body and $\mathbf{F}$ is the force exerted on the body by the fluid.

11. Calculate the far-field acoustic pressure when

$$\left(\frac{1}{c_o^2} \frac{\partial^2}{\partial t^2} - \nabla^2 \right) p = -f(t) \frac{\partial}{\partial x_2} \left[\delta(x_1 - L)\delta(x_2)\delta(x_3) \right], \quad L > a,$$

in the presence of the rigid airfoil $-a < x_1 < a$, $x_2 = 0$, $-\infty < x_3 < \infty$ of *compact chord* $2a$ (see §2.17.2). Show that

$$p(\mathbf{x}, t) \approx \frac{\rho_o \cos\Theta}{4\pi c_o |\mathbf{x}|} \frac{L}{\sqrt{L^2 - a^2}} \frac{\partial^2 f}{\partial t^2} \left(t - \frac{|\mathbf{x}|}{c_o} \right), \quad |\mathbf{x}| \to \infty,$$

where $\Theta = \cos^{-1}(x_2/|\mathbf{x}|)$ is the angle between the normal to the airfoil and the radiation direction (as in Figure 2.17.3).

12. A plane, time harmonic sound wave $p_I = p_0 e^{i(k_o x_1 - \omega t)}$ is incident upon an acoustically compact, rigid sphere of radius R with centre at the origin. Show that at large distances the scattered sound is given by

$$p_s \approx \frac{p_0 (k_o R)^2 R}{3|\mathbf{x}|} \left(\frac{3}{2} \cos\theta - 1 \right) e^{-i\omega(t - |\mathbf{x}|/c_o)}, \quad |\mathbf{x}| \to \infty, \quad \cos\theta = \frac{x_1}{|\mathbf{x}|}.$$

13. Show in the linearised approximation that the displacement $\xi(t)$ of the fluid slug in the neck of the Helmholtz resonator of Figure 6.6.4 satisfies

$$\frac{d^2\xi}{dt^2} + \frac{\rho_1 c_1^2 A_0}{\rho_0 V \ell} \xi = 0.$$

14. In Figure 6.6.1 suppose that the three ducts of cross sections A_0, A_1, A_2 meeting at the junction O are closed at their outer ends. Let the corresponding duct lengths be ℓ_0, ℓ_1, ℓ_2 and suppose that $\rho_0 = \rho_1 = \rho_2$ and $c_0 = c_1 = c_2$. Show that the low-frequency resonance frequencies of the closed system satisfy

$$\sum_{n=0}^{2} A_n \tan(k_0 \ell_n) = 0.$$

If the outer end of the duct of length ℓ_0 is open, show that the resonance frequencies are determined by

$$A_0 \cot(k_0 \ell_0) = A_1 \tan(k_0 \ell_1) + A_2 \tan(k_0 \ell_2).$$

15. Use the results of §6.7.1 to calculate the pressure and acoustic particle velocity at large distance r from the open end of the duct of Figure 6.7.1 produced by an acoustic wave $p = p'e^{ik_ox}$ incident from within upon the open end when $k_oR \ll 1$, where R is the duct radius. Calculate the power radiated through the surface of a large sphere centred on the open end, and verify that a fraction $(k_oR)^2$ of the incident power radiates from the duct.

16. Verify for the problem of §6.7.3 that the acoustic power radiated from the open end of the duct is given by

$$\Pi_{rad} = \frac{1}{2}|p|^2\text{Re}\,Y \equiv (k_oR)^2\Pi_I, \quad \text{where} \quad Y = \frac{iA_o}{\rho_o\omega(\ell_E + ik_oA_o/4\pi)},$$

p is the complex amplitude of the pressure at $x = 0$, and $\Pi_I = \frac{1}{2}|p'|^2\bar{Y}_o$ is the incident acoustic power.

17. Show that the resonance frequencies of the double-bulb Helmholtz resonator in the figure satisfy

$$\omega^4 - \omega^2 c_o^2 A_o\left[\frac{1}{V_1}\left(\frac{1}{\ell_1} + \frac{1}{\ell_2}\right) + \frac{1}{V_2\ell_2}\right] + \frac{c_o^4 A_o^2}{\ell_1\ell_2 V_1 V_2} = 0,$$

where V_1, V_2 are the bulb volumes, both necks have the same cross-sectional area A_o, and the lengths ℓ_1, ℓ_2 are inclusive of end corrections

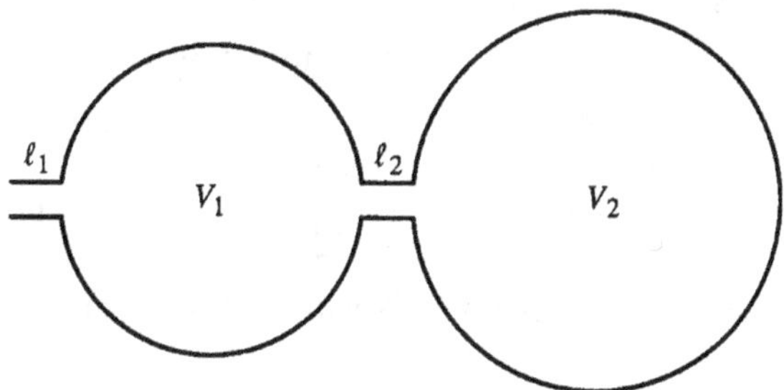

18. A low-frequency plane acoustic wave $p = p_I(t - x/c_o)$ propagates within a semi-infinite, circular-cylindrical duct of cross-sectional area A towards the open end at $x = 0$ (Figure 6.7.1). By use of Equations (6.7.3) and (6.7.5), show that the sound radiated out of the open end is given at a large distance r from the exit by

$$p \approx \frac{A}{2\pi c_o r}\frac{\partial p_I}{\partial t}\left(t - \frac{r}{c_o}\right).$$

19. Use the method of §6.7, Example 3, to show that the end correction of a flanged semi-infinite duct of square cross section of side a is $\ell_E \approx 0.813R$, where R is the radius of a circular duct of the same cross-sectional area ($R = a/\sqrt{\pi}$). Assume that in the square exit plane ($x = 0$) of the duct

$$\frac{\partial\varphi^*}{\partial x} = \alpha\left[1 + \frac{\mu(y^4 + z^4)}{a^4}\right], \quad -\frac{a}{2} < y,\ z < \frac{a}{2}.$$

20. Show that for a plane sound wave propagating in a smooth-walled duct (satisfying the conditions of §6.10) the acoustic power dissipated per unit area of the duct wall is

$$\Pi = \frac{\sqrt{\omega}}{2\sqrt{2}\rho_o c_o^2}|p'|^2\left(\sqrt{\nu} + \frac{\beta c_o^2}{c_p}\sqrt{\chi}\right),$$

where p' is the complex amplitude of the plane wave of frequency $\omega > 0$.

21. Show that for a plane sound wave $\propto e^{ikx}$ propagating in free space the complex wavenumber is given approximately by

$$k = \pm k_o\left\{1 + \frac{i\omega}{2\rho_o c_o^2}\left[\frac{4\eta}{3} + \eta' + \frac{\kappa}{c_p}(\gamma - 1)\right]\right\},$$

where γ is the ratio of the specific heats.

22. A plane time harmonic sound wave ($\propto e^{-i\omega t}$) of amplitude p' impinges on a plane rigid wall at an angle of incidence θ to the normal. If the heat capacity of the wall is very large (so that its temperature remains constant during the interaction) show that the acoustic power dissipated per unit area of the wall is

$$\frac{\sqrt{\omega}}{2\sqrt{2}\rho_o c_o^2}|p'|^2\left(\sqrt{\nu}\sin^2\theta + \frac{\beta c_o^2}{c_p}\sqrt{\chi}\right).$$

23. Energy equation for sound propagation in a duct. Consider the problem of plane-wave propagation in the duct of Figure 6.10.1. By multiplying the wave equation $\left(\partial^2/c_o^2\partial t^2 - \nabla^2\right)\varphi = 0$ by $\rho_o\partial\varphi/\partial t$ and integrating over the cross section of the irrotational 'core' of the duct, obtain the acoustic energy equation in the form

$$\frac{\partial}{\partial t}\left[A\left(\frac{\rho_o v^2}{2} + \frac{p^2}{2\rho_o c_o^2}\right)\right] + \frac{\partial}{\partial x_1}(Apv) = \ell_p p v_\delta,$$

where $v = \partial\varphi/\partial x_1$, $p = -\rho_o\partial\varphi/\partial t$. Explain the physical interpretation of the terms in this equation.

Bibliography

Abramowitz, M. and Stegun, I. A. (eds.) 1970 *Handbook of Mathematical Functions* (9th corrected printing), U.S. Department of Commerce, National Bureau of Standard Applied Mathematics Series No. 55.

Batchelor, G. K. 1967 *An Introduction to Fluid Dynamics*, Cambridge University Press.

Birkhoff, G. and Zarantonello, E. H. 1957 *Jets, Wakes and Cavities*. New York, Academic.

Copson, E. T. 1947 *Proc. Edinburgh Math. Soc.* **8**, 14–19. On the problem of the electrified disc.

Debye, P. 1909 *Math. Ann.* **67**, 535–558. Näherungsformeln für die Zylinderfunktionen für grosse Werte des Arguments und unbeschränkt veränderliche Werte des Index.

Dowling, A. P. and Ffowcs Williams, J. E. 1983 *Sound and Sources of Sound*, Chichester, Ellis Horwood.

Durand, W. F. (ed.) 1934–36. *Aerodynamic Theory*, 6 volumes. Berlin, Springer.

Fried, B. D. and Conte, S. D. 1961 *The Plasma Dispersion Function*, New York, Academic.

Glauert, H. 1929 Technical Report of the Aeronautical Research Committee, R. & M. No. 1242.

Goldstein, M. E. 1976 *Aeroacoustics*, New York, McGraw-Hill.

Goldstein, S. 1960 *Lectures on Fluid Mechanics*, New York, Interscience.

Gurevich, M. I. 1965 *Theory of Jets in Ideal Fluids*, New York, Academic.

Hadamard, J. 1952 *Lectures on Cauchy's Problem in Linear Partial Differential Equations*, New York, Dover.

Havelock, T. H. 1908 *Proc. R. Soc. London Ser. A* **81**, 398–430. The propagation of groups of waves in dispersive media, with application to waves on water produced by a travelling disturbance.

Havelock, T. H. 1914 *The Propagation of Disturbances in Dispersive Media*. Cambridge Tracts in Mathematics and Mathematical Physics, No. 17, Cambridge University Press.

Howe, M. S. 1995 *Q. J. Mech. Appl. Math.* **48**, 401–426. On the force and moment exerted on a body in an incompressible fluid, with application to rigid bodies and bubbles at high and low Reynolds numbers.

Howe, M. S. 1998 *Proc. R. Soc. London Ser. A* **454**, 1523–1534 The compression wave produced by a high-speed train entering a tunnel.

Howe, M. S. 2003 *Theory of Vortex Sound*, Cambridge University Press.

Howe, M. S. and Iida, M. 2003 *Int. J. Aeroacoust.* **2**, 13–33. Influence of separation on the compression wave generated by a train entering a tunnel.

Howe, M. S., Iida, M., Fukuda, T., and Maeda, T. 2000 *J. Fluid Mech.* **425**, 111–132. Theoretical and experimental investigation of the compression wave generated by a train entering a tunnel with a flared portal.

Howe, M. S., Lauchle, G. C., and Wang, J. 2001 *J. Fluid Mech.* **436**, 41–57. Aerodynamic lift and drag fluctuations of a sphere.

John, F. 1949 *Commun. Pure Appl. Math.* **2**, 13–57. On the motion of floating bodies I.

Keller, J. B. 1957 *J. Appl. Phys.* **28**, 859–864. Teapot effect.

Kelvin, Lord 1867a *Philos. Mag.* **33**, 511–512. The translatory velocity of a circular vortex ring.

Kelvin, Lord 1867b *Philos. Mag.* **34**, 15–24. On vortex atoms.

Kelvin, Lord 1887 *Proc. R. Soc.* **XLII**, 80. On the waves produced by a single impulse in water of any depth, or in a dispersive medium.

Khrabrov, A. and Ol, M. 2004 *J. Aircraft* **41**, 944–948. Effects of flow separation on aerodynamic loads in linearized thin airfoil theory.

Lamb, Horace 1932 *Hydrodynamics* (6th ed.), Cambridge University Press.

Landau, L. D. and Lifshitz, E. M. 1987 *Fluid Mechanics* (2nd ed.), Oxford, Pergamon.

Levich, V. 1962 *Physico-Chemical Hydrodynamics*, Englewood Cliffs, NJ, Prentice-Hall.

Levine, H. and Schwinger, J. 1948 *Phys. Rev.* **73**, 383–406. On the radiation of sound from an unflanged circular pipe.

Lighthill, M. J. 1958 *An Introduction to Fourier Analysis and Generalised Functions*, Cambridge University Press.

Lighthill, M. J. 1960 *Philos. Trans. R. Soc. London Ser. A* **252**, 397–430. Studies on magneto-hydrodynamic waves and other anisotropic wave motions.

Lighthill, M. J. 1963 Chapters 1 and 2 of *Laminar Boundary Layers* (editor, L. Rosenhead), Oxford University Press.

Lighthill, Sir James 1975 *Mathematical Biofluiddynamics*, Philadelphia, Society of Industrial and Applied Mathematics.

Lighthill, James 1978 *Waves in Fluids*, Cambridge University Press.

Lighthill, J. 1986 *An Informal Introduction to Theoretical Fluid Mechanics*, Oxford, Clarendon.

Miles, J. W. and Lee, Y. K. 1975 *J. Fluid Mech.* **67**, 445–464. Helmholtz resonance of harbours.

Milne-Thomson, L. M. 1968 *Theoretical Hydrodynamics* (5th ed.), London, Macmillan.

Noble, B. 1958 *Methods Based on the Wiener–Hopf Technique*, London, Pergamon.

Pierce, A. D. 1989 *Acoustics, An Introduction to Its Principles and Applications*, New York, American Institute of Physics.

Prandtl, L. 1952 *Essentials of Fluid Dynamics*, London, Blackie and Sons.

Rayleigh, Lord 1870 *Philos. Trans. R. Soc. London* **161**, 77–118. On the theory of resonance.

Rayleigh, Lord 1945 *Theory of Sound*, Volumes 1 and 2, New York, Dover.

Saffman, P. G. 1993 *Vortex Dynamics*, Cambridge University Press.

Sedov, L. I. 1965 *Two Dimensional Problems in Hydrodynamics and Aerodynamics*, New York, Wiley.

Simmons, N. 1939 *Q. J. Math.* (Oxford) **10**, 281–298. Free stream-line flow past a vortex.

Stoker, J. J. 1957 *Water Waves*, New York, Interscience.

Wagner, H. 1925 *Z. Angew. Math. Mech.* **5** Part 1. Uber die Einstehung des dynamischen Auftriebes von Tragflügeln.

Whitham, G. B. 1961 *Commun. Pure Appl. Math.* **14**, 675–691. Group velocity and energy propagation for three-dimensional waves.

Whitham, G. B. 1965 *J. Fluid Mech.* **22**, 273–283. A general approach to linear and nonlinear dispersive waves using a Lagrangian.

Index

For EU product safety concerns, contact us at Calle de José Abascal, 56–1°,
28003 Madrid, Spain or eugpsr@cambridge.org.